SECOND EDITION

Toxic Constituents of Plant Foodstuffs

FOOD SCIENCE AND TECHNOLOGY

A Series of Monographs

A complete list of titles in this series appears at the end of this volume.

SECOND EDITION

TOXIC CONSTITUENTS OF PLANT FOODSTUFFS

EDITED BY

IRVIN E. LIENER

Department of Biochemistry
College of Biological Sciences
University of Minnesota
St. Paul, Minnesota

ACADEMIC PRESS 1980
A Subsidiary of Harcourt Brace Jovanovich, Publishers
New York London Toronto Sydney San Francisco

ACADEMIC PRESS, INC.
111 Fifth Avenue, New York, New York 10003

United Kingdom Edition published by
ACADEMIC PRESS, INC. (LONDON) LTD.
24/28 Oval Road, London NW1 7DX

Library of Congress Cataloging in Publication Data
Main entry under title:

Toxic constituents of plant foodstuffs.

Includes bibliographies and index.
1. Food contamination. 2. Food poisoning.
3. Food--Composition. I. Liener, Irvin E.
RA1260.T69 1979 615.9'54 79-51681
ISBN 0-12-449960-0

PRINTED IN THE UNITED STATES OF AMERICA

80 81 82 83 9 8 7 6 5 4 3 2 1

Contents

List of Contributors

Numbers in parentheses indicate the pages on which the authors' contributions begin.

Leah C. Berardi (183), Southern Regional Research Center, Agricultural Research Service, U.S. Department of Agriculture, New Orleans, Louisiana 70179

Yehudith Birk (161), Department of Agricultural Biochemistry, Faculty of Agriculture, The Hebrew University of Jerusalem, Rehovot, Israel

William F. Busby, Jr. (329), Department of Nutrition and Food Science, Massachusetts Institute of Technology, Cambridge, Massachusetts 02139

M. Chevion (265), Department of Cellular Biochemistry, The Hebrew University, Hadassah Medical School, Jerusalem, Israel

M. E. Daxenbichler (103), Northern Regional Research Center, Agricultural Research Service, U.S. Department of Agriculture, Peoria, Illinois 61604

G. Glaser (265), Department of Cellular Biochemistry, The Hebrew University, Hadassah Medical School, Jerusalem, Israel

Leo A. Goldblatt (183), Southern Regional Research Center, Agricultural Research Service, U.S. Department of Agriculture, New Orleans, Louisiana 70179

Werner G. Jaffé (73), Facultad de Ciencias, Universidad Central de Venezuela, Escuela de Biologia, Caracas, Venezuela

Madhusudan L. Kakade* (7), Land O'Lakes, Inc., Minneapolis Minnesota 55440

Irvin E. Liener (1, 7, 429), Department of Biochemistry, College of Biological Sciences, University of Minnesota, St. Paul, Minnesota 55108

J. Mager (265), Department of Cellular Biochemistry, The Hebrew University, Hadassah Medical School, Jerusalem, Israel

*Present address: Cargill, Inc., Minneapolis, Minnesota 55440.

R. D. Montgomery (143), East Birmingham Hospital, Bordsley Green East, Birmingham B9 5 ST, England

G. Padmanaban (239), Department of Biochemistry, Indian Institute of Science, Bangalore 560012, India

Irena Peri (161), Department of Agricultural Biochemistry, Faculty of Agriculture, The Hebrew University of Jerusalem, Rehovot, Israel

Frank Perlman (295), Allergy Clinic, Portland Medical Center, Portland, Oregon 97205

H. L. Tookey (103), Northern Regional Research Center, Agricultural Research Service, U.S. Department of Agriculture, Peoria, Illinois 61604

C. H. VanEtten (103), Northern Regional Research Center, Agricultural Research Service, U.S. Department of Agriculture, Peoria, Illinois 61604

Gerald N. Wogan (329), Department of Nutrition and Food Science, Massachusetts Institute of Technology, Cambridge, Massachusetts 02139

Shmuel Yannai (371), Department of Food Engineering and Biotechnology, Technion—Israel Institute of Technology, Haifa, Israel

Preface

The warm reception the first edition of this book received has been most encouraging to the authors and has provided the stimulus needed to undertake the writing of a second edition. More importantly, however, since the first edition was published a decade ago, interest in the use of plant proteins, particularly in the human diet, has continued to grow. Several reasons can be given for this interest: (1) the grim realization that the gap between the protein needed for feeding a rapidly expanding world population and the amount of protein available for meeting this need continues to widen, and that plant protein offers the most promise for bridging this gap, (2) the growing body of evidence that appears to link heart disease to the excessive consumption of animal protein, containing as it does, such a high proportion of saturated fats, and (3) the beneficial effect that the crude fiber and other complex polysaccharides provided by crude sources of plant protein are believed to have in the prevention of colon cancer. These positive effects, however, must be weighed against the fact that plant materials are known to contain substances which, because of their toxic properties, may limit the nutritional potential of proteins of plant origin unless eliminated by suitable processing techniques.

Information regarding the physicochemical properties and physiological effects of these toxic components has increased dramatically since the first appearance of this book, a fact that has given added impetus for updating our fund of knowledge in this important area of food toxicology. The general format of the earlier edition has been followed in this edition with minor changes in the authors of some of the chapters. A more major change has been the replacement of the chapter on Cycads with one on Carcinogens by G. N. Wogan; this, no doubt, provides a much broader coverage of naturally occurring substances that have been shown to produce carcinogenic effects in animals. In addition, the untimely death of L. Friedman has necessitated the replacement of his chapter on Adventitious Toxic Factors in Processed Foods with a similar chapter by S. Yannai bearing the title, Toxic Factors Induced by Processing.

It should be appreciated that the final chapter in this book must remain unwritten. As more and more novel sources of plant proteins are introduced into the diet of man and animals, constant surveillance for the possible presence of heretofore unrecognized toxic substances will be necessary. It can only be hoped that our experiences of the past will alert us as to possible courses of action that should be taken to effect their elimination if such sources of protein are to reach their maximum nutritional potential.

IRVIN E. LIENER

Preface to the First Edition

One of the most dramatic developments in the food industry has been the use of chemical additives for improving the functional properties, nutritive value, shelf-life, and esthetic qualities of processed foodstuffs. Less desirable has been the inadvertent introduction of certain chemicals at some point of the food chain, whether it be in the harvesting of the raw material or in the final distribution of the processed food to the consumer. Regardless of their origin, these "foreign" chemicals have been closely scrutinized by public health officials who are concerned with the potential hazards which such chemicals might constitute to the health of the consumer public. Important as these considerations are, a fact which is sometimes not fully appreciated is that certain foods, particularly those of plant origin, may contain *natural* chemical constituents which may also be potentially harmful to man or animals. Unlike chemicals which are deliberately or inadvertently added to foods, these natural toxicants pose a special problem since their elimination is not readily amenable to legislative action. The purpose of this book is to direct the attention of those who are in any way involved in one of the many phases of food production to the existence of this problem. For it is only after these toxic constituents have been recognized and studied that effective measures can be taken to reduce their harmful effects.

The material presented in this work represents the joint effects of more than a dozen experts in their respective fields. It is their hope that the information which they have assembled will be of interest and value to all scientists who are concerned in one way or another with the safety of our food supply, whether it be the plant geneticist who seeks to develop new nontoxic varieties of plants, the nutritionist who is asked to make recommendations regarding the nutritive properties of certain foods, the technologist who must develop methods of processing which will destroy these toxic constituents, the toxicologist who must be able to detect these noxious substances wherever they may have been implicated in cases of poisoning, and, finally, the public health official who is entrusted with the task of enforcing whatever regulations may be necessary to insure the safety of a particular food. Since these toxic substances are normal, genetically determined components of plants, their function in the metabolism of the living cell may also be of basic interest to the biochemist and plant physiologist. The reader could undoubtedly add others to the list.

A subject as complex and diverse as the one dealt with here poses certain problems with respect to the manner in which the material should be presented. A toxic substance may be classified in one of several ways, according to (1) the physiological manifestations which it evokes in the animal body (i.e., an inhibition of growth, glandular hypertrophy, carcinogenetic effect, etc.), (2) some kind of activity by which it can be recognized in the test-tube (i.e., inhibition of trypsin, agglutination of red blood cells, etc.), (3) its chemical structure (i.e., glycoside, protein, amino acid, etc.), or (4) the foodstuff from which it is derived (i.e., soybeans, cottonseed, fava bean, etc.). Because of the incompleteness of much of the information that is currently available, no one scheme of classification seemed appropriate. Nevertheless, nearly all of the known naturally occurring toxicants have sufficient properties in common so that they may be grouped together according to at least one of the aforementioned categories. It is felt that this arbitrary manner of organization allows for a more uniform treatment of the subject matter and that the need for cross references is thus held to a minimum.

The work does not cover all of the known naturally occurring toxic substances but deals with some of the more important topics which have not heretofore received the detailed review and evaluation which they deserve. An indication of the subject matter covered as well as the sources of supplementary references for suggested reading can be found in Chapter 1.

IRVIN E. LIENER

CHAPTER 1

Introduction

IRVIN E. LIENER

As the discrepancy between the world's supply of protein and the growth of the global population continues to widen, ways and means of breaching this gap have become a matter requiring immediate action. On the basis of available statistics (Abbott, 1966; Altschul, 1967), it can be estimated that by the end of the century there will be approximately six billion people on the face of the earth and that a twofold increase in the protein supplied by plant materials and a fourfold increase in the protein of animal origin will be needed to maintain the same level of nutrition that we have at present and that is admittedly even now suboptimal in many parts of the world. Most experts feel that to expect such an increase in the supply of animal protein in many of the poorer countries is unrealistic. Although cost of production and religious and cultural practices are important factors leading to this conclusion, in the final analysis the simple but stark fact emerges that man and beast will ultimately be forced to compete with one another for the same living space and the same available food supply (Bonner, 1961).

We are thus forced to consider plant proteins as the major source of dietary protein in the future, particularly in those parts of the world where poverty and high birth rate always seem to go hand in hand. Although the production of cereal grains such as wheat, corn, and rice could be conceivably expanded to provide a sufficient amount of the total protein needed, nutritionists are well aware of the fact that cereal proteins are in general of poor quality because of an inherent deficiency of certain amino acids, particularly lysine. On the other hand, the protein derived from such oil-bearing seeds as soybean, cottonseed, and the peanut or from legumes such as peas and beans, although somewhat deficient in the sulfur-containing amino acids, can be combined with cereal proteins to produce a protein mixture of high nutritive value (Altschul, 1967). Such all-vegetable protein diets not only have been successfully used in child-feeding

TOXIC CONSTITUENTS OF PLANT FOODSTUFFS, SECOND EDITION

ISBN 0-12-449960-0

programs but have also produced dramatic cures of the dreaded disease of protein malnutrition known as kwashiorkor (Swaminathan, 1967).

Although the nutritional merits of the protein from oilseeds and legumes are of prime importance, there are a number of other overriding considerations that place these plant proteins foremost on the list of potential candidates for expanding our future world supply of protein. In contrast to the low level of protein in most cereal grains (6–14%), the oilseeds and legumes contain 20–50% protein. This means that the yield of protein per unit acre is much greater, certainly an important consideration where arable land is at a premium, as in Japan and certain parts of Asia. Moreover, the oilseeds and legumes may be readily adapted to grow under a wide variety of climatic conditions. Sold on the retail market, they are relatively cheap and usually within the purchasing capabilities of the average family in the poorer countries. The very fact that they are already a part of the diet in many parts of the world greatly simplifies efforts to increase their consumption in such countries. The protein of the oilseeds can also be isolated and converted to beverages and textured foods that simulate the appearance and flavor of animal protein products (Smith and Circle, 1972).

Having thus extolled the virtues of the plant proteins, the author must now admit that there are certain liabilities associated with their use as food which cannot be dismissed lightly. For reasons that scientists have yet to fathom, nature has seen fit to endow many plants with the capicity to synthesize a wide variety of chemical substances that are known to exert a deleterious effect when ingested by man or animals. It did not require much experience for primitive man to learn to avoid the consumption of those plants that produced an immediate, unpleasant reaction and that may even have proved fatal. More importantly, man learned that by the simple expedience of cooking plant products, most of these toxic constituents could be destroyed to produce highly palatable and nutritious foods. To modern-day man this may sound like a very simple and obvious thing to do, yet it is estimated that man survived for 98% of the proved 2.6 million years of his history without learning how to use fire for cooking his food. It was only after man learned to use fire for cooking, about 40,000 years ago, that it became possible for him to take advantage of a greatly expended food supply in the form of cooked vegetable foods. As pointed out by Leopold and Arudrey (1972), the rapid expansion of the population of modern *Homo sapiens* did in fact coincide with the development of cooking, which, by destroying the toxic substances of plant foods, brought a major evolutionary advantage to man. What could not have been realized by early man, and even today may not be fully appreciated, is the fact that, although there might not be an immediate violent reaction to a certain food, there still might be a slow cumulative effect resulting in frank disease or less than optimal health. It is this aspect of the problems that will pose the greatest challenge in the future, since knowledge of this nature is accumulated very slowly and with difficulty, particularly if the causative principles remain unidentified.

The subject of naturally occurring toxicants in foods has been comprehensively covered in a symposium sponsored by the American Chemical Society in 1968, in a monograph published by the National Research Council of the National Academy of Sciences in 1973, and, more recently, in a review by Salunkhe and Wu (1977). The subject matter of this volume may in a sense be considered to be complementary to these publications in that its purpose is to provide an in-depth coverage of what have been judged to be some of the more important topics that have heretofore not received the detailed review and evaluation they merit.

No attempt has been made in this book to cover all of the natural toxic substances known to be present in plant materials. A considerable body of information has been amassed over the years which deals with the distribution of poisons in plants that have been implicated in losses in livestock in various parts of the world (Garner, 1957; Kingsbury, 1964; Radeleff, 1964). Although such losses may at times be of economic significance, most of these plants have rarely been used as human food, except in cases of accidental poisoning, especially of children (O'Leary, 1964). Nor are we concerned here with toxins of bacterial origin, although we do not mean to minimize the importance of foodborne infections (Tanner and Tanner, 1953; Riemann, 1969). The contamination of foods by toxic substances produced by molds (mycotoxins) is deemed to be of sufficient relevance and importance that it is briefly discussed in Chapter 11, which deals with naturally occurring carcinogens. Although poisoning by foods of marine origin is of importance in such countries as Japan, whose seafood contributes about 10% of the total food supply of the Japanese, this subject has been dealt with elsewhere (Schantz, 1974). The principal concern of this book is foods of plant origin, which account for 70% of the world's supply of protein (Abbott, 1966), particularly those plant foodstuffs that are at present commonly consumed by man or animals or that may have potential value in the future, as the search for new sources of protein becomes more acute.

The nondescript terms "toxic factor" or "toxicant" are commonly used to refer to those substances found in foods that produce a deleterious effect when ingested by man or animals. These terms can be misleading since, strictly speaking, they imply that the substance in question is lethal beyond a given level of intake, and the toxicologist may in fact assess its toxicity in terms of its LD_{50}, that is, that dose which causes the death of 50% of the animals tested. In fact, although some plants are known to produce a violent expression of poisoning (Kingsbury, 1964), much more subtle effects, produced only by prolonged ingestion of a given plant, are more commonly observed. Such effects might include an inhibition of growth, a decrease in food efficiency, a goiterogenic response, pancreatic hypertrophy, hypoglycemia, and liver damage. Other factors that should be taken into consideration include the species of animal, its age, size, and sex, its state of health and plane of nutrition, and any stress factors that might be superimposed on these variables. The reader will therefore be asked to give the term "toxic," as used throughout this volume, a most liberal interpretation to mean nothing more or less than an adverse physiological response produced in man or animals by a particular food or a substance derived therefrom.

It should perhaps be emphasized that the evidence that a particular food constitutes a hazard to the health of man is frequently only presumptive. As might be expected, much of the research relating to the toxicity of foodstuffs has been done with plants commonly used as food by farm animals. Under these conditions the animals consume large quantities of a particular plant over a long period of time, a situation quite foreign from the normal eating patterns of man. Thus, a toxic substance, if present, might produce symptoms of poisoning that might not otherwise be apparent. If the causative factor has been isolated, its toxicity is frequently evaluated by using a route of administration (such as intraperitoneal or subcutaneous injection) that is physiologically unrelated to the normal mode of ingestion. The final link in the chain of evidence, namely, the demonstration that the ingestion of the purified toxin will produce some physiological damage to man at a level comparable to that which would be present in the quanitity of food that he normally consumes, must be left undone. In effect, therefore, the only evidence for toxicity to man is often only the knowledge that a substance known to be toxic to animals under a given set of conditions is present in a food that he consumes. This information is nevertheless of sufficient importance to at least alert one to the possible hazards involved in the consumption of such foods by man.

As has already been pointed out, man has through trial and error learned which foods to avoid and which foods require some form of processing to make them safe for consumption. It is also true that some margin of safety is provided by the fact that man's diet is generally of such a varied nature that any naturally occurring toxicants in a particular food would be diluted to the point where the normal detoxification mechanisms of the body could adequately cope with them. It is conceivable, however, that abnormal patterns of food consumption might bring to the surface toxic manifestations that would not otherwise be apparent. A good example of this is the periodic eruption of lathyrism in certain parts of India, which is associated with the consumption of certain varieties of plants during times of famine when cereal grains are in short supply (see Chapter 8). It is this sort of occurrence that may be expected to become more commonplace as the shortage of protein foods becomes more acute and people are forced to become more indiscriminate in their choice of life-sustaining plant foods.

The fact that some sources of plant protein are capable of producing harmful effects in animals is in itself of importance with respect to man's food supply.

Man's dependence on animal protein can be expected to increase in proportion to population growth in those countries where higher standards of living will continue to prevail (Altschul, 1967). Were it not for the fact that such oilseeds as soybeans and cottonseed can be processed so as to inactivate their toxic constituents, these rich sources of plant protein would not occupy such a position of importance as they now command in the feeding of animals (Altschul, 1958). Thus, the role of naturally occurring toxicants as it relates to their effects on animals cannot be dismissed as irrelevant to the problem of feeding the world of tomorrow.

References

Abbott, J. C. (1966). *Adv. Chem. Ser.* **57,** 13.

Altschul, A. M., ed. (1958). "Processed Plant Protein Foodstuffs." Academic Press, New York.

Altschul, A. M. (1965). "Proteins: Their Chemistry and Politics," pp. 289–290. Basic Books, New York.

Altschul, A. M. (1967). *Science* **158,** 221–226.

American Chemical Society (1968). *J. Agric. Food Chem.* **17,** 413–538.

Bonner, J. (1961). *Fed. Proc.* **20** (Suppl. 7), 369–372.

Garner, R. J. (1957). "Veterinary Toxicology." Baillière, London.

Kingsbury, J. M. (1964). "Poisonous Plants of the United States and Canada." Prentice-Hall, Englewood Cliffs, New Jersey.

Leopold, A. C., and Audrey, R. (1972). *Science* **176,** 512–514.

National Academy of Sciences, National Research Council (1973). "Toxicants Occurring Naturally in Foods," 2nd ed., N.A.S.—N.R.C., Washington, D.C.

O'Leary, S. B. (1964). *Arch. Environ. Health* **9,** 216–242.

Radeleff, R. D. (1964). "Veterinary Toxicology." Lea & Febiger, Philadelphia, Pennsylvania.

Riemann, H., ed. (1969). "Food-Borne Infections and Intoxications." Academic Press, New York.

Salunkhe, D. K., and Wu, M. T. (1977). *Crit. Rev. Food Sci. Nutr.* **9,** 265–324.

Schantz, E. J. (1974). *In* "Toxic Constituents of Animal Foodstuffs" (I. E. Liener, ed.), pp. 111–130. Academic Press, New York.

Smith, A. K., and Circle, S. J., eds. (1972). "Soybeans: Chemistry and Technology." Avi Publ. Co., Westport, Connecticut.

Swaminathan, M. (1967). *Newer Methods Nutr. Biochem.* **3,** 197–241.

Tanner, F. W., and Tanner, L. P. (1953). "Food-borne Infections and Intoxications." Garrand Press, Champaign, Illinois.

CHAPTER 2

Protease Inhibitors

IRVIN E. LIENER AND MADHUSUDAN L. KAKADE

I. INTRODUCTION

Substances that have the ability to inhibit the proteolytic activity of certain enzymes are found throughout the plant kingdom, particularly among the legumes. These so-called protease inhibitors have attracted the attention of scientists in many disciplines—nutritionists because of their possible effect on the nutritive value of plant proteins, protein chemists because the reaction of these inhibitors, which are proteins, with enzymes provides a simple model system for studying protein–protein interaction, and members of the medical profession because the unique pharmacological properties of these inhibitors hold considerable promise for clinical application in the field of medicine.

TOXIC CONSTITUENTS OF PLANT FOODSTUFFS, SECOND EDITION

ISBN 0-12-449960-0

General reviews on the subject of protease inhibitors are available (Werle and Zickgraf-Rüdel, 1972; Ryan, 1973; Tschesche, 1974; Birk, 1976; Richardson, 1977). More specialized aspects of this subject relating to nutrition (Couch and Hooper, 1972; Rackis, 1974; Liener, 1977), molecular properties (Laskowski and Sealock, 1971), and immunology (Ryan, 1977) have also been reviewed. The proceedings of an international symposium on protease inhibitors (Fritz *et al.*, 1974) is highly recommended to the reader interested in an update of the progress of research in this area. Papers dealing with protease inhibitors continue to appear in ever-increasing numbers, so that the present review can only portray the situation as it exists at the time of writing. Many facets of the subject are still controversial and unexplained, so that a final evaluation of the information at hand is still not possible. Within these limits, an attempt will be made to review the chemical and physical properties of the protease inhibitors, their possible significance in the diet of man and animals, and their physiological role in the plant in which they are found.

II. DISTRIBUTION IN THE PLANT KINGDOM

A. Distribution among Plants

Read and Haas (1938) appear to have been the first to recognize the presence of an inhibitor in plant material. They reported that an aqueous extract of soybean flour inhibited the ability of trypsin to liquefy gelatin. The fraction of soybean protein responsible for this effect was partially purified by Bowman (1944, 1946, 1948) and Ham and Sandstedt (1944) and subsquently isolated in crystalline form by Kunitz (1945, 1946). The existence of an inhibitor of trypsin in soybeans that could be inactivated by heat seemed to offer, at the time, a reasonable explanation for the observation made many years before by Osborne and Mendel (1917) that heat treatment improved the nutritive value of soybean protein. The realization that protease inhibitors might be of nutritional significance in plant foodstuffs, particularly in such important dietary sources of protein as legumes and cereals, stimulated a search for similar factors in other plant materials. The list of protease inhibitors found in various plants is now a long and growing one. Table I is a summary of our current knowledge of the distribution of these inhibitors in the plant kingdom. More detailed information concerning individual inhibitors that have been more thoroughly studied with respect to their physicochemical properties and possible mode of action can be found in Section III.

B. Distribution within the Plant

It will be noted from Table I that the most frequent source of protease inhibitors is the seed of the plant, but their location is not necessarily restricted to this part of the plant. For example, in some legumes, such as the mung bean (Hon-

TABLE I

DISTRIBUTION AND PROPERTIES OF PLANT PROTEASE INHIBITORS

Botanical name	Common name	Part of plant	Enzymes inhibited[a]	References
Abrus precatorius	Guinea pea	Seed	T,C,S	Sumathi and Pattabiraman (1976)
Acacia arabica	Acacia	Seed	T,C	Sumathi and Pattabiraman (1976)
Adenanthera pavonia	Red wood	Seed	T,C	Sumathi and Pattabiraman (1976)
Allium cepa	Onion	Bulb	T	Sumathi and Pattabiraman (1975)
Allium sativum	Garlic	Bulb	T	Sumathi and Pattabiraman (1975)
Amorphophallus companulatus	Amorphophallus	Tuber	T,C	Sumathi and Pattabiraman (1975)
Anacardium occidentale	Cashew nut	Nut	T,Pl	Xavier-Filho and Ainouz (1977)
Ananus sativus	Pineapple	Stem	B,Pa,F	See Section III,C,6
Arachis hypogaea	Peanut, groundnut	Nut	T,C,Pl,K	See Section III,A,7
Artocarpus integrifolia	Jack fruit	Seed	T	Siddappa (1957)
Avena sativa	Oats	Endosperm	T	Laporte and Trémoliéres (1962); Mikola and Kirsi (1972)
Bambusa arundinaria	Bamboo	Sprouts	T	Sugiura (1975)
Bauhinia pupurea		Seed	T,C,E,Pr	See Section III,A,14
Beta vulgaris	Beet, beetroot	Tuber	T	Vogel and Hartwig (1968)
Brassica campestris	Rapeseed	Seed	T	Hata *et al.* (1967), Ogawa *et al.* (1971c)
Brassica juncea	Indian mustard	Seed	T	Hata *et al.* (1967)
Brassica oleracea	Broccoli, brussels sprouts	Seed	T	Chen and Mitchell (1973)
Brassica rapa	Turnip, rutabaga	Seed	T	Sohonie and Bhandarkar (1954)
Cajanus cajan	Pigeon pea, red gram	Seed	T	See Section III,A,12
Canavalia ensiformis	Jack bean, sword bean	Seed	T,C,S	Ubatuba (1955), Sumathi and Pattabiraman (1976)

(Continued)

TABLE I (*Continued*)

Botanical name	Common name	Part of plant	Enzymes inhibited[a]	References
Ceratonia siliqua	Carob bean	Seed	T	Borchers *et al.*, (1947b)
Cercis canandensis	Redbud tree, wild lilac	Seed	T	Borchers *et al.*, (1947b)
Chamacrista fasiculata	Partridge pea	Seed	T	Borchers *et al.*, (1947b)
Cicer arietinum	Chick-pea, Bengal gram, garbanzo	Seed	T,C	See Section III,A,13
Clittoria ternatea	Butterfly pea	Seed	T,C,S	Sumathi and Pattabiraman (1976)
Colocasia esculenta	Taro	Tuber	T,C	Sohonie and Bhandarkar (1954)
Cucumis sativus	Cucumber	Seed, leaves	T,C	Chen and Mitchell (1973), Walker-Simmons and Ryan (1977a)
Cucurbita maxima	Squash	Leaves	T	Walker-Simmons and Ryan (1977a)
Curcuma amada	Mango ginger	Seed	T	Sumathi and Pattabiraman (1975)
Cyamopsis psoraloides	Guar bean	Seed	T	Couch *et al.* (1966)
Cyamopsis tetragonoloba	Cluster bean	Seed	T,C,S	Sumathi and Pattabiraman (1976)
Dolichos biflorus	Horse gram	Seed	T	Subbulakrishini *et al.* (1976)
Dolichos lablab	Hyacinth bean, field bean, hakubenzu bean	Seed	T,C,Th	See Section III,A,11
Eleusine coracana	Finger millet	Seed	T,C	Veerabhadrappa *et al.* (1978)
Faba vulgaris	Double bean	Seed	T	Sohonie *et al.* (1959)
Fagopyrum esculentum	Buckwheat	Endosperm	T	Laporte and Trémoliéres (1962)
Fragaria virginiana	Strawberry	Leaves	T	Walker-Simmons and Ryan (1977a)
Gleditsia tricanthos	Honey locust	Seed	T	Borchers *et al.* (1947b)
Glycine max	Soybean	Seed	See Section III,A,1	See Section III,A,1
Glymocladus dioica	Kentucky coffee bean	Seed	T	Borchers *et al.* (1947b)

Hordeum vulgare	Barley	Endosperm, embryo	T,S, fungal proteases, endopeptidases	See Section III,B,1
Helianthus annuus	Sunflower	Seed	T	Bielorai and Bondi (1963), Agren and Liedén (1968), Roy and Bhat (1974)
Ipomoea batata	Sweet potato	Tuber	T	See Section III,C,2
Lablab purpureus		Seed	T	Miége *et al*. (1976)
Lactuca sativa	Lettuce	Seed	T	Shain and Mayer (1965)
Larix occidentalis	Larch	Leaves	T	Walker-Simmons and Ryan (1977a)
Lathyrus odoratus	Sweet pea	Seed	T	Weder and Belitz (1969)
Lathyrus sativus	Chickling vetch, grass pea	Seed	T,C	See Section III,A,10
Lens esculenta (culinaris)	Lentil	Seed	T	Sohonie and Bhandarkar (1955)
Lespedeza stipularea	Lespedeza	Seed	T	Borchers *et al*. (1947b)
Lupinus albus	Lupine	Seed	T	Gallardo *et al*. (1974)
Lycopersicum esculentum	Tomato	Leaves	T,C	Walker-Simmons and Ryan (1977a,b), Gustafson and Ryan (1976)
Malus pumila	Apple	Leaves	T	Walker-Simmons and Ryan (1977a)
Manihot utilissima	Cassava, tapioca	Tuber	T,C	Sumathi and Pattabiraman (1975)
Medicago sativa	Alfalfa, lucerne	Meal	T	See Section III,A,16
Mucana deeringianum	Florida velvet bean	Seed	T	Borchers *et al*. (1947b)
Nicotiana tobacum	Tobacco	Leaves	T,C	Ryan and Shumway (1970), Walker-Simmons and Ryan (1977a)
Oryza sativum	Rice	Grain, embryo	T,S,endopeptidase	See Section III,C,5
Phaseolus aconitifolius	Moth bean	Seed	T	Subbulakshani *et al*. (1976)
Phaseolus angularis	Adzuki bean	Seed	T,C	See Section III,A,6
Phaseolus aureus	Mung bean, green gram	Seed	T,endopeptidase	See Section III,A,5

(Continued)

TABLE I (*Continued*)

Botanical name	Common name	Part of plant	Enzymes inhibited[a]	References
Phaseolus coccineus	Scarlet runner bean, French bean	Seed	T,C	See Section III,A,4
Phaseolus lunatus	Lima bean, butter bean	Seed	T,C	See Section III,A,2
Phaseolus mungo (radiatus)	Black gram	Seed	T,C,S	Sohonie and Bhandarkar (1955), Sumathi and Pattabiraman (1976)
Phaseolus vulgaris	Navy bean, kidney bean, pinto bean, French bean, white bean, wax bean, haricot bean, garden bean	Seed	T,C,E	See Section III,A,3
Phleum pratense		Seed	T,C	Mejbaum-Katzenellengoben and Lorenc-Kubis (1969)
Pisum sativum	Field pea, garden pea	Seed	T	See Section III,A,15
Poa praetensis		Seed	T,C	Mejbaum-Katzenellenbogen and Lorenc-Kubis (1969)
Prosopis tamarugo	Tamarugo	Seed	T	Gallardo *et al*. (1974)
Psophocarpis tetragonolobus	Winged bean, gao bean, asparagus pea	Seed	T	Sohonie and Bhandarkar (1954), Jaffé and Korte (1976)
Raphanus sativus	Radish	Seed	T	See Section III,C,5
Scopolia japonica		Callus	T,C,Pl,Pe,K	See Section III,C,4
Secale cereale	Rye	Flour, embryo, endosperm	T,C	See Section III,B,2

Sesbania aculenta	Dhaincha seeds	Seed	T	Katoch and Chopora (1974)
Solanum melongena	Eggplant	Exocarp	T,Pr	See Section III,C,3
Solanum tuberosum	Potato	Tuber, leaves	T,C,CPase A and B, K, cathepsins A and D, microbial proteases	See Section III,C,1
Sophora japonica	Japanese pagoda tree	Seed	T	Borchers *et al.* (1947b)
Sorghum bicolor	Sorghum	Grain	T	Xavier-Filho (1974)
Spinacia oleracea	Spinach	Leaves	T	Chen and Mitchell (1973)
Stizobolium deeringianum	Velvet bean	Seed	T	Borchers *et al.* (1947b)
Tamarindus indica	Tamarind	Seed	T,C	Sumathi and Pattabiraman (1976)
Trifolium repens	Clover	Leaves	T	Kendall (1951), Walker-Simmons and Ryan (1977a)
Triticale	Triticale	Grain	T,C	Madl and Tsen (1973)
Triticum durum	Durum	Flour	T,C	Camus and Laporte (1973)
Triticum vulgare	Wheat	Whole, wheat flour, embryo	T,C,larval protease inhibitor	See Section III,B,3
Typhonium trilobatum	Typhinium	Bulb	T,C	Sumathi and Pattabiraman (1975)
Vicia cracca		Seed	T	Sundberg *et al.* (1970)
Vicia faba	Broad bean, field bean, fava bean	Seed, leaves	T,C,Th,Pr,Pa, microbial proteases	See Section III,A,9; Walker-Simmons and Ryan (1977a)
Vigna unguiculata (sinensis)	Cowpea, black-eyed pea, Southern pea, serido pea	Seed	T,C	See Section III,A,8
Vitis vinifera	Grape	Leaves	T	Walker-Simmons and Ryan (1977a)
Voandzeia subterranea	Bambara beans	Seed	T	Owusu-Domfeh (1973)
Zea mays	Corn, maize	Kernel	T	See Section III,B,4

[a]Key to abbreviations: B, bromelain; C, chymotrypsin; CPase, carboxypeptidase; E, elastase; F, ficin; K, kallikrein; Pa, papain; Pe, pepsin; Pl, plasmin; Pr, Pronase; S, subtilisin; T, trypsin; Th, thrombin.

avar and Sohonie, 1955) and field bean (Ambe and Sohonie, 1956a,b), a fairly high level of trypsin inhibitor activity is found in the leaves as well. In tuberous plants, such as the white potato (Ryan and Huisman, 1967) and sweet potato (Honavar and Sohonie, 1955), protease inhibitors are found in the leaves as well as the tuber. Among the cereals, including corn, barley, wheat, and rye, protease inhibitors are located primarily in the endosperm and to a lesser extent in the germ (Melville and Scandalios, 1972; Mikola and Kirsi, 1972; Halim *et al.*, 1973a; Kirsi, 1974; Kirsi and Mikola, 1977). Those inhibitors present in the endosperm, however, appear to be different from those found in the embryo (Kirsi and Mikola, 1971, 1977).

Within the cotyledon of such legumes as the soybean, kidney bean, and mung bean, a greater concentration of the protease inhibitors is found in the outer part of the cotyledon mass (Zimmerman *et al.*, 1967; Collins and Sanders, 1976). In *Vicia faba* there is a twofold greater concentration of trypsin inhibitor in the hull than in the cotyledon (Marquardt *et al.*, 1975). As for the intracellular localization of the inhibitors, they do not appear to be associated with the protein bodies but rather with the cytosol, at least in the case of mung beans (Baumgarten and Chrispeels, 1976), peas (Hobday *et al.*, 1973), and *Lablab purpureus* (Miége *et al.*, 1976). When protease inhibitors were induced in tomato leaves by wounding, however, they were localized in the protein bodies in the cell vacuoles (Walker-Simmons and Ryan, 1977b; Shumway *et al.*, 1976). The relationship between the intracellular location of the protease inhibitors and their physiological function in the plant is not well understood (for further discussion, see Section IV).

III. PHYSICOCHEMICAL PROPERTIES OF INHIBITORS FROM VARIOUS PLANTS

Much of the data recorded in Table I with respect to the protease inhibitors found in different plants are based on the use of crude extracts or partially purified preparations. In every instance, however, in which such inhibitors have been isolated in a high state of purity, the active substance has proved to be a protein or a mixture of proteins having inhibitory activity toward proteases (these will be referred to as "isoinhibitors"). Those inhibitors that have been purified to the extent that meaningful measurements of their chemical and physical properties can be made will be treated individually in more detail in the paragraphs that follow.

A. Legumes

1. *Glycine max* (Soybean)

Broadly speaking, the protease inhibitors that have been isolated from soybeans fall into two main categories: those that have a molecular weight of

20,000–25,000 with relatively few disulfide bonds and a specificity directed primarily toward trypsin, and those that have a molecular weight of only 6000–10,000 with a high proportion of disulfide bonds and capable of inhibiting trypsin and chymotrypsin at independent binding sites. Although the major components in these two classes of inhibitors are referred to as the Kunitz and Bowman–Birk inhibitors, respectively, there appear to be a number of related isoinhibitors in each of these categories that differ in some minor respects from the Kunitz and Bowman–Birk inhibitors.

a. Kunitz Inhibitor. *i. Isolation and Properties.* The original method whereby the Kunitz inhibitor was purified (Kunitz, 1945, 1946) has been largely supplemented by techniques involving chromatography (Rackis *et al.*, 1959; Rackis and Anderson, 1964; Frattali and Steiner, 1968; Yamamoto and Ikenaka, 1967; Obara *et al.*, 1970; Obara and Watanabe, 1971; Kassell, 1970), gel electrophoresis (Eldridge and Wolf, 1969; Frattali and Steiner, 1969), and isoelectric focusing (Catsimpoolas, 1969; Catsimpoolas *et al.*, 1969; Wang, 1971). Regardless of the methodology employed, one invariably observes a heterogeneous array of components exhibiting protease inhibitor activity which differ somewhat in molecular weight (18,000–24,000), isoelectric point (pH 3.5–4.4), amino acid composition, and heat stability. Since each investigator used his own designation for each of these components, there is considerable confusion in the literature as to the exact identity and relationship of each of these inhibitors. Nevertheless, there can be little doubt that the major component of those inhibitors that have molecular weights of approximately 20,000 is what is commonly referred to as the Kunitz soybean inhibitor.

In addition to the presence of isoinhibitors within the same variety of soybeans, there appear to be at least two genetic variants of the major component itself among different varieties of soybeans. Certain varieties of soybeans contain a major inhibitory component that exhibits a slower mobility when examined by disc gel electrophoresis at pH 8.3 than the more common major inhibitor component found in most soybean varieties (Singh *et al.*, 1969; Clark *et al.*, 1970; Hymowitz and Hadley, 1972). In general, those varieties of soybeans possessing the slower-moving inhibitor showed lower trypsin inhibitor activity than those with the fast-moving variant (Clark and Hymowitz, 1972). This slower-moving genetic variant of the Kunitz inhibitor, however, has not yet been purified and characterized.

The complete amino acid sequence of the Kunitz inhibitor as reported by Koide and Ikenaka (1973a,b) and Koide *et al.* (1973) is shown in Fig. 1. It consists of 181 amino acid residues and two disulfide bonds, with the reactive site being located at residues Arg 63 and Ile 64 (see further discussion of active site below). On the basis of this sequence and theoretical considerations, Koide *et al.* (1973) predicted that this molecule should have very little α-helical structure and is largely in the form of a random coil, thus confirming earlier physicochemical

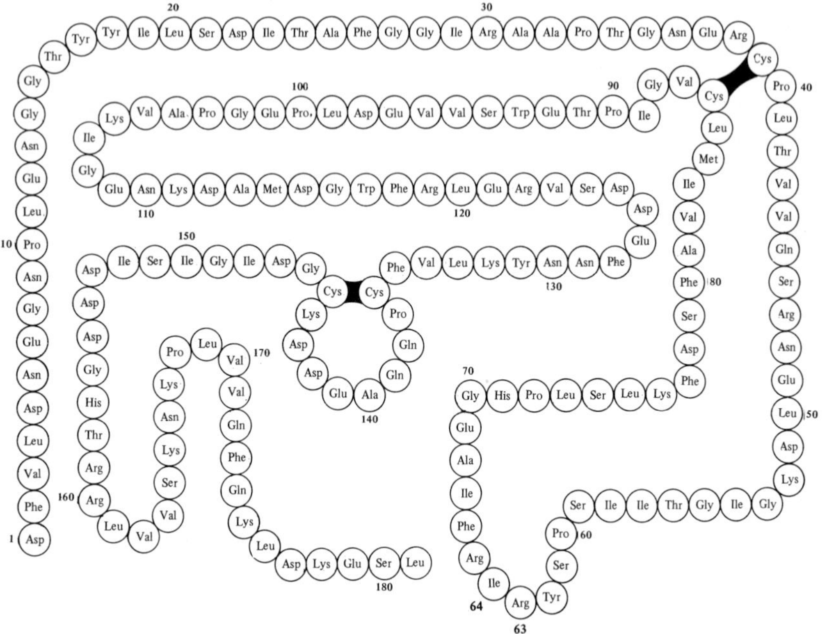

FIG. 1. Amino acid sequence of the Kunitz soybean inhibitor. (From Koide *et al.*, 1973, reproduced with the permission of the *European Journal of Biochemistry*.)

measurements (Wu and Scheraga, 1962; Edelhoch and Steiner, 1963; Jirgensons, 1967); this was later verified by X-ray crystallographic data (Blow *et al.*, 1974). This may account for the fact that the Kunitz inhibitor, unlike most globular proteins, is quite resistant to such denaturing agents as urea (Edelhoch and Steiner, 1963) and guanidinium chloride (Leach and Fish, 1977). Although the disulfide bond encompassing the small loop of the molecule, Cys 136 and Cys 145, may be selectively cleaved without loss in activity (DiBella and Liener, 1969), rupture of both disulfide bonds causes inactivation (Steiner, 1964, 1965).

ii. Specificity and kinetic parameters. Kunitz (1947) was the first to show that his inhibitor combined with trypsin in a stoichiometric fashion, i.e., 1 mole of inhibitor inactivated 1 mole of enzyme. The reaction with trypsin is almost instantaneous, the half-life of this reaction being about 4 sec with a second-order velocity constant of 2×10^7 L/mole/sec (Green, 1957). The complex that forms is a very tight one, the most recent measurements yielding a value of 10^{-11} M for the dissociation constant at pH 6.5 (Laskowsi and Sealock, 1971). At low pH values (pH < 3) the complex completely dissociates to regenerate the active enzyme and inhibitor.

The Kunitz inhibitor is capable of inhibiting trypsins derived from a wide variety of sources, including the cow, pig, salmon, stingray, barracuda, and turkey (Kassell, 1970). Bovine chymotrypsin, human plasmin, cocoonase, and plasma kallikrein are also inhibited to various degrees. With the exception of chymotrypsin, the stoichiometry and kinetic parameters for interaction of the latter enzymes with the Kunitz inhibitor have not been studied in great detail. Among the enzymes not inhibited by the Kunitz inhibitor are the thiol proteases, pepsin, thrombin, collagenase, and the carboxypeptidases. In the case of bovine chymotrypsin, the Kunitz inhibitor forms a highly dissociable complex with a dissociation constant of 10^{-6} *M* (Kasche, 1973). Despite their differences in substrate specificity, the reactive site of the inhibitor for chymotrypsin was shown to be the same as that for trypsin (Bidlingmeyer *et al.*, 1972), although Quast and Steffen (1975) presented evidence for a second, weak binding site for chymotrypsin.

The question whether the soybean inhibitors are capable of inhibiting human trypsin deserves special comment in view of the expanding use of soybeans in human diets. Human trypsin is known to exist in two forms: a cationic species, which constitutes the major component, and an anionic species, which comprises only about 10–20% of the total trypsin activity (Robinson *et al.*, 1972; Mallory and Travis, 1973; Figarella *et al.*, 1975). While the latter is fully inactivated by the Kunitz inhibitor (Mallory and Travis, 1973, 1975), the cationic form of trypsin is only weakly inhibited (Feeney *et al.*, 1969; Travis and Roberts, 1969; Figarella *et al.*, 1975).

iii. Mechanism of action. Early attempts to elucidate the model of interaction of the Kunitz inhibitor with trypsin involved the chemical modification of specific amino acid residues of either the inhibitor or the enzyme to see if such modifications affected the ability of these components to interact with each other. Such studies revealed the essentiality of tryptophan (Steiner, 1966; Odani *et al.*, 1971), tyrosine (Gorbunoff, 1970; Papaioannou and Liener, 1970), cystine (Steiner, 1964, 1965), and arginine (Fritz *et al.*, 1969; DeLarco and Liener, 1973) residues of the Kunitz inhibitor for interaction with trypsin. Although trypsin that had been rendered catalytically inactive by treatment with diisopropyl fluorophosphate or TLCK (1-chloro-3-tosylamido-7-amino-2-heptanone) was no longer capable of binding the Kunitz inhibitor (Green, 1953; Feinstein and Feeney, 1966), anhydrotrypsin, in which the serine residue of the active site had been converted to dehydroalanine, was strongly bound to the inhibitor (Ako *et al.*, 1972a, 1974). The reason for this paradoxical situation did not become apparent until X-ray crystallography revealed the true manner whereby trypsin is inhibited by the Kunitz inhibitor (see below).

Major contributions to our understanding of the mechanism of action of protease inhibitors has come from the laboratory of Laskowski and his group. They observed that treatment of the Kunitz inhibitor with catalytic levels of trypsin under acidic conditions led to a cleavage of an Arg-Ile bond (subsequently identified as Arg 63-Ile 64 in Fig. 1) without producing any loss in inhibitory activity (Finkenstadt and Laskowski, 1965; Ozawa and Laskowski, 1966). Equilibrium between the original, unmodified inhibitor (''virgin'' inhibitor) and the modified inhibitor is reached when 86% of the virgin inhibitor has been thus converted to the modified inhibitor (Niekamp *et al.*, 1969). A complex of the modified inhibitor with trypsin can be dissociated by a sudden drop in pH (Hixson and Laskowski, 1970) or by exposure to 6 *M* guanidinium chloride (Sealock and Laskowski, 1969), yielding predominantly the virgin inhibitor; this implies a resynthesis of the Arg-Ile bond in the modified inhibitor. The conclusion that this peptide bond constitutes the reactive site of the inhibitor is further supported by the fact that removal of the newly exposed arginine residue of the modified inhibitor with carboxypeptidase B (Finkenstadt and Laskowski, 1965) or chemical modification of its new N-terminal isoleucine residue (Haynes and Feeney, 1968; Kowalski and Laskowski, 1972) leads to irreversible inactivation of the modified inhibitor.

The fact that the Arg 63-Ile 64 bond of the Kunitz inhibitor can be cleaved and then resynthesized without loss of activity has permitted enzymatic and chemical replacement of amino acid residues at positions 63 and 64. Thus, it was possible by enzymatic manipulations to replace Arg 63 with lysine without loss of specificity toward trypsin (Sealock and Laskowski, 1969), whereas replacement of Arg 63 with tryptophan led to the formation of an inhibitor of chymotrypsin (Leary and Laskowski, 1973). Clearly, then, it is the nature of the amino acid located at position 63 that determines the specificity of the Kunitz inhibitor. Chemical–enzymatic manipulations were also employed to replace Ile 64 with alanine, leucine, or glycine without affecting its specificity toward trypsin (Kowalski and Laskowski, 1976a). The insertion of an additional amino acid to the N-terminal side of Ile 64, however, produced an inactive derivative (Kowalski and Laskowski, 1976b). These results indicate that although the nature of the amino acid located at position 64 is not critical, the spatial distance separating this residue from arginine (or lysine) at position 63 is decisive.

Although the studies described above have identified in an elegant fashion the reactive site of the Kunitz inhibitor, the exact nature of the forces that bind the inhibitor with the enzyme has proved more difficult to elucidate. Laskowski and Sealock (1971) had earlier postulated that the stability of the complex might be due to the formation of an ester bond between Arg 63 of the modified inhibitor and the active-site serine residue of trypsin. This type of bond would be analogous to the acyl enzyme intermediate that has been postulated to exist on the pathway leading to the hydrolysis of peptide bonds by serine proteinases. That this could not be the case in this situation was dictated by the fact that the catalytically inactive form of trypsin, wherein the active-site serine had been converted to dehydroalanine (Ako *et al.*, 1972a,1974), retained the capacity to complex with the Kunitz inhibitor. The most convincing evidence for the absence of an acyl enzyme intermediate has come from the X-ray crystallographic studies reported from Blow's laboratory (Blow *et al.*, 1974; Janin *et al.*, 1974; Sweet *et al.*, 1974). These studies have revealed that the Kunitz inhibitor–trypsin complex exists not as an acyl enzyme intermediate but as a tetrahedral adduct between the active-site serine residue of trypsin and the carbonyl group of Arg 63. Moreover, the tight affinity is actually due to the close complementary fit of these two interacting molecules reinforced by a large number of weak noncovalent hydrophobic and hydrogen bonds. Most of these bonds are localized near the active site of the enzyme and the recognition site (Arg 63-Ile 64) of the inhibitor, the actual zone of contact involving a very limited region of the inhibitor molecule (see Fig. 2). Only about 12 amino acids out of the 181 that comprise the inhibitor actually make contact with the trypsin molecule. The tetrahedral adduct, the formation of which would depend on a catalytically active form of the enzyme, does not appear to be necessary for complex formation and in fact would be more likely to decrease its stability (Means *et al.*, 1974). This would explain why anhydrotrypsin is still capable of binding to the Kunitz inhibitor.

From the evidence at hand, the sequence of reactions shown in Fig. 3 has been proposed as a model for the mechanism whereby the Kunitz inhibitor interacts with trypsin (Means *et al.*, 1974; Huang and Liener, 1977). Trypsin (E) initially interacts with the virgin inhibitor (I) in much the same way as trypsin would interact with a substrate containing a lysine- or arginine-containing peptide bond to form a tetrahedral adduct $(EI)_t$. Unlike a natural substrate, however, the adduct so formed is very stable due to a very close complementary fit, which is reinforced by an accumulation of weak, noncovalent bonds at the contact zone. This is the situation that exists when trypsin and the Kunitz inhibitor interact in

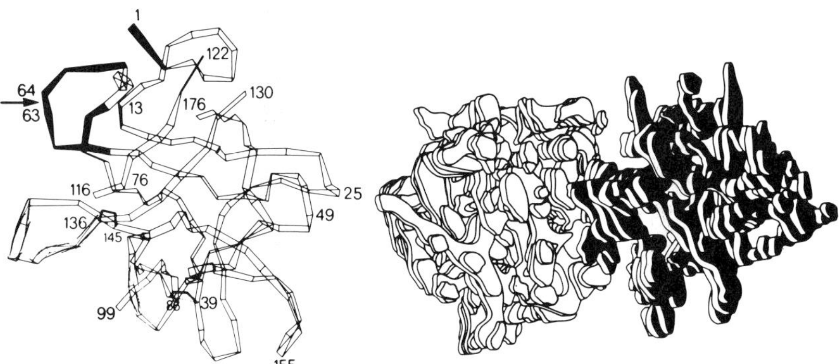

FIG. 2. Folding of the polypeptide backbone chain of the Kunitz inhibitor is shown on the left Amino acid residues in intimate contact with trypsin are shown in black. Shown on the right is a model of the Kunitz inhibitor–trypsin complex. That part representing trypsin is less heavily shaded. Reprinted with permission from R. M. Sweet, H. T. Wright, J. Janin, C. H. Chothia, and D. M. Blow, 1974, *Biochemistry* **13**, 4212–4228. Copyright by the American Chemical Society.

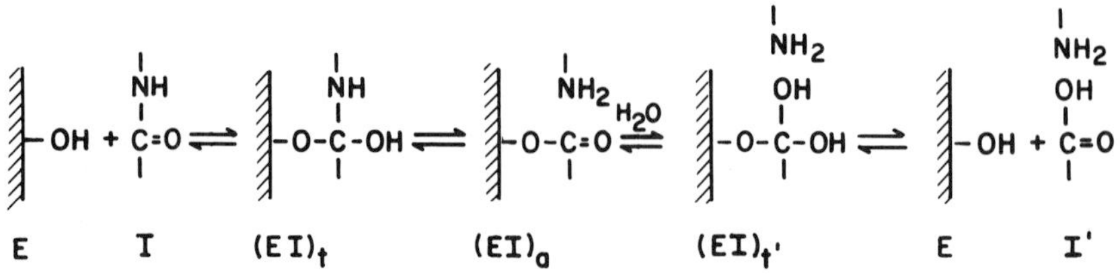

$$E\text{-OH} + \underset{|}{\overset{|}{\underset{|}{NH}}}\text{C=O} \rightleftharpoons (EI)_t \rightleftharpoons (EI)_a \overset{H_2O}{\rightleftharpoons} (EI)_{t'} \rightleftharpoons E\text{-OH} + I'$$

E T (EI)$_t$ (EI)$_a$ (EI)$_{t'}$ E I'

FIG. 3. Postulated mechanism of inhibition of trypsin by the Kunitz inhibitor (E, trypsin; I, virgin inhibitor; (EI)$_t$, tetrahedral adduct in stable complex; (EI)$_a$, acyl enzyme intermediate; (EI)$_{t'}$, tetrahedral adduct between enzyme and enzymatically modified inhibitor, I'). See text for further explanation.

equimolar proportions at neutral pH. When treated with catalytic levels of trypsin under acidic conditions, however, the inhibitor behaves more like a normal substrate so that (EI)$_t$ undergoes peptide cleavage to form the acyl enzyme (EI)$_a$. The latter then leads to the formation of a second tetrahedral intermediate (EI)$_{t'}$, composed of the enzyme and the modified inhibitor I'. Then (EI)$_{t'}$ rapidly collapses to regenerate the enzyme and the modified inhibitor.

b. Bowman–Birk Inhibitor. A protease inhibitor that has properties quite different from those of the Kunitz inhibitor was first described by Bowman (1944) and subsequently purified and characterized by Birk and co-workers (Birk, 1961; Birk *et al.*, 1963b) and Frattali and Steiner (1969). The so-called Bowman–Birk inhibitor differs from the Kunitz inhibitor in the following important respects:

1. It is a relatively small molecule with a molecular weight of 8000 (Frattali, 1969; Kakade *et al.*, 1970a). Earlier estimates indicated a molecular weight of approximately 20,000, but this was apparently due to the fact that this molecule undergoes self-association, which is concentration dependent (Harry and Steiner, 1969; Miller *et al.*, 1969).
2. It is especially rich in cystine residues in that it has seven disulfide bonds, but it is devoid of glycine and tryptophan (Odani and Ikenaka, 1972).
3. It is a "doubled-headed" inhibitor with independent binding sites for chymotrypsin as well as trypsin (see below).
4. It exhibits marked stability toward heat, acid, and alkali (Bowman, 1946; Birk, 1961), a property most likely attributable to the stabilizing effect of the disulfide bonds on the structure of the protein.

The complete amino acid sequence of the Bowman–Birk inhibitor is shown in Fig. 4. It is a single polypeptide chain with 71 amino acids including seven disulfide bonds. Limited proteolysis with trypsin and chymotrypsin in conjunction with chemical modification studies has clearly established the existence of two independent binding sites for trypsin and chymotrypsin (Seidel and Liener, 1971, 1972a,b,c; Odani and Ikenaka, 1972). The trypsin-reactive site (Lys

16-Ser 17) as well as the chymotrypsin-reactive site (Leu 44-Ser 45) lie within nonapeptide loops formed by a single disulfide bond. The sequences of amino acids surrounding these two reactive sites are in fact remarkably similar not only to each other but to the corresponding active sites of other legume inhibitors (see Table II). By taking advantage of a single methionine residue at position 27, which could be cleaved with CNBr, Odani and Ikenaka (1973b) succeeded in separating the inhibitor into two active fragments: one (38 residues) contained the trypsin-reactive site and inhibited only trypsin, and the other (29 residues) contained the chymotrypsin-reactive site and was effective only toward chymotrypsin. Nishino *et al.*, (1977) reported the synthesis of the nonapeptide loop encompassing the trypsin-reactive site (Cys 14 to Cys 22). Although this peptide did not form a stable complex with trypsin, it did display significant inhibitory activity toward the esterase and peptidase activities of this enzyme.

There have been several recent reports that there are present in soybeans as many as five protease inhibitors that have properties quite similar to that of the Bowman–Birk inhibitor (Mamiha *et al.*, 1972; Odani and Ikenaka, 1976; Hwang *et al.*, 1977a). One of these (designated as CII) was a double-headed inhibitor that inhibited elastase as well as trypsin (Odani and Ikenaka, 1976). The trypsin- and elastase-reactive sites contained arginine and alanine residues, respectively, in regions of the molecule that were homologous to that of the Bowman–Birk inhibitor (see Table II). Another inhibitor (DII) was inhibitory only toward trypsin and had arginine residues at each of two reactive sites. It was suggested by these authors that these double-headed inhibitors had evolved from an ancestral gene by internal gene duplication and subsequent mutation, causing differences in the reactive site. Inhibitor DII probably represents a primitive form of

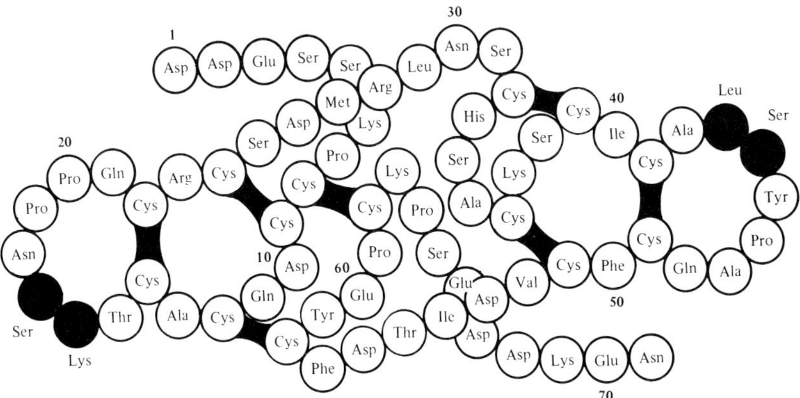

FIG. 4. Amino acid sequence of the Bowman–Birk inhibitor. The reactive sites of trypsin (Lys 16-Ser 17) and chymotrypsin (Leu 44-Ser 45), are shown in black. [From Odani and Ikenaka, 1973a, reproduced with the permission of the *Journal of Biochemistry (Tokyo)*.]

TABLE II

SEQUENCE HOMOLOGY AROUND REACTIVE SITES OF DOUBLE-HEADED INHIBITORS IN LEGUMES[a]

Source				Reactive site ↓						Inhibitor type
Soybeans										
Bowman-Birk[b]	Cys	Ala	Cys-Thr	16 Lys	17 Ser	Asn	Pro	Pro	Gln-Cys	Trypsin inhibitors
CII[c]	Cys	Ala	Cys-Thr	49 Arg	50 Ser	Met	Pro	Gly	Gln-Cys	Trypsin inhibitors
DII[c]	Cys	Met	Cys-Thr	24 Arg	25 Ser	Met	Pro	Pro	Gln-Cys	Trypsin inhibitors
	Cys	Met	Cys-Thr	51 Arg	52 Ser	Gln	Pro	Gly	Gln-Cys	Trypsin inhibitors
Lima bean[d] LBI-I,IV	Cys	Leu / Ala	Cys-Thr	26 Lys	27 Ser	Ile	Pro	Pro	Gln-Cys	Trypsin inhibitors
Garden bean[e] Isoinhibitor II	Cys	Met	Cys-Thr	53 Arg	54 Ser	Met	Pro	Gly	Lys-Cys	Trypsin inhibitors
Chick pea[f]	Cys	Val	Cys-Thr	14 Lys	15 Ser	Ile	Pro	Pro	Gln-Cys	Trypsin inhibitors
Runner bean[g] PCI-3			Ile-Tyr	Lys	Ser	Gln	Pro			Trypsin inhibitors
Soybeans Bowman-Birk[b]	Cys	Ile	Cys-Ala	43 Leu	44 Ser	Tyr	Pro	Ala	Gln-Cys	Chymotrypsin inhibitors
Lima bean[d] LBI-I,IV	Cys	Ile	Cys-Thr	53 Leu	54 Ser	Ile	Pro	Ala	Gln-Cys	Chymotrypsin inhibitors
LBI-IV′	Cys	Ile	Cys-Thr	53 Phe	54 Ser	Ile	Pro			Chymotrypsin inhibitors
Runner bean[g] PCI-3		Asp	Val-Ala	Leu	Ser	Pro				Chymotrypsin inhibitors
Soybean[c] CII	Cys	Met	Cys-Thr	22 Ala	23 Ser	Met	Pro	Pro	Gln-Cys	Elastase inhibitors
Garden bean[e] Isoinhibitor II			Cys-Thr	26 Ala	27 Ser	Ile	Pro	Pro	Gln-Cys	Elastase inhibitors

[a] Identical sequences are enclosed by solid lines.
[b] Odani and Ikenaka (1972).
[c] Odani and Ikenaka (1976).
[d] Stevens *et al.* (1974).
[e] Wilson and Laskowski (1975).
[f] Belew and Eaker (1976).
[g] Hory and Weder (1976); Hory *et al.* (1976).

the double-headed inhibitor, and the prototype inhibitor in legumes was most likely a single-headed inhibitor of trypsin with an arginine or lysine at its active site.

Hwang *et al.* (1977a,b) reported the crystallization of a low molecular weight (6800), double-headed trypsin/chymotrypsin inhibitor that differs from the Bowman–Birk inhibitor in composition and in its failure to cross-react with antisera to the latter. The availability of a double-headed inhibitor in crystalline form affords a unique opportunity for X-ray crystallographic studies of ternary as well as binary complexes of this inhibitor with two different proteases.

c. Inhibitors of Insect Proteases. A soybean fraction was described that inhibited the larval growth and digestive proteases of the flour beetles *Tribolium confusum* (Lipke *et al.,* 1954) and *Tribolium castaneum* (Birk and Applebaum, 1960). This effect, however, could not be duplicated with either the Kunitz or the Bowman–Birk inhibitor (Birk *et al.,* 1962). Birk *et al.* (1963a) were able to purify the larval growth inhibitor by chromatography on hydroxylapatite, and although this growth inhibitor fraction was also capable of inhibiting the activity of larval proteases, it was devoid of trypsin and chymotrypsin inhibitor activity. On the other hand, the proteolytic enzymes from the larval midgut of the yellow mealworm (*Tenebrio molitor*) were inhibited by the Kunitz and Bowman–Birk inhibitors but not by the protein fraction which was active against the *Tribolium* gut proteases (Applebaum *et al.,* 1964)

d. Papain Inhibitor. Learmonth (1951, 1958) reported the presence of a heat-labile inhibitor of papain in the germ of the soybean. A heat-stable inhibitor of papain was also shown to be present in soybean flour (Radoeva, 1968). Since papain is not inhibited by any of the previously described protease inhibitors, it would appear that the papain inhibitor cannot be identical to any of the known soybean inhibitors. Because the papain inhibitor also inactivated the proteases of wheat, which are believed to be responsible for the development of poor loaf volume and texture in bread, the addition of finely ground soybean germ to bread doughs was recommended (Learmonth, 1957).

2. *Phaseolus lunatus* (Lima Bean)

The crystallization of a trypsin inhibitor from lima beans was first reported by Tauber *et al.* (1949), but these crystals were subsequently shown to be an inactive protein to which the inhibor had been adsorbed (Fraenkel-Conrat *et al.,* 1952). The inhibitor isolated by the latter workers was found to be a low molecular weight protein (10,000) rich in cystine (16.5%) and relatively resistant to heat, acid, and alkali. These are properties very similar to those of the Bowman–Birk inhibitor, a fact that was later verified by the striking homology of amino acid sequences between these two inhibitors (see below).

The lima bean inhibitor was subsequently shown to be composed of at least four active components with molecular weights ranging from 8000 to 10,000. All were capable of inhibiting trypsin and chymotrypsin in the fashion of double-headed inhibitors (Jones *et al.*, 1963; Ferdinand *et al.*, 1965; Haynes and Feeney, 1967; Krahn and Stevens, 1970, 1971, 1972a,b, 1973). Partial sequences of two of these inhibitors, LB-I and LB-IV, were determined and the reactive sites for trypsin and chymotrypsin identified (Tan and Stevens, 1971a,b; Stevens *et al.*, 1974). As shown in Table II, the trypsin- and chymotrypsin-reactive sites in both inhibitors are Lys 26-Ser 27 and Leu 53-Ser 54, respectively. A variant of LB-IV (LB-IV′) was also found in which a phenylalanine replaced leucine in the chymotrypsin site. As in the case of the Bowman–Birk inhibitor, there is not only a striking homology between the two sites, but also a striking homology with the equivalent sites in the Bowman–Birk inhibitor.

Little is known concerning the mechanism of interaction between double-headed inhibitors, such as the lima bean inhibitor and trypsin and chymotrypsin. Since, however, the lima bean can form a complex with the catalytically inactive enzymes anhydroptrypsin (Ako *et al.*, 1972a, 1974) and anhydrochymotrypsin (Ako *et al.*, 1972b), it can be concluded that, as in the case of tbe Kunitz inhibitor, the active-site serine residue of these enzymes is not necessary for the binding process.

3. *Phaseolus vulgaris* (Kidney Bean, Navy Bean, French Bean, Garden Bean, Pinto Bean)

The large number of different varieties of beans that are classified as *Phaseolus vulgaris* makes it very difficult to compare the results of investigators who have studied protease inhibitors from this particular species. In the discussion that follows, an attempt is made to compare the properties of the different varieties of *P. vulgaris*.

a. Kidney Beans. Pusztai (1966) appears to have been the first to claim the isolation of an inhibitor from kidney beans that inhibited chymotrypsin as well as trypsin; in a later paper (Pusztai, 1968), he described some of its physicochemical properties. Like most legume inhibitors, its molecular weight was rather low (10,000–15,000), and it was rich in cystine (14%).

b. Navy Beans. Wagner and Riehm (1967) reported the isolation of a trypsin inhibitor from navy beans that had a molecular weight of 23,000 and cystine content of 14%. Bowman (1971) and Whitley and Bowman (1975) more recently reported the presence of a family of protease inhibitors in navy beans and succeeded in purifying one of these to homogeneity. It was similar in size and composition to the Bowman–Birk and lima bean inhibitors.

c. French Beans. Multiple forms of doubled-headed inhibitors of trypsin and chymotrypsin were isolated from the French bean by Belitz and co-workers (Ortanderl and Belitz, 1971; Belitz *et al.*, 1972). All had molecular weights of about 9000 but showed minor differences in amino acid composition, specific activity, and N-terminal groups. The trypsin- and chymotrypsin-reactive sites in one of these were identified as Lys 23-Ser 24 and Phe 52-Ser 53, respectively (Belitz and Fuchs, 1972, 1973). As in the case of the Bowman–Birk inhibitor (Odani and Ikenaka, 1973b), it was possible by appropriate enzymatic and chemical treatments to cleave this inhibitor into two separate active fragments (Weder, 1973).

d. Garden Beans. A number of isoinhibitors were isolated from the garden bean, all having molecular weights in the range 8000–9000 but differing somewhat in amino acid composition and specificity (Wilson and Laskowski, 1973, 1975). Although they all appeared to inhibit trypsin and chymotrypsin to various degrees, one of these, isoinhibitor II, was also a strong inhibitor of elastase. Isoinhibitors I and II had lysine and arginine residues at their respective trypsin-reactive sites (the chymotrypsin-reactive sites were not identified). The elastase-reactive site of isoinhibitor II was identified as an alanine residue. When the partial sequences surrounding the trypsin- and elastase-reactive sites of isoinhibitor II are compared with other double-headed inhibitors (see Table II), their homology to such inhibitors becomes evident.

e. Pinto Beans. Wang (1975) reported the isolation of a crystalline protease inhibitor from pinto beans that had a molecular weight of 19,000 (gel filtration) and was highly active toward trypsin and chymotrypsin. This preparation, however, could be resolved by gel electrophoresis into two components having minimum molecular weights of 9100 and 10,000 (based on amino acid analysis only) and differing markedly in amino acid composition. If these values are doubled, the corresponding molecular weights, 18,200 and 20,000, agree with the molecular weight of the original crystalline protein. Thus, the pinto bean inhibitor appears to have the characteristics of the Kunitz inhibitor with respect to size but resembles most of the other legume inhibitors in terms of its dual specificity. Because of the known tendency of some of these low molecular weight inhibitors to undergo self-association (Harry and Steiner, 1969), the value of 19,000 for the true molecular weight of this inhibitor should be accepted with reservation.

4. Phaseolus coccineus (Runner Bean)

As many as five double-headed isoinhibitors of trypsin and chymotrypsin have been isolated and characterized from the runner bean (Ortanderl and Belitz,

1971; Belitz *et al.*, 1971a, 1972; Weder *et al.*, 1975; Weder and Hory, 1976). Although their isoelectric points ranged from pH 4.46 to pH 5.68 and their molecular weights from 8200 to 11,600, only minor differences in amino acid composition were noted. The dissociation constant for trypsin inhibition ranged from 1×10^{-7} *M* to 2×10^{-9} *M* and that for chymotrypsin inhibition from 1×10^{-6} *M* to 2×10^{-8} *M*. The trypsin- and chymotrypsin-reactive sites of the major isoinhibitor, PCI-3 were identified as Lys-Ser and Leu-Ser, respectively (Hory and Weder, 1976; Hory *et al.*, 1976). Partial sequences surrounding these sites are recorded in Table II.

5. *Phaseolus aureus* (Mung Bean, Green Gram)

The presence of a trypsin inhibitor in green gram was first reported by Honavar and Sohonie (1959). Chu and associates (Chu and Chi, 1965a,b, 1966; Chu *et al.*, 1965) subsequently isolated two inhibitors, A and B, from the mung bean in crystalline form. They established the fact that inhibitor A was derived from inhibitor B as a result of deamination produced by treatment with trichloroacetic acid during the isolation procedure. Estimates of the molecular weight of inhibitor A gave values between 8000 and 9000. This inhibitor was capable of combining with 1 or 2 moles of trypsin, and both of these complexes were crystallized. Chymotrypsin was only weakly inhibited. It was postulated that this inhibitor was in fact a double-headed inhibitor with two trypsin-containing sites. In this respect, therefore, it is quite similar to the soybean inhibitor, DII, described by Odani and Ikenaka (1976).

Baumgartner and Chrispeels (1976) isolated from the cotyledons of the mung bean two protease inhibitor fractions. One of these had a molecular weight of 12,000 and was inhibitory toward trypsin and the seed endopeptidase. The other inhibitor was quite small, having a molecular weight of only 2000, and exhibited inhibitor activity only toward the seed endopeptidase. The larger of these two inhibitors was further resolved by affinity chromatography on Sepharose–trypsin into a trypsin inhibitor component and a component that inhibited only the endopeptidase. None of these inhibitors has been further characterized. The possibility that the endopeptidase inhibitors has an important physiological function in the plant is considered in Section IV.

6. *Phaseolus angularis* (Adzuki Bean)

Two protease inhibitors, designated as inhibitors I and II, were isolated from the Adzuki bean (Yoshida and Yoshikawa, 1975). Inhibitor I, with a molecular weight of 8000, combined with two molecules of trypsin, suggesting that it might be a double-headed inhibitor with two trypsin-containing sites analogous to inhibitor A from the mung bean and inhibitor DII from the soybean. Inhibitor II was somewhat larger than inhibitor I, its molecular weight being 8600, but, unlike the latter, it inhibited both trypsin and chymotrypsin stoichiometrically, indicative of two independent binding sites. The identity of the reactive sites of these inhibitors has not yet been reported.

7. *Arachis hypogaea* (Peanut, groundnut)

Although there were earlier reports of trypsin inhibitor activity in the peanut (Borchers *et al.*, 1947b; Lord and Wakelam, 1950; Cama *et al.*, 1956), Tixier (1968) appears to have been the first to purify and partially characterize this inhibitor, which had a molecular weight of 17,000 and proved to be an inhibitor not only of trypsin, but also of chymotrypsin and plasmin. The antiplasmin activity is of particular interest in view of earlier reports that peanut fractions with antitryptic activity are effective hemostatic agents in the treatment of patients with hemophilia (Boudreaux and Frampton, 1960; Astrup *et al.*, 1960, 1962; Bisordi, 1964). Cepelak *et al.* (1963) also reported the isolation of a protease inhibitor from the skin of defatted peanuts which shortened the bleeding time of experimental animals, presumably due to its antiplasmin activity.

More recent investigations by Hochstrasser *et al.* (1969b) and Hochstrasser *et al.* (1970b) revealed that there are actually two isoinhibitors in peanuts, both of which have a molecular weight of approximately 17,000. These were believed to exist as tetramers of a subunit containing 48 amino acids. Some uncertainty exists regarding the exact location of the trypsin- reactive site of these inhibitors, although it is fairly certain that arginine is one of the constituent amino acid residues. Birk's group (Tur-Sinai *et al.*, 1972; Birk, 1974) also reported the isolation of a trypsin/chymotrypsin inhibitor from peanuts, which differed from those previously reported in that it had a molecular weight of only 8000. Rather unexpectedly, the trypsin-reactive site, which contained arginine, did not prove to be the trypsin-inhibitor site since removal of the new C-terminal arginine residue, which became exposed after trypsin modification, did not cause a loss in trypsin inhibitor activity (Birk, 1974). Birk is of the opinion that the chymotrypsin-inhibiting site coincides with, or is in vicinity of, the trypsin-inhibiting site, and hence this inhibitor may not be a true double-headed inhibitor. Similar to the Kunitz soybean inhibitor, the peanut inhibitor apparently can complex with catalytically inactive trypsinogen or anhydrochymotrypsin (Stewart and Doherty, 1971, 1973).

8. *Vigna sinensis (unguiculata)* (Cowpea, Black-eyed Pea)

The trypsin inhibitor first noted in crude extracts of this legume by Borchers *et al.* (1974b) was subsequently purified and characterized by Ventura and coworkers (Ventura and Filho, 1967; Ventura *et al.*, 1971, 1972a,b). It proved to be a low molecular weight (10,000) trypsin/chymotrypsin inhibitor rich in cystine. Somewhat later Royer's group (Royer *et al.,* 1974; Royer, 1975), using immobilized trypsin, demonstrated the presence of five isoinhibitors having similar properties.

More recently, Gennis and Cantor (1976a–e; Gennis, 1976) described the properties of two double-headed inhibitors from the Redbow variety of black-eyed peas that were isolated by sequential chromatography on immobilized trypsin and chymotrypsin. Both had molecular weights of about 8000 and were particularly rich in cystine, which no doubt accounted for their unusual resistance to denaturation by heat and urea. One of these, BEPCI, had an isoelectric point of 5.1 and inhibited both trypsin and chymotrypsin, whereas the other, BEPTI, was effective only against trypsin. These inhibitors were found to exist in solution primarily as an equilibrium between dimeric and tetrameric forms of the 8000 molecular weight monomer. Unlike the Bowman–Birk and lima bean double-headed inhibitors, which complex with trypsin and chymotrypsin in their monomeric forms (Seidel and Liener, 1972c; Krahm and Stevens, 1971), it is the dimeric form of these inihbitors that interacts with trypsin and/or chymotrypsin at two independent binding sites. In the case of BEPTI, the dimer reacts with one molecule of trypsin at one site on one of the two subunits (so-called "half-site reactivity") and with a second molecule of trypsin at a different site on the other subunit. The net result is that the dimer can bind one or two molecules of trypsin. In the case of BEPCI, the dimer can bind one molecule of chymotrypsin at a site on one of the two subunits and another molecule of trypsin at a second site on the other subunit. Thus, it is possible to have complexes of dimer + trypsin, dimer + chymotrypsin, and dimer + trypsin + chymotrypsin.

During the course of their investigations, Gennis and Cantor (1976a) also isolated an endogenous seed protease from black-eyed peas by affinity chromatography on immobilized Kunitz soybean inhibitor. The fact that this protease apparently exists in the seed as a complex with BEPCI suggests that the latter may have some physiological function in the plant (see Section IV).

9. *Vicia faba* (Field Bean, Fava Bean)

The initial observation by Wilson *et al.* (1972a) that *Vicia faba* contains a trypsin inhibitor was shortly followed by reports describing the isolation and characterization of this inhibitor (Warsy and Stein, 1973; Warsy *et al.,* 1974). Not unexpectedly, several isoinhibitors were noted, two of which were purified. Both inhibitors had a molecular weight of 11,000 but differed in isoelectric points. Affinity chromatography on carrier-bound trypsin was also employed by Ortanderl and Belitz (1977) for the isolation of inhibitors from this legume. Three isoinhibitors were isolated, but since their molecular weights were about 6000 it is not clear whether these inhibitors are the same as those reported earlier by Warsy *et al.* (1974). One of the features that distinguishes the *Vicia faba* inhibitors from most other legume inhibitors is their very broad specificity. They inhibit not only trypsin and chymotrypsin but also thrombin, Pronase, papain, and several serine proteinases of microbial origin.

10. Lathyrus sativus (Field Pea, Chick-pea)

The improved digestibility of *L. sativus* subjected to heat treatment had been attributed to the destruction of a trypsin inhibitor (Esh and Som, 1952), but the actual isolation of the inhibitor and its partial purification was first reported by Roy and Rao (1971). This fraction was subsequently separated into five isoinhibitors by ion-exchange chromatography and gel electrophoresis (Roy, 1972). Sumathi and Pattabiraman (1976) showed that extracts of *L. sativus* were equally potent against trypsin and chymotrypsin and succeeded in isolating two fractions, both of which were active against these two enzymes (Bhat *et al.*, 1976). These inhibitors are most likely double-headed, but this cannot be concluded unequivocally, since these two fractions were not completely homogeneous. The presence of protease inhibitors in *L. sativus* may be of some nutritional significance, since the heat treatment used in the preparation of unleavened bread, which is commonly used in the diet in India, results in a destruction of only about 50% of the inhibitor (Roy and Rao, 1971).

11. Dolichos lablab (Hyacinth Bean, Field Bean, Hakubenzu Bean)

An inhibitor of trypsin in this legume was originally reported by Gaitonde and Sohonie (1951) and subsequently purified and characterized as a protein having a molecular weight of about 24,000 which inhibited not only trypsin but also chymotrypsin and thrombin (Banerji and Sohonie, 1969; Sohonie and Ambe, 1955). A more recent report by Furusawa *et al.* (1974) described the purification of a trypsin/chymotrypsin inhibitor from this bean which had a molecular weight of only 9500, and, unlike most plant inhibitors, it contained more than 20% carbohydrate.

12. Cajanus cajan (Pigeon Pea, Red Gram)

Although the trypsin inhibitor activity of this legume is quite low compared to that of other legumes (Honavar *et al.*, 1962), there have been attempts to purify this inhibitor (Sohonie and Bhandarkar, 1955; Tawde, 1961). Although rigorous proof of purity was lacking, this inhibitor was reported to have a molecular weight of 15,660 and was quite resistant to denaturation by heat.

13. Cicer arietinum (Chick-pea, Bengal Gram)

Earlier reports of trypsin inhibitory activity in this legume (Borchers *et al.*, 1947a; Chattapadhgay and Bannerjee, 1953; Sohonie and Bhandarkar, 1954) have been followed by more thorough studies on this inhibitor. Smirnoff *et al.* (1976) described the isolation of a trypsin/chymotrypsin inhibitor with a molecular weight of about 10,000. Limited proteolysis with these two enzymes indicated that a Lys-X and Tyr-X were the reactive sites for trypsin and chymotrypsin, respectively. Cleavage of the single methionine residue of this molecule with CNBr followed by pepsin yielded two fragments—A, which had a molecular weight of 4382, inhibited only trypsin, and B, the molecular weight of which was 3268, inhibited only chymotrypsin. The specific trypsin inhibitory activity of fragment A was twice that of the native inhibitor, suggesting the unmasking of another trypsin site as a result of the cleavage. The specific chymotrypsin inhibitory activity of fragment B was about one-half that of the native inhibitor, indicating possible conformational damage.

Affinity chromatography on Sepharose-bound trypsin was used by Belew *et al.* (1975) to isolate a trypsin/chymotrypsin fraction that could be resolved into six isoinhibitors by ion-exchange chromatography. All of these isoinhibitors had a molecular weight of about 10,000, and, although they all inhibited chymotrypsin to the same degree, their inhibitory activity toward trypsin differed somewhat. It was subsequently established (Belew, 1977) that four of these isoinhibitors were actually limited degradation products of a common precursor molecule that arose as a result of interaction with matrix-bound trypsin during the course of isolation. These studies emphasize the fact that caution should be exercised regarding the existence of isoinhibitors when matrix-bound trypsin is employed for their purification.

The amino acid sequence surrounding the trypsin-reactive site, Lys 14-Ser 15, as shown in Table II, is strikingly homologous to that of other double-headed legume inhibitors (Belew and Eaker, 1976).

14. Bauhinia purpurea

An inhibitor showing an unusually broad range of specificity toward various proteases was isolated from the seeds of this plant (Goldstein *et al.*, 1973). This inhibitor, which had a molecular weight of 24,300, inhibited trypsin, chymotrypsin, pepsin, and the thiol proteases papain, ficin, and bromelain. Exhaustive dialysis of the inhibitor against mercaptoethanol caused a complete loss of activity toward the thiol proteases without affecting its activity toward trypsin and chymotrypsin. The trypsin inhibitor complex did not inhibit chymotrypsin, indicating that this inhibitor is probably not double-headed. Pinsky and Schwimmer (1974) also isolated an inhibitor of serine proteases from *Bauhinia,* but, unlike the one described by Goldstein *et al.*, this one did not inhibit thiol proteases. Since chymotrypsin interfered with the inhibitor's activity toward trypsin, these authors were of the opinion that the inhibitor had two active centers, the latter being so situated that one enzyme could not complex with the inhibitor after complex formation with the other enzyme. However, one could just as well explain these results by a single active site for which both enzymes would compete. Sufficient evidence is not yet available to permit a conclusion as to whether the two inhibitors isolated by these two groups are in fact the same.

15. Pisum sativum (Garden Pea, Field Pea)

The trypsin inhibitory activity of extracts of the garden pea (Chattapadhgay and Bannerjee, 1953; Learmonth, 1958; Mansfeld *et al.*, 1959) has been attributed to as many as nine isoinhibitors (Weder and Hory, 1972a,b). Their molecular weights fall into a rather narrow range (10,000–12,000), but they differ somewhat in amino acid composition and in their isoelectric points. Since a trypsin matrix was used for the demonstration of these isoinhibitors, it is not certain how much of this heterogeneity may have been due to limited proteolysis.

16. Medicago sativa (Alfalfa)

Ramirez and Mitchell (1960) described the partial purification of a trypsin inhibitor from alfalfa that they ascribed to a nondialyzable polypeptide or noncoagulable protein. More recently, Chien and Mitchell (1970) reported the purification of two inhibitor fractions, but no evidence of the purity of their fractions was presented. The alfalfa inhibitors appeared to be quite heat-stable since extracts of commercial alfalfa meal dehydrated at high temperatures were also inhibitory toward trypsin (Beauchene and Mitchell, 1957). Moorijman (1965) is of the opinion that the alfalfa trypsin inhibitor is a saponin–peptide or saponin–amino acid complex. In this connection it may be mentioned that the saponins of the soybean are known to be effective inhibitors of trypsin by virtue of their ability to interact nonspecifically with proteins (Ishaaya and Birk, 1965).

B. Cereals

1. Hordeum vulgare (Barley)

The amount of protein responsible for the protease inhibitory activity of barley is relatively high and has been estimated to be approximately 0.45 g/kg, or 5–10% of all the water-soluble protein present in barley (Mikola and Suolinna, 1969). The inhibitors present in aqueous extracts of barley are active against a wide variety of proteases including trypsin (Belohlawek, 1964; Mikola and

Suolinna, 1969), the alkaline proteases produced by certain fungi and bacteria (Matsushima, 1955; Mikola and Suolinna, 1971), and the endopeptidases of germinating barley (Enari *et al.*, 1964). The inhibitors responsible for these various activities, however, are not identical and may be found in different parts of the plant. Thus, one of the inhibitors of trypsin, which has a molecular weight of about 14,000, is found in the endosperm. It is quite specific for trypsin and can be immunologically distinguished from a trypsin inhibitor that is present in the embryo and that has a molecular weight of 18,500 (Kirsi and Mikola, 1971; Mikola and Kirsi, 1972). The inhibitor fraction responsible for the inhibition of microbial alkaline proteases and chymotrypsin is found primarily in the endosperm and is actually composed of as many as five isoinhibitors having molecular weights of about 25,000 but possessing isoelectric points ranging from 4.6 to 5.4 (Mikola and Suolinna, 1971). The inhibitor that is active toward the endogenous proteinases of barley is found in small amounts primarily in the endosperm of resting grain and disappears at the onset of germination (Mikola and Enari, 1970). The fact that the disappearance of this inhibitor is followed by an increase in endopeptidase activity suggests that this inhibitor plays a regulatory role in the metabolism of the plant (see also Section IV).

A number of other investigators have also reported the isolation of various protease inhibitors from barley, although it is not clear whether they are identical to those already described. For example, Ogiso *et al.* (1975) isolated a trypsin inhibitor from a Japanese variety of barley that had a molecular weight of 14,200 and in this respect resembled the endospermal trypsin inhibitor of Kirsi and Mikola (1971). This enzyme showed a remarkable stability toward denaturation by urea and guanidine, which was attributed to the stabilizing effect of disulfide bonds (Ogiso *et al.*, 1976). This inhibitor was also shown to have an arginine residue at its reactive site (Ogiso *et al.*, 1975; Boisen, 1976). Warchalewski and Shupin (1973) isolated an inhibitor from the albumin of barley grits, which appeared to differ from previously reported inhibitors in that it inhibited chymotrypsin as well as trypsin. Yoshikawa *et al.* (1976) reported the isolation of a still another inhibitor which, unlike the microbial alkaline protease inhibitors previously described, was strictly specific for subtilisin and had no activity toward any other microbial protease.

2. *Secale cereale* (Rye)

Hochstrasser *et al.* (1969a) used matrix-bound trypsin to isolate two protease inhibitor fractions from rye flour. One was a single molecular species with a molecular weight of 17,000, and the other was a mixture of three isoinhibitors with molecular weights of about 12,000. The larger of the inhibitors was found to be composed of two chains linked by disulfide bonds. It was believed, however, that these two chains arose as a result of the cleavage of an Arg-Ala bond during interaction with the insolubilized trypsin used for the isolation. The low molecular weight inhibitors were not further characterized. Mikola and Kirsi (1972) also reported inhibitors of trypsin to be present in the embryo and endosperm of rye. The inhibitors, however, were not identical in that the one from the embryo had a molecular weight of 18,500, whereas the one derived from the endosperm had a somewhat lower molecular weight of 14,000. The latter exhibited partial immunological identity to the en;dospermal trypsin inhibitor of barley (see previous section). Polanowski (1974) isolated a protease inhibitor from rye endosperm that had a molecular weight of only 10,000 and was active against chymotrypsin as well as trypsin. These inhibitors seem to resemble, in size at least, the low molecular weight isoinhibitors previously reported by Hochstrasser *et al.* (1969).

3. Triticum vulgare (Wheat)

A heat-sensitive trypsin inhibitor was first isolated from wheat flour by Shyamala and Lyman (1964). A trypsin inhibitor in wheat germ, first noted by Creek and Vasaitas (1962), was later isolated by Hochstrasser *et al.* (1969a) and was found to consist of three inhibitors, one with a molecular weight of 17,000–18,300 and two others with a molecular weight of 12,000. The larger of the two inhibitors was shown to have a trypsin-sensitive site at Arg 35-Ala 36.

Little work seems to have been done on protease inhibitors in the endosperm and bran of wheat. Mikola and Kirsi (1972) described a trypsin inhibitor in the endosperm that is similar in properties to that found in endosperm of barley and rye (see previous sections). Mistunaga (1974) found that their endospermal protease inhibitor was quite active against chymotrypsin. This author also reported that an extract of the bran was capable of inhibiting pepsin.

An inhibitor of papain was shown to be present in aqueous extracts of wheat flour and bran (Hites *et al.*, 1951), but this inhibitor does not appear to have been further characterized. Likewise, little is known concerning the relation of an inhibitor of the larval proteases of the flour beetle, *Tribolium castaneum,* present in wheat flour (Applebaum and Konijn, 1966), to some of the other characterized protease inhibitors of wheat.

4. Zea mays (Corn, Maize)

Insolubilized trypsin was employed by Hochstrasser *et al.* (1967, 1968) to isolate a trypsin inhibitor from corn. The purified inhibitor had a molecular weight of 21,000 and appeared to be a polymer of identical subunits containing 168 amino acid residues. The trypsin-reactive site was identified as Arg 24-Leu 25 (Hochstrasser *et al.,* 1970a). Chen and Mitch;ell (1973) isolated a trypsin inhibitor from sweet corn that also displayed moderate activity toward chymotrypsin, but, in the absence of any reported physicochemical properties, its identity to Hochstrasser's inhibitor is not certain. Multiple trypsin inhibitor components have also been reported (Halim *et al.*, 1973b). It is of interest that opaque-2 corn contains about twice as much inhibitor as normal corn (Mitchell *et al.*, 1976).

5. *Oryza sativa* (Rice)

A trypsin inhibitor was isolated from the embryo of rice seed by Horiguchi and Kitagishi (1971). This inhibitor was active toward rice seed protease as well as trypsin. Kato *et al.* (1974) described an inhibitor from rice that inhibited subtilisin but not trypsin and chymotrypsin. This inhibitor contained 200 amino acid residues, including 2 cystine residues, and the sequence of the first 20 N-terminal residues suggests that it may be homologous to the Kunitz soybean inhibitor. This inhibitor may be identical to an inhibitor of *Aspergillus* protease reported earlier by Matsushima (1955) to be present in the bran of rice seed.

6. *Sorghum bicolor* (Sorghum)

In a brief communication Xavier-Filho (1974) reported that an acid extract of sorghum seeds contained a heat-stable inhibitor of trypsin that had a molecular weight of about 15,000. Over one-half of the inhibitor activity in the acid extract, however, was due to low molecular weight components (<6000), which were thought to be tannins, the latter binding nonspecifically to proteins.

C. Other Plants

1. *Solanum tuberosum* (Potato)

Protease inhibitors are believed to account for as much as 15–25% of the soluble proteins of the potato tuber (Ryan *et al.*, 1976). As many as 13 different species of inhibitors have been identified (Belitz *et al.*, 1971a), of which about 10 have been purified and at least partially characterized (Solyom *et al.*, 1964; Hochstrasser and Werle, 1969; Iwasaki *et al.*, 1971, 1972; Kaiser and Belitz, 1971, 1972, 1973; Melville and Ryan, 1970, 1972; Santarius and Belitz, 1972, 1975; Hojima *et al.*, 1973a,b; Kiyohara *et al.*, 1973a,b; Belitz *et al.*, 1974; Ryan *et al.*, 1974; Worowski, 1974; Rouleau and Lamy, 1975; Bryant *et al.*, (1976). Specificity of inhibition varies from one inhibitor to another but includes the serine proteases, carboxypeptidase, papain, microbial proteases, and kallikreins. Specificity apart, those potato inhibitors that have been purified and most extensively characterized fall into three main categories: (1) inhibitor I, molecular weight of 39,000; (2) inhibitor II, molecular weight of 21,000; and (3) the carboxypeptidase inhibitor, molecular weight 4100.

a. Inhibitor I. This inhibitor, because its specificity is directed most strongly toward chymotrypsin, has been referred to as chymotrypsin inhibitor I, although it inhibits subtilisin, Pronase, and other alkaline microbial proteases as well and is a weak inhibitor of trypsin (Melville and Ryan, 1970, 1972; Kiyohara *et al.*, 1973a). This inhibitor exists in solution as a tetramer of four different subunits, each of which has a molecular weight of about 10,000. Even these individual subunits can exhibit microheterogeneity due to additional or substitute amino acids (Richardson *et al.*, 1976). The tetramer inhibits 4 moles of chymotrypsin, which indicates that each of the subunits probably possesses a chymotrypsin-binding site. The amino acid sequences of each of these subunits were determined (Richardson, 1974; Richardson and Cossins, 1974) and were found to display extensive homology with each other but not with other known protease inhibitors. The major chymotrypsin-reactive site is also heterogeneous and has been identified as either Met 47-Asp 48 or Leu 47-Asp 48 (Richardson *et al.*, 1977). There was also some indication of another binding site for trypsin, which was Lys 28-Gln 29 or Arg 55-Leu 56. Due to a disagreement as to the exact number of amino acid residues in inhibitor I, Leu 56-Asx 57 was also proposed as the chymotrypsin-reactive site (Kiyohara *et al.*, 1973a,b). From a mechanistic viewpoint, it is of interest that inhibitor I is capable of forming a complex with anhydrochymotrypsin (Ako *et al.*, 1974).

b. Inhibitor II. This inhibitor, a major component of the heat-stable proteins of the Russet Burbank potato tuber, has a molecular weight of 20,000 and is composed of dimers of four distinctly different protomers (Bryant *et al.*, 1976). The latter are likewise not homologous to the protomers that make up inhibitor I. Inhibitor II must therefore be regarded as a family of dimeric hybrids composed of a heterogeneous mixture of four possible protomers. Since each dimer inhibits two molecules of chymotrypsin, each protomer probably possesses a binding site for this enzyme. Although each of the various protomers is a potent inhibitor of chymotrypsin, the activity of each toward trypsin varies.

Inhibitor II is similar in many respects to inhibitors IIa and IIb isolated from a Japanese variety of potatoes (Kiyohara *et al.*, 1973a; Iwasaki *et al.*, 1971, 1972), to inhibitor A5 from a German variety (Belitz *et al.*, 1971b; Santarius and Belitz, 1972), and to an inhibitor isolated from a Russian variety (Mosolov *et al.*, 1974). These varietal differences were also reflected in the studies by Kasier *et al.* (1974), who found that different varieties of potatoes exhibited different but characteristic inhibitor patterns upon isoelectric focusing, and by Ryan *et al.*, (1976), who noted a sevenfold variation in inhibitor II content among different varieties of potatoes. Despite such varietal differences, preparations of inhibitor II from Japanese, American, and German potato varieties were immunochemically related (Ryan and Santarius, 1976).

Inhibitors IIa and IIb differ somewhat with respect to their specificity, inhibitor II showing the opposite inhibitory pattern (Iwasaki *et al.*, 1972). Iwasaki *et al.* (1974, 1975) succeeded in obtaining active fragments from these two inhibitors. The fragment from inhibitor IIa (molecular weight, 4800) was a strong inhibitor of trypsin, whereas the fragment from inhibitor IIb (molecular weight, 4300) was a potent inhibitor of chymotrypsin and subtilisin. The trypsin-reactive site of the active fragment from inhibitor IIa was identified as Lys 32-Ser 33 (Iwasaki *et al.*, 1976). The chymotrypsin site remains to be identified, but, from studies on the intact inhibitor IIb, this site would also appear to involve a lysine residue (Iwasaki *et al.*, 1973a,b, 1974). In this respect, this inhibitor would resemble the Kunitz soybean trypsin inhibitor, which binds chymotrypsin, albeit weakly, at the same arginine residue that is part of the trypsin-binding site (see Section III,A,′).

c. Carboxypeptidase Inhibitor. An inhibitor of carboxypeptidases A and B, having a molecular weight of approximately 3500, was isolated and characterized by Ryan and co-workers (Rancour and Ryan, 1968; Ryan, 1971; Ryan *et al.*, 1974). This inhibitor proved to be a mixture of two polypeptide chains having 38 and 39 amino acid residues, respectively; these chains had the same amino acid sequence, except that one had an additional N-terminal amino acid (Hass *et al.*, 1975; Nau and Biemann, 1976). From chemical modification studies, Hass *et al.* (1976a) concluded that the carboxyl terminal region of the inhibitor molecule was in contact with the enzyme in the complex. The parallel effects of chemical modification on the inhibitor activity toward carboxypeptidases A and B suggested that both enzymes utilize the same binding site of the inhibitor. Catalytic activity is not required for complex formation since the apoenzyme of carboxypeptidase A still forms a complex with inhibitor (Ako *et al.*, 1976).

d. Other Inhibitors. Mention should be made of several other protease inhibitors that have been purified from potato tubers and that possess properties suggesting that they are probably different from those already discussed. For example, Sawada *et al.* (1974) described a family of isoinhibitors having a molecular weight of approximately 5500 with broad inhibitory specificity toward chymotrypsin, elastase, kallikrein, plasmin, and thrombin. These isoinhibitors appear to resemble, at least in size, a polypeptide inhibitor of chymotrypsin isolated and characterized by Hass *et al.* (1976b). From an evolutionary aspect, it is interesting that this inhibitor has extensive sequence homology with the carboxypeptidase inhibitor and the protomers of inhibitors II, suggesting that all three of the inhibitors may have evolved from a common polypeptide precursor (Ryan and Hass, 1977). Also reported to be present in potato extracts are inhibitors of elastase (Solyom *et al.*, 1964), papain (Hoff *et al.*, 1972), the kallikreins (Werle *et al.*, 1959; Moriya *et al.*, 1970; Hojima *et al.*, 1971a,b,

1973a,b), and cathepsins (Keilova and Tomasek, 1976a,b; Worowski, 1974, 1975; Busse and Belitz, 1976). All of these inhibitors possess properties which suggest that they are not identical to the other protease inhibitors already described here.

2. *Ipomoea batata* (Sweet Potato)

The trypsin inhibitor activity of sweet potatoes, first noted by Sohonie and Bhandarkar (1954), was later attributed to two fractions with antitryptic activity (Sohonie and Honavar, 1956). More recent studies (Sugiura *et al.*, 1973; Ogiso *et al.*, 1974) have shown these inhibitors to have a molecular weight of 23,000–24,000, to be relatively weak inhibitors of plasmin and kallikrein, and to be inactive toward chymotrypsin. Chemical modification studies indicated arginine to be the reactive site, although modification of one tyrosine and one histidine residue also led to a significant loss of activity.

3. *Solanum melongena* (Eggplant)

A trypsin inhibitor fraction isolated from the exocarp of the eggplant was shown by isoelectric focusing to be composed of three components (Kanamori *et al.*, 1975a,b, 1976a). It appeared to be a single-headed inhibitor, the specificity of which was directed primarily toward trypsin and the reactive site of which was identified as Arg-Ser (Yamada *et al.*, 1976). The isolation of a trypsin inhibitor that was dialyzable has also been described (Kanamori *et al.*, 1976b). Since its properties were indistinguishable from those of the nondialyzable inhibitor which had a molecular weight of 5000–10,000, this inhibitor probably represents the same inhibitor which escaped during dialysis.

4. *Scopolia japonica*

During the course of screening various plant cultures for antiplasmin activity, Sakato *et al.* (1975) noted that the calli from *Scopolia japonica* exhibited strong inhibition of trypsin and chymotrypsin and somewhat weaker activity toward plasmin, kallikrein, and pepsin. This inhibitor proved to be a mixture of five components with molecular weights ranging from 4000 to 6000.

5. *Raphanus sativus* (Japanese Radish Seed)

A low molecular weight inhibitor of trypsin was purified by gel filtration from the seeds of the Japanese radish (Hata *et al.*, 1967) and later further resolved by ion-exchange chromatography into three components (Ogawa *et al.*, 1968). Two of these components with molecular weights of 8000 and 12,000 were further characterized (Ogawa *et al.*, 1971a). One of these inhibitors had arginine at its reactive site, whereas the other had lysine. Complexes of these inhibitors were also isolated; these could be dissociated in the pH range of 2–5 or by heat and protein denaturants without loss of activity (Ogawa *et al.*, 1971b).

6. *Ananus sativus* (Pineapple)

During the course of the purification of bromelain from pineapple stem, Perlstein and Kezdy (1975) noted an unexpected increase in enzyme activity, suggesting the presence of an endogenous protease inhibitor. This inhibitor was subsequently shown to consist of a family of seven isoinhibitors which, although differing in charge, had very similar molecular weights (about 5600) and amino acid composition. Within the pH range of 4–5 these inhibitors very effectively inhibited not only bromelain but also the related thiol enzymes papain and ficin and inhibited trypsin slightly. Amino acid sequence studies of one of these inhibitors (fraction VII) (Reddy *et al.*, 1975) revealed that is was composed of two chains, one of which had 41 residues and the other 10 to 11 residues, held together by disulfide bonds. Amino acid substitutions were noted at four sites in the sequence, thus giving rise to further microheterogeneity within fraction VII. Speculations regarding the reactive site of this inhibitor and its possible homology to other protease inhibitors were discussed by Heinrikson and Kezdy (1976).

IV. PHYSIOLOGICAL ROLE IN THE PLANT

In contrast with our detailed knowledge regarding the properties of the protease inhibitors, our knowledge as to their function in the plant, at best, is rather meager and speculative. Whatever role these inhibitors might play in the plant may be reasonably ascribed to their ability to inhibit proteases. Whether their action is directed primarily toward the endogenous proteases of the plant or toward exogenous proteases produced by invading organisms is a question that has yet to be fully resolved. The inhibition of mammalian enzymes, which is so frequently observed, is also a puzzle and may represent an interesting side effect devoid of any true physiological significance as far as the plant is concerned.

A. Role as Inhibitor of Endogenous Proteases

In the case of some plants, such as barley (Kirsi and Mikola, 1971), lettuce (Shain and Mayer, 1965), cowpea (Xavier-Filho, 1973; Royer *et al.*, 1974), rice seed (Horiguchi and Kitagishi, 1971), and mung bean (Baumgarten and Chrispeels, 1976), it was shown that inhibitors that are active against endogenous proteases disappear during germination. Thus, the role of these protease inhibitors was presumed to be one that favors protein catabolism during germination or prevents degradation of storage protein during seed maturation. Attractive as this hypothesis may appear, there are a number of observations that do not support the generality of this concept.

1. In some cases, the protease inhibitors isolated from plants are in fact incapable of inhibiting endogeneous proteases of the same plant. This was shown to be true in the case of soybeans (Birk, 1968), peas (Hobday *et al.*, 1973), and corn (Melville and Scandalios, 1972). Moreover, no change in the protease inhibitor activity of soybeans occurs during germination (Collins and Sanders, 1976), and, in the case of kidney beans (Palmer *et al.*, 1973), there may be an actual increase.

2. In some seeds, such as the pea (Hobday *et al.*, 1973) and mung bean (Baumgarten and Chrispeels, 1976), most of the protease inhibitor is located outside the protein bodies. This would appear to preclude the inhibitor from playing a role in regulating the proteolysis of storage protein in these organelles during germination.

3. A kinetic study of the rise in endopeptidase activity and decline in inhibitory activity in the mung bean revealed no causal relationship between these two phenomena (Baumgarten and Chrispeels, 1976). That the inhibitors might function to protect the cytoplasm from accidental rupture of the protease-containing bodies was suggested as a possibility.

B. Role in Defense Mechanism

Extensive studies by Ryan and co-workers (Ryan, 1968, 1974; Green and Ryan, 1972, 1973; Ryan and Green, 1974; Ryan and Huisman, 1970; Shumway *et al.*, 1970) strongly support the concept that the protease inhibitors may have evolved as part of the defense mechanism of plants toward invading microbes and insects. Many plant leaves, after mechanical wounding or detachment or following an attack by insects, undergo an accumulation of protease inhibitors not only at the site of damage but also in adjacent tissues. For example, the attack of potato leaves by the potato beetle leads to an accumulation of protease inhibitors I and II (Green and Ryan, 1972, 1973); the crown gall tissue of tobacco, tomato, and potato plant induced by tumor bacteria contains high levels of protease inhibitor I (Wong *et al.*, 1976); and the protease inhibitor activity of tomato plants rapidly increases following infection with pathogenic fungi (Peng and Black, 1976). This inhibitor response is believed to be mediated by a wound hormone, or protease inhibitor-inducing factor (PIIF), which triggers the synthesis of protease inhibitors in other parts of the plant following leaf damage. PIIF from one species of plant can frequently induce protease inhibitor accumulation in another plant species (McFarland and Ryan, 1974; Gustafson and Ryan, 1976; Walker-Simmons and Ryan, 1977a).

Although plants are known to produce inhibitors of a wide variety of microbial proteases (see, for example, Sections III,B,1, and III,C,1, and Mosolov *et al.*, 1976) and of the gut proteases of insects (see Sections III,A,1 and III,B,3), in only a few cases has it actually been shown that these inhibitors are effective toward the proteases elaborated by the organisms that attack the host plant. Only in the case of the gut proteases of insects, to which reference has already been made, does this seem to be true. In addition, Senser *et al.* (1974) showed that the extracellular proteases of microbes from spoiled potatoes were, in many cases, inhibited by inhibitors from the same variety of potato. Halim *et al.* (1973c) likewise showed that the trypsin inhibitor of corn suppressed the growth of fungi capable of invading corn, and Weiel and Hapner (1976) noted that those varieties of barley least susceptible to grasshopper attack contained the highest levels of trypsin inhibitor. In the latter two instances, however, it is not certain whether the inhibition of fungal growth or resistance to insect attack was actually due to a curtailment of proteolytic activity.

V. NUTRITIONAL SIGNIFICANCE OF PROTEASE INHIBITORS

A. Soybeans

The fact that protease inhibitors are so widely distributed among the very plants that are an important source of dietary protein throughout the world has stimulated a vast amount of research regarding their possible nutritional significance. Because of the important role that the soybean plays in animal feeding and its potential contribution to human nutrition, the protease inhibitors of this plant have received particular attention. Unfortunately, the literature dealing with this subject is fraught with inconsistencies, claims, and counterclaims so that a clear-cut picture of the role of the protease inhibitors in animal nutrition has yet to emerge. Some of the reasons for this discomforting situation, especially from a reviewer's point of view, can be attributed to variations in experimental conditions involving such factors as the species of the experimental animals as well as their strain, age, and sex, composition of the diets employed, and the failure to use well-defined preparations of the protease inhibitors. This situation is best summarized in the words of one reviewer (Anonymous, 1962), "In spite of the many experimental approaches with a wide variety of animal subjects, it appears that the considerable research effort (in this field) has done more to demonstrate the complexity of the problem than to elucidate the mechanisms involved." This statement is just as applicable today as it was over 15 years ago.

Osborne and Mendel (1917) are generally credited with having been the first to observe that soybeans, unless cooked for several hours, would not support the growth of rats, an observation that has since been extended to many other experimental animals (Liener, 1958). In accordance with the classic concept that the nutritive value of a protein is determined by its amino acid composition, numerous studies were undertaken to determine whether supplementation of the raw protein with various amino acids would achieve the same effect as heating. Such experiments showed that the addition of methionine or cystine to unheated soybean meal improves protein utilization to essentially the same extent as proper heating (Hayward and Hafner, 1941; Evans and McGinnis, 1948; Barnes *et al.*, 1962; Borchers, 1962a). It is important to note, however, that additional methionine will not raise the nutritive value of raw soy to the level of heated soybean similarly supplemented with methionine (Liener *et al.*, 1949).

Although experiments with rats have shown that heated soybean oil meal is somewhat more digestible than the unheated meal, these differences have generally been too small to account for the marked difference in biological value between the two sources of protein (Melnick *et al.*, 1946; Carroll *et al.*, 1952). In fact, the *net* absorption of nitrogen and sulfur (Johnson *et al.*, 1939; Carroll *et al.*, 1952) and methionine itself (Melnick *et al.*, 1946; Liener and Wada, 1953) from the digestive tract of the rat is essentially the same for both raw and heated soybeans. Differences in nitrogen absorption in the rat seem to be confined to the terminal 20% of the small intestine, where over twice as much nitrogen is absorbed from the heated soybean as from the raw (Carroll *et al.*, 1952). It follows, therefore, that a considerable portion of the nitrogen from raw soy that escapes hydrolysis in the small intestine must be absorbed from the large intestine and has little utility for growth. In the chick, however, the net absorption of protein is significantly less with the unheated meal (Evans and McGinnis, 1948; Bouthilet *et al.*, 1950; Saxena *et al.*, 1963d; Nitsan, 1965; Nesheim and Garlich, 1966). Thus, the basic difference between the rat and chick, with respect to the absorption of nitrogen and sulfur from raw and heated soybeans, is that less net nitrogen and sulfur is absorbed from raw soy by the chick than by the rat but the site of absorption rather than the net absorption is of significance in the rat.

Melnick *et al.* (1946), on the basis of experiments involving the *in vitro* release of amino acids from soybean protein by pancreatin, suggested that the methionine of raw soybean was liberated more slowly by the proteolytic enzymes of the intestines than the other essential amino acids. As a result the absorption of methionine is delayed, and it is not available for mutual supplementation of the remainder of the other amino acids. This concept was supproted both by Kunitz's discovery of a heat-labile trypsin inhibitor in raw soybeans (see Section III,A,1) and by the fact that active antitryptic fractions from raw soybeans were capable of inhibiting the growth of rats (Klose *et al.*, 1946; Liener *et al.*, 1949; Borchers *et al.*, 1948b), chicks (Ham *et al.*, 1945), and mice (Westfall and Hauge, 1948). There is in fact direct evidence for the inhibition of proteolysis in the intestinal tract of chicks (Alumot and Nitsan, 1961; Bielorai and Bondi, 1963), an effect that is much less pronounced in older animals (Bornstein and Lipstein, 1963; Saxena *et al.*, 1963c; Nitsan and Alumot, 1964). There have been conflicting reports, however, concerning the effectiveness of supplementing the diet with trypsin in order to overcome the growth inhibitory properties of the trypsin inhibitor (Almquist and Merritt, 1953a; Brambila *et al.,* 1961).

Although Melnick's theory would explain why the trypsin inhibitor interferes with the availability of methionine from raw soybeans, other observations are not in accord with his hypothesis. *In vitro* studies have shown that the trypsin inhibitor does not specifically retard the enzymatic release of methionine but appears to affect all of the amino acids to approximately the same extent (Riesen *et al.,* 1947; Ingram *et al.,* 1949; Liener and Fevold, 1949; Hou *et al.,* 1949; Clandinin and Robblee, 1972; Nehring *et al.,* 1963). Goldberg and Guggenheim (1962) observed that several amino acids, including lysine, tryptophan, as well as methionine, were all more slowly absorbed from the gut in the case of rats receiving raw soybeans versus those fed the heated product. Almquist and Merritt (1951, 1953b) therefore questioned the necessity of postulating a specific interference with the enzymatic release of methionine in order to explain the methionine deficiency provoked by raw soybeans. They believe that the action of this inhibitor involves a general interference with digestion so that a substantial amount of the most limiting acid, which, in the case of soybeans is methionine, is excreted unabsorbed (in the case of chicks) or absorbed too late to be of value to the animal (in the case of rats). In support of the concept, these authors showed that the addition of the trypsin inhibitor, in the form of raw soybean meal, to rations containing proteins with marginal levels of lysine, arginine, isoleucine, or tryptophan caused the experimental animals (chicks) to become markedly deficient in these amino acids.

There is compelling evidence that the growth-retarding effect of the trypsin inhibitor may have little to do with its ability to inhibit protein digestion in the intestines, at least in the case of rodents. Thus, active antitryptic preparations were shown to retard the growth of rats (Desikachar and De, 1947; Liener *et al.,* 1949; DeMuelenaere, 1964; Khayambashi and Lyman, 1966) and mice (Westfall *et al.,* 1948) when incorporated into diets containing predigested protein or free amino acids. There have also been reports that, in contrast to chicks, raw soybeans or trypsin inhibitor preparations do not depress proteolytic activity in the intestinal tract of rats and mice (Lyman, 1957; Brendenkamp and Luck, 1969; Lyman and Lepkovsky, 1957; Haines and Lyman, 1961; Nitsan and Bondi, 1965). These observations raise the question as to how the trypsin inhibitor can inhibit the growth of rats without affecting intestinal proteolysis and at the same time cause an enhanced requirement for methionine.

Chernick *et al.* (1948) were the first to report that chicks fed raw soybeans developed hypertrophy of the pancreas. This observation has since been confirmed not only for the chick (Singh *et al.,* 1964; Lepkovsky *et al.,* 1965; Nitsan and Alumot, 1964, 1965; Garlich and Nesheim, 1966) but also for the rat (Booth *et al.,* 1960, 1964; DeMuelenaere, 1964; Rackis, 1965; Gertler *et al.,* 1967).

Pancreatic hypertrophy has not been observed, however, in other species of animals fed raw soybeans, including calves (Gorrill and Thomas, 1967; Kakade *et al.*, 1976), pigs (Jensen *et al.*, 1974), dogs (Patten *et al.*, 1971), and adult guinea pigs (Patten *et al.*, 1973). Pancreatic hypertrophy can be produced by feeding animals highly purified preparations of the Kunitz inhibitor (Nesheim *et al.*, 1962; Rackis, 1965; Garlich and Nesheim, 1966; Sambeth *et al.*, 1967), although a direct relationship between trypsin inhibitor activity of various soybean fractions and their ability to produce pancratic hypertrophy cannot always be established (Rackis *et al.*, 1963; Saxena *et al.*, 1963a; Pubols *et al.*, 1964). Histological and biochemical examination of the enlarged pancreas reveals true hyperplasia, which is characterized by an increase in the number of cells in the pancreatic tissue (Applegarth *et al.*, 1964; Singh *et al.*, 1964; Kakade *et al.*, 1967; Salman *et al.*, 1968; Saxena *et al.*, 1963b) but which at the same time shows evidence that the zymogen granules have been depleted (Applegarth *et al.*, 1964; Salman *et al.*, 1967). Konijn and Guggenheim (1967) believe that pancreatic enlargement is the result of an increase in cell size rather than the number of cells. However, recent reports tend to indicate hyperplasia as well as hypertrophy of the pancreas in rats fed soybean trypsin inhibitor (Melmed *et al.*, 1976; Yanatori and Fujita, 1976). The reports from the same authors indicate, furthermore, that the soybean trypsin inhibitor has a stimulating effect on the endocrine secretion of rat pancreas as well (Melmed *et al.*, 1973; Yanatori and Fujita, 1976).

Lyman and Lepkovsky (1957) suggested that the growth depression caused by the trypsin inhibitor may be the result of the endogenous loss of essential amino acids derived from a hyperactive pancreas that is responding in a compensatory fashion to the effects of the trypsin inhibitor. The loss of methionine and cystine in this fashion would be particularly acute since soybean protein is notoriously deficient in these amino acids. In agreement with this hypothesis is the observation that amino acid supplementation effectively counteracts the growth depression despite the persistence of pancreatic hypertrophy (Booth *et al.*, 1960; Khayambashi and Lyman, 1966). The nature of this endogenous loss most likely differs in the chick and the rat. In the chick, much of this endogeneous nitrogen and sulfur probably ends up in the feces, as evidenced by the fact that the proteolytic activity of the feces from chicks fed raw soybean is almost entirely of pancreatic origin (Lepkovsky *et al.*, 1959). In the rat, where the fecal loss of nitrogen and sulfur is relatively small, these losses could be incurred as a result of bacterial degradation in the lower part of the intestinal tract (Carroll *et al.*, 1953) or because the cystine remains bound in a form that is not available for utilization by the rat (Barnes *et al.*, 1965a,b). It is entirely possible that the beneficial effect of antibiotics on diets containing raw soybeans (Borchers *et al.*, 1957; Braham *et al.*, 1959; Barnes *et al.*, 1965a,b) may be due to the suppression of microflora responsible for the degradation of the sulfur-containing amino acids in the colon.

Kakade *et al.* 1970b), on the other hand, believe that the action of antibiotics is to increase the intestinal absorption of amino acids, specifically that of the sulfur-containing amino acids. This would meet the demand for the increased synthesis of pancreatic enzymes in response to dietary trypsin inhibitor. This would explain the supplementary effect of antibiotics in overcoming the growth inhibition of rats fed trypsin inhibitor (Kakade *et al.*, 1970b) or the lack of growth depression in germfree chicks fed the trypsin inhibitor (Hewitt *et al.*, 1973), which is observed despite the fact that the pancreas of these experimental animals were enlarged. DeMuelenaere (1964) pointed out that a sloughing-off of the intestinal mucosa may also contribute to the endogenous loss of nitrogen when rats are fed the soybean trypsin inhibitor.

Studies by Geratz (1968, 1969) indicate that *p*-aminobenzamidine (*p*-ABA), a synthetic, low molecular weight compound capable of inhibiting trypsin, is also capable of producing in animals effects similar to those reported for the soybean trypsin inhibitors, namely, growth inhibition, pancreatic enlargement, and increased secretion of pancreatic enzymes. The apparent similarity in the action of *p*-ABA and naturally occurring trypsin inhibitors must be attributed to their ability to inhibit trypsin, a conclusion that was also reached by Kakade *et al.* (1970b).

The most direct proof of the role of trypsin inhibitors in causing the deleterious effects in rats came from the studies of Kakade *et al.* (1973). These workers used a unique approach, which involved feeding rats diets containing unheated soybean protein from which the trypsin inhibitors had been selectively removed by affinity chromatography on insolubilized trypsin. It was concluded that approximately 40% of the pancreatic hypertrophic effect of the original extract could be accounted for by the trypsin inhibitors. The remaining 60% growth-inhibiting and pancreatic hypertrophic effects of the unheated extract apparently were due to the resistance of native protein to attack by digestive enzymes.

The mechanism whereby the trypsin inhibitor causes pancreatic enlargement is not clear. Khayambashi and Lyman (1969) reported that rat pancreas which had been perfused with blood plasma from a rat fed the trypsin inhibitor secreted excessive amounts of enzymes. This plasma factor was found to be relatively unstable at 4°C and had a molecular weight of 10,000 or less. They concluded that soybean trypsin inhibitor enhances the formation and/or release of a humoral pancreozymin-like substance from the intestinal wall, which markedly stimulates the external enzyme secretion of the rat pancreas. Support for such a contention came from the experiments with rats in which repeated injections of pancreozymin (PZ) or cholecystokinin–pancreozymin (CCK–PZ) cause an increase in pancreatic size and enzyme secretion (Rothman and Wells, 1967; Snook, 1969; Mainz *et al.*, 1973). That the trypsin inhibitor-induced pancreatic changes are mediated through the gastrointestinal tract has been verified experimentally by a number of investigators (Melmed and Bouchier, 1969; Ihse *et al.*, 1974; Ihse,

1976; Schneeman *et al.*, 1977). Pancreatic enlargement is thus the consequence of repeated formation and/or release of humoral agent in response to the dietary trypsin inhibitors for a period of time.

An interesting hypothesis was advanced concerning the regulatory role of trypsin inhibitor in causing pancreatic changes (Green and Lyman, 1972; Green *et al.*, 1973; Lyman *et al.*, 1974; Schneeman and Lyman, 1975; Schneeman *et al.*, 1977). These authors suggested that pancreatic enzyme secretion, at least in the rat, is regulated by a negative feedback mechanism mediated by intestinal trypsin and chymotrypsin; i.e., the presence of these enzymes in the small intestine suppresses the secretion of pancreatic enzymes. According to their hypothesis, the soybean trypsin inhibitor evokes a pancreatic enzyme response by effectively removing active trypsin and chymotrypsin from the intestine, thereby preventing their normal feedback regulation of secretion. Dietary protein could also act as a regulator of pancreatic enzyme secretion in much the same way as soybean trypsin inhibitor by complexing with trypsin to form a protein–trypsin complex (Green *et al.*, 1973). This would explain the growth inhibition and pancreatic enlargement of rats fed native proteins (Kakade *et al.*, 1973; Thompson and Liener, 1978). It should be pointed out that the exact nature of the manner in which the humoral factors are released by removal of trypsin or chymotrypsin from the intestine is not yet known.

Melamed and Bouchier (1969) suggested another mode of pancreatic stimulation, which is determined by the level of pancreatic trypsin inhibitor released into the intestines. By this so-called positive feedback mechanism the pancreatic inhibitor itself provides an important stimulation for the production and repletion of digestive enzymes by the acinar cells of the pancreas. In other words, while the negative feedback mechanism may initiate the stimulation of the pancreas leading to an increased production of the pancreatic inhibitor, the latter reinforces and sustains this stimulatory effect by a positive feedback mechanism. The pancreatic inhibitor would also be expected to lower the level of intestinal trypsin in much the same way as the soybean trypsin inhibitor; so it could likewise be a participant in the negative feedback effect. The net effect of this rather complicated series of events is an accentuated loss of the sulfur-containing amino acids, resulting in growth inhibition when the dietary supply of these amino acids is marginal.

In an attempt to establish a more direct relationship between the soybean trypsin inhibitor and the utilization of methionine, Barnes and his group (Kwong *et al.*, 1962; Kwong and Barnes, 1963; Barnes and Kwong, 1965) investigated the effect of the administration of the Kunitz inhibitor in the metabolism of methionine. By using radioactive methionine, these workers demonstrated an increased conversion of methionine to cystine and metabolic CO_2 following a dose of the trypsin inhibitor. Most of the increased synthesis of cystine took place in the pancreas. It was postulated that the effect of the trypsin inhibitor was to increase the metabolic conversion of methionine to cystine, which would intensify the body's need for methionine for tissue protein synthesis. Similar experiments conducted by Frost and Mann (1966) led them to a somewhat different hypothesis, namely, that the trypsin inhibitor interfered with the incorporation of cystine into protein by blocking the enzyme cystathionine synthetase. Borchers *et al.* (1965) studied the effect of raw soybeans on the metabolism of several labeled amino acids and concluded that some factor in raw soybeans interferes with the catabolism of threonine and valine. These are the same amino acids which, in combination with methionine, produce maximal growth response in rats when added to diets containing raw soybeans (Booth *et al.*, 1960; Borchers, 1961, 1962b).

Looking at the evidence as a whole, it seems questionable whether it is necessary to postulate that the soybean trypsin inhibitors play a direct role in the metabolism of certain amino acids. Certainly the increased secretory activity of the pancreas that accompanies the ingestion of trypsin inhibitors would be expected to result in an increase in the synthesis of pancreatic enzymes, which are known to be rich in cystine (Neurath, 1961). This would account for the disproportionate amount of cystine found in the small intestines of rats receiving raw soybeans (Carroll *et al.*, 1953). Since this cystine is most likely derived from methionine, the preferential synthesis of pancreatic enzymes would accentuate the need for methionine required for the synthesis of other body tissues. It is of interest that homoserine, produced as a by-product of the conversion of methionine to cysteine, and valine and threonine can all undergo the same metabolic fate through propionate as a common intermediate (Fig. 5). Since the concentration of propionate would be increased when methionine is converted to cysteine, the rates at which threonine and valine would be metabolized by the same pathway might be expected to be repressed.

In the final analysis, the relative proportions of trypsin inhibitor and proteolytic enzymes present in the small intestine at any given time will determine the manner in which nitrogen is lost from the body (see Fig. 5). In the case of the young chicken the hypertrophic response of the pancreas is delayed (Nitsan and Alumot, 1964) so that the amount of trypsin produced by the pancreas is not sufficient to counteract the trypsin inhibitor (TI $>$ T), and an inhibition of intestinal proteolysis results. It has been suggested that the higher level of trypsin inhibitor in the intestines of chicks may be due, in part, to the fact that little peptic inactivation of these inhibitors occurs in the gizzard of the chicken compared to the rat (Nitsan and Bondi, 1965). As a result of inhibited proteolysis, a substantial portion of the dietary protein is excreted into the feces and represents an exogenous loss of nitrogen. In the case of the rat and the older chicken, the quantity of trypsin and other enzymes poured out from the pancreas is sufficient to prevent an inhibition of proteolysis (T $>$ TI) so that much of the nitrogen found in the intestinal tract is of endogenous origin and is particularly rich in

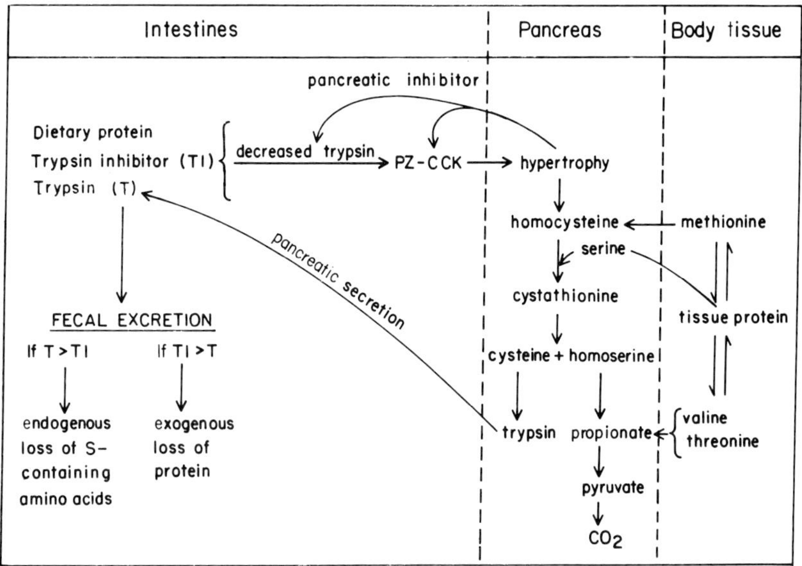

FIG. 5. A scheme proposed to explain the experimental evidence relating to the effect of the soybean protease inhibitor on the nutritive value of the protein. See text for explanatory details. Detailed information concerning the metabolic pathways of methionine, valine, and threonine can be found in West *et al.* (1966).

cystine. The latter is apparently lost to the organism because of subsequent bacterial destruction or because it exists in a bound, unavailable form, although fecal losses of nitrogen and methionine may also occur to some extent (Kwong *et al.,* 1962). It can be concluded, therefore, that the trypsin inhibitor may cause both an exogenous and endogenous loss of nitrogen and that the relative importance that each of these pathways assumes depends on the experimental animal and the conditions selected by the investigator.

The deleterious effects produced in rats and chicks by feeding soybean trypsin inhibitor prompted a number of investigators to examine its nutritional role in other animals. Trypsin inhibitor failed to show any deleterious effects in calves (Kakade *et al.,* 1976) and growing pigs (Yen *et al.,* 1977). The feeding of raw soybeans to pigs (Jensen *et al.,* 1974), calves (Gorrill and Thomas, 1967; Kakade *et al.,* 1976), dogs (Patten *et al.,* 1971), and adult guinea pigs (Patten *et al.,* 1973) likewise failed to produce pancreatic enlargement and excessive secretion of enzymes similar to those observed in rats and chicks fed raw soybeans. To help explain these differences among various species, Kakade *et al.* (1976) discussed an interesting relationship between the size of the pancreas and the hypertrophic response to raw soybeans or the inhibitor purified therefrom. As shown in Table III, there appears to be a direct relationship between the size of the pancreas and the sensitivity of response to raw soybeans or the trypsin

TABLE III

RELATIONSHIP BETWEEN SIZE OF PANCREAS OF VARIOUS SPECIES OF ANIMALS AND RESPONSE OF PANCREAS TO RAW SOYBEANS OR TRYPSIN INHIBITOR

Species	Size of pancreas (% of body weight)	Pancreatic hypertrophy	Reference
Mouse	0.6–0.8	+	Schingoethe *et al.* (1970b)
Rat	0.5–0.6	+	Kakade *et al.* (1973)
Chick	0.4–0.6	+	Lepkovsky *et al.* (1959)
Guinea pig	0.29	±[a]	Patten *et al.* (1973)
Dog	0.21–0.24	−	Patten *et al.* (1971)
Pig	0.10–0.12	−	Yen *et al.* (1977)
Human being	0.09–0.12[b]	(−)[c]	
Calf	0.06–0.08	−	Kakade *et al.* (1976)

[a]Observed in young guinea pigs but not in adults.
[b]Taken from Long (1961).
[c]Predicted response.

inhibitor. The pancreas of those species of animals whose weights exceeded 0.3% of the body weight became hypertrophic, whereas those whose weights were below this value did not respond to the hypertrophic effects of trypsin inhibitor. The guinea pig would appear to be on the borderline in that the adult guinea pig, in contrast to young ones, was insesitive in its pancreatic response (Patten *et al.*, 1973). One would predict from this relationship that the human pancreas would not be sensitive to the trypsin inhibition and that the soybean trypsin inhibitor would play a minor role, if any, in human nutrition.

One can only speculate as to the reason for this apparent relationship between pancreas size and sensitivity to the hypertrophic effect of protease inhibitors. A possible explanation may lie in the fact that there appears to be a direct correlation between the size of the pancreas and the level of proteolytic activity that is produced (Goss, 1966; Schingoethe *et al.*, 1970b). It follows from this that those animals with a larger pancreas (as a percentage of the body weight), such as the rat and the chick, have a higher requirement for the sulfur amino acids needed for the synthesis of pancreatic enzymes than do animals with a smaller pancreas such as the dog, pig, and calf. Thus, animals with a larger pancreas might be expected to be more responsive to stimuli, such as the soybean trypsin inhibitor, which increase the production of these enzymes by causing pancreatic hypertrophy.

The structural features of trypsin inhibitors themselves seem to have some nutritional significance. Since many of the protease inhibitors present in legumes are rich in cystine (see Section III,A), Kakade *et al.* (1969) hypothesized that a dietary loss of cystine derived from the inhibitor itself could contribute in a significant fashion to the poor protein quality of these legumes. It can be estimated that the protease inhibitors, although comprising only about 2.5% of the bean protein, contribute approximately 32 and 40% of the total cystine of the protein of lima and navy beans, respectively. Experimental results designed to test the above-mentioned hypothesis support the contention that in the case of the navy bean protease inhibitor, the cystine that it contains is not available to chicks for growth. It was further demonstrated that the unavailability of cystine from the navy bean protease inhibitor was due to its resistance to enzymatic attack. This resistance to enzyme attack is probably due to the highly compact structure of the molecule produced by a large number of disulfide bonds.

Preoccupation with attempts to elucidate the nutritional significance of the soybean trypsin inhibitors should not obscure the fact that other growth inhibitors may also be present in raw soybeans (Schingoethe *et al.*, 1970a, 1974; Tidemann and Schingoethe, 1974). Several reports have appeared in the literature in which soybean fraction possessing high levels of antitryptic activity were shown to have little or no growth-depressing activity (Rackis *et al.*, 1963; Saxena *et al.*, 1963a; Garlich and Nesheim, 1966; Sambeth *et al.*, 1967). Conversely, the water-soluble residue of soybeans with little or no antitryptic activity was reported to inhibit the growth of rats (Borchers *et al.*, 1948a; Birk and Gertler, 1961; Rackis *et al.*, 1963; Saxena *et al.*, 1963a; Rackis, 1965). It was also reported that the Bowman–Birk inhibitor failed to produce growth depression in rats and/or chicks despite the fact the pancreas of these animals were hypertrophic (Gertler *et al.*, 1967; Nitsan and Gertler, 1972). Some workers reported that the growth inhibition in raw soybeans cannot be completely overcome by supplementation with amino acids (Hill *et al.*, 1953; Saxena *et al.*, 1962), which suggests that an interference with the availability of amino acids, such as would be expected from the effects of a trypsin inhibitor, is not the only factor involved. Also unexplained is the fact that the nutritive value of germinated soybeans is superior to that of the raw meal, despite the fact that the trypsin inhibitor content is unchanged (Desikachar and De, 1947, 1950). Aside from the possible role that the hemagglutinin might play in the nutritive value of raw soybeans and other legumes (see Chapter 3), little is known about the nature of such growth inhibitors.

B. Other Plants

The widespread distribution of trypsin inhibitors in legumes provides the most likely explanation for the observation that heating increases the *in vitro* (Waterman and Johns, 1921; Waterman and Jones, 1921; Jones and Waterman, 1922; Kakade and Evans, 1963) and *in vivo* (Jaffé, 1950a) digestibility of many leguminous proteins. It should be noted, however, that not all legumes that have trypsin inhibitors have their nutritive value enhanced by heating (Borchers and Ackerson, 1950; Jaffé, 1950b).

Since the growth of animals fed diets containing the lima bean inhibitor is depressed (Klose *et al.*, 1949; Tauber *et al.*, 1949), it is generally believed that the beneficial effect of heat treatment on the nutritive value of lima beans (Johns and Finks, 1920a; Finks and Johns, 1921a; Everson and Heckert, 1944; Richardson, 1948) is due to the destruction of the trypsin inhibitor. As in the case of soybeans, it would appear that the trypsin inhibitor does not exert its deleterious effect on growth solely by an inhibition of intestinal proteolysis, since lima bean fractions high in antitryptic activity inhibited the growth of rats fed acid-hydrolyzed casein (Klose *et al.*, 1948). Pancreatic hypertrophy was observed in rats fed the purified lima bean inhibitor (Lyman *et al.*, 1962).

Although the beneficial effect of heat on the nutritive value of many varieties of *Phaseolus vulgaris* is well known (Johns and Finks, 1920a,b; Everson and Heckert, 1944; Richardson, 1948; Evans and Bandemer, 1967), only recently have attempts been made to assess the role of the trypsin inhibitors. Jaffé and Lette (1968) reported that certain varieties of *Phaseolus vulgaris* having very low levels of trypsin inhibitor activity nevertheless had poor growth-promoting properties for rats unless autoclaved. Kakade and co-workers (Kakade and Evans, 1965a,b) isolated fractions of the navy bean that were high in antitryptic activity and inhibited the growth of rats and chicks. Pancreatic hypertrophy accompanied the ingestion of kidney beans (Wagh *et al.*, 1963) and navy beans (M. L. Kakade, T. L. Barton, and P. J. Schaibel, unpublished observations, 1967) by chicks. Protein fractions of the kidney bean that are devoid of antitryptic activity but high in hemagglutinating activity also inhibit the growth of rats (Honavar *et al.*, 1962; see also Chapter 3). It is clear that it is difficult to define the precise role of the trypsin inhibitors of *Phaseolus vulgaris* in view of the multiple nature of the growth inhibitors that exist in this species.

Purified trypsin inhibitors from the double bean (*Faba vulgaris*) and the field bean (*Dolichos lablab*) do not inhibit the growth of rats (Apte and Sohonie, 1957; Sohonie *et al.*, 1958; Phadke and Sohonie, 1962). Studies of Wilson *et al.* (1972a,b) also indicate that broad bean (*Vicia faba* L.) trypsin inhibitor had no effect on the growth of chicks, although pancreatic hypertrophy and reduced feed efficiency were observed in the same group of chicks. Impairment of growth and pancreatic hypertrophy were observed in rats fed the trypsin inhibitor from peanuts (Kwaan *et al.*, 1968) and from *Lathyrus sativus* (Roy, 1972). A trypsin inhibitor isolated from alfalfa was found to be growth inhibitory when fed to mice (Delic *et al.*, 1974). Mitchell *et al.* (1976) isolated a trypsin inhibitor from opaque-2 corn but failed to find any deleterious effects on growth or weight of pancreas of rats fed the inhibitor. Creek and Vasaitis (1962) suggested that the beneficial effect of heat treatment on wheat germ (Creek *et al.*, 1961; Parish and

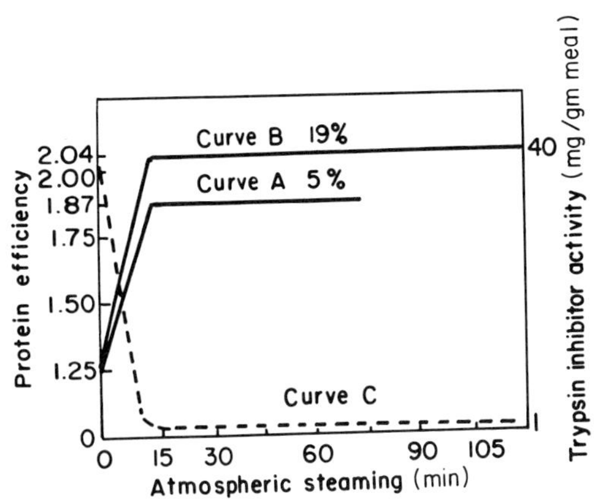

FIG. 6. Effect of autoclaving on protein efficiency and trypsin inhibitor activity of raw soybean meal. Conditions: live steam at atmospheric pressure, 100°C; curve A, protein efficiency (grams of weight gained per gram of protein consumed) of meals with 5% moisture before autoclaving; curve B, protein efficiency of meals with 19% moisture prior to autoclaving; curve C, decrease in trypsin inhibitor activity with time of autoclaving. (From Rackis, 1972, reproduced by permission of the Avi Publishing Co., Westport, Connecticut.)

trypsin inhibitor content and nutritive value of soybean dhal (dehusked, split soybeans) that had been soaked for half a minute (moisture content 31.3%) or overnight (moisture content 60.4%). In the case of beans soaked by either method, steaming for 60 min destroyed the inhibitor completely, whereas it was necessary to autoclave the unsoaked dhal (moisture content 6.5%) at 15 psi pressure for 30 min to obtain the same result. The improvement in nutritive value (protein efficiency using rats) obtained under these conditions was somewhat better in the case of the soaked beans than with the unsoaked sample.

Only partial inactivation of the trypsin inhibitor in cotyledons, chips, or whole soybeans was obtained by steaming for 20 min, presumably due to larger particle size (Rackis, 1966). The trypsin inhibitor of whole soybeans with an initial moisture content of 20% was completely destroyed by atmospheric steaming for 15 min, and, when the moisture content was raised to 60% or more by overnight soaking, boiling for only 5 min sufficed to inactivate the inhibitor (Albrecht *et al.*, 1966). A simple method involving extrusion cooking has been developed for producing a full-fat soybean flour for use in underdeveloped countries (Albrecht *et al.*, 1967; Mustakas *et al.*, 1964). The final product is free of trypsin inhibitors and is reported to be nutritionally equivalent to good-quality toasted soybean flour.

Most investigators are in general agreement that, when properly processed, soybean milk, which is essentially a water extract of the soybean, has a nutritive value almost equivalent to that of cow's milk (Shurpalikar *et al.*, 1961). The

Bolt, 1963) is due to the destruction of a trypsin inhibitor. Sh (1964) pointed out that, since whole wheat flour contains only inhibitor activity of raw soybean meal, it is unlikely that th could exert any influence on its nutritive value even if it we heated. Tsai (1976) attributed the observed growth depressio hypertrophy in chicks fed unheated rice bran to its trypsin inh

VI. EFFECT OF PROCESSING

A. Heat Treatment

Most of the plant protease inhibitors are destroyed by heat, an ef ally accompanied by a general enhancement of the nutritive value (Liener, 1962). It is the relative ease with which deleterious compo the trypsin inhibitors, can be inactivated by appropriate methods o has no doubt contributed to the popularity of legumes as a staple the diet in many countries of the world. The various methods legumes for human consumption were described by Dean (1958).

The soybean has received the most attention with respect to the treatment on trypsin inhibitory activity, and, in general, the extent t activity is destroyed by heat is a function of the temperature, duratio particle size, and moisture conditions—variables that are closely cont industrial processing of soybean oil meal in order to obtain a pro maximal nutritive value. The economic importance of controlling th was emphasized by McKinney and Cowan (1956), who pointed ou percentage of increase in feed efficiency that can be effected by prop ing is worth about $4 million per year to American agriculture.

Borchers *et al.* (1947a) reported that the trypsin inhibitor activity o extracted soybean meal was destroyed by exposure to flowing steam f or by autoclaving under the following conditions: 5 psi for 45 min, 10 min, 15 psi for 20 min, or 20 psi for 10 min. It is generally agreed that th inhibitor is associated with soybean meals that have been inadequately h achieve optimal nutritive value (Borchers *et al.*, 1948a; Westfall and 1948). The effect of atmospheric steaming (100°C) on the trypsin inhibi tent and protein efficiency of dehulled and defatted raw soybean flak studied at two levels of moisture content, 5 and 19% (Smith *et al.*, Rackis, 1965). These results, shown in Fig. 6, reveal that, regardless of m content, over 95% of the trypsin inhibitor content of the meals is des within 15 min. The improvement in protein efficiency effected by atmos steaming at a level of 19% moisture was somewhat higher than at the 5% Similar results were also reported for full-fat soybean flakes (Rackis, 196

Krishnamurthy *et al.* (1958) studied the effect of heat processing o trypsin inhibitor activity inherently present in soy milk can be effectively eliminated by heating the milk at 93°C for 30–75 min, 5–10 min at 121°C, or by spray-drying the milk for 30 min at 121°C (Hackler *et al.*, 1965; Van Buren *et al.*, 1964). In a detailed study of the effect of varying degrees of heat treatment to which soy milk was subjected, maximal protein efficiency ratios were obtained when about 90% of the trypsin inhibitor had been destroyed (Hackler *et al.*, 1965).

Churella *et al.* (1976) measured the trypsin inhibitor activity of commercially available soy infant formulas and concluded that low residual trypsin inhibitor activity remaining in the formula had no nutritional significance for the rat, as judged by growth and size of the pancreas of the animals. Indeed, the nutritional quality of soybean milk for infant feeding is quite comparable to that of cow's milk (Kay *et al.*, 1960; Fomon, 1962).

The trypsin inhibitor activity contributed by soybeans used in the preparation of Mexican tortilla is destroyed by treatment with 1% $Ca(OH_2$ at 80°C for 1 hr (Cravioto *et al.*, 1951). Trypsin inhibitory is reduced to virtually zero in bread that has been prepared from formulas containing 3% soybean flour (J. W. Hayward, personal communication). Kotter *et al.* (1970) reported that canned frankfurter-type sausages containing 1.5% soy isolate were essentially devoid of any trypsin inhibitor activity after the canning process. Furthermore, Nordal and Fossum (1974) reported that the trypsin inhibitor activity provided by soybean protein in meat products was actually made more labile to heat inactivation due to some component in the meat ingredients. They postulated that this factor increased the sensitivity of the trypsin inhibitors to heat inactivation by causing a rupture of disulfide bonds in the inhibitor molecule, particularly the Bowman–Birk inhibitor, which is rich in disulfide bonds.

It was pointed out that rats fed commercially toasted samples of soybean flour (steam-treated at 100°C for 20 min with 18% moisture) (Kakade and Liener, 1973) with PDI's (protein dispersibility index, which measures the amount of protein that can be dispersed in water) ranging from 10 to 30 have consistently grown as well as those fed casein. The destruction of trypsin inhibitor in the toasted soy flour is estimated to be 90–92% (Kakade *et al.*, 1974). However, several reports in the literature indicate that the nutritive value of fully toasted soy flour can be enhanced further by acid or alkali treatment (Colvin and Ramsey, 1969; Delobez *et al.*, 1971), an effect that was attributed to the complete or greater degree of destruction of residual trypsin inhibitor. Although azeotrope extraction of defatted soy flakes with hexane–alcohol destroys lipoxygenase activity and thus improves the flavor score, this treatment must be combined with live steam in order to inactivate the trypsin inhibitor fully (Rackis *et al.*, 1975a). Thus, a combination of azeotrope extraction and live steam enhances the nutritive value as well as the flavor of edible soy flours.

An examination of trypsin inhibitor activity of several commercially available soy products, e.g., soy protein concentrate, soy isolate, textured soy protein (extruded as well as spun), and soy infant formula, indicates about 90% destruction of the inhibitory activity (Kakade *et al.*, 1974; Liener, 1973; Churella *et al.*, 1976; M. L. Kakade and S. M. Martin, unpublished observations, 1976). Since Rackis *et al.* (1975b) found that only 70–80% of the trypsin inhibitory activity needed to be destroyed in order to achieve maximal gains in weight and in PER's with rats, the above-mentioned soy products should be high in nutritional quality.

Borchers *et al.* (1972) used dielectric heating to bring about rapid improvement in nutritional quality of soybeans. These authors found 72% loss of inhibitory activity due to dielectric heating at 168°C for 2 min. The heating of soybean meals by microwave radiations has also been used (Wang and Alexander, 1971).

Aside from the soybean, comparatively few studies have dealt with the effect of variable heat treatment on the trypsin inhibitor content of other crude plant materials. Kakade and Evans (1965b) reported that autoclaving navy beans for 5 min at 121°C destroyed about 80% of the trypsin inhibitory activity, and the growth performance of rats that were fed beans subjected to this heat treatment was considerably improved. Longer heating periods, however, were detrimental. The destruction of the inhibitor and improvement in nutritive value were paralleled by an increase in the enzymatic liberation of lysine as measured *in vitro* (Kakade and Evans, 1966a). Chemical determination of available lysine did not prove to be a very reliable index of the effect of heat on the quality of inadequately heated navy bean protein.

Ryan (1966) pointed out that the chymotrypsin inhibitor of the potato is quickly destroyed by heating in the intact potato even though the purified inhibitor is quite stable. Bessho and Kurosawa (1966) also studied the effect of varying degrees of heat treatment on the inactivation of the *Bacillus* protease inhibitor contained in homogenates and slices of potatoes.

Marquardt *et al.* (1976) observed that about 90% of trypsin inhibitor activity of broad beans (*Vicia faba*) was destroyed by autoclaving (120°C for 20 min), extrusion cooking (152°C), or microwave radiation at 107°C for 30 min. The growth of chicks fed heated broad bean improved significantly over those fed unheated beans, although autoclaving gave the best weight gain in chicks. The studies of Elias *et al.* (1976), on the other hand, indicated that autoclaving (120°C for 15 min), cooking at 90–95°C for 45 min, toasting at 210°C for 30 min, or extrusion cooking destroyed only 10–50% of the inhibitory activity of cowpea (*Vigna sinensis*).

Dry heat at 125°C for 5 hr or 140°C for 2 hr or 150°C for 1 hr destroyed 90% of the inhibitory activity of peanut (*Arachis hypogaea*), whereas moist heat at 100° for 15 min was sufficient to destroy 100% of the inhibitory activity (Woodham and Dawson, 1968).

B. Germination

Germination is known to result in an improvement in the nutritive value of soybeans (Everson *et al.*, 1944; Mattingly and Bird, 1945; Viswanatha and De, 1951), although the level of the trypsin inhibitor does not appear to change during germination (Desikachar and De, 1947; Collins and Sanders, 1976). Germination also improves the biological value of black gram (*Phaseolus mungo*) without any significant change in antitryptic activity (Chattapadhgay and Bannerjee, 1953). Conflicting results, however, were reported for lentil (*Lens esculenta*) and chick-pea (*Cicer arietin;um*); an improvement in nutritive value of these legumes after germination was observed in one instance (Chattapadhgay and Bannerjee, 1953) but not in another (Devadatta *et al.*, 1951). Kakade and Evans (1966b) reported that the germination of navy beans (*Phaseolus vulgaris*) did not improve its nutritive value, despite the fact that there was a slight transitory decrease in trypsin inhibitor activity during the first two days of germination. Curiously enough, the biological value of the field pea (*Pisum sativum*) decreased following germination without any observable change in trypsin inhibitor content (Chattapadhgay and Bannerjee, 1953). Hobday *et al.* (1973), in fact, reported a decrease in the trypsin inhibitory activity of germinated field peas. The results of Jaya *et al.* (1976) and Venkatamaran *et al.* (1976) indicate that germination of green gram (*Phaseolus aureus*) and cowpea (*Vigna sinensis*) did not improve their nutritive value when fed to rats. Chick-pea germinated for 24 hr showed an improvement in PER as compared to that of ungerminated chick-pea. However, the germination of all three legumes for 72 hr reduced their protein quality. It is unfortunate that in no instance were trypsin inhibitor activity measurements made. Germinating horse gram (*Dolichos biflorus*) and moth bean (*Phaseolus aconitifolius*) for 72 hr increased the *in vitro* protein digestibility of these legumes and also decreased the trypsin inhibitory activity, although these changes were more marked in germinated moth bean than in germinated horse gram (Subbulakshmi *et al.*, 1976).

Palmer *et al.* (1973) found a gradual improvement in the weight gain of rats fed kidney beans germinated for various periods of time. Since a significant improvement in the nutritive value of the beans was obtained despite a twofold increase in trypsin inhibitor activity after 8 days of germination, the authors ruled out the trypsin inhibitors as a major toxic component of raw beans.

In general, there would appear to be little correlation between the effect of germination on the protease inhibitors of seeds and their nutritive value, and other factors affecting growth response are presumably involved.

C. Fermentation

Fermented soybean preparations known as *"tempeh"* or *"natto"* are popular dietary items in the Orient (Dean, 1958; Smith, 1963). Van Veen and Schaeffer

(1950) reported that the nitrogen of tempeh is readily available, which they attributed to the destruction of the trypsin inhibitor during the preparation of this food. Other investigators (Hackler *et al.*, 1964; Smith *et al.*, 1964b), however, noted little if any improvement in the nutritive value of soybeans fermented by the tempeh mold. Steinkraus *et al.* (1962) in fact found that prolonged fermentation actually caused a decrease in the growth-promoting quality of tempeh fed to rats. Since a decrease in lysine and methionine took place during fermentation (Steinkraus *et al.*, 1962; Stillings and Hackler, 1965), the lower nutritive value sometimes observed for tempeh may be due to amino acid deficiencies. There was no evidence of pancreatic enlargement in rats fed diets containing tempeh (Smith *et al.*, 1964b), which would indicate that the trypsin inhibitors of soybean had been destroyed, presumably due to the heat treatment involved in the preparation of tempeh (the beans are boiled for 30 min prior to the fermentation). Several reports suggest that the nutritive value and digestibility of natto are somewhat better than that of the unfermented bean (Cheong *et al.*, 1959; Arimoto *et al.*, 1962) but less than that of autoclaved soybeans (Hayaishi and Ariyama, 1960) or casein (Standal, 1963). Van Veen and Steinkraus (1970) claim that the digestibility of soybean products for human beings is greatly enhanced by fermentation. Studies by Wang *et al.* (1968, 1969) indicate that the use of *Rhizopus oligosporus* improves the nutritional quality of fermented soy and/or wheat products. None of these studies, however, have actually involved measurements of the trypsin inhibitor content of the fermented products.

D. Protein Isolates

The term "protein isolate" does not necessarily imply that, since a protein has been isolated, it is pure or homogeneous according to currently accepted criteria for protein purity. In most instances, a protein isolate does represent, however, in the major protein component of a particular plant from which other minor components, including the protease inhibitors, have been removed during purification and, as such, might be expected to exhibit improved nutritional qualities.

Although the raw soybean is an unsatisfactory source of protein for growth, the isolated protein, *"glycinin,"* is capable of supporting good growth in rats (Osborne and Mendel, 1912; De and Ganguly, 1947). Soybean curd or *tofu,* which is a popular dish in the Orient, is, in a sense, a protein isolate since it is the protein that is precipitated with calcium salt from a hot-water extract of the whole bean (Smith, 1963). The biological value of tofu is equivalent to that of properly processed soybean meal (Pian, 1930) or casein (Standal, 1963). Since the preparation of tofu involves the cooking or steaming of the beans prior to extraction with water, tofu is believed to be free of the trypsin inhibitor (Dean, 1958), although no specific data on this point are available.

In recent years a great deal of interest has centered around the use of soybean isolates in the formulation of textured foods (Smith and Circle, 1972). Soybean protein, isolated by extraction and precipitation, may be spun into fibers which can then be manipulated to give products simulating the texture and flavor of meat foodstuffs. In a comprehensive study of the protein quality of a soybean protein-textured food using rats, dogs, and children as experimental subjects, Bressani *et al.* (1967) found that the quality of the final textured food product was almost equivalent to that of casein or beef. The protein efficiency (as measured with rats) of the original protein isolate, however, was very low but could be improved by heat treatment. The authors concluded that the low protein efficiency ratio of the protein isolate was most likely due to the presence of residual growth inhibitors and that, during the process of changing the isolate into fiber, these inhibitors were removed. These results are in agreement with the findings of Longenecker *et al.* (1964), who had previously reported that heat treatment improved the nutritional value of a number of commercially available soybean isolates, which again suggests that growth inhibitors were not completely removed during the isolation of these proteins.

Only limited data are available on the trypsin inhibitory activity of proteins isolated from plants other than the soybean. It has been reported that cooking enhances the nutritive value of the globulin-like protein of the navy bean (Johns and Finks, 1920a,b) and the velvet bean (Finks and Johns, 1921b), designated as phaseolin and stizolobin, respectively. *In vitro* digestibility studies on stizolobin (Waterman and Jones, 1921) demonstrated that cooking likewise enhanced the susceptibility of this protein to the combined action of pepsin and trypsin. By today's standards, the methods by which these proteins were prepared would be considered quite crude, and it is entirely possible that the poor nutritive value of these proteins may have been due to the presence of residual amounts of a growth inhibitor such as the trypsin inhibitors. Alternatively, the beneficial effect of heat treatment on the digestibility of these protein isolates may be a consequence of the denaturation of the protein as shown with soybean protein (Kakade *et al.* (1973).

REFERENCES

Agren, G., and Liedén, S. A. (1968). *Acta Chem. Scand.* **22,** 1981–1988.

Ako, H., Foster, R. J., and Ryan, C. A. (1972a). *Biochem. Biophys. Res. Commun.* **41,** 1402–1407.

Ako, H., Ryan, C. A., and Foster, R. J. (1972b). *Biochem. Biophys. Res. Commun.* **46,** 1639–1645.

Ako, H., Foster, R. J., and Ryan, C. A. (1974). *Biochemistry* **13,** 132–139.

Ako, H., Hass, G. M., Grahn, D. T., and Neurath, H. (1976). *Biochemistry* **15,** 2573–2578.

Albrecht, W. J. Mustakas, G. C., and McGhee, J. E. (1966). *Cereal Chem.* **43,** 400–407.

Albrecht, W. J. Mustakas, G. C., McGhee, J. E., and Griffin, E. L. (1967). *Cereal Sci. Today* **12,** 81–83.

Almquist, H. J., and Merritt, J. B. (1951). *Arch. Biochem. Biophys.* **31,** 450–453.

Almquist, H. J., and Merritt, J. B. (1953a). *Proc. Soc. Exp. Biol. Med.* **83,** 269.
Almquist, H. J., and Merritt, J. B. (1953b). *Proc. Soc. Exp. Biol. Med.* **84,** 333–334.
Alumot, E., and Nitsan, Z. (1961). *J. Nutr.* **73,** 71–77.
Ambe, K. S., and Sohonie, K. (1956a). *Experientia* **12,** 302–303.
Ambe, K. S., and Sohonie, K. (1956b). *J. Sci. Ind. Res., Sect. C* **15,** 136–140.
Anonymous (1962). *Nutr. Rev.* **21,** 19–21.
Applebaum, S. W., and Konijn, A. M. (1966). *J. Insect Physiol.* **12,** 665–669.
Applebaum, S. W., Birk, Y., Harpaz, I., and Bondi, A. (1964). *Comp. Biochem. Ciophys.* **11,** 85–103.
Applegarth, A., Furata, F., and Lepkovsky, S. (1964). *Poult. Sci.* **43,** 733–739.
Apte, U., and Sohonie, K. (1957). *J. Sci. Ind. Res., Sect. C* **16,** 225–227.
Arimoto, K., Tamura, E., Nishihara, A., Tamura, A., and Kobatake, Y. (1962). *Kokuritsu Eiyo Kenkyusho Kenkyu Hokoku* pp. 40–45; cited in *Chem. Abstr.* **66,** 1687q (1966).
Astrup, T., Brakman, P., Ollendorf, P., and Rasmussen, J. (1960). *Thromb. Diath. Haemorrh.* **5,** 329–340.
Astrup, T., Brakman, P., and Sjölin, K. E. (1962). *Nature (London)* **194,** 980–981.
Banerji, A. P., and Sohonie, K. (1969). *Enzymologia* **36,** 137–152.
Barnes, R. H., and Kwong, E. (1965). *J. Nutr.* **86,** 245–252.
Barnes, R. H., Fiala, G., and Kwong, E. (1962). *J. Nutr.* **77,** 278–284.
Barnes, R. H., Kwong, E., and Fiala, G. (1965a). *J. Nutr.* **85,** 123–126.
Barnes, R. H., Fiala, G., and Kwong, E. (1965b). *J. Nutr.* **85,** 127–131.
Baumgartner, B., and Chrispeels, M. J. (1976). *Plant Physiol.* **58,** 1–6.
Beauchene, R. E., and Mitchell, H. L. (1957). *J. Agric. Food Chem.* **5,** 762–765.
Belew, M. (1977). *Eur. J. Biochem.* **73,** 411–420.
Belew, M., and Eaker, D. (1976). *Eur. J. Biochem.* **62,** 499–508.
Belew, M., Porath, J., and Sundberg, L. (1975). *Eur. J. Biochem.* **60,** 247–258.
Belitz, H.-D., and Fuchs, A. (1972). *Chem., Mikrobiol., Technol. Lebensm.* **1,** 132–134.
Belitz, H.-D., and Fuchs, A. (1973). *Z. Lebensm.-Unters. -Forsch.* **152,** 129–133.
Belitz, H.-D., Fuchs, A., and Grimm, L. (1971a). *Lebensm.-Wiss.* + Technol. **4,** 89–92.
Belitz, H.-D., Kaiser, K. P., and Santarius, K. (1971b). *Biochem. Biophys. Res. Commun.* **42,** 420–427.
Belitz, H. -D., Fuchs, A., Nitsche, G., and Al-Sulton, T. (1972). *Z. Lebensm.-Unters. -Forsch.* **150,** 215–220.
Belitz, H.-D., Kaiser, K. P., and Santarius, K. (1974). *Z. Lebensm.-Unters. -Forsch.* **154,** 206–209.
Belohlawek, L. (1964). *Ber. Getreidechem.-Tag., Detmold* Pp. 19–25.
Bessho, H., and Kurosawa, S. (1966). *Eiyo To Shokuryo* **18,** 365–368.
Bhat, P. G., Karnik, S. S., and Pattabiraman, T. N. (1976). *Indian J. Biochem. & Biophys.* **13,** 339–343.
Bidlingmeyer, U. D., Leary, T. R., and Laskowski, M., Jr. (1972). *Biochemistry* **11,** 3303–3310.
Bielorai, R., and Bondi, A. (1963). *J. Sci. Food Agric.* **14,** 124–132.
Birk, Y. (1961). *Biochim. Biophys. Acta* **54,** 378–381.
Birk, Y. (1968). *Ann. N.Y. Acad. Sci.* **146,** 388–399.
Birk, Y. (1974). *Proteinase Inhibitors, Boyer Sump., 5th, 1973* pp. 355–361.
Birk, Y. (1976). *In* "Methods in Enzymology" (L. Lorand, ed.), Vol. 45, Part B, pp. 695–751. Academic Press, New York.
Birk, Y., and Applebaum, S. W. (1960). *Enzymologia* **22,** 318–326.
Birk, Y., and Gertler, A. (1961). *J. Nutr.* **75,** 379–387.
Birk, Y., Harpaz, I., Ishaaya, I., and Bondi, A. (1962). *J. Insect Physiol.* **8,** 417–429.
Birk, Y., Gertler, A., and Khalef, S. (1963a). *Biochim. Biophys. Acta* **67,** 326–328.
Birk, Y., Gertler, A., and Khalef, S. (1963b). *Biochem. J.* **87,** 281–284.

Bisordi, M. (1964). *Lancet* **2,** 476.
Blow, D. M., Janin, J., and Sweet, R. M. (1974). *Nature (London)* **249,** 54–57.
Boisen, S. (1976). *Phystochemistry* **15,** 641–642.
Booth, A. N., Robbins, D. J., Ribelin, W. E., and DeEds, F. (1960). *Proc. Soc. Exp. Biol. Med.* **104,** 681–683.
Booth, A. N., Robbins, D. J., Ribelin, W. E., DeEds, F., Smith, A. K., and Rackis, J. J. (1964). *Proc. Soc. Exp. Biol. Med.* **116,** 1067–1069.
Borchers, R. (1961). *J. Nutr.* **75,** 330–334.
Borchers, R. (1962a). *J. Nutr.* **77,** 309–311.
Borchers, R. (1962b). *J. Nutr.* **78,** 330–332.
Borchers, R., and Ackerson, C. W. (1950). *J. Nutr.* **41,** 339–345.
Borchers, R., Ackerson, C. W., and Sandstedt, R. M. (1947a). *Arch. Biochem. Biophys.* **12,** 367–374.
Borchers, R., Ackerson, C. W., and Kimmett, L. (1947b). *Arch. Biochem. Biophys.* **13,** 291–293.
Borchers, R., Ackerson, C. W., and Mussehl, F. E. (1948a). *Poult. Sci.* **27,** 601–604.
Borchers, R., Ackerson, C. W., and Mussehl, F. E. (1948b). *Arch. Biochem. Biophys.* **19,** 317–322.
Borchers, R., Mohammed-Abadi, D.-G., and Weaver, J. M. (1957). *J. Agric. Food Chem.* **5,** 371–373.
Borchers, R., Andersen, S. M., and Spelts, J. (1965). *J. Nutr.* **86,** 253–255.
Borchers, R., Manage, L. D., Nelson, S. O., and Stetsen, L. E. (1972). *J. Food Sci.* **37,** 331–334.
Bornstein, S., and Lipstein, B. (1963). *Poult. Sci.* **48,** 61–70.
Boudreaux, H. B., and Frampton, V. L. (1960). *Nature (London)* **185,** 469–470.
Bouthilet, R. J., Hunter, W. L., Luhman, C. A., Ambrose, D., and Lepkovsky, S. (1950). *Poult. Sci.* **29,** 837–840.
Bowman, D. E. (1944). *Proc. Soc. Exp. Biol. Med.* **57,** 139–140.
Bowman, D. E. (1946). *Proc. Soc. Exp. Biol. Med.* **63,** 547–550.
Bowman, D. E. (1948). *Arch. Biochem. Biophys.* **16,** 109–113.
Bowman, D. E. (1971). *Arch. Biochem. Biophys.* **144,** 541–548.
Braham, J. E., Bird, H. R., and Baumann, C. A. (1959). *J. Nutr.* **67,** 149–158.
Brambila, S., Nesheim, M. C., and Hill, F. W. (1961). *J. Nutr.* **75,** 13–20.
Bredenkamp, B. L. F., and Luck, D. N. (1969). *Proc. Soc. Exp. Biol. Med.* **132,** 537–539.
Bressani, R., Vitari, F., Elias, L. G., DeZaghi, S., Alvardo, J., and Odell, A. O. (1967). *J. Nutr.* **93,** 349–360.
Bryant, J., Green, T. R., Gurusaddaiah, J., and Ryan, C. A. (1976). *Biochemistry* **15,** 3418–3423.
Busse, T., and Belitz, H.-D. (1976). *Z. Lebensm.-Unters. -Forsch.* **162,** 357–364.
Cama, H. R., Balasudaram, S., and Malik, A. A. (1956). *Proc. Int. Congr. Biochem., 3rd, 1955* Resumes Communs., p. 113.
Camus, M. C., and Laporte, J. C. (1973). *Ann. Technol. Agric.* **22,** 111–120.
Carroll, R. W., Hensley, G. W., and Graham, W. R., Jr. (1952). *Science* **115,** 36–39.
Carroll, R. W., Hensley, G. W., Sittler, C. L., Wilcox, E. L., and Graham, W. R., Jr. (1953). *Arch. Biochem. Biophys.* **45,** 260–269.
Catsimpoolas, N. (1969). *Sep. Sci.* **4,** 483–492.
Catsimpoolas, N., Ekenstam, C., and Meyer, E. W. (1969). *Biochim. Biophys. Acta* **175,** 76–81.
Cepelak, V., Horakova, Z., and Padr, Z. (1963). *Nature (London)* **198,** 295.
Chattapadhgay, H., and Bannerjee, S. (1953). *Indian J. Med. Res.* **41,** 185–189.
Chen, I., and Mitchell, H. L. (1973). *Phytochemistry* **12,** 327–330.
Cheong, J. S., Ke, S. Y., and Yoon, D. S. (1959). *Bull. Sci. Res. Inst. (Seoul)* **4,** 41–45.
Chernick, S. S., Lepkovsky, S., and Chaikoff, I. L. (1948). *Am. J. Physiol.* **155,** 33–41.
Chien, J. F., and Mitchell, H. L. (1970). *Phytochemistry* **9,** 717–720.

Chu, H.-M., and Chi, C.-W. (1965a). *Acta Biochim. Biophys. Sin.* **5,** 519–528.
Chu, H.-M., and Chi, C.-W. (1965b). *Sci Sin.* **14,** 1441–1453.
Chu, H.-M., and Chi, C.-W. (1966). *Acta Biochim. Biophys. Sin.* **6,** 22–31.
Chu, H.-M., Lo, S.-S., and Jen, M.-H. (1965). *Sci. Sin.* **14,** 1454–1463.
Churella, H. R., Yao, B. C., and Thomason, W. A. B. (1976). *J. Agric. Food Chem.* **24,** 393–397.
Clandinin, D. R., and Robblee, A. R. (1952). *J. Nutr.* **46,** 525–530.
Clark, R. W., and Hymowitz, T. (1972). *Biochem. Genet.* **6,** 169–182.
Clark, R. W., Mies, D. W., and Hymowitz, J. (1970). *Crop Sci.* **10,** 486–487.
Collins, J. L., and Sanders, G. G. (1976). *J. Food Sci.* **41,** 168–172.
Colvin, B. M., and Ramsey, H. A. (1969). *J. Dairy Sci.* **52,** 270–273.
Couch, J. R., and Hooper, F. G. (1972). *Newer Methods Nutr. Biochem.* **5,** 183–193.
Couch, J. R., Creger, C. R., and Bakshi, Y. K. (1966). *Proc. Soc. Exp. Biol. Med.* **123,** 263–265.
Cravioto, R., Guzman, J. G., and Massieu, H. G. (1951). *Ciencia (Mexico City)* **11,** 81–82.
Creek, R. D., and Vasaitas, V. (1962). *Poult. Sci.* **41,** 1351–1352.
Creek, R. D., Vasaitas, V., and Schumaier, G. (1961). *Poult. Sci.* **40,** 822–824.
De, S. S., and Ganguly, J. (1947). *Nature (London)* **159,** 341–342.
Dean, R. F. A. (1958). *In* "Processed Plant Protein Foodstuffs" (A. M. Altschul, ed.), pp. 205–247. Academic Press, New York.
DeLarco, J., and Liener, I. E. (1973). *Biochim. Biophys. Acta* **303,** 274–283.
Delić, I., Stojisavljević, T., and Stojanović, S. (1974). *Acta Vet. (Belgrade)* **24,** 1–5.
Delobez, R., Duterte, R., and Rambard, M. (1971). *Rev. Fr. Gras* **18,** 381–389.
DeMuelenaere, H. J. H. (1964). *J. Nutr.* **82,** 197–205.
Desikachar, H. S. R., and De, S. S. (1947). *Science* **106,** 421–422.
Desikachar, H. S. R., and De, S. S. (1950). *Biochim. Biophys. Acta* **5,** 285–289.
Devadatta, S. C., Acharya, B. N., and Nadkarni, S. B. (1951). *Proc. Indian Acad. Sci., Sect. B* **33,** 150–158.
DiBella, F. P., and Liener, I. E. (1969). *J. Biol. Chem.* **244,** 2824–2829.
Edelhoch, H., and Steiner, R. F. (1963). *J. Biol. Chem.* **238,** 931–938.
Eldridge, A. C., and Wolf, W. J. (1969). *Cereal Chem.* **46,** 470–478.
Elias, L. G., Hernandez, M., and Bressani, R. (1976). *Nutr. Rep. Int.* **14,** 385–403.
Enari, J.-J., Mikola, J., and Linko, M. (1964). *J. Inst. Brew., London* **70,** 405–411.
Esh, G. C., and Som, J. M. (1952). *Indian J. Physiol. Allied Sci.* **6,** 61–65.
Evans, R. J., and Bandemer, S. (1967). *Cereal Chem.* **44,** 417–426.
Evans, R. J., and McGinnis, J. (1948). *J. Nutr.* **35,** 477–488.
Everson, G., and Heckert, A. (1944). *J. Am. Diet. Assoc.* **20,** 81–82.
Everson, G., Steinbock, H., Cederquist, D. C., and Parsons, H. R. (1944). *J. Nutr.* **27,** 225–229.
Feeney, R. E., Means, G. E., and Bigler, J. C. (1969). *J. Biol. Chem.* **244,** 1957–1960.
Feinstein, G., and Feeney, R. E. (1966). *J. Biol. Chem.* **241,** 5183–5189.
Ferdinand, W., Moore, S., and Stein, W. H. (1965). *Biochim. Biophys. Acta* **96,** 524–527.
Figarella, C., Negri, G. A., and Guy, O. (1975). *Eur. J. Biochem.* **53,** 457–463.
Finkenstadt, W. R., and Laskowski, M., Jr. (1965). *J. Biol. Chem.* **240,** PC962–PC963.
Finks, A. J., and Johns, C. O. (1921a). *Am. J. Physiol.* **56,** 505–507.
Finks, A. J., and Johns, C. O. (1921b). *Am. J. Physiol.* **57,** 61–67.
Fomon, S. J. (1962). *Proc. Conf. Soybean Prod. Protein Human Foods, 1961* pp. 175–178.
Fraenkel-Conrat, H., Bean, R. C., Ducay, E. D., and Olcott, H. S. (1952). *Arch. Biochem. Biophys.* **37,** 393–407.
Frattali, V. (1969). *J. Biol. Chem.* **244,** 274–280.
Frattali, V., and Steiner, R. F. (1968). *Biochemistry* **7,** 521–530.
Frattali, V., and Steiner, R. F. (1969). *Biochem. Biophys. Res. Commun.* **34,** 480–487.

Fritz, H., Fink, E., Gebhardt, M., Hochstrasser, K., and Werle, E. (1969). *Hoppe-Seyler's Z. Physiol. Chem.* **350,** 933–944.
Fritz, H., Tschesche, H., Greene, L. J., and Truscheit, E., eds. (1974). "Proteinase Inhibitors," Bayer Symp. V. Springer-Verlag, Berlin and New York.
Frost, A. B., and Mann, G. V. (1966). *J. Nutr.* **89,** 49–54.
Furusawa, Y., Kurosawa, Y., and Chuman, I. (1974). *Agric. Biol. Chem.* **38,** 1157–1164.
Gaitonde, M. K., and Sohonie, K. (1951). *Curr. Sci.* **20,** 217–218.
Gallardo, F., Araya, H., Pak, N., and Tagle, M. A. (1974). *Arch. Latinoam. Nutr.* **2,** 183–189.
Garlich, J. D., and Nesheim, M. C. (1966). *Proc. Soc. Exp. Biol. Med.* **118,** 1022–1025.
Gennis, L. S. (1976). *J. Biol. Chem.* **251,** 763–768.
Gennis, L. S., and Cantor, C. R. (1976a). *J. Biol. Chem.* **251,** 734–740.
Gennis, L. S., and Cantor, C. R. (1976b). *J. Biol. Chem.* **251,** 741–746.
Gennis, L. S., and Cantor, C. R. (1976c). *J. Biol. Chem.* **251,** 747–753.
Gennis, L. S., and Cantor, C. R. (1976d). *J. Biol. Chem.* **251,** 754–762.
Gennis, L. S., and Cantor, C. R. (1976e). *J. Biol. Chem.* **257,** 769–775.
Geratz, J. D. (1968). *Am. J. Physiol.* **214,** 595–600.
Geratz, J. D. (1969). *Am. J. Physiol.* **216,** 812–820.
Gertler, A., Birk, Y., and Bondi, A. (1967). *J. Nutr.* **91,** 358–370.
Goldberg, A., and Guggenheim, K. (1962). *Biochem. J.* **83,** 129–135.
Godlstein, Z., Trop, M., and Birk, Y. (1973). *Nature (London), New Biol.* **246,** 29–31.
Gorbunoff, M. J. (1970). *Biochim. Biophys. Acta* **221,** 314–325.
Gorrill, H. D. L., and Thomas, J. W. (1967). *J. Nutr.* **92,** 215–223.
Goss, R. J. (1966). *Science* **153,** 1615–1616.
Green, G. M., and Lyman, R. L. (1972). *Proc. Soc. Exp. Biol. Med.* **140,** 6–12.
Green, G. M., Olds, B. A., and Lyman, R. L. (1973). *Proc. Soc. Exp. Biol. Med.* **142,** 162–167.
Green, N. M. (1953). *J. Biol. Chem.* **205,** 535–551.
Green, N. M. (1957). *Biochem. J.* **66,** 407–415.
Green, T. R., and Ryan, C. A. (1972). *Science* **175,** 776–777.
Green, T. R., and Ryan, C. A. (1973). *Plant Physiol.* **51,** 19–21.
Gustafson, G., and Ryan, C. A. (1976). *J. Biol. Chem.* **251,** 7004–7010.
Hackler, L. R., Steinkraus, K. H., Van Buren, J. P., and Hand, D. B. (1964). *J. Nutr.* **82,** 452–456.
Hackler, L. R., Van Buren, J. P., Steinkraus, K. H., El-Raroi, I., and Hand, D. B. (1965). *J. Food Sci.* **30,** 723–728.
Haines, P. C., and Lyman, R. L. (1961). *J. Nutr.* **74,** 445–452.
Halim, A. H., Wassom, C. E., and Mitchell, H. L. (1973a). *Crop Sci.* **13,** 405–409.
Halim, A. H., Mitchell, H. L., and Wassom, C. E. (1973b). *Trans. Kans. Acad. Sci.* **76,** 289–293.
Halim, A. H., Wassom, C. E., Mitchell, H. L., and Edmunds, L. K. (1973c). *J. Agric. Food Chem.* **21,** 1118–1119.
Ham, W. E., and Sanstedt, R. M. (1944). *J. Biol. Chem.* **154,** 505–506.
Ham, W. E., Sandstedt, R. M., and Mussehl, F. E. (1945). *J. Biol. Chem.* **161,** 635–642.
Harry, J. B., and Steiner, R. F. (1969). *Biochemistry* **8,** 5060–5064.
Hass, G. M., Nau, H., Biemann, K., Grahn, D. T., Ericsson, L. H., and Neurath, H. (1975). *Biochemistry* **14,** 1334–1342.
Hass, G. M., Ako, H., Grahn, D. T., and Neurath, H. (1976a). *Biochemistry* **15,** 93–100.
Hass, G. M., Venkatakrishnan, R., and Ryan, C. A. (1976b). *Proc. Natl. Acad. Sci. U.S.A.* **73,** 1941–1944.
Hata, T., Ogawa, T., and Hayashi, R. (1967). *Bull, Res. Inst. Food Sci., Kyoto Univ.* **30,** 51–60.
Hayashi, Y., and Ariyama, H. (1960). *Tokoku J. Med. Res.* **11,** 1171–1185.
Haynes, R., and Feeney, R. E. (1967). *J. Biol. Chem.* **242,** 5378–5385.
Haynes, R., and Feeney, R. E. (1968). *Biochim. Biophys. Acta* **159,** 209–211.

Hayward, J. W., and Hafner, F. H. (1941). *Poult. Sci.* **20,** 139–150.

Heinrikson, R. L., and Kezdy, F. J. (1976). *In* "Methods in Enzymology" (L. Lorand, ed.), Vol. 45, part B, pp. *45,* 740–751. Academic Press, New York.

Hewitt, D., Coates, M. E., Kakade, M. L., and Liener, I. E. (1973). *Br. J. Nutr.* **29,** 423–435.

Hill, C. H., Borchers, R., Ackerson, C. W., and Mussehl, F. E. (1953). *Arch. Biochem. Biophys.* **43,** 286–288.

Hites, B. D., Sandstedt, R. M., and Schaumburg, L. (1951). *Cereal Chem.* **28,** 1–18.

Hixson, H. F., Jr., and Laskowski, M., Jr. (1970). *J. Biol. Chem.* **245,** 2027–2035.

Hobday, S. M., Thurman, D. A., and Barber, D. J. (1973). *Phytochemistry* **12,** 1041–1046.

Hochstrasser, K., Muss, M., and Werle, E. (1967). *Hoppe-Seyler's Z. Physiol. Chem.* **348,** 1337–1340.

Hochstrasser, K., Schwarz, S., Illchmann, K., and Werle, E. (1968). *Hoppe-Seyler's Z. Physiol. Chem.* **349,** 1449–1455.

Hochstrasser, K., Werle, E., Schwarz, S., and Siegelmann, R. (1969a). *Hoppe-Seyler's Z. Physiol. Chem.* **350,** 249–254.

Hochstrasser, K., Illchmann, K., and Werle, E. (1969b). *Hoppe-Seyler's Z. Physiol. Chem.* **350,** 929–932.

Hochstrasser, K., Illchmann, K., and Werle, E. (1970a). *Hoppe-Seyler's Z. Physiol. Chem.* **351.** 721–728.

Hochstrasser, K., Illchmann, K., Werle, E., Hösse, R., and Schwarz, S. (1970b). *Hoppe-Seyler's Z. Physiol. Chem.* **351,** 1503–1512.

Hoff, J. E., Jones, C. M., Sosa, M. P., and Rodis, P. (1972). *Biochem. Biophys. Res. Commun.* **49,** 1525–1527.

Hojima, Y., Moriya, H., and Moriwaki, C. (1971a). *J. Biochem. (Tokyo)* **69,** 1019–1025.

Hojima, Y., Moriyama, H., and Moriwaki, C. (1971b). *J. Biochem. (Tokyo)* **69,** 1027–1032.

Hojima, Y., Horiwaki, C., and Moriya, H. (1973a). *J. Biochem.* **73,** 923–932.

Hojima, Y., Horiwaki, C., and Moriya, H. (1973b). *J. Biochem. (Tokyo)* **73,** 933–943.

Honavar, P. M., and Sohonie, K. (1955). *J. Univ. Bombay, Sect. B* **24,** 64–69.

Honavar, P. M., and Sohonie, K. (1959). *Ann. Biochem. Exp. Med.* **19,** 57–58.

Honavar, P. M., Shih, C.-V., and Liener, I. E. (1962). *J. Nutr.* **77,** 109–114.

Horiguchi, T., and Kitagishi, K. (1971). *Plant Cell Physiol.* **12,** 907–915.

Hory, H. D., and Weder, J. K. P. (1976). *Z. Lebensm.-Unters. -Forsch.* **162,** 349–356.

Hory, H. D., Weder, J. K. P., and Belitz, H. D. (1976). *Z. Lebensm.-Unters. -Forsch.* **162,** 341–347.

Hou, H. C., Riesen, W. H., and Elvehjem, C. A. (1949). *Proc. Soc. Exp. Biol. Med.* **70,** 416–419.

Huang, J.-S., and Liener, I. E. (1977). *Biochemistry* **16,** 2474–2478.

Hwang, D. L., Foard, D. E., and Wei, C. H. (1977a). *J. Biol. Chem.* **252,** 1099–1101.

Hwang, D. L., Lin, K.-T. D., Yang, W.-K., and Foard, D. E. (1977b). *Biochem. Biophys. Acta* **495,** 369–382.

Hymowitz, T., and Hadley, H. H. (1972). *Crop Sci.* **12,** 197–198.

Ihse, I. (1976). *Scand. J. Gastroenterol.* **11,** 11–15.

Ihse, I., Arnesjö, B., and Lundquist, I. (1974). *Scand. J. Gastroenterol.* **9,** 719–724.

Ingram, G. R., Riesen, W. H., Cravens, W. W., and Elvehjem, C. A. (1949). *Poult. Sci.* **28,** 898–902.

Ishaaya, I., and Birk, Y. (1965). *J. Food Sci.* **30,** 118–120.

Iwasaki, T., Kiyohara, T., and Yoshikawa, M. (1971). *J. Biochem. (Tokyo)* **70,** 817–826.

Iwasaki, T., Kiyohara, T., and Yoshikawa, M. (1972). *J. Biochem. (Tokyo)* **72,** 1029–1035.

Iwasaki, T., Kiyohara, T., and Yoshikawa, M. (1973a). *J. Biochem. (Tokyo)* **73,** 1039–1048.

Iwasaki, T., Kiyohara, T., and Yoshikawa, M. (1973b). *J. Biochem. (Tokyo)* **74,** 335–340.

Iwasaki, T., Kiyohara, T., and Yoshikawa, M. (1974). *J. Biochem. (Tokyo)* **75,** 843–851.

Iwasaki, T., Wada, J., Kiyohara, T., and Yoshikawa, M. (1975). *J. Biochem. (Tokyo)* **78,** 1267–1274.

Iwasaki, T., Kiyohara, T., and Yoshikawa, M. (1976). *J. Biochem. (Tokyo)* **79,** 381–391.

Jaffé, W. G. (1950a). *Proc. Soc. Exp. Biol. Med.* **75,** 219–220.

Jaffé, W. G. (1950b). *Acta Cient. Venez.* **1,** 62–64.

Jaffé, W. G., and Korte, R. (1976). *Nutr. Rep. Int.* **14,** 449–455.

Jaffé, W. G., and Lette, C. L. V. (1968). *J. Nutr.* **94,** 203–210.

Janin, J., Sweet, R. M., and Blow, D. M. (1974). *in Protease Inhibitors, Bayer Symp., 5th, 1973* pp. 513–520.

Jaya, T. V., Krishnamurthy, K. S., and Venkatamaran, L. V. (1976). *Nutr. Rep. Int.* **12,** 175–184.

Jensen, A. H., Yen, J. T., and Hymowitz, T. (1974). *J. Anim. Sci.* **38,** 304–309.

Jirgensons, B. (1967). *J. Biol. Chem.* **242,** 912–918.

Johns, C. O., and Finks, A. J. (1920a). *J. Biol. Chem.* **41,** 379–389.

Johns, C. O., and Finks, A. J. (1920b). *Science* **52,** 414–415.

Johnson, L. M., Parsons, H. T., and Steinbock, H. (1939). *J. Nutr.* **18,** 423–434.

Jones, D. B., and Waterman, H. C. (1922). *J. Biol. Chem.* **52,** 351–366.

Jones, G., Moore, S., and Stein, W. H. (1963). *Biochemistry* **2,** 66–71.

Kaiser, K.-P., and Belitz, H. D. (1971). *Chem., Mikrobiol., Technol. Lebensm.* **1,** 1–7.

Kaiser, K.-P., and Belitz, H. D. (1972). *Chem., Mikrobiol., Technol. Lebensm.* **1,** 191–194.

Kaiser, K.-P., and Belitz, H. D. (1973). *Z. Lebensm.-Unters. -Forsch.* **151,** 18–22.

Kaiser, K.-P., Bruhn, L. C., and Belitz, H. D. (1974). *Z. Lebensm.-Unters. -Forsch.* **154,** 339–347.

Kakade, M. L., and Evans, R. J. (1963). *Mich., Agric. Exp. Stn., Q. Bull.* **46,** 87–91.

Kakade, M. L., and Evans, R. J. (1965a). *J. Agric. Food Chem.* **13,** 450–452.

Kakade, M. L., and Evans, R. J. (1965b). *Br. J. Nutr.* **19,** 269–276.

Kakade, M. L., and Evans, R. J. (1966a). *Can. J. Biochem.* **44,** 648–650.

Kakade, M. L., and Evans, R. J. (1966b). *J. Food Sci.* **31,** 781–783.

Kakade, M. L., and Liener, I. E. (1973). *In* "Man, Food, and Nutrition" (M. Rechneigl, ed.), pp. 231–242. CRC Press, Cleveland, Ohio.

Kakade, M. L., Barton, T. L., and Schaibel, P. J. (1967). *Poult. Sci.* **46,** 1578–1580.

Kakade, M. L., Arnold, R. L., Liener, I. E., and Waibel, P. E. (1969). *J. Nutr.* **99,** 34–41.

Kakade, M. L., Simons, N. R., and Liener, I. E. (1970a). *Biochim. Biophys. Acta* **200,** 168–169.

Kakade, M. L., Simons, N. R., and Liener, I. E. (1970b). *J. Nutr.* **100,** 1003–1008.

Kakade, M. L., Hoffa, D. E., and Liener, I. E. (1973). *J. Nutr.* **103,** 1772–1778.

Kakade, M. L., Rackis, J. J., McGhee, E., and Puski, G. (1974). *Cereal Chem.* **51,** 376–382.

Kakade, M. L., Thompson, R. D., Englestad, W. D., Behrens, G. C., Yoder, R. O., and Crane, F. M. (1976). *J. Dairy Sci.* **59,** 1484–1489.

Kanamori, M., Ibuki, F., Tashiro, M., Yamada, M., and Miyoshi, M. (1975a). *J. Nutr. Sci. Vitaminol.* **21,** 421–428.

Kanamori, M., Ibuki, F., Yamada, M., Tashiro, M., and Miyoshi, M. (1975b). *J. Nutr. Sci. Vitaminol.* **21,** 429–436.

Kanamori, M., Ibuki, F., Tashiro, M., Yamada, M., and Miyoshi, M. (1976a). *Biochim. Biophys. Acta* **439,** 398–405.

Kanamori, M., Ibuki, F., Yamada, M., Tashiro, M., and Miyoshi, M. (1976b). *Agric. Biol. Chem.* **40,** 839–844.

Kasche, V. (1973). *Stud. Biophys.* **35,** 45–56.

Kassell, B. (1970). *In* "Methods in Enzymology" (G. Perlmann and L. Lorand, eds.). Vol. 19, pp. 853–871. Academic Press, New York.

Kato, I., Tominaga, N., and Kihara, H. (1974). *In* "Protease Inhibitors" (H. Fritz, H. Tschesche, L. J. Greene, and E. Truscheit, eds.), Bayer Symp. V, pp. 599–600. Springer-Verlang, Berlin and New York.

Katoch, B. S., and Chopra, A. K. (1974). *Indian J. Anim. Sci.* **44,** 551–562.
Kay, J. L., Daeschner, C. W., Jr., and Desmond, M. M. (1960). *Am. J. Dis. Child.* **100,** 264–276.
Keilova, H., and Tomasek, V. (1976a). *Collect. Czech. Chem. Commun.* **41,** 489–497.
Keilova, H., and Tomasek, V. (1976b). *Collect. Czech. Chem. Commun.* **41,** 2440–2447.
Kendall, K. A. (1951). *J. Dairy Sci.* **34,** 499–500.
Khayambashi, H., and Lyman, R. L. (1966). *J. Nutr.* **89,** 455–464.
Khayambashi, H., and Lyman, R. L. (1969). *Am. J. Physiol.* **217,** 646–651.
Kirsi, M. (1974). *Physiol. Plant.* **32,** 89–93.
Kirsi, M., and Mikola, J. (1971). *Planta* **96,** 281–291.
Kirsi, M., and Mikola, J. (1977). *Physiol. Plant* **39,** 110–114.
Kiyohara, T., Iwasaki, T., and Yoshikawa, M. (1973a). *J. Biochem. (Tokyo)* **73,** 89–95.
Kiyohara, T., Fujii, M., Iwasaki, T., and Yoshikawa, M. (1973b). *J. Biochem. (Tokyo)* **74,** 675–682.
Klose, A. A., Hill, B., and Fevold, H. L. (1946). *Proc. Soc. Exp. Biol. Med.* **62,** 10–12.
Klose, A. A., Greaves, J. D., and Fevold, H. L. (1948). *Science* **108,** 88–89.
Klose, A. A., Hill, B., Greaves, J. D., and Fevold, H. L. (1949). *Arch. Biochem.* **22,** 215–223.
Koide, T., and Ikenaka, T. (1973a). *Eur. J. Biochem.* **32,** 401–407.
Koide, T., and Ikenaka, T. (1973b). *Eur. J. Biochem.* **32,** 417–431.
Koide, T., Tsunasawa, S., and Ikenaka, T. (1973). *Eur. J. Biochem.* **32,** 408–416.
Konijn, A. M., and Guggenheim, K. (1967). *Proc. Soc. Exp. Biol. Med.* **126,** 65–67.
Kotter, L., Palitzsch, A., Belitz, H.-D., and Fischer, K.-H. (1970). *Fleischwirtschaften* **8,** 1063–1064.
Kowalski, D., and Laskowski, M., Jr. (1972). *Biochemistry* **11,** 3451–3459.
Kowalski, D., and Laskowski, M., Jr. (1976a). *Biochemistry* **15,** 1300–1309.
Kowalski, D., and Laskowski, M., Jr. (1976b). *Biochemistry* **15,** 1309–1315.
Krahn, J., and Stevens, F. C. (1970). *Biochemistry* **9,** 2646–2652.
Krahn, J., and Stevens, F. C. (1971). *FEBS Lett.* **13,** 339–341.
Krahn, J., and Stevens, F. C. (1972a). *Biochemistry* **11,** 1804–1815.
Krahn, J., and Stevens, F. C. (1972b). *FEBS Lett.* **28,** 313–316.
Krahn, J., and Stevens, F. C. (1973). *Biochemistry* **12,** 1330–1335.
Krishnamurthy, K., Taskar, P. K., Ramakrishnan, J. N., Rajogopolan, R., Swaminathan, M., and Subrahmanyan, V. (1958). *Ann. Biochem. Exp. Med.* **18,** 153–156.
Kunitz, M. (1945). *Science* **101,** 668–669.
Kunitz, M. (1946). *J. Gen. Physiol.* **29,** 149–154.
Kunitz, M. (1947). *J. Gen. Physiol.* **30,** 311–320.
Kwaan, H. C., Kok, P., and Astrup, T. (1968). *Experientia* **24,** 1125–1126.
Kwong, E., and Barnes, R. H. (1963). *J. Nutr.* **81,** 392–398.
Kwong, E., Barnes, R. H., and Fiala, G. (1962). *J. Nutr.* **77,** 312–316.
Laporte, J., and Trémoliéres, J. (1962). *C.R. Seances Soc. Biol. Ses Fil.* **156,** 1261–1263.
Laskowski, M., Jr., and Sealock, R. W. (1971). *In* "The Enzymes" (P. D. Boyer, ed.), 3rd ed., Vol. 3, pp. 376–473. Academic Press, New York.
Leach, B. S., and Fish, W. W. (1977). *J. Biol. Chem.* **252,** 5239–5243.
Learmonth, E. M. (1951). *Nature (London)* **167,** 820.
Learmonth, E. M. (1957). British patent 784,830.
Learmonth, E. M. (1958). *J. Sci. Food Agric.* **9,** 269–273.
Leary, J. R., and Laskowski, J., Jr. (1973). *Fed. Proc., Fed. Am. Soc. Exp. Biol.* **32,** 1363.
Lepkovsky, S., Bingham, E., and Pencharz, R. (1959). *Poult. Sci.* **38,** 1289–1295.
Lepkovsky, S., Furuta, F., and Dimick, M. K. (1971). *Br. J. Nutr.* **25,** 235–241.
Liener, I. E. (1958). *In* "Processed Plant Protein Foodstuffs" (A. M. Altschul, ed.), pp. 79–129. Academic Press, New York.

Liener, I. E. (1962). *Am. J. Clin, Nutr.* **11,** 281–298.
Liener, I. E. (1973). *In* "Proteins in Human Foods" (J. W. G. Porter and B. A. Rolls, eds), pp. 481-500. Academic Press, New York.
Liener, I. E. (1977). *In* "Evaluation of Proteins for Human" (C. E. Bodwell, ed.), pp. 284–303. Avi Publ. Co., Westport, Connecticut.
Liener, I. E., and Fevold, H. L. (1949). *Arch. Biochem. Biophys.* **21,** 395–407.
Liener, I. E., and Wada, S. (1953). *Proc. Soc. Exp. Biol. Med.* **82,** 484–486.
Liener, I. E., Deuel, H. J., Jr., and Fevold, H. L. (1949). *J. Nutr.* **39,** 325–339.
Lipke, H., Fraenkel, G. S., and Liener, I. E. (1954). *J. Agric. Food Chem.* **2,** 410–414.
Long, C. (1961). *In* "Biochemists Handbook," p. 675. Van Nostrand-Reinhold, Princeton, New Jersey.
Longnecker, J. B., Martin, W. H., and Sarrett, H. P. (1964). *J. Agric. Food Chem.* **12,** 411–412.
Lord, J. W., and Wakelam, J. A. (1950). *Br. J. Nutr.* **4,** 154–160.
Lyman, R. L. (1957). *J. Nutr.* **62,** 285–294.
Lyman, R. L., and Lepkovsky, S. (1957). *J. Nutr.* **62,** 269–284.
Lyman, R. L., Wilcox, S. S., and Monsen, E. R. (1962). *Am. J. Physiol.* **106,** 104–111.
Lyman, R. L., Olds, B., and Green, G. M. (1974). *J. Nutr.* **104,** 105–110.
McFarland, D., and Ryan, C. A. (1974). *Plant Physiol.* **54,** 706–708.
McKinney, L. L., and Cowan, J. C. (1956). *Soybean Dig.* **16,** 14–16 and 18.
Madl, R. L., and Tsen, C. C. (1973). *In* "Triticale: First Man-Made Cereal, Symposium" (C. C. Tsen, ed.), pp. 168–182. Am. Assoc. Cereal Chem., St. Paul, Minnesota.
Mainz, D. L., Black, D., and Webster, P. D. (1973). *J. Clin. Invest.* **52,** 5300–5310.
Mallory, P. A., and Travis, J. (1973). *Biochemistry* **12,** 2847–2851.
Mallory, P. A., and Travis, J. (1975). *Am. J. Clin. Nutr.* **28,** 823–830.
Mamiya, Y., Totsuka, K., Shoji, K., and Aso, K. (1972). *Jpn. Soc. Food Sci. Technol.* **19,** 12–23.
Mansfeld, V., Ziegelhoffer, A., Horakova, Z., and Hladovec, J. (1959). *Naturwissenschaften* **46,** 172–173.
Marquardt, R. R., McKirdy, J. A., Ward, T., and Campbell, L. D. (1975). *Can. J. Anim. Sci.* **55,** 421–429.
Marquardt, R. R., Campbell, L. D., and Ward, T. (1976). *J. Nutr.* **106,** 275–284.
Mattingly, J. P., and Bird, H. R. (1945). *Poult. Sci.* **24,** 344–352.
Matsushima, K. (1955). *Nippon Nogei Kagaku Kaishi* **29,** 883–887.
Means, G. E., Ryan, D. S., and Feeney, R. E. (1974). *Acc. Chem. Res.* **7,** 315–320.
Mejbaum-Katzenellenbogen, N., and Lorenc-Kubis, I. (1969). *Acta Bot. Pol.* **38,** 664–669.
Melmed, R. N., and Bouchier, I. A. D. (1969). *J. Br. Soc. Gastroenterol.* **10,** 973–979.
Melmed, R. N., Turner, R. C., and Holt, S. J. (1973). *J. Cell Sci.* **13,** 279–285.
Melmed, R. N., El-Aeser, A. A. A., and Holt, S. J. (1976). *Biochim. Biophys. Acta* **421,** 280–288.
Melnick, D., Oser, B. L., and Weiss, S. (1946). *Science* **103,** 326–329.
Melville, J. C., and Ryan, C. A. (1970). *Arch. Biochem. Biophys.* **138,** 700–702.
Milville, J. C., and Ryan, C. A. (1972). *J. Biol. Chem.* **247,** 3445–3453.
Melville, J. C., and Scandalios, J. G. (1972). *Biochem. Genet.* **7,** 15–31.
Miége, M.-N., Mascherpa, J.-M., Royer-Spierer, A., Grange, A., and Miége, J. (1976). *Planta* **131,** 81–86.
Mikola, J., and Enari, J. M. (1970). *J. Inst. Brew. London* **76,** 182–188.
Mikola, J., and Kirsi, M. (1972). *Acta Chem. Scand.* **26,** 787–795.
Mikola, J., and Suolinna, E. M. (1969). *Eur. J. Biochem.* **9,** 555–560.
Mikola, J., and Suolinna, E. M. (1971). *Arch. Biochem. Biophys.* **144,** 566–575.
Millar, D. B. S., Willick, G. E., Steiner, R. F., and Frattali, V. (1969). *J. Biol. Chem.* **244,** 281–284.
Mistunaga, T. (1974). *J. Nutr. Sci. Vitaminol.* **20,** 153–159.

Mitchell, H. L., Parrish, D. B., Cormey, M., and Wassom, C. E. (1976). *J. Agric. Food Chem.* **24,** 1254–1255.

Moorijman, J. G. J. (1965). *Diss. Abstr.* **25,** 4390.

Moriya, H., Hojima, Y., Moriwaki, C., and Tajima, T. (1970). *Experientia* **26,** 720–721.

Mosolov, V. V., Shulmina, A. I., and Malova, E. L. (1974). *Biokhimiya* **39,** 793–798.

Mosolov, V. V., Loginova, M. D., Fedurkina, N. V., and Benken, I. I. (1976). *Plant Sci. Lett.* **7,** 77–80.

Mustakas, G. C., Griffin, E. L., Jr., Allen, L. E., and Smith, O. B. (1964). *J. Am. Oil Chem. Soc.* **41,** 607–614.

Nau, H., and Biemann, K. (1976). *Anal. Biochem.* **73,** 175–186.

Nehring, K., Bock, H.-D., and Wunsche, J. (1963). *Nahrung* **3,** 233–242.

Nesheim, M. C., and Garlich, J. D. (1966). *J. Nutr.* **88,** 187–192.

Nesheim, M. C., Garlich, J. D., and Hopkins, D. T. (1962). *J. Nutr.* **78,** 89–94.

Neurath, H. (1961). *In* "The Exocrine Pancreas" (A. V. S. DeRueck and M. P. Cameron, eds.), p. 76. Little,Brown, Boston, Massachusetts.

Niekamp, C. W., Hixson, H. F., Jr., and Laskowski, M., Jr. (1969). *Biochemistry* **8,** 16–22.

Nishino, N., Aoyogi, H., Kato, T., and Izumiya, N. (1977). *J. Biochem. (Tokyo)* **82,** 901–909.

Nitsan, Z. (1965). *Poult. Sci.* **44,** 1036–1043.

Nitsan, Z., and Alumot, E. (1964). *J. Nutr.* **84,** 179–184.

Nitsan, Z., and Alumot, E. (1965). *Poult. Sci.* **44,** 1210–1214.

Nitsan, Z., and Bondi, A. (1965). *Br. J. Nutr.* **19,** 177–187.

Nitsan, Z., and Gertler, A. (1972). *Nutr. Metab.* **14,** 371–376.

Nordal, J., and Fossum, K. (1974). *Z. Lebens.-Unters. -Forsch.* **154,** 144–150.

Obara, T., and Watanabe, Y. (1971). *Cereal Chem.* **48,** 523–527.

Obara, T., Kimua-Kobayashi, M., Kobayashi, T., and Watanabe, Y. (1970). *Cereal Chem.* **47,** 597–606.

Odani, S., and Ikenaka, T. (1972). *J. Biochem. (Tokyo)* **71,** 839–848.

Odani, S., and Ikenaka, T. (1973a). *J. Biochem. (Tokyo)* **74,** 697–715.

Odani, S., and Ikenaka, T. (1973b). *J. Biochem. (Tokyo)* **74,** 857–860.

Odani, S., and Ikenaka, T. (1976). *J. Biochem. (Tokyo)* **80,** 649–643.

Odani, S., Koide, T., and Ikenaka, T. (1971). *J. Biochem. (Tokyo)* **70,** 925–936.

Ogawa, T., Higasa, Y., and Hata, T. (1968). *Agric. Biol. Chem.* **32,** 484–491.

Ogawa, T., Higasa, T., and Hata, T. (1971a). *Agric. Biol. Chem.* **35,** 712–716.

Ogawa, T., Higasa, T., and Hata, T. (1971b). *Agric. Biol. Chem.* **35,** 717–723.

Ogawa, T., Higasa, T., and Hata, T. (1971c). *Mem. Res. Inst. Food Sci., Kyoto Univ.* **32,** 1–6.

Ogiso, T., Tamura, S., Kato, Y., and Suguira, M. (1974). *J. Biochem.* **76,** 147–156.

Ogiso, T., Noda, T., Sako, Y., Kato, Y., and Aoyama, M. (1975). *J. Biochem. (Tokyo)* **78,** 9–17.

Ogiso, T., Aoyama, M., Watanabe, M., and Kato, Y. (1976). *J. Biochem. (Tokyo)* **79,** 321–328.

Ortanderl, H. H., and Belitz, H.-D. (1971). *Lebensm.-Wiss. + Technol.* **4,** 85–88.

Ortanderl, H. H., and Belitz, H. D. (1977). *Z. Lebensm.-Unters. -Forsch.* **163,** 31–34.

Osborne, T. B., and Mendel, L. B. (1912). *Hoppe-Seyler's Z. Physiol. Chem.* **80,** 307–309.

Osborn, T. B., and Mendel, L. B. (1917). *J. Biol. Chem.* **32,** 369–387.

Owusu-Domfeh, K. (1973). *Ghana J. Agric. Sci.* **5,** 99–102.

Ozawa, K., and Laskowski, M., Jr. (1966). *J. Biol. Chem.* **241,** 3955–3961.

Palmer, R., McIntosh, A., and Pusztai, A. (1973). *J. Sci. Food Agric.* **24,** 937–944.

Papaioannou, S. E., and Liener, I. E. (1970). *J. Biol. Chem.* **245,** 4931–4938.

Parish, R. D., and Bolt, R. J. (1963). *Nature (London)* **199,** 398–399.

Patten, J. R., Richards, E. A., and Pope, H., II (1971). *Proc. Soc. Exp. Biol. Med.* **137,** 59–63.

Patten, J. R., Patten, J. A., and Pope, H., II (1973). *Food Cosmet. Toxicol.* **11,** 577–583.

Peng, J.-H., and Black, L. L. (1976). *Phytopathology* **66,** 958–963.

Perlstein, S. H., and Kezdy, F. J. (1973). *J. Supramol. Struct.* **1,** 249–254.
Phadke, K., and Sohonie, K. (1962). *J. Sci. Ind. Res., Sect. C* **21,** 272–275.
Pian, J. H. C. (1930). *Chin. J. Physiol.* **4,** 431–436.
Pinsky, A., and Schwimmer, V. H. (1974). *Phytochemistry* **13,** 779–783.
Polanowski, A. (1974). *Acta Biochim. Pol.* **43,** 27–37.
Pubols, M. H., Saxena, H. C., and McGinnis, J. (1964). *Proc. Soc. Exp. Biol. Med.* **117,** 713–717.
Pusztai, A. (1966). *Biochem. J.* **101,** 379–384.
Pusztai, A. (1968). *Eur. J. Biochem.* **5,** 252–259.
Quast, U., and Steffen, E. (1975). *Hoppe-Seyler's Z. Physiol. Chem.* **356,** 617–620.
Rackis, J. J. (1965). *Fed. Proc., Fed. Am. Soc. Exp. Biol.* **24,** 1488–1493.
Rackis, J. J. (1966). *Food Technol.* **20,** 102–104.
Rackis, J. J. (1972). *In* "Soybeans: Chemistry and Technology" (A. K. Smith and S. J. Circle, eds.), pp. 158–202. Avi Publ. Co., Westport, Connecticut.
Rackis, J. J. (1974). *J. Am. Oil Chem. Soc.* **51,** 161A–174A.
Rackis, J. J., and Anderson, R. L. (1964). *Biochem. Biophys. Res. Commun.* **15,** 230–235.
Rackis, J. J., Sasame, H. A., and Smith, A. K. (1959). *J. Am. Chem. Soc.* **81,** 6265–6270.
Rackis, J. J., Smith, A. K., Nash, A. M., Robbins, D. J., and Booth, A. N. (1963). *Cereal Chem.* **40,** 531–538.
Rackis, J. J., McGhee, J. E., Honig, D. H., and Booth, A. N. (1975a). *J. Am. Oil Chem. Soc.* **52,** 249A–253A.
Rackis, J. J., McGhee, J. E., and Booth, A. N. (1975b). *Cereal Chem.* **52,** 85–93.
Radoeva, A. (1968). *Nauch. Tr., Vissh. Inst. Kranit. Vkusova Prom., Plovdiv* pp. 371–380; cited in *Chem. Abstr.* **78,** 56647u (1973).
Ramirez, J. S., and Mitchell, H. L. (1960). *J. Agric. Food Chem.* **8,** 393–395.
Rancour, J. M., and Ryan, C. A. (1968). *Arch. Biochem. Biophys.* **125,** 380–383.
Read, J. W., and Haas, L. W. (1938). *Cereal Chem.* **15,** 59–68.
Reddy, M. N., Keim, P. S., Heinrikson, R. L., and Kezdy, F. J. (1975). *J. Biol. Chem.* **250,** 1741–1750.
Richardson, L. R. (1948). *J. Nutr.* **36,** 451–462.
Richardson, M. (1974). *Biochem. J.* **137,** 101–112.
Richardson, M. (1977). *Phytochemistry* **16,** 159–169.
Richardson, M., and Cossins, L. (1974). *FEBS Lett.* **45,** 11–13.
Richardson, M., McMillan, R. T., and Barker, R. D. J. (1976). *Biochem. Soc. Trans.* **4,** 1107–1108.
Richardson, M., Barker, R. D. J., McMillan, R. T., and Cossins' L. M. (1977). *Phytochemistry* **16,** 837–839.
Riesen, W. H., Clandinin, D. R., Elvehjem, C. A., and Cravens, W. W. (1947). *J. Biol. Chem.* **167,** 143–150.
Robinson, L. A., Kim, W. J., White, T. T., and Hadorn, B. (1972). *Scand. J. Gastroenterol.* **7,** 43–45.
Rothman, S. S., and Wells, H. (1967). *Am. J. Physiol.* **213,** 215–218.
Rouleau, M., and Lamy, F. (1975). *Can. J. Biochem.* **53,** 958–974.
Roy, D. N. (1972). *J. Agric. Food Chem.* **20,** 778–780.
Roy, D. N., and Bhat, R. V. (1974). *J. Sci. Food Agric.* **25,** 765–769.
Roy, D. N., and Rao, S. P. (1971). *J. Agric. Food Chem.* **19,** 251–259.
Royer, A. (1975). *Phytochemistry* **14,** 915–919.
Royer, A., Miége, M. N., Grange, A., Miége, J., and Mascherpa, J. M. (1974). *Planta* **119,** 1–16.
Ryan, C. A. (1966). *Biochemistry* **5,** 1592–1596.
Ryan, C. A. (1968). *Plant Physiol.* **43,** 1880–1881.
Ryan, C. A. (1971). *Biochem. Biophys. Res. Commun.* **44,** 1265–1270.
Ryan, C. A. (1973). *Annu. Rev. Plant Physiol.* **24,** 173–196.

Ryan, C. A. (1974). *Plant Physiol.* **54,** 328–332.

Ryan, C. A. (1977). *In* "Immunological Aspects of Foods" (N. Catsimpoolas, ed.), pp. 182–198. Avi Publ. Co., Westport, Connecticut.

Ryan, C. A., and Green, T. R. (1974). *Recent Adv. Phytochem.* **8,** 123–140.

Ryan, C. A., and Hass, G. M. (1977). *Fed. Proc., Fed. Am. Soc. Exp. Biol.* **36,** 2603.

Ryan, C. A., and Huisman, O. C. (1967). *Nature (London)* **214,** 1047–1049.

Ryan, C. A., and Huisman, W. (1970). *Plant Physiol.* **45,** 484–489.

Ryan, C. A., and Santarius, K. (1976). *Plant Physiol.* **58,** 683–685.

Ryan, C. A., and Shumway, L. K. (1970). *Plant Physiol.* **45,** 512–514.

Ryan, C. A., Hass, G. M., and Kuhn, R. W. (1974). *J. Biol. Chem.* **249,** 5495–5503.

Ryan, C. A., Kuo, T., Pearce, G., and Kunkel, R. (1976). *Am. Potato J.* **53,** 433–455.

Sakato, K., Tanaka, H., and Misawa, M. (1975). *Eur. J. Biochem.* **55,** 211–219.

Salman, A. J., Dal Bargo, G., Pubols, M. H., and McGinnis, J. (1967). *Proc. Soc. Exp. Biol. Med.* **126,** 694–698.

Salman, A. J., Pubols, M. H., and McGinnis, J. (1968). *Proc. Soc. Exp. Biol. Med.* **128,** 258–261.

Sambeth, W., Nesheim, M. C., and Serafin, J. A. (1967). *J. Nutr.* **92,** 479–490.

Santarius, K., and Belitz, H.-D. (1972). *Chem., Mikrobiol., Technol. Lebensm.* **2,** 56–62.

Santarius, K., and Belitz, H.-D. (1975). *Chem., Mikrobiol., Technol. Lebensm.* **3,** 180–184.

Sawada, J., Uasui, H., Armamoto, T., Yamada, M., Okazaki, T., and Tanaka, I. (1974). *Agric. Biol. Chem.* **38,** 2559–2561.

Saxena, H. C., Jensen, L. S., and McGinnis, J. (1962). *J. Nutr.* **77,** 241–244.

Saxena, H. C., Jensen, L. S., and McGinnis, J. (1963a). *Proc. Soc. Exp. Biol. Med.* **112,** 101–105.

Saxena, H. C., Jensen, L. S., McGinnis, J., and Lauber, J. K. (1963b). *Proc. Soc. Exp. Biol. Med.* **112,** 390–393.

Saxena, H. C., Jensen, L. S., and McGinnis, J. (1963c). *J. Nutr.* **80,** 391–396.

Saxena, H. C., Jensen, L. S., and McGinnis, J. (1963d). *Poult. Sci.* **42,** 788–790.

Schingoethe, D. J., Aust, S. D., and Thomas, J. W. (1970a). *J. Nutr.* **100,** 739–748.

Schingoethe, D. J., Gorrill, A. D. L., Thomas, J. W., and Yang, M. G. (1970b). *Can. J. Physiol. Pharmacol.* **48,** 43–49.

Schingoethe, D. J., Tideman, L. J., and Uckert, J. R. (1974). *J. Nutro.* **104,** 1304–1312.

Schneeman, B. O., and Lyman, R. L. (1975). *Proc. Soc. Exp. Biol. Med.* **148,** 897–903.

Schneeman, B. O., Chang, I., Smith, L. B., and Lyman, R. L. (1977). *J. Nutr.* **107,** 281–288.

Sealock, R. W., and Laskowski, M., Jr. (1969). *Biochemistry* **8,** 3703–3710.

Seidel, D. S., and Liener, I. E. (1971). *Biochim. Biophys. Acta* **251,** 83–93.

Seidel, D. S., and Liener, I. E. (1972a). *Biochim. Biophys. Acta* **258,** 303–309.

Seidel, D. S., and Liener, I. E. (1972b). *Biochem. Biophys. Res. Commun.* **42,** 1101–1107.

Seidel, D. S., and Liener, I. E. (1972c). *J. Biol. Chem.* **247,** 3533–3538.

Senser, F., Belitz, H.-D., Kaiser, K.-P., and Santarius, K. (1974). *Z. Lebens.-Unters. -Forsch.* **155,** 100–101.

Shain, Y., and Mayer, A. M. (1965). *Physiol. Plant.* **18,** 853–859.

Shumway, L. K., Rancour, J., and Ryan, C. A. (1970). *Planta* **93,** 1–14.

Shumway, L. K., Yang, V. V., and Ryan, C. A. (1976). *Planta* **129,** 161–165.

Shurpalikar, S. R., Chandrasekhara, M. R., Swaminathan, M., and Subrahmanyan, V. (1961). *Food Sci.* **11,** 52–64.

Shyamala, G., and Lyman, R. L. (1964). *Can. J. Biochem. Physiol.* **42,** 1825–1832.

Siddappa, G. S. (1957). *J. Sci. Ind. Res., Sect. C* **16,** 199–201.

Singh, H., Schaible, P. J., Zindel, H. C., and Ringer, R. K. (1964). *Mich., Agric. Exp. Stn., Q. Bull.* **47,** 17–23.

Singh, L., Wilson, C. M., and Hadley, H. H. (1969). *Crop Sci.* **9,** 489–491.

Smirnoff, P., Khalef, S., Birk, Y., and Applebaum, S. W. (1976). *Biochem. J.* **157,** 745–751.

Smith, A. K. (1963). *Cereal Sci. Today* **8,** 196, 200, and 210.
Smith, A. K., and Circle, S. J., eds. (1972). "Soybeans: Chemistry and Technology." Avi Publ. Co., Westport, Connecticut.
Smith, A. K., Rackis, J. J., McKinney, L. L., Robbins, D. J., and Booth, A. N. (1964a). *Feedstuffs* **36,** 46–47.
Smith, A. K., Rackis, J. J., Hesseltine, C. W., Smith, M., Robbins, D. J., and Booth, A. N. (1964b). *Cereal Chem.* **41,** 173–181.
Snook, J. T. (1969). *J. Nutr.* **97,** 286–294.
Sohonie, K., and Ambe, K. S. (1955). *Nature (London)* **175,** 508–509.
Sohonie, K., and Bhandarkar, A. P. (1954). *J. Sci. Ind. Res., Sect. B* **13,** 500–503.
Sohonie, K., and Bhandarkar, A. P. (1955). *J. Sci. Ind. Res., Sect. C* **14,** 100–104.
Sohonie, K., and Honavar, P. M. (1956). *Sci. Cult.* **21,** 538.
Sohonie, K., Apte, U., and Ambe, K. S. (1958). *J. Sci. Ind. Res., Sect. C* **17,** 42–46.
Sohonie, K., Huprikar, S. V., and Joshi, M. R. (1959). *J. Sci. Ind. Res., Sect. C* **18,** 95–98.
Solyom, A., Borsy, J., and Tolnay, P. (1964). *Biochem. Pharmocal.* **13,** 391–394.
Standal, B. R. (1963). *J. Nutr.* **81,** 279–285.
Steiner, R. F. (1964). *Nature (London)* **204,** 579.
Steiner, R. F. (1965). *Biochim. Biophys. Acta* **100,** 111–121.
Steiner, R. F. (1966). *Arch. Biochem. Biophys.* **115,** 257–270.
Steinkraus, K. H., Hand, D. B., Van Buren, J. P., and Hackler, L. R. (1962). *Proc. Conf. Soybean Prod. Protein Human Foods, 1961* pp. 83–92.
Stevens, F. C., Wuerz, C., and Krahn, J. (1974). *In* "Protease Inhibitors" (H. Fritz, H. Tschesche, L. J. Greene, and E. Truscheit, eds.), Bayer Symp. V, pp. 344–353. Springer-Verlag, Berlin and New York.
Stewart, K. K., and Doherty, R. F. (1971). *FEBS Lett.* **16,** 226–228.
Stewart, K. K., and Doherty, R. F. (1973). *Fed. Proc., Fed. Am. Soc. Exp. Biol.* **32,** 1598.
Stillings, B. R., and Hackler, L. R. (1965). *J. Food Sci.* **30,** 1043–1048.
Subbulakshmi, G., Ganeshkumar, K., and Venkataraman, L. V. (1976). *Nutr. Rep. Int.* **13,** 19–31.
Suguira, M. (1974). Japanese Patent, No. 74 37,275; cited in *Chem. Abstr.* **82,** 134930e. (1975).
Suguira, M., Ogiso, T., Takeuti, K., Tamura, S., and Ito, A. (1973). *Biochim. Biophys. Acta* **328,** 407–417.
Sumathi, S., and Pattabiraman, T. N. (1975). *Indian J. Biochem. Biophys.* **12,** 383–385.
Sumathi, S., and Pattabiraman, T. N. (1976). *Indian J. Biochem. Biophys.* **13,** 52–56.
Sundberg, L., Porath, J., and Aspberg, K. (1970). *Biochim. Biophys. Acta* **221,** 394–395.
Sweet, R. M., Wright, H. T., Janin, J., Chothia, C. H., and Blow, D. M. (1974). *Biochemistry* **13,** 4212–4228.
Tan, C. G. L., and Stevens, F. C. (1971a). *Eur. J. Biochem.* **18,** 503–514.
Tan, C. G. L., and Stevens, F. C. (1971b). *Eur. J. Biochem.* **18,** 515–000.
Tauber, H., Kershaw, B. B., and Wright, R. D. (1949). *J. Biol. Chem.* **179,** 1155–1161.
Tawde, S. (1961). *Ann. Biochem. Exp. Med.* **21,** 359–366.
Thompson, R. M., and Liener, I. E. (1978). *Fed. Proc., Fed. Am. Soc. Exp. Biol.* **37,** 264.
Tidemann, L. J., and Schingoethe, D. J. (1974). *J. Agric. Food Chem.* **22,** 1059–1062.
Tixier, R. (1968). *Hebd. Seances Acad. Sci., Ser. D* **266,** 2498–2500.
Travis, J., and Roberts, R. (1969). *Biochemistry* **8,** 2884–2889.
Tsai, Y.-C. (1976). *J. Chin. Agric. Chem. Soc.* **14,** 187–194.
Tschesche, H. (1974). *Angew. Chem., Int. Ed. Engl.* **13,** 10–28.
Tur-Sinai, A., Birk, Y., Gertler, A., and Rigbi, M. (1972). *Biochim. Biophys. Acta* **263,** 666–672.
Ubatuba, F. B. (1955). *Rev. Bras. Biol.* **5,** 1–8.
Van Buren, J. P., Steinkraus, K. H., Hackler, L. R., El-Rawi, I., and Hand, D. B. (1964). *J. Agric. Food Chem.* **12,** 524–528.

Van Veen, A. G., and Schaeffer, G. (1950). *Doc. Neerl. Indones. Morbis Trop.* **2,** 270–281.
Van Veen, A. G., and Steinkraus, K. H. (1970). *J. Agric. Food Chem.* **18,** 576–578.
Veerabhadrappa, P. S., Majunath, N. H., and Virupaksha, T. K. (1978). *J. Sci. Food Agric.* **29,** 353–358.
Venkatamaran, L. V., Haya, T. V., and Krishnamurthy, K. S. (1976). *Nutr. Rep. Int.* **13,** 197–205.
Ventura, M. M., and Filho, J. X. (1967). *An. Acad. Bras. Cienc.* **38,** 553–566.
Ventura, M. M., Filho, J. X., Moreira, R. A., Aquino, A. deM., and Pinheiro, P. A. (1971). *An. Acad. Bras. Cienc.* **43,** 233–242.
Ventura, M. M., de Araijo, G. B., Cagnin, A. H., and Martin, C. deO. (1972a). *An. Acad. Bras. Cienc.* **44,** 581–582.
Ventura, M. M., Aragão, J. B., and Ikenoto, H. (1972b). *An. Acad. Bras. Cienc.* **44,** 582–583.
Viswanatha, J., and De, S. S. (1951). *Indian J. Physiol. Allied Sci.* **5,** 51–58.
Vogel, R., and Hartwig, G. (1968). *In* "Nature Protease Inhibitors" (R. Vogel, I. Trautschold, and E. Werle, eds.), p. 37. Academic Press, New York.
Wagh, P. V., Klaustermeier, D. F., Waibel, P. E., and Liener, I. E. (1963). *J. Nutr.* **80,** 191–195.
Wagner, L. P., and Riehm, J. P. (1967). *Arch. Biochem. Biophys.* **121,** 672–677.
Walker-Simmons, M., and Ryan, C. A. (1977a). *Plant Physiol.* **59,** 437–439.
Walker-Simmons, M., and Ryan, C. A. (1977b). *Plant Physiol.* **60,** 61–63.
Wang, A. L., Ruttle, D. I., and Hesseltine, C. W. (1968). *J. Nutr.* **96,** 109–114.
Wang, A. L., Ruttle, D. I., and Hesseltine, C. W. (1969). *Fed. Proc., Fed. Am. Soc. Exp. Biol.* **28,** 304.
Wang, D. (1975). *Biochim. Biophys. Acta* **393,** 583–596.
Wang, L. C. (1971). *Cereal Chem.* **48,** 303–312.
Wang, R. W., and Alexander, J. C. (1971). *Nutr. Rep. Int.* **4,** 387–391.
Warchalewski, J. R., and Shupin, J. (1973). *J. Sci. Food Agric.* **24,** 995–1009.
Warsy, A. S., and Stein, M. (1973). *Qual. Plant.—Plant Foods Hum. Nutr.* **23,** 157–169.
Warsy, A. S., Norton, G., and Stein, M. (1974). *Phytochemistry* **13,** 2481–2486.
Waterman, H. C., and Johns, C. O. (1921). *J. Biol. Chem.* **46,** 9–17.
Waterman, H. C., and Jones, D. B. (1921). *J. Biol. Chem.* **47,** 285–295.
Weder, J. K. P. (1973). *Z. Lebensm.-Unters. -Forsch.* **153,** 83–86.
Weder, J. K. P., and Belitz, H.-D. (1969). *Dtsch. Lebensm.-Rundsch.* **65,** 78–79.
Weder, J. K. P., and Hory, H. D. (1972a). *Lebensm.-Wiss. + Technol.* **5,** 54–63.
Weder, J. K. P., and Hory, H. D. (1972b). *Lebensm.-Wiss. + Technol.* **5,** 86–90.
Weder, J. K. P., and Hory, H. D. (1976). *Chem., Mikrobiol., Technol. Lebensm.* **4,** 161–169.
Weder, J. K. P., Kassubek, A., and Hory, H. D. (1975). *Chem., Mikrobiol., Technol. Lebensm.* **4,** 79–84.
Weiel, J., and Hapner, K. D. (1976). *Phytochemistry* **15,** 1885–1887.
Werle, E., and Zickgraf-Rüdel, G. (1972). *Z. Klin. Chem. Klin. Biochem.* **10,** 139–150.
Werle, E., Appel, W., and Hopp, E. (1959). *Z. Vitam.-Horm.- Fermentforsch.* **10,** 127–136.
West, E. S., Todd, W. R., Mason, H. S., and Van Bruggen, J. T. (1966). "Textbook of Biochemistry," pp. 1203 and 1246. Macmillan, New York.
Westfall, R. J., and Hauge, S. M. (1948). *J. Nutr.* **35,** 374–389.
Westfall, R. J., Bosshardt, D. K., and Barnes, R. H. (1948). *Proc. Soc. Exp. Biol. Med.* **68,** 498–500.
Whitley, E. J., Jr., and Bowman, D. E. (1975). *Arch. Biochem. Biophys.* **169,** 42–50.
Wilson, B. J., McNab, J. M., and Bentley, H. (1972a). *J. Sci. Food Agric.* **23,** 679–684.
Wilson, B. J., McNab, J. M., and Bentley, H. (1972b). *Poult. Sci.* **13,** 521–523.
Wilson, K. A., and Laskowski, M., Sr. (1973). *J. Biol. Chem.* **248,** 756–762.
Wilson, K. A., and Laskowski, M., Sr. (1975). *J. Biol. Chem.* **250,** 4261–4267.
Wong, P. P., Kuo, T., Ryan, C. A., and Kado, C. I. (1976). *Plant Physiol.* **57,** 214–217.

Woodham, A. A., and Dawson, R. (1968). *Br. J. Nutr.* **22,** 589–599.
Worowski, K. (1974). *Thromb. Diath. Haemorrh.* **32,** 617–632.
Worowski, K. (1975). *Experientia* **31,** 637–638.
Wu, Y. V., and Scheraga, H. A. (1962). *Biochemistry* **1,** 698–705.
Xavier-Filho, J. (1973). *Physiol. Plant.* **28,** 149–154.
Xavier-Filho, J. (1974). *J. Food Sci.* **39,** 422–423.
Xavier-Filho, J., and Ainouz, I. L. (1977). *Biol. Plant.* **19,** 183–189.
Yamada, M., Tashiro, M., Yamaguchi, H., Yamada, H., Ibuki, F., and Kanamori, M. (1976). *J. Biochem. (Tokyo)* **80,** 1293–1297.
Yamamoto, M., and Ikenaka, T. (1967). *J. Biochem. (Tokyo)* **62,** 141–149.
Yanatori, Y., and Fujita, T. (1976). *Arch. Histol. Jpn.* **39,** 67–78.
Yen, J. T., Jensen, A. A., and Simon, J. (1977). *J. Nutr.* **107,** 156–165.
Yoshida, C., and Yoshikawa, M. (1975). *J. Biochem. (Tokyo)* **78,** 935–945.
Yoshikawa, M., Iwasaki, T., Fujii, M., and Ogaki, M. (1976). *J. Biochem. (Tokyo)* **79,** 765–773.
Zimmerman, G., Weissmann, S., and Yannai, S. (1967). *J. Food Sci.* **32,** 129–130.

CHAPTER 3

Hemagglutinins (Lectins)

WERNER G. JAFFÉ

I. INTRODUCTION

The use of fire for cooking by primitive man must have had an important impact on the supply of edible plant materials. Starchy plant products became available for human consumption through cooking (since the digestive system is not adapted to hydrolyze raw starch), and a number of seeds that are toxic in the raw state became edible food after cooking.

The presence of heat-labile toxic factors in plant products, mainly legume seeds, makes them unsuitable as food for human beings and higher animals unless they are properly cooked. Nonetheless, there is archeological evidence

TOXIC CONSTITUENTS OF PLANT FOODSTUFFS, SECOND EDITION

ISBN 0-12-449960-0

that they have been used by human beings since prehistoric times. The wild, ancestral bean *Phaseolus aborigineus* has been found in many parts of the American continent from Argentina to Mexico, and some of the local cultivars of the garden bean are very closely related to this wild species. It can only be imagined how man has learned to prepare edible food from these wild seeds, since they contain thermolabile, toxic constituents, as do modern cultivated beans. Among the antinutritional factors that are inactivated by the cooking process are the lectins, or hemagglutinins, which are the subject of this chapter.

Lectins have been detected in many plant families, from slime molds (Rosen *et al.*, 1975), fungi, and lichens to flowering plants (Krüpe, 1956). They have also been detected in animals (Cohen, 1974), such as sponges, crustaceans, (Pauly, 1974), mollusks, e.g., snails (Uhlenbruck and Steinhausen, 1972), fish blood serum, the eggs of amphibians, and even mammalian tissue (Stockert *et al.*, 1974).

The first description of a phytohemagglutinin, or plant lectin, was presented by Stillmark (1889), who studied the toxicity of castor beans and press cake from the production of castor oil. From his very thorough investigation, he concluded that the toxic action was due to a protein, which he called "ricin" and which was capable of agglutinating red blood cells from human beings and animals.

Several other toxic plant proteins were described in the years following the discovery of ricin, although many have never been studied in much detail. The relative facility with which castor beans can be obtained and the strong toxicity of ricin probably account for the fact that more investigators were attracted to this plant rather than to similar but less available plants. The early literature on the toxicological properties of the plant hemagglutinins was reviewed by Ford (1913) and Brocq-Rousseu and Fabre (1947).

Landsteiner and Raubitschek (1908) observed that extracts from many edible crude legume seeds agglutinated red cells, but no toxic action was detected in these seeds at that time. They also established that the relative hemagglutinating activities of various seeds were quite different when tested with red blood cells from different animals, and they compared this specificity with that of the antibodies of animal blood serum (Table I). The name "agglutinin" was first proposed by Elfstrand (1897) for the phytohemagglutinins, and only later was its use extended to immunoagglutinins.

The observations of Renkonen (1948) and Boyd and Reguera (1949) on the different activities of some plant agglutinins toward the human blood groups aroused the interest of immunologists and prompted an intensive, systematic investigation of a great number of plants with regard to their potential use as blood typing reagents. As a result, a distinction was made between "specific" and "nonspecific" agglutinins. These terms have produced some confusion because they obscure the fact that the nonspecific agglutinins, although acting on the cells of any human blood group, nevertheless show very characteristic dif-

TABLE I

AGGLUTINATION OF BLOOD FROM DIFFERENT ANIMALS BY LEGUME EXTRACTS[a]

Blood	Seed extracts			
	Beans	Peas	Lentils	Sweet peas
Human	800	40	30	20
Horse	16,000	128	64	128
Rabbit	8,000	1,000	2,000	200
Sheep	1,600	4	—	—
Pigeon	32,000	—	—	400
Carp	800	400	200	10
Frog	400	80	—	8

[a] Landsteiner and Raubitschek (1908). The highest dilutions of the respective extracts still active in the agglutination test are reported.

ferences when brought into contact with erythrocytes from different animal species. Only a fraction of the known agglutinins exhibit blood group specificity. They are of special interest to the hematologist, and several excellent reviews have been published on this aspect (Krüpe, 1956; Bird, 1959; Saint-Paul, 1961; Boyd, 1963, 1974; Tobiska, 1964). Blood group-specific and -nonspecific agglutinins may exist together in some plants, such as navy beans (*Phaseolus vulgaris*) (Toms and Turner, 1965) and *Vicia cracca* (Asberg *et al.*, 1968).

The application of this erythroagglutinating action to the separation of red blood cells from leukocytes led Nowell (1960) to the important observation of the mitosis-inducing activity of the kidney bean agglutinin. A new field of investigation is based on this discovery, which led to the development of simple methods for the study of human chromosomes and its clinical application (see Ling, 1968). It also proved to be very useful for the examination of the biochemical events involved in the conversion of resting cells to actively growing ones. Several other plant lectins have been found to stimulate mitosis in lymphocytes.

Another observation that proved to be extremely stimulating for research on lectins was that on the differential activity in normal and cancerous cells (see Lis and Sharon, 1973). It is therefore not surprising that, since the first edition of this book was published, the number of papers published on lectins and on scientific investigations in which lectins were used has grown enormously. They are tools for the study of cell membranes under normal and pathological conditions (Walborg *et al.*, 1975), the structure of glycolipids (Lis and Sharon, 1973), and membrane dynamics (Sachs, 1974), and are used in affinity chromatography. Their chemical (Lis and Sharon, 1973; Sharon and Lis, 1972), botanical (Liener,

1976), taxonomic (Toms and Western, 1971), and immunological (Jaffé, 1977) aspects have been reviewed. Compared with this extensive literature, the number of papers dealing with the antinutritional properties of lectins is rather small. Liener provided a condensed review of this aspect in 1974.

Of 147 plant families from which several species were tested, 79 were shown to contain hemagglutinin-positive species (Toms and Western, 1971). In the Leguminosae positive species are common, and many of the edible pulses contain lectins (Landsteiner and Raubitschek, 1908). Owing to their occurrence in edible plant products and the toxic properties of some of them, they are of special interest to the food scientist. This chapter will primarily discuss lectins from edible plants.

II. PURIFICATION

The purification of lectins to homogeneity poses problems not commonly encountered in the purification of other proteins. Lectins may appear in multiple forms that possess more or less similar biological activities and differ only slightly in their chemical and physical properties. Many lectins are composed of subunits, which may undergo different association–dissociation reactions. The chromatographic behavior and other characteristics of lectins may depend on the experimental conditions, especially the presence of certain metal ions. Moreover, not only are lectins from various sources different, but even those obtained from one plant species may differ from one another according to the exact variety or cultivar used for fractionation. Therefore, results from different laboratories frequently are difficult to compare, because it cannot be determined whether differences in the properties described are due to the fact that different fractions were obtained or that the homologous fractions from various seed materials have different properties.

The specific affinity for certain sugar residues can be used for the purification of lectins. Liener (1976) compiled a number of pertinent examples. Therefore, only a few will be mentioned here. Frequently, lectins can be obtained in a single step in relatively pure form and in excellent yield by affinity chromatography. Sephadex and Sepharose have been used with some lectins, which are bound to the glucosyl or galactosyl residues of these polymers. Reductively aminated disaccharides bound to aminoethyl polyacrylamide can be used for the preparation of many different lectins (Baues and Gray, 1977). A rapid method for derivatizing agarose with a variety of carbohydrates for use in affinity chromatography was described by Bloch and Burger (1974a). Biospecific adsorbents were prepared by a one-step reaction between epoxy-activated Sepharose 6B and lectin-specific sugars (Vretblad, 1976). Formalized erythrocytes can also be used for the isolation of lectins from crude plant extracts (Reitherman *et al.*, 1974). Advantage may be taken of the fact that most lectins are glycoproteins and can therefore interact with some other lectins. Thus, concanavalin A covalently bound to Sepharose can be used for the removal of different lectins from crude plant extracts (Bessler and Goldstein, 1973). Many lectins and Sepharose-bound lectins are now commercially available.

III. GENERAL PROPERTIES

The lectins, or hemagglutinins, are proteins that possess a specific affinity for certain sugar molecules. Since carbohydrate moieties exist in most animal cell membranes, they may attach to these so-called receptor groups if the specific structure of the latter fits the former. The hemagglutination reaction, which has been studied with some plant extracts since the last century, is the most readily observable effect of this attachment and will occur only if the lectin molecule has at least two active groups.

As is indicated by their names, hemagglutinins, or lectins, can be characterized and detected by their action on red blood cells. Other types of cells can be agglutinated as well. In some cases, the binding may occur without visible agglutination. The term "lectin," originally proposed by Boyd and Shapleigh (1954) to denote plant hemagglutinins, was chosen to call attention to the specificity of this reaction (from the Latin *legere,* to choose). Liener (1976) proposed distinguishing between "phytolectins," "zoolectins," mycolectins," etc., on the basis of their respective origin.

The receptor site of the cell surface must be exposed in order to react with the specific lectin. Moreover, the degree of site mobility in the cell membrane determines agglutinability (Sachs, 1974). Activation of erythrocytes by digestion with papain, trypsin, or Pronase is probably due to exposure of receptor sites, rendering them accessible to the lectin. Malignant transformation and other physiological or pathological changes in the cell surface structure may also influence the ability to interact with lectins (Etzler, 1974).

IV. STRUCTURE

The only common characteristic of the plant lectins is that all are proteins. Many bear covalently bound sugars and can thus be classified as glycoproteins. Concanavalin A and probably a few other lectins are exceptions and are devoid of sugar residues. The carbohydrates found in different lectins, even those from different varieties of a single plant species and isolectins derived from the same seed lot, may be different. Glycopeptides bearing glucosamine and mannose have been obtained by proteolytic digestion of several lectins.

The structure of some lectins has been elucidated in detail (see Liener, 1976). Concanavalin A at neutral pH consists of a tetramer, which is composed of four identical subunits of molecular weight 26,000 formed by a single polypeptide chain consisting of 237 amino acids (Cunningham *et al.*, 1975). Each subunit contains two metal-binding sites, one for Mn^{2+} and one for Ca^{2+}, and another site for a sugar residue (Fig. 1). There is evidence that the tetrameric form of concanavalin A has been elucidated (Edelman *et al.*, 1972), and the structural features of the binding sites for sugar molecules and for Ca^{2+} and Mn^{2+} are known. Evidence for hydrophobic sites has also been presented (Hardman and Ainsworth, 1973). Neutral lipids can be extracted in small amounts, causing the loss of activity (Jaffé and Palozzo, 1971).

Analogous structural conditions prevail in many legume lectins; i.e., they consist of two to four subunits and contain bivalent metal ions. In many cases, several similar isolectins exist in the corresponding seeds, which can be separated by electrophoresis or chromatographic methods.

Foriers *et al.* (1977) observed extensive sequence homologies in the first 25 residues of the lectins from lentils, peas, soybeans, kidney beans, and peanuts. However, immunological cross-reactions among lectins derived from different legume seeds have not been found in most of the cases in which pertinent studies have been performed, except in very closely related species (Jaffé, 1977).

Interesting insight into the structural relationship between the highly toxic ricin and the *Ricinus* agglutinin from the castor bean (*Ricinus communis*) has been obtained. The toxic and hemagglutinating properties of the castor bean have been

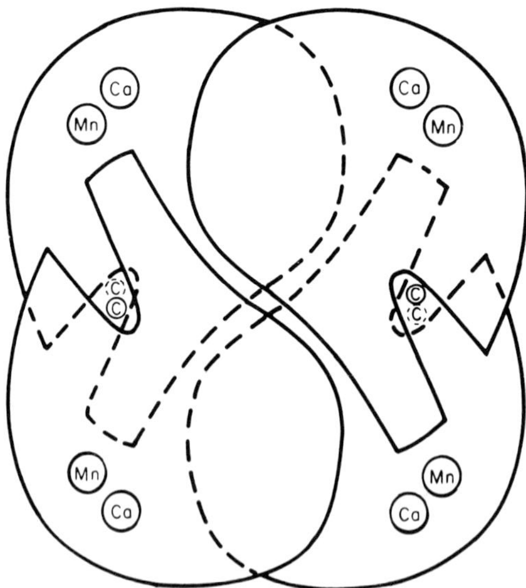

FIG. 1. Schematic representation of the tetrameric structure of concanavalin A viewed down the Z axis. The proposed binding sites for transition metals are indicated by Mn and Ca, and those for saccharides are indicated by C. (From Edelman *et al.*, 1972, reproduced by permission.)

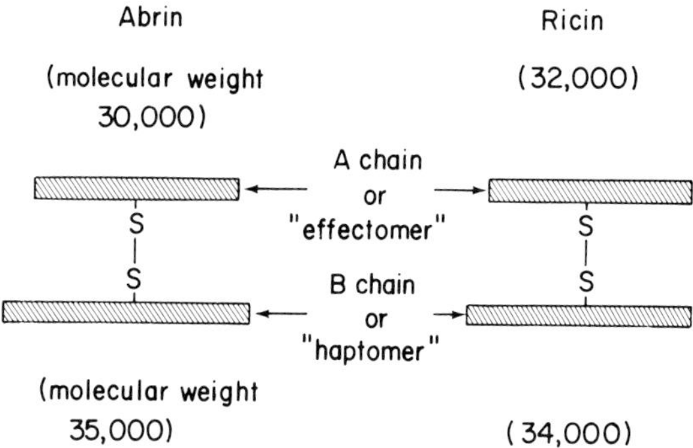

FIG. 2. Schematic structure of ricin and abrin. (From Olsnes *et al.*, 1974, reproduced by permission.)

studied since the last century (see Ford, 1913). Funatsu and his group (1972) were the first to separate a nonagglutinating toxic component from a nontoxic agglutinin. The toxic component is now generally called ricin and is composed of two polypeptide chains held together by disulfide bonds (Fig. 2). The smaller A chain, or "effectomer," was found to inhibit protein synthesis in a cell-free system of rabbit reticulocytes. The B chain, or "haptomer," functions as a carrier moiety, which serves to anchor the toxin to the cell surface, a binding that probably involves galactose-containing receptor sites (Olsnes and Pihl, 1973). Studies on the structure of the *Ricinus* agglutinin have revealed that it is a tetramer composed of two different subunits. The heavier of the two chains is probably identical to the B chain of ricin, whereas the lighter chain is homologous but not identical to the A chain of ricin. Since the B chain contains the sugar-binding site, the agglutinin has two binding sites (Olsnes *et al.*, 1974). Very similar conditions prevail for the hemagglutinating and toxic fractions of abrin from the jequirity bean (*Abrus precatorius*) (Olsnes, 1975).

V. LOCALIZATION

Most hemagglutinins of higher plants are found in seeds, but tubers and plant saps are also sources of lectins. In some cases their presence in leaves, stems, and bark has been demonstrated (Tobiška, 1964). According to Renkonen (1960), unripe seeds contain lectins that may be partially or completely bound to inhibitors. These inhibitors are thermostable and are not precipitated by acetone. During maturation the hemagglutinin titer may rise very rapidly. This may be caused by a release of the lectins from their inhibitors.

By the use of immunological methods, Mialonier *et al.* (1973) localized the bean lectin in the cytoplasm of the cotyledon and the embryo. It appeared during ripening and disappeared during germination. These authors therefore ascribed a certain role to the lectin in seed germination. Similar localization was found by Howard *et al.* (1972) for lentil seeds. During the early growth of the seed of lima beans, lectin activity increases rapidly in comparison with the dry weight. During germination, lectin decreases rapidly, but small quantities can be detected in various plant parts (Martin *et al.*, 1964). Lectins in seeds can be found concentrated in the protein bodies of the endosperm, where they appear during maturation and disappear rapidly during germination (Youle and Huang, 1976).

Clarke *et al.* (1975) localized concanavalin A in jack bean cotyledons and bean lectin phytohemagglutinin in the cotyledons of bean seeds using a fluorescent method with a fluorescein isocyanate-labeled globulin preparation. In these seeds they also found lectins with specifity for β-galactosyl and β-glucosyl artificial antigens located in the intercellular spaces of the cotyledon parenchyma, which were associated with the cell membrane. This localization is considered as evidence of self-recognition factors of the β-lectins in plant cells.

VI. ACTIVITY AND PHYSIOLOGICAL FUNCTION IN THE PLANT

Very important differences in agglutination activity can be observed between the ripe seeds of different varieties of many lectin-containing plants or other plant organs, such as potato tubers. (Tobiška, 1964). Besides the possible presence of inhibitors, other explanations are also likely. Application of fertilizers can raise the lectin content of beans (Tobiška and Lhotecka-Brázdová, 1960). Lectin formation in plants may also be controlled by genetic factors. Schertz *et al.* (1960) studied the heredity of the anti-A activity in lima beans by crossing varieties with high or low agglutinating activity. They were able to demonstrate genetic control of lectin activity. The gene for high activity is a dominant factor. Similarly, in beans, the lectin type A, which is the most active in agglutination of blood cells and the most toxic, is inherited as a single dominant factor when crossed with D-type beans (Brücher *et al.*, 1969; Jaffé *et al.*, 1972). This fact can be used to outbreed this toxicity factor. Klozowa and Turkova (1978a,b) studied bean extracts of cultivars of different hematological behavior by immunological methods. Their results are in accordance with the above observations. Several soybean lines lacking hemagglutinin activity have been detected by Pull *et al.* (1978). The presence of lectin is controlled by a single dominant gene (Orf *et al.*, 1978). The occurrence of a special lectin in small-seeded lentil subspecies (Fliolová *et al.*, 1975) also points to a genetic influence, as does the relationship between chromosome number in wheat and the number of isolectins present (Rice, 1976).

Lectin-free potato varieties were observed by Ortega and Cardona (1970). These authors found differences among varieties or cultivars not only in the strength of hemagglutinating action but also in some physicochemical parameters, like such as heat resistance and solubility. Similar observations exist for soybeans (Turner and Liener, 1975) and for kidney beans (Jaffé *et al.*, 1972). On the other hand, very similar if not identical lectins may exist in closely related species. *Canavalia gladiata* and *Canavalia maritima* contain variants of concanavalin A that differ by only one amino acid residue from that of *Canavalia ensiformis* (Hague, 1975).

Speculations as to the function of lectins in the plant have been expressed in earlier reviews (Tobiška, 1964; Toms and Western, 1971; Liener, 1976). These include the following: they (1) act as antibodies to conteract soil bacteria, (2) protect plants against fungal attack, (3) attach glycoprotein enzymes in organized multienzyme systems, (4) play a role in the development and differentiation of embryonic cells, and (5) transport or store sugars.

A bruchid beetle (*Callosobruchus maculatus*) that can eat lectin-free cowpeas (*Vigna ungulata*) will die when the isolated black bean lectin is added to its diet (Janzen *et al.*, 1976). Wild and cultivated *Phaseolus* are attacked by several species of bruchids. These dietary requirements of these insects are extremely specific. Part of the adaptive significance of lectins may be protection of the seeds from attack by insect seed predators.

Certain bacteria such as *Rhizobium* form nodules in the roots of legumes and fix nitrogen, which is vital for the plant. A successful symbiotic relationship between bacteria and plants is characterized by a high degree of host–symbiont selectivity. There are several observations that point to the involvement of lectins in the molecular basis for this selectivity. The presence of lectins in the root nodules of some legumes was first observed by Eisler and Portheim (1926). Hamblin and Kent (1973) found that lectins were present in the roots of bean plants and that a strain of *Rhizobium phaseoli,* which had been treated with these lectins, was capable of agglutinating erythrocytes. Bohlool and Schmidt (1974) presented evidence that soybean lectin is capable of binding to 22 of 25 infective strains of *R. japonicum,* the symbiont of soybeans, while it does not bind to any of the tested strains of *Rhizobium* that do not nodulate soybeans. The lipopolysaccharides of the O antigens of four *Rhizobium* strains were isolated by Wolpert and Albersheim (1976). These compounds could bind to the lectin of the symbiont, which had been covalently attached to agarose. However, they did not react with the lectins of the other legumes tested. The establishment of a successful symbiotic relationship was therefore attributed to selective interaction between the lectins of the host with the O antigens of the symbiont. A lectin from clover agglutinates specifically infective strains of *Rhizobium trifolii* and binds to the capsular antigen from this nodule bacterium (Dazzo and Hubbel, 1975). Since lectin is found on the surface of clover root hairs, its position aids in the binding of the symbiont. In this case, too, the reaction is specific and not shared by other noninfective bacterial strains. Planqué and Kijino (1977) reported the presence of a new type of polysaccharide in the rhizobial liposaccharides after isolation of the latter from bacterial walls and the interaction with a lectin isolated from pea seeds. Rougé and Labroue (1977), however, could not detect any lectins in the roots of beans, peas, or field beans 1 week after germination and concluded that infection of young plants with *Rhizobium* occurs at a time when the roots are totally devoid of phytohemagglutinins. The lectin-free soybeans described by Pull *et al.* (1978) were effectively nodulated by *Rhizobium japonicum*. The role that the lectins play in the establishment of the symbiosis between roots and nitrogen-fixing bacteria must therefore be studied further. The claim of Brücher *et al.* (1969) that no significant selection pressure can be attributed to the presence of lectins was based on the detection of mixed populations of wild beans (*Phaseolus aborigeneus*), including plants that gave positive or negative reaction in the agglutination test using rabbit blood cells and activated cow red blood cells. It was established later that bean seed extracts inactive toward both of these blood preparations will agglutinate hamster erythrocytes and, hence, do contain lectins (Jaffé *et al.,* 1974). The conclusion that no vital role exists for the bean agglutinins cannot, therefore, be sustained. The fact that no lectins could be detected in a number of legumes could be due to the methodology used for screening.

VII. TOXICITY SYMPTOMS PRODUCED BY LECTINS

Not all lectins produce toxic effects in experimental animals. Some raw legume seeds cause growth retardation and sometimes even death when incorporated into the diet. Only when the same seeds are nontoxic after cooking can lectins be suspected to be one of the toxic constituents. Moreover, toxicity also depends on the animal species and even on the special animal strains used.

Early death of weanling rats is observed when they are fed diets prepared with raw kidney beans (Jaffé and Vega Lette, 1968), field beans (Jaffé, 1950), and winged beans (Jaffé and Korte, 1976). Raw soybeans will cause the death of weanling guinea pigs, but not rats (Patten *et al.,* 1973). Different strains of mice show different mortality rates when injected with an extract from raw beans (Jaffé, 1962). Honavar *et al.* (1962) observed that kidney bean lectin inhibits growth and causes death of rats when given at a 0.5% level in the diet. In a similar experiment with chicks, Wagh *et al.* (1963) noted that the decrease in growth was much less than that in rats, and no lethal action was detected.

In order to exhibit a toxic action when ingested by the oral route, a lectin must resist digestion. The lethal doses of soybean and kidney bean lectins administered by intraperitoneal injection are similar (Liener and Pallansch, 1952; Jaffé, 1960). Raw soybean meal is much less toxic when fed to rats than are raw kidney beans (Jaffé and Vega Lette, 1968). The purified lectin from beans causes the death of young rats (Honavar *et al.,* 1962), but the soybean lectin had no lethal action when administered by stomach tube at a level of 500 mg/kg (Liener and Rose, 1953). Differences in susceptibility to digestion could be related to these variations; this is suggested by observations on the action of pepsin, which inactivates soybean lectin rapidly (Liener, 1958). The kidney bean lectin, however, is not completely inactivated after 6 days of digestion by this enzyme (Goddard and Mendel, 1929). Hemagglutinating activity can be detected in the feces of rats after the ingestion of raw kidney beans but not after the ingestion of soybeans, indicating that the bean lectin is less susceptible to the action of rat digestive enzyme *in vivo* than is the lectin from soybean. In the feces of mice, agglutinating action can be observed after the administration of raw soybean flour (W. G. Jaffé, unpublished results). Indeed, the growth of mice is much more affected by raw soybeans than that of rats. The detection of biologically active wheat germ lectin in fecal collections of humans after the consumption of wheat germ is of special interest is this context (Brady *et al.*, 1978).

Crude preparations of lectins obtained from the immature seeds of soybeans, groundnuts, cowpeas (*Vigna unguiculata*), lima beans (*Phaseolus lunatus*), pigeon peas (*Cajanus cajan*), and rice beans (*Vigna umbrellata, Ivum*) when injected into young rats produced liver damage and death (Ikegwuonu and Bassir, 1977). Oral toxicity of most of these seeds has not been reported, however.

It has been shown that bean lectin administered orally to rats reduces the absorption of all nutrients (Jaffé, 1960). *In vitro* experiments with intestinal loops taken from rats fed the lectin revealed a significant decrease in the rate of absorption of glucose across the intestinal wall compared to controls (Jaffé and Camejo, 1961). Bean lectin is taken up from a saline solution when treated with minced rat intestine (Jaffé, 1960). It can be concluded that the action of that lectin is to combine with the cells lining the intestinal wall, thus causing nonspecific interference with the absorption of nutrients. Support for this hypothesis comes from studies of Etzler and Branstrator (1974), who found that a number of different lectins react with crypts and villi of the intestine but at different regions of the intestine, depending on the specifity of the lectin. Binding of concanavalin A by the epithelial cells of human colon and colonic tumor was observed by Brattain *et al.* (1976). A decrease in active and passive accumulation of nutrients through the gut, as detected by the everted-sac technique, was found in rats after having eaten a diet containing raw soy flour (Pope *et al.*, (1975). The amount of water movement and the transmural potentials of the soy-fed animal groups were smaller than those of the controls. All these results point to the binding of some hemagglutinins to the intestinal wall, thus causing nonspecific interference with the absorption of nutrients. The simultaneous existence of proteinase inhibitors and their stimulation of the excretion of pancreatic juice may also lead to an excessive loss of endogenous protein (Booth *et al.*, 1960), obscuring the effect of lectins on intestinal absorption, since both will result in increased fecal nitrogen excretion.

On the other hand, in the case of soybean, Turner and Liener (1975) discussed the possibility that trypsin inhibitor may suppress the adverse effect of the soybean lectin. Indeed, Sambeth *et al.* (1967) concluded from their experiments on the biological activity of various fractions of soybean protein that they may be the result of interactions among the various components contained therein.

Intradermal injection of lectins may produce hemorrhagic, Arthus-like lesions with edema and necrosis. This effect was first observed with crude ricin solutions by Madson and Walborn (1904). The preparations used by these authors probably contained both the toxic and hemagglutinating compounds of the castor bean. Kind and Peterson (1968) studied the local reaction of concanavalin A and pokeweed lectin in the production of skin lesions in mice. They observed that the intensity of reaction paralleled the ability of these lectins to precipitate serum proteins and that previous treatment with nitrogen mustard, which suppresses polymorphonuclear leukocytes, reduced the local skin reaction significantly. Jaffé and Gómez (1975) used the skin test in mice to compare partially purified lectins from different bean cultivars. The extension and severity of the lesions paralleled the oral and intraperitoneal toxicity of these materials. Shier *et al.* (1974) used the inflammation induced by concanavalin A and other lectins to test the effect of antiinflammatory drugs.

Interesting results were obtained when germfree animals were used in oral toxicity tests. Raw soybean meal caused less growth depression in germfree than in conventional chicks (Coates *et al.*, 1970). Fractions prepared from navy beans that were free of trypsin inhibitor activity but rich in hemagglutinating action depressed the growth of conventional chicks to about 26% but did not affect that of germfree animals (Hewitt *et al.*, 1973). Jayne-Williams (1973) observed growth depression, low body temperature, and a high mortality rate in Japanese quail fed diets containing raw jack bean meal or isolated concanavalin A. Germfree birds performed normally under the same experimental conditions, and all survived the 14-day experimental period. These experiments point to the decisive role of lectins in producing the toxicity in conventional but not in germfree birds and lend weight to the suggestion of Miller and Coates (1966) that the microflora of the alimentary tract plays an important role in the growth-depressing action of raw soy meal and other toxic legumes. The possibility advanced by Coates *et al.* (1970) that the lectins or other compounds may be converted to toxins by microbial action in the gut is not supported by their intraperitoneal and intradermic toxicity reactions.

Jayne-Williams and Burgess (1974) studied further the toxic effect of raw navy beans on the Japanese quail. The toxicity was associated with the presence of high concentrations of lectin. Birds given raw beans showed a high incidence of liver infection, but bacteriological and chemical examination of gut contents of birds fed raw or autoclaved beans revealed no significant differences. In experiments with germfree birds, several strains of coliform bacteria were capable of causing death when the raw bean diet was fed. The authors suggested that the toxicity of raw navy beans toward conventional quail was due to impairment of body defense mechanisms by the lectins, leading to tissue invasion by normally innocuous strains of the intestinal microflora.

Some cases of human intoxication through the ingestion of raw or partially cooked bean products have been observed. Griebel (1950) reported an outbreak of massive poisoning after the consumption of partially cooked bean flakes. Faschingbauer and Kofler (1929) reported on intoxication by raw runner beans. More recently, several cases of human intoxication by red kidney beans were described in Great Britain (Anonymous, 1976). Mixtures of locally grown ground beans and cereals are being used in infant feeding programs in some developing countries. Korte (1972) reported that lectins are not completely destroyed by the cooking of these mixtures under field conditions prevailing in Africa. As a consequence, diarrhea and poor growth response of the children may result.

VIII. DETOXIFICATION

The detoxification of lectins and other potentially toxic factors is usually achieved by the traditional methods of household cooking. Nevertheless, under special conditions complete detoxification may not always be achieved, especially if ground seeds are used or industrial processes for quick cooking products are applied. Some of the pertinent aspects have been reviewed by Stein (1976). When beans are used in animal feed, cooking is usually omitted and poor feed utilization due to antinutritional factors may not be easily recognized. Knowledge about heat denaturation of these factors is therefore of considerable practical importance. Obtaining strains of legumes with low levels or none of these factors through plant breeding methods offer interesting possibilities.

The resistance of lectins to inactivation by dry heat (de Muelenaere, 1964) deserves special emphasis. A reduction of the boiling point of water in mountainous regions could result in incomplete destruction of toxicity. The addition of kidney bean flour to wheat flour for the manufacture of bread (Anonymous, 1948) and the use of legume flour for making baked goods (Morcos and Boctor, 1959; D'Appolonia, 1977) should be viewed with caution. Preliminary soaking prior to cooking is required for complete elimination of toxicity of kidney beans (Jaffé, 1949) and field beans (Phadke and Sohonie, 1962). The most toxic bean lectins are also the most heat resistant; after 30 min of cooking they are not completely destroyed (Jaffé and Flores, 1975).

IX. ASSAY METHODS

The detection of lectins in plant extracts is still performed mostly by a serial dilution technique with visual estimation of the end point. A microdilution technique has found widespread acceptance because of the minute amounts of sample needed. It can be performed with a single seed or part of a seed. Commercial microtitration kits are available for this purpose and are suitable for routine testing of multiple samples. The presence of sodium chloride or some other salt is required for agglutination. The washed red blood cells should be activated by 0.5-hr treatment with a suitable proteinase, Pronase, trypsin, or papain, since the sensitivity of the agglutinin reaction is usually enhanced considerably by this treatment. A control with a plant extract known to contain a lectin of specificity similar to the one being investigated must always be included in the experiment. Not all blood samples of one animal species will react in an identical manner owing to the existence of different blood groups. Therefore, a positive control is always required with samples of unknown activity. In screening tests, activated blood cells from several different animal should be used, because the presence of lectins could be overlooked due to their inactivity in some types of blood (see Tables I and II). Human and rabbit blood are most frequently used but may not be suitable for the detection of some specific lectins. The agglutination–dilution test is only semiquantitative and has been critically evaluated by Burger (1974). A quantitative method devised by Liener (1955) is based on the photometric measurement of the density of a suspension of erythrocytes that are not agglutinated by the lectin. Several more sophisticated methods have been proposed for investigating the lectin–cell surface interaction (see Bittiger and Schnebli, 1975). For example, Hwang *et al.* (1974) and Kaneko *et al.* (1975) have applied spectrophotometric measurements to the study of the binding of lectins to cancer cells. Radioactive labeling of lectin molecules has likewise been frequently used.

The mitogenic activity of lectins is best measured by their ability to stimulate the *in vitro* DNA synthesis of peripheral human lymphocytes or of mouse spleen lymphocytes as determined by the uptake of labeled thymidine. The result of this test greatly depends on the optimal concentration of the lectins (Rigas and Tisdale (1969).

The toxicological assays are especially useful for the elucidation of the nutritive properties of lectins. Since lectins frequently exist in plant material together with trypsin inhibitors, the corresponding experimental diets may be supplemented with a tryptic casein digest and with methionine, the limiting essential amino acid of legume seeds (Jaffé and Vega Lette, 1968). Control diets prepared with heat-activated material should always be included in the test.

The use of affinity chromatography has permitted the purification of many lectins. Their possible toxicity can be assayed by intraperitoneal or intradermic application, since only minute amounts of the pure compounds are required. When toxic reactions are observed, oral toxicity should also be tested.

X. LECTINS OF EDIBLE PLANTS

A. Bean Lectins

The hemagglutinating action of crude extracts from legume seeds, particularly those of the *Phaseolus* genus, has been studied more than that of any other plant group because of the different hemagglutinating and mitogenic activities found in these seeds. Tobiška (1964) quoted about 200 *Phaseolus* species and varieties that have been examined for hemagglutinating activity, and Toms and Western (1971) presented a list containing over 500 entries. Three groups among the *Phaseolus* species can be distinguished according to the hemagglutination test. The first, represented by the common or garden bean *P. vulgaris,* contains nonspecific lectins, which act on practically all types of animal blood and all human blood groups. They are not inhibited by simple sugars in the conventional concentration range. They are sometimes called panagglutinins. The second group is represented by *P. lunatus,* the lima bean, which contains blood-group-specific lectins characterized by a high anti-A_1 activity. The seeds of a third group of *Phaseolus* plants show no hemagglutinating activity at all, at least with the blood samples used in the respective tests.

Beans constitute an important source of dietary protein for large segments of the world's population, and many reports exist in the literature concerning the toxic effects that may accompany the ingestion of raw beans classified as *Phaseolus vulgaris.* The feeding of raw kidney beans as part of a diet for experimental animals causes rapid loss of weight and death, but the heated seeds have no similar effect. Johns and Finks (1920) and Berczeller (1922) explained this by the low digestibility of the raw beans. The existence of a bean agglutinin had been reported by Landsteiner and Raubitschek (1908), and Wienhaus (1909) studied it in some detail but failed to detect significant toxic action, probably because he supposed that it was similar to ricin in poisonous activity. He proposed the name of "phasin" for this agglutinin, a term that has since been occasionally used to designate nontoxic plant agglutinins. The abbreviated term PHA is often used for the bean hemagglutinin. Lüning and Bartels (1926) were probably the first to relate the toxic action of beans with the agglutinin content. Jaffé (1949) again observed the toxic action of raw beans and showed that it could not be explained by poor digestibility or by the presence of trypsin inhibitors, because enzymatically digested casein when added to the toxic diet did not improve the performance of the experimental animals. Further experiments by Jaffé (1960) and Honavar *et al.* (1962) clearly established that the hemagglutinins are most likely responsible for this toxicity. Both groups of workers found that when the hemagglutinins isolated from the black bean were fed to rats at levels as low as 0.5% of the diet, a definite retardation of growth was observed. The kidney bean hemagglutinin was somewhat more toxic, a level of 0.5% being sufficient to cause the death of rats within 2 weeks (Honavar *et al.,* 1962). Higher levels of this hemagglutinin merely hastened the onset of death. Wagh *et al.* (1963) showed that the kidney bean hemagglutinin also inhibits the growth of chicks.

Several investigators have observed that more than one hemagglutinating fraction may be present in kidney bean extracts. Pierkarski (1957) prepared three, and Jaffé and Gaede (1959) isolated two fractions possessing hemagglutinating and toxic properties. Prager and Speer (1959) separated a crude bean extract into three agglutinating fractions by chromatography on DEAE-cellulose. Jaffé and Hanning (1965) demonstrated by immunoelectrophoresis that at least two different hemagglutinins existed in a black bean extract, which were separated by ammonium sulfate precipitation and free-flow electrophoresis. Takahashi *et al.* (1967) separated two hemagglutinating proteins from wax beans by DEAE-cellulose chromatography.

Rigas *et al.* (1966) demonstrated that a bean agglutinin that was homogeneous by several criteria, including different types and conditions of electrophoresis, gel filtration, chromatography, and ultracentrifugation, and that had only one N-terminal amino acid, alanine, dissociated slowly into subunits when kept in 8 *M* urea. These could be separated by gel electrophoresis. Chromatography of the lectin on the cation-exchange resin IRC-50 resulted in the appearance of several fractions differing in their physical and biological properties, amino acid composition, and subunit ratio. It was suggested by the authors that dissociation into subunits and recombination occurs during chromatography, that different subunits may be responsible for the hemagglutinating, mitogenic, and toxic activities, and that inactive subunits may also be present.

A protein fraction that inhibits the growth of rats and has hemagglutinating and mitogenic properties was isolated from navy beans of the Sanilac variety by Evans *et al.* (1973). Andrew (1974) isolated two proteins in a highly purified state from a toxic fraction of navy (haricot) beans, one of which had very strong agglutinating activity. Using ammonium sulfate precipitation and absorption onto DEAE-cellulose, Andrew and Jayne-Williams (1974) obtained a homogeneous protein from navy beans with both agglutinating and toxic properties for the Japanese quail. The proteins of two bean varieties were fractionated by continuous high-voltage electrophoresis (Pusztai *et al.*, 1975). Thus, a toxic albumin fraction related to the isolectins was obtained. A similar isolectin was found in the globulin fraction. Kidney bean lectins were purified by affinity chromatography on a fetuin–Sepharose-4B column by Pusztai and Palmer (1977). The pure lectins depressed the growth of rats when added to a 5% casein-containing diet and was identified as the toxic principle of this seed. Hamaguchi *et al.* (1977) likewise isolated a highly toxic lectin from a bean cultivar.

Yachnin and Svenson (1972) found five heterogeneous proteins in a commercial bean phytohemagglutinin. Each consists of isomeric noncovalently bound tetramers made up of two different subunits designated L and R, which differ slightly in their amino acid sequence and also in their affinity to erythrocytes and lymphocytes (Fig. 3). The tetramer with four L subunits (L-PHAP) is a potent leukoagglutinin with little hemagglutinating activity. The hybrid tetramers with two or more R subunits exhibit potent hemagglutinating activity but modest

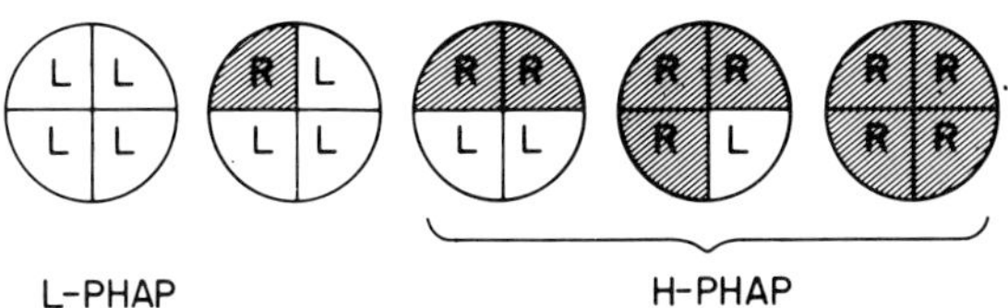

FIG. 3. Schematic representation of the tetrameric structure of the five lectins in the kidney bean; L and R are the subunits responsible for leukoagglutinating and hemagglutinating activities, respectively. (From Miller *et al.*, 1975, reproduced by permission.)

leukoagglutinating activity. The hybrid 1R–3L is devoid of hemagglutinating activity, presumably because it is monovalent with respect to its interaction with erythrocytes (Miller *et al.*, 1975). Intact lectins can be obtained by recombination of the subunits. This model has not yet been checked with the various bean lectin preparations possessing different activities. It may account for at least some of these differences.

Harms-Ringdahl and Jornvall (1974) isolated a mitogenic factor from red kidney beans that stimulated RNA synthesis in mouse spleen lymphocytes and in a bacterial system but had only a slight effect on chicken spleen lymphocytes. Unlike the mitogens previously described, which have molecular weights of about 120,000, it had a molecular weight of only about 10,000 and an unusually high content of cystine.

When the agglutinating activity of extracts of seeds from different bean cultivars were tested, it was established that, according to the specifity of agglutination of blood from different animal species, the seeds could be classified into four groups: one active on all blood samples, called A; one active on all except activated cow blood, called B; one very active on activated cow blood but inactive on rabbit or human blood, called C; and one with little activity except on activated blood from the hamster and the mouse, called D (Jaffé *et al.*, 1972, 1974). In all cases, treatment of the red blood cells with Pronase enhanced the sensitivity of the reaction, with the exception of cow blood, which should be activated by trypsin.

When the extracts of the different bean types were tested by injection into mice (Jaffé and Brücher, 1972) or by feeding the ground seeds to rats (Jaffé and Vega Lette, 1968; Jaffé and Brücher, 1972), it was established that only the A- and C-type beans, i.e., those that displayed agglutinating activity toward trypsinated cow blood, were also toxic (Table II), whereas those varieties that agglutinated only activated hamster and/or rabbit blood were of low toxicity. These results emphasize the importance of testing the hemagglutinating activity of seeds against several species of blood cells before concluding that a particular bean is toxic. The use of trypsinated cow blood cells would appear to offer an important tool for detecting beans that are potentially toxic. Commercial lots of beans

TABLE II

CORRELATION OF SPECIFIC HEMAGGLUTINATING ACTIVITY WITH THE INTRAPERITONEAL TOXICITY IN MICE OF EXTRACTS OF DIFFERENT VARIETIES AND CULTIVARS OF *P. vulgaris* AND THEIR MITOGENIC ACTION[a]

Variety	Rabbit blood	Trypsinated cow blood	Toxicity (No. of injected mice/ No. of dead mice)	Mitogenic activity
Balin de Albenga	+	+	5/4	+
Merida	+	+	9/9	+
Negro Nicoya	+	+	5/4	+
Saxa	+	+	5/5	+
Peruvita	+	−	5/0	−
Palleritos	+	−	6/0	−
Juli	+	−	5/0	−
Cubagua	+	−	5/0	−
Porillo	−	+	5/5	+
Negra No. 584	−	+	5/3	+
Vainica Saavegra	−	+	10/6	+
Hallado	−	−	5/0	−
Madrileño	−	−	5/0	−
Alabaster	−	−	5/0	−
Triguito	−	−	6/0	−

[a]From Jaffé and Brücher (1972) and Jaffé *et al.* (1972).

frequently are composed of seeds belonging to different types. A number of single seeds should therefore be tested with different blood samples to establish the homogeneity of the lot. At least two isolectins can be obtained from each of the four bean types (Jaffé *et al.*, 1972). Only those of types A and C exhibit significant mitogenic activity.

The partial separation of toxic and hemagglutinating fractions, as reported by Kakade and Evans (1965) and by Stead *et al.* (1966), could perhaps be explained if toxic C-type beans, which do not agglutinate rabbit or human blood, and nontoxic B-type beans with high agglutinating activity were both present in the seed lots used for the fractionation work. Bean varieties without hemagglutinating activity have been described in several papers (Krüpe *et al.*, 1968; Brücher *et al.*, 1969). In these cases, trypsinated cow blood or Pronase-activated hamster blood was not tested. Therefore, it is likely that the corresponding bean varieties belonged to the C or D types. No authentic agglutinin-free bean varieties have yet been discovered. Leukoagglutinin bean lectins that agglutinate human leukocytes and induce mitosis have been claimed not to agglutinate red blood cells (Weber *et al.*, 1971; Oh and Conrad, 1972; Yachnin and Svenson, 1972). Pure, commercial leukoagglutinin preparations, however, strongly agglutinate trypsin-treated cow blood and Pronase-treated hamster blood but are inactive on human and rabbit blood and are therefore probably C-type lectins (Jaffé *et al.*, 1972). The fact that only the extracts from bean varieties active as agglutinins on trypsinized cow blood are strong mitogens permits a classification of the beans used by different investigators for the purification of the respective fractions. The substances described by Rigas *et al.* (1972) and by Evans *et al.* (1973) were evidently A-type lectins because they had strong mitogenic and hemagglutinating action on human blood. The isolectins studied by Pusztai and Watt (1974) and by Felsted *et al.* (1975) were apparently of the B type with very little mitogenic activity. The so-called leukoagglutinins are probably C-type lectins, as already mentioned. D-Type lectins were obtained by Jaffé *et al.* (1974).

The major bean seed proteins differ significantly in their physicochemical properties when they are obtained from different bean cultivars (Kloz, 1971; Adriaanse *et al.*, 1969). Therefore, it should not be surprising that the properties of the lectins from different bean lots may also differ.

It must be emphasized that such terms as kidney beans, wax beans, navy beans, and garden beans as well as the seed colors are of no significance in relation to toxicity, since one of the different lectin types may be present in any of these beans, although the toxic A type is found frequently in red and black beans and the nontoxic D type in white beans (Jaffé and Brücher 1972).

B. Broad Bean Lectins

Extracts from the broad bean, horse bean, or faba bean (*Vicia faba*) agglutinate human and animal red blood cells. Ingestion of broad beans may cause an acute hemolytic anemia in some sensitive persons bearing certain serum and erythrocyte abnormalities. This disease is called favism and is related to a hereditary deficiency of the enzyme glucose-6-phosphate dehydrogenase (G6PD). (See Chapter 9). Perera and Frumin (1965) reported the occurrence in broad beans of a heat-labile lectin that would not agglutinate erythrocytes of homozygous G6PD-deficient American blacks, whereas erythrocytes from heterozygous G6PD-deficient and normal donors were strongly agglutinated. These results are at variance with the previously published observation by Greenberg and Wong (1961) that the lectins from broad beans do not discriminate between normal and G6PD-deficient red blood cells.

A glycoprotein lectin that is mitogenic for mouse spleen lymphocytes was obtained in crystalline form from the seeds by Wang *et al.* (1974). Allen *et al.* (1976) purified the *Vicia faba* lectin by affinity chromatography and compared its specifity toward different sugars and sugar derivatives with that of lentils and peas and with that of concanavalin A. The lectin molecule consists of two apparently identical subunits. There also exist fragments of subunits in the seed extract.

C. Soybean Lectins

The nutritional value of soybeans is enhanced by heat treatment. This fact, which was already observed by Osborne and Mendel in 1917, is attributed to the destruction of heat-labile growth inhibitory factors. Trypsin inhibitors, poor digestibility of the native, undenatured proteins, lectins, and possibly other growth inhibitors could be responsible. Most growth studies on the action of raw soybeans have been performed with rats or chicks. In rats and chicks fed a diet prepared with raw soybean meal, growth performance is reduced by at least 25% as compared with that of animals receiving a diet of heated soybeans. The growth of mice is more depressed than that of rats and chicks. Weanling guinea pigs lose weight and suffer a high mortality rate when fed a diet containing raw soybeans (Patten *et al.*, 1973).

Evidence that part of the toxic action of raw soybeans is related to the lectin came from the work of Liener and co-workers, who isolated a hemagglutinating protein from raw soy flour and extensively studied its physical, chemical, and biological aspects. In the earlier work this protein was called soyin, but in the later papers the name soybean hemagglutinin was preferred since a proteolytic enzyme had been called soyin at an earlier date. The agglutinin was homogenous as determined by chromatography on DEAE-cellulose, electrophoresis at various pH values, starch gel electrophoresis, and ultracentrifugation (Pallansch and Liener, 1953). From the investigation of the N-terminal amino acids it appeared to consist of two peptide chains (Wada *et al.*, 1958). In addition to the hemagglutinin described by Liener and his co-workers, three other minor components were separated from a crude soybean extract by Lis *et al.* (1966); all four have agglutinating properties and are similar in electrophoretic and chromatographic behavior. Separation was achieved by chromatography on DEAE-cellulose under strictly controlled conditions. The fractions are similar in amino acid composition and contain neutral sugars and glucosamine in different amounts. The soybean lectins were further studied by Catsimpoolas and Meyer (1969), who used isoelectric focusing, and later by Gordon *et al.* (1973), who applied affinity chromatography, taking advantage of its specific binding to galactopyranosyl residues. At least four isolectins have thus been separated from soybean whey proteins. Catsimpoolas and Meyer (1969) discussed the possibility that they arise through the hydrolysis of labile amide groups. Lotan *et al.* (1975a,b) detected four subunits of two different types in soybean lectin.

Soybean extracts agglutinate rabbit blood cells rapidly. Rat erythrocytes are agglutinated only by large amounts of the agglutinin, and blood cells from sheep and calves are completely refractory to agglutination (Liener and Pallansch, 1952). Bird (1953) stated that soybean extracts contain cold agglutinins active on human blood cells at low temperature.

The toxicity of the soybean agglutinin was studied under various conditions by Liener (1951) and Liener and Pallansch (1952). When it was injected intraperitoneally, the LD_{50} for young rats was about 50 mg/kg. When added to a diet containing autoclaved soybean meal at a level of 1%, it depressed growth to about 75% of the controls, and food intake was reduced. About one-half of the growth depression caused by raw meal was attributed to the action of the lectin. Its destruction by heat is associated with an improvement in the nutritive value of soybean (Liener and Hill, 1953). Liener (1953) had shown that, when the purified lectin was added to a diet containing heated soybean meal at a level approximating its occurrence in the raw meal, the growth of rats was significantly retarded. Growth inhibition, however, was not observed when the food intake of the control diet containing heated soybean meal was restricted to the food intake of the same diet containing soybean lectin. Stead *et al.* (1966) performed a chromatographic separation of soybean proteins on DEAE-cellulose. The hemagglutinating activity and toxicity were concentrated in one of the peaks in the eluate, but the material from another peak was also toxic when injected into rats, although it was devoid of significant hemagglutinating action. Birk and Gertler (1961) observed that most of the hemagglutinating material can be extracted from raw soybean meal, but the nutritional quality of the residue is not much improved.

When hemagglutinating activities in extracts from different soybean cultivars were compared, very significant quantitative differences were observed (Kakade *et al.,* 1972). However, no correlation was found between lectin activity and toxicity for growing rats, nor were the trypsin inhibitor action and growth depression quantitatively related. More recently, Turner and Liener (1975) removed the lectin from a crude extract of unheated soybean flour by affinity chromatography. When fed to young rats, the hemagglutinin-free extract, obtained by passage through a column of Sepharose-bound concanavalin A, produced a rate of growth only slightly higher than that obtained with the original extract from which the hemagglutinin had not been removed. Thus, under these experimental conditions, there was no clear indication for the toxic action of the soybean lectin.

Probably there are still other, incompletely defined factors in soybeans, such as the polypeptide toxic for mice described by Schingoethe *et al.* (1970). The interaction of the multiple biologically active factors might account for the growth-depressing and other toxic effects of raw soybeans. Little attention has been given by the majority of investigators interested in the nutritional properties of soybeans to the possible nutritional implications of lectins. Trypsin inhibitors and consequent pancreatic enlargement are frequently considered the main cause of growth depression and even death of experimental animals fed raw soybean seeds. Inclusion of research on the soybean lectin in these investigations could significantly advance our understanding of the causative agents of these events. Considering the simplicity of the necessary tests, it is rather surprising that this aspect is so often neglected.

D. Lentil Lectins

The presence of hemagglutinin activity in extracts from lentils (*Lens culinaris* or *Lens esculenta*) has been known since the observations of Landsteiner and Raubitschek in 1908. Its purification was described by Howard and Sage (1969) and Tichá *et al.* (1970). It is absorbed on Sephadex and can be purified in this way. The activity depends on the presence of bivalent ions Ca^{2+}, Zn^{2+}, and $Mn^{2}+$. In acidic solution it dissociates into subunits. Two isoagglutinins can be separated by elution from a Sephadex column. The relative abundance differs between different seed lots. Some authors found small amounts of carbohydrates, but others could not detect any sugar in their preparations.

Strosberg *et al.* (1976) found three components in a purified lectin preparation that could be separated by gel electrophoresis in the presence of sodium dodecyl sulfate. The amino acid squences of its two polypeptide chains showed striking homology with the corresponding chains from pea lectin but were considerably different from each other.

From a small-seed subspecies (*Lens esculenta microsperma*) a lectin that is not bound to Sephadex was obtained by Flialová *et al.* (1975). However, it differs little from that of the common lentil in composition and with respect to the inhibitory effect of sugars on its agglutination activity.

Growth-promoting activity of raw and cooked lentils is quite similar in the rat growth test. This may indicate that the lentil lectin is nontoxic or that its concentration in the seeds is too low to produce a significant effect.

E. Pea Lectins

The hemagglutinating activity of pea seeds was detected by Landsteiner and Raubitschek (1908). A hemagglutinin was purified from unspecified white peas by Huprikar and Sohonie (1965). When incorporated into an experimental rat diet at a level of 1%, it had no significant effect on animal growth. When injected into rats, the white pea lectin had an LD_{50} of 143 mg/kg (Manage *et al.*, 1972).

Entlicher *et al.* (1970), Marik *et al.* (1974), and Trowbridge (1974) isolated two lectins from garden peas (*Pisum sativum*). A third isolectin, which was detected by isoelectric focusing (Entlicher and Kocourek, 1975), might be the result of hybridization of the other two lectins. Despite the fact that they are probably tetramers, the pea lectins have only two binding sites. The specificity and properties are similar to those of the lectin from lentils and concanavalin A. Homology was observed between residues 13–39 of one chain of the pea lectin and residues 82–98 of concanavalin A (Van Driessche *et al.*, 1976).

F. Field Bean Lectins

Ground field beans (*Dolichos lablab*) incorporated into a rat diet reduced growth significantly and may have caused the death of the animals (Jaffé, 1950). Autoclaving, but not germination of the seeds, abolished this toxic action (Salgarkar and Sohonie, 1965b). Liver necrosis and diminished levels of several liver enzymes were found in animals fed a raw field bean diet. The LD_{50} of injected field bean lectin was about 80 mg/kg for both rats and mice (Manage *et al.*, 1972).

Two lectin fractions were obtained from the ripe seeds by Salgarkar and Sohonie (1965a). One was homogeneous electrophoretically, and the other contained three components. The major lectin when fed to rats in a casein diet produced growth inhibition and death in these animals. The growth depression caused by the purified lectin was less than that observed with an amount of field bean meal containing equivalent hemagglutinating activity. It was investigated by Rao *et al.* (1976), who obtained subunits in the presence of 8 *M* urea. The native lectin was resistant to trypsin digestion and contained 1–2% sugars but did not require metal ions for its hemagglutinating activity. Erythrocytes of different animal species and human A, B, and O group individuals were agglutinated.

In contrast, the lectin from *Dolichos biflorus,* the horse gram, shows specifity toward type-A human erythrocytes and blood group A substance. This was used to advantage by Carter and Etzler (1975) in the isolation of this lectin using insoluble polyleucyl hog blood group A + M substance as an affinity column. The preparation thus obtained was further fractionated into at least two components by chromatography on concanavalin A–Sepharose. Both consisted of four subunits and had a similar specifity toward blood group A human red blood cells. Horse gram lectin was nontoxic to mice and rats when injected in doses up to 250 mg/kg. Oral administration restricted the growth of young rats (Manage *et al.*, 1972).

G. Other Legume Lectins

The toxic and hemagglutinating properties of the runner bean (*P. coccineus*) were observed by de Muelenaere (1965). Two lectins were isolated from the scarlet runner bean. Both are glycoproteins and have a molecular weight of 20,000. One of these lectins consists of four identical subunits and contains Ca^{2+}, Zn^{2+}, and Mn^{2+}. They are nonspecific toward human erythrocytes. The so-called lectin I was not mitogenically active in the tested concentration. Lectin II stimulated mitosis of mouse spleen cells (Nowaková and Kocourek, 1974).

The lectin of the lima or double bean (*P. lunatus*) has aroused special interest since Renkonen (1948) and Boyd and Reguera (1949) independently discovered its blood group specifity for human A erythrocytes. Two active components were separated by specific absorption to insolubilized type A substance, a tetramer and a dimer composed of identical subunits (Galbraith and Goldstein, 1972). They are glycoproteins containing Ca^{2+} and Mn^{2+}. Crude double bean lectin restricted the growth of young rats and was toxic when injected into mice or rats (Manage *et al.*, 1972).

Uhlenbruck *et al.* (1969) used the lectin from peanuts (*Arachis hypogoea*) to study the nature of the T antigen and T-transformed (neuramidase-treated) cells, since it gave the same immunological reaction as anti-T antibodies responsible for T polyagglutination occurring in several bacterial and viral infections. The lectin was inhibited by galactose and lactose. Lotan *et al.* (1975c) purified an anti-T lectin from peanuts by an affinity chromatographic procedure.

H. Potato Lectins

The hemagglutinating activity of the press juice of potatoes was first observed by Marcusson-Begun (1925). A lectin from potatoes was obtained by Krüpe and Ensgraber (1962) and by Marinkovich (1964). Allen and Neuberger (1973) further purified the potato lectin and studied its chemical properties in some detail.

The hemagglutinating activity in different potato cultivars can vary considerably (Tobiška, 1964). Delmotte *et al.* (1975) were able to purify this lectin by affinity chromatography on Sepharose, which had been substituted with *p*-aminobenzyl-1-theoacetylglucosamide. Ortega and González (1970) found that some potato varieties are devoid of hemagglutinating action while some others have high agglutinin titers. After 45 min of cooking, the extracts of some samples were still able to agglutinate human erythrocytes. One wonders how these lectins resist the short heating used for the preparation of potato chips and other quickly cooked potato products and what their possible nutritive effect may be.

I. Lectins of Cereals

A lectin present in a lipase preparation from wheat germ is of some interest because of its capacity to agglutinate tumor cells (Aub *et al.*, 1963). It was obtained in crystalline form and has a molecular weight of 35,000 for the dimer, with an unusually high content of half-crystine residues (Levine *et al.*, 1972; Nagata and Burger, 1974). The monomer has two binding sites specific for *N*-acetylglucosamine. At neutral pH it exists as the dimer. Its purification can be achieved by affinity chromatography of chitin (Bloch and Burger, 1974b), a method that could be useful for large-scale purification. Another simple affinity chromatographic method proposed by Vretblad (1976) uses epoxy-activated Sepharose 6B treated with *N*-acetylglucosamine for the purification of the three wheat germ isolectins.

Raw wheat germ was found to depress the growth and fat utilization of chicks. Attia and Creek (1965) compared these results with similar experiments performed with raw soybeans and discussed the possible involvement of the wheat germ lectin in the growth-depressing effect.

Rice (1976) found that germ from hexaploid wheat (*Triticum aestivum*) contained three forms of agglutinin separable by ion-exchange chromatography, whereas germ from tetraploid wheat (*Triticum turgidum*) contained only two such forms. Thomas *et al.* (1977) resolved purified wheat germ agglutinin into four homogeneous isolectins and studied saccharide-induced conformational transitions. They concluded that the binding site is composed of several subsites.

A lectin from barley germ was purified by Fortiers *et al.* (1976). It could be separated into two compounds by isolectric focusing. A rice seed lectin prepared by Takahashi *et al.* (1973) has an unusually low molecular weight of about 10,000 and contained about 27% carbohydrate. It displays mitogenic activity toward human lymphocytes.

XI. CONCLUSIONS

Compared with the extensive research carried out on the chemical, physical, and biochemical characteristics of the lectins, our knowledge of their toxicological and nutritive properties is very scant. Nevertheless, this is an important aspect, since the consumption of legumes as a cheap protein source for the ever-growing number of human beings in developed and less developed countries is becoming increasingly encouraged by nutrition planners. New species and cultivars are envisaged for use in human and farm animal nutrition. Careful toxicological studies are required in all these cases. Also, the trend to produce protein isolates or quick cooking food products requires the consideration of possible antinutritional effects. It should be emphasized that obvious toxic reactions may not always be easily observed if the degree of toxicity is low or if toxic constituents are only partly destroyed or eliminated, resulting in slightly reduced food utilization that is difficult to detect. Moreover, the absence of toxic manifestations in one animal species does not preclude possible toxic effects in others, including human beings. The modern methods of protein fractionation, especially the technique of affinity chromatography, have opened new approaches, which should be useful for pursuing research in the field of toxicology of plant lectins.

References

Adriaanse, A., Klop, W., and Robbers, J. E. (1969). *J. Sci. Food Agric.* **20,** 647–650.

Allen, A. K., and Neuberger, A. (1973). *Biochem. J.* **135,** 307–314.

Allen, A. K., Desai, N. N., and Neuberger, A. (1976). *Biochem. J.* **155,** 127–135.

Andrew, A. T. (1974). *Biochem J.* **13,** 421–429.

Andrew, A. T., and Jayne-Williams D. J. (1974). *Br. J. Nutr.* **32,** 181–188.

Anonymous (1948). *Chem. Ind. & Eng. News* **26,** 2516.

Anonymous (1976). *Br. Med. J.* **2,** 1268.

Asberg, K., Holmen, H., and Porath, J. (1968). *Biochim. Biophys. Acta* **160,** 116–117.

Attia, F., and Creek, R. D. (1965). *Cereal Chem.* **42,** 494–497.

Aub, J. C., Tieslan, C., and Lankester, A. (1963). *Proc. Natl. Acad. Sci. U.S.A.* **50,** 613.

Baues, R. J., and Gray, G. R. (1977). *J. Biol. Che.* **252,** 57–60.

Berczeller, L. (1922). *Biochem. Z.* 129–239.

Bessler, W., and Goldstein, I. J. (1973). *FEBS Lett.* **34,** 58–62.

Bird, G. W. G. (1959). *Br. Med. J.* **15,** 165.

Birk, Y., and Gertler, A. (1961). *J. Nutr.* **75,** 379–387.

Bittiger, H., and Schnebli, H. P., eds. (1975). "Concanavalin A Tool." Wiley, New York.

Bloch, R., and Burger, M. M. (1974a). *FEBS Lett.* **44,** 286–289.

Bloch, R., and Burger, M. M. (1974b). *Biochem. Biophys. Res. Commun.* **58,** 13–19.

Bohlool, B. B., and Schmidt, E. L. (1974). *Science* **185,** 269–271.

Booth, A. N., Robbins, D. J., Ribelin, W. E., and DeEds, F. (1960). *Proc. Soc. Exp. Biol. Med.* **104,** 681–683.

Boyd, W. C. (1963). *Vox Sang.* **8,** 1–32.

Boyd, W. C. (1974). *Ann. N.Y. Acad. Sci.* **234,** 396–408.

Boyd, W. C., and Reguera, R. M. (1949). *J. Immunol.* **62,** 333.

Boyd, W. C., and Shapleigh, E. (1954). *Science* **119,** 419.

Brady, P. G., Vannier, A. M., and Banwell, J. G. (1978). *Gastroenterology* **75,** 236–239.

Brattain, M. G., Pretlow, T. G., Pittman, J. M., and Weiler, A. (1976). *Br. J. Cancer* **33,** 659–663.

Brocq-Rousseu, D., and Fabre, R. (1947). "Les toxines végétales." Hermann, Paris.

Brücher, O., Wechsler, M., Levy, A., Palozzo, A., and Jaffé, W. G. (1969). *Phytochemistry* **8,** 1739–1743.

Burger, M. M. (1974). (S. Fleischer and L. Packer, eds.), *In* "Methods in Enzymology," Vol. 32, pp. 615–621. Academic Press, New York.

Carter, W. G., and Etzler, M. E. (1975). *J. Biol. Chem.* **250,** 2756–2762.

Catsimpoolas, N., and Meyer, E. W. (1969). *Arch. Biochem. Biophys.* **132,** 279–285.

Clarke, A. E., Knox, R. B., and Jermyn, M. A. (1975). *J. Cell Sci.* **19,** 157–167.

Coates, M. E., Hewitt, D., and Golob, P. (1970). *Br. J. Nutr.* **24,** 213–225.

Cohen, E., ed. (1974). "Biomedical Perspectives of Agglutinins of Invertebrate and Plant Origin," Ann. N.Y. Acad. Sci. No. 234. N.Y. Acad. Sci., New York.

Cunningham, B. A., Wang, J. L., Waxdale, M. J., and Edelman, G. M. (1975). *J. Biol. Chem.* **250,** 1503–1512.

D'Appolonia, B. L. D. (1977). *Cereal Chem.* **54,** 53–63.
Dazzo, F. B., and Hubbel, D. H. (1975). *App. Microbiol.* **30,** 1017–1033.
Delmotte, F., Kieda, C., and Monsigny, M. (1975). *FEBS Lett.* **53,** 324–330.
de Muelenaere, H. J. H. (1964). *Nature (London)* **201,** 1029–1030.
de Muelenaere, H. J. H. (1965). *Nature (London)* **206,** 827–828.
Edelman, G. M., Cunningham, B. A., Recke, G. N., Jr., and Becker, J. L. (1972). *Proc. Natl. Acad. Sci. U.S.A.* **69,** 2580–2584.
Eisler, M., and Portheim, L. (1926). *Z. Immumitätsforsch.* **47,** 59–82.
Elfstrand, W. (1897). *Görbersdorfer Veröff.* **1,** 1.
Entlicher, G., and Kocourek, J. (1975). *Biochim. Biophys. Acta* **393,** 165–169.
Entlicher, G., Kostir, J. U., and Kocourek, J. (1970). *Biochim. Biophys. Acta* **221,** 272–281.
Etzler, M. E. (1974). *Ann. N.Y. Acad. Sci.* **224,** 260–275.
Etzler, M., and Branstrator, M. L. (1974). *J. Cell Biol.* **62,** 329–343.
Evans, R. J., Pusztai, A., Watt, W. B., and Bauer, D. H. (1973). *Biochim. Biophys. Acta* **303,** 175–184.
Faschingbauer, H., and Kofler, L. (1929). *Wien. Klin. Wochenschr.* **42,** 1069–1073.
Felsted, R. L., Leavitt, R. D., and Bachur, N. (1975). *Biochim. Biophys. Acta* **405,** 72–81.
Felsted, R. L., Egotin, M. J., Laevitt, R. D., and Bachur, N. R. (1977). *J. Biol. Chem.* **252,** 2967–2971.
Flialová, D., Tichá, M., and Kocourek, J. (1975). *Biochim. Biophys. Acta* **393,** 170–180.
Ford, W. W. (1913). *Zentralbl. Bakteriol., Parasitenkd., Infektionskr. Hyg., Abt. 1: Ref.* **58,** 129–161 and 193–222.
Foriers, A., De Neve, R., and Kanarek, L. (1976). *Arch. Int. Physiol. Biochim.* **84,** 617–618.
Foriers, A., Wuilmart, C., Sharon, N., and Strosberg, A. D. (1977). *Biochim. Biophys. Acta* **75,** 980–986.
Funatsu, M. (1972). *In* "Proteins: Structure and Function" (M. Funatsu *et al.*, eds.), Vol. 2, pp. 103–140. Wiley, New York.
Galbraight, W., and Goldstein, I. J. (1972). *Biochemistry* **11,** 3976–3984.
Goddard, V. R., and Mendel, L. B. (1929). *J. Biol. Chem.* **82,** 447–463.
Gordon, J. A., Blumberg, S., Lis, H., and Sharon, N. (1973). *FEBS Lett.* **21,** 193–196.
Greenberg, M. S., and Wong, H. (1961). *J. Lab. Clin. Med.* **57,** 733.
Griebel, C. (1950). *Z. Lebensm.-Unters.-Forsch.* **90,** 191–199.
Hague, D. R. (1975). *Plant Physiol.* **55,** 636–642.
Hamaguchi, Y., Yagi, N., Nishino, A., Mochizuki, T., Mizukami, T., and Miyoshi, M. (1977). *J. Nutr. Sci. Vitaminol.* **23,** 525–534.
Hamblin, J., and Kent, S. P. (1973). *Nature (London), New Biol.* **245,** 28–30.
Hardman, K. D., and Ainsworth, C. F. (1973). *Biochemistry* **12,** 4442–4448.
Harms-Ringdahl, M., and Jornvall, H. (1974). *Eur. J. Biochem.* **48,** 541–547.
Hassing, G. S., and Goldstein, I. J. (1970). *Eur. J. Biochem.* **16,** 549–556.
Hewitt, D., Coates, M. E., Kakade, M. L., and Liener, I. E. (1973). *Br. J. Nutr.* **29,** 423–435.
Honavar, P. M., Shih, C. V., and Liener, I. E. (1962). *J. Nutr.* **77,** 109–114.
Howard, I. K., and Sage, H. J. (1969). *Biochemistry* **8,** 2436–2441.
Howard, I. K., Sage, H. J., and Morton, C. B. (1972). *Arch. Biochem. Biophys.* **149,** 323–326.
Huprikar, S. V., and Sohonie, K. (1965). *Enzymologia* **28,** 333–345.
Hwang, K. M., Murphree, S. A., and Sartorelli, A. C. (1974). *Cancer Res.* **34,** 3396–3402.
Ikeqwuonu, F. I., and Bassir, O. (1976). *Toxicol. Appl. Pharmacol.* **40,** 217–226.
Jaffé, W. G. (1949). *Experientia* **5,** 81–84.
Jaffé, W. G. (1950). *Acta Cient. Venez.* **1,** 62–64.
Jaffé, W. G. (1960). *Arzneim. Forsch.* **10,** 1012–1016.

Jaffé, W. G. (1977). *In* "Immunological Aspects of Foods" (N. Catsimpoolas, ed.), Chapter 7, pp. 170–181. Avi Publ., Westport, Connecticut.
Jaffé, W. G., and Brücher, O. (1972). *Arch. Latinoam. Nutr.* **22,** 267–281.
Jaffé, W. G., and Camejo, H. (1961). *Acta Cient. Venez.* **12,** 59–63.
Jaffé, W. G., and Flores, M. E. (1975). *Arch. Latinoam. Nutr.* **25,** 79–90.
Jaffé, W. G., and Gaede, K. (1959). *Nature (London)* **183,** 1329–1331.
Jaffé, W. G., and Gómez, M. J. (1975). *Qual. Plant.—Plant Foods Hum. Nutr.* **24,** 359–365.
Jaffé, W. G., and Hanning, K. (1965). *Arch. Biochem. Biophys.* **109,** 80–91.
Jaffé, W. G., and Korte, R. (1976). *Nutr. Rep. Int.* **14,** 449–455.
Jaffé, W. G., and Palozzo, A. (1971). *Acta Cient. Venez.* **22,** 102–105.
Jaffé, W. G., and Vega Lette, C. L. (1968). *J. Nutr.* **94,** 203–210.
Jaffé, W. G., Brücher, O., and Palozzo, A. (1972). *Z. Immunitaetsforsch.* **142,** 439–447.
Jaffé, W. G., Levy, A., and González, I. D. (1974). *Phytochemistry* **13,** 2685–2693.
Janzen, D. H., Juster, H. B., and Liener, I. E. (1976). *Science* **192,** 795–796.
Jayne-Williams, D. J. (1973). *Nature (London), New Biol.* **243,** 150–151.
Jayne-Williams, D. J., and Burgess, C. D. (1974). *J. Appl. Bacteriol.* **37,** 149–169.
Johns, C. O., and Finks, A. J. (1920). *J. Biol. Chem.* **41,** 379–389.
Kakade, M. L., and Evans, R. J. (1965). *J. Agric. Food Chem.* **13,** 450–452.
Kakade, M. L., Simons, N. R., Liener, I. E., and Lambert, J. W. (1972). *J. Agric. Food Chem.* **20,** 87–96.
Kaneko, I., Hayatsu, H., and Ukita T. (1975). *Biochim. Biophys. Acta* **392,** 131–140.
Kind, L. S., and Peterson, W. A. (1968). *Science* **160,** 312–313.
Kloz, J. (1971). *In* "Chemotaxonomy of the Leguminosae" (J. B. Harborne, D. Boulter, and B. L. Turner, eds.), Chapter 9, pp. 309–365. Academic Press, New York.
Klozowá, E., and Turkova, V. (1978a). *Biol. Plant. (Praha)* **20,** 129–134.
Klozowá, E., and Turkova, V. (1978b). *Biol. Plant. (Praha)* **20,** 373–376.
Korte, R. (1972). *Ecol. Food Nutr.* **1,** 303–307.
Krüpe, M. (1956). "Blutgruppen-spezifische Eiweisskörper (Phytagglutinine)." Enke, Stuttgart.
Krüpe, M., and Ensgraber, A. (1962). *Behringwerk-Mitt.* **42,** 48–54.
Krüpe, M., Wirth, W., Nils, D., and Ensgraber, A. (1968). *Z. Immumitaetsforsch.* **135,** 19–42.
Landsteiner, K., and Raubitschek, H. (1908). *Zentralbl. Bakteriol., Parasitenkd., Infektionskr. Hyg., Abt. 2: Orig.* **45,** 660.
Levine, D., Kaplan, M. J., and Greenaway, P. I. (1972). *Biochem. J.* **129,** 847–856.
Liener, I. E. (1951). *J. Biol. Chem.* **193,** 183–191.
Liener, I. E. (1953). *J. Nutr.* **49,** 527–539.
Liener, I. E. (1955). *Arch. Biochem. Biophys.* **54,** 223–231.
Liener, I. E. (1958). *J. Biol. Chem.* **233,** 401–405.
Liener, I. E. (1974). *J. Agric. Food Chem.* **22,** 17–22.
Liener, I. E. (1976). *Annu. Rev. Plant Physiol.* **27,** 291–319.
Liener, I. E., and Hill, E. G. (1953). *J. Nutr.* **49,** 609–615.
Liener, I. E., and Pallansch, M. J. (1952). *J. Biol. Chem.* **197,** 29–36.
Liener, I. E., and Rose, I. E. (1953). *Proc. Soc. Exp. Biol. Med.* **83,** 539–544.
Ling, N. R. (1968). "Lymphocyte Stimulation." North-Holland Publ., Amsterdam.
Lis, H., and Sharon, N. (1973). *Annu. Rev. Biochem.* **42,** 541–574.
Lis, H., Fridman, C., Sharon, N., and Katchalski, E. (1966). *Arch. Biochem. Biophys.* **117,** 301–309.
Lotan, R., Cacan, R., Cacan, M., Debray, H., Carter, W. G., and Sharon, N. (1975a). *FEBS Lett.* **57,** 100–103.
Lotan, R., Lis, H., and Sharon, N. (1975b). *Biochem. Biophys. Res. Commun.* **62,** 144–150.
Lotan, R., Skutelsky, E., Canon, D., and Sharon, N. (1975c). *J. Biol. Chem.* **250,** 8518–8523.

Lüning, O., and Bartels, W. (1926). *Z. Lebensm.-Unters.-Forsch.* **51,** 220–228.
Madson, T., and Walbum, L. (1904). Toxines et antitoxines de la ricine. *Zentralbl. Bakteriol., Parasitenkd., Infektimskr. Hyg., Abt. 2* **36,** 242.
Manage, C., Joshi, A., and Sohonie, K. (1972). *Toxicon* **10,** 89–90.
Marcusson-Begum, H. (1925). *Z. Immunitaetsforsch.* **45,** 49–73.
Marik, T., Entlicher, G., and Kocourek, J. (1974). *Biochim. Biophys. Acta* **336,** 53–61.
Marinkovich, A. (1964). *J. Immunol.* **93,** 732–741.
Martin, F. W., Waszenko-Zacharczenko, E., Boyd, W. C., and Schertz, K. F. (1964). *Ann. Bot. (London)* **28,** 319–324.
Mialonier, G., Privat, J. P., Monsigny, M., Kahlem, G., and Durand, R. (1973). *Physiol. Veg.* **11,** 519–537.
Miller, J. B., Hsu, R., Heinrikson, R., and Yachnin, S. (1975). *Proc. Natl. Acad. Sci. U.S.A.* **72,** 1388–1391.
Miller, W. S., and Coates, M. E. (1966). *Proc. Nutr. Soc.* **25,** iv.
Morcos, S. R., and Boctor, A. M. (1959). *Br. J. Nutr.* **13,** 163–167.
Nagata, Y., and Burger, M. M. (1974). *J. Biol. Chem.* **249,** 3116–3122.
Nowaková, N., and Kocourek, J. (1974). *Biochim. Biophys. Acta* **359,** 320–333.
Nowell, P. (1960). *Cancer Res.* **20,** 462.
Oh, J. H., and Conrad, R. A. (1972). *Arch. Biochem. Biophys.* **146,** 525–530.
Olsnes, S. (1975). *Bull. Inst. Pasteur, Paris* **74,** 85–99.
Olsnes, S., and Pihl, A. (1973). *Biochemistry* **12,** 3121–3126.
Olsnes, S., Saltvedt, E., and Pihl, A. (1974). *J. Biol. Chem.* **249,** 803–810.
Ortega, M., and González, M. T. (1970). *Ann. Bromatol. (Madrid)* **22,** 353–375.
Osborne, T. B., and Mendel, L. B. (1917). *J. Biol. Chem.* **32,** 369–387.
Pallansch, M. J., and Liener, I. E. (1953). *Arch. Biochem. Biophys.* **45,** 366–374.
Patten, J. R., Patten, J. A., and Pope, H. (1973). *Food Cosmet. Toxicol.* **11,** 577–584.
Pauly, G. B. (1974). *Ann. N.Y. Acad. Sci.* **234,** 145–160.
Perera, C. B., and Frumin, A. M. (1965). *Bibl. Haematol. (Basel)* **23,** 589.
Phadke, K., and Sohonie, K. (1962). *J. Sci. Ind. Res., Sect. C* **21,** 178–181.
Pierkarski, L. (1957). *Diss. Pharm.* **9,** 255–263.
Planqué, K., and Kijno, J. W. (1977). *FEBS Lett.* **73,** 64–66.
Pope, H. O., Patten, J. R., and Lawrence, A. L. (1975). *Br. J. Nutr.* **33,** 117–125.
Prager, M. D., and Speer, R. J. (1959). *Proc. Soc. Exp. Biol. Med.* **100,** 68–70.
Pull, S. P., Pueppke, S. G., Hymowitz, T., and Orf, J. H. (1978). *Science* **200,** 1277–1279.
Pusztai, A., and Palmer, R. (1977). *J. Sci. Food Agric.* **28,** 620–623.
Pusztai, A., and Watt, W. B. (1974). *Biochim. Biophys. Acta* **365,** 57.
Pusztai, A., Grant, G., and Palmer, R. (1975). *J. Sci. Food Agric.* **26,** 149–156.
Orf, J. H., Hymowitz, T., Pull, S. P., and Pueppke, S. G. (1978). *Crop Sci.* **18,** 899–900.
Rao, D. N., Hariharan, K., and Rajagopal, R. D. (1976). *Lebensm. Wiss. u. Technol.* **9,** 246–250.
Reitherman, R. W., Rosen, S. D., and Barondes, S. H. (1974). *Nature (London)* **248,** 599–600.
Renkonen, K. O. (1948). *Ann. Med. Exp. Biol. Fenn.* **26,** 66–72.
Renkonen, K. O. (1960). *Ann. Med. Exp. Biol. Fenn.* **38,** 26.
Rice, R. H. (1976). *Biochim. Biophys. Acta* **444,** 175–180.
Rigas, D. A., and Tisdale, V. V. (1969). *Experientia* **25,** 399–401.
Rigas, D. A., Johnson, E. A., Jones, R. T., McDermed, I. D., and Tisdale, V. V. (1966). *Chromatogr. Methods Immed. Sep., Proc. Meet., 1965* pp. 151–223.
Rigas, D. A., Head, C., and Eginitis-Rigas, C. (1972). *Physio. Chem. Phys.* **4,** 153–165.
Rosen, S. D., Reitherman, R. W., and Barondes, S. H. (1975). *Exp. Cell Res.* **95,** 159–166.
Rougé, P., and LaBroue, L. (1977). *C. R. Hebd. Seances Acad. Sci., Ser. D* **284,** 2423–2426.

Sachs, L. (1974). *In* "The Cell Surface in Development" (A. A. Moscona, ed.), Chapter 7, p. 127. Wiley, New York.
Saint-Paul, M. (1961). *Transfusion* **4,** 3–37.
Salgarkar, S., and Sohonie, K. (1965a). *Indian J. Biochem.* **2,** 193–196.
Salgarkar, S., and Sohonie, K. (1965b). *Indian J. Biochem* **2,** 197–199.
Sambeth, W., Nesheim, M. C., and Serafin, J. A. (1967). *J. Nutr.* **92,** 479–490.
Schertz, K. F., Jurgelky, W., and Boyd, W. C. (1960). *Proc. Natl. Acad. Sci. U.S.A.* **46,** 529–532.
Schingoethe, D. J., Aust, S. D., and Thomas, J. W. (1970). *J. Nutr.* **100,** 739–748.
Sharon, N., and Lis, H. (1972). *Science* **177,** 949–959.
Shier, W. T., Trotter, J. I., III, and Reading, C. (1974). *Proc. Soc. Exp. Biol. Med.* **146,** 590–593.
Stead, R. H., de Muelenaere, H. J. H., and Quicke, G. V. (1966). *Arch. Biochem. Biophys.* **113,** 703–708.
Stein, M. (1976). *Qual. Plant.—Plant Foods Hum. Nutr.* **26,** 227–243.
Stillmark, H. (1889). *Arch. Pharmakol. Inst. Dorpat* **3,** 59.
Stockert, R. J., Morell, A. G., and Scheinberg, I. H. (1974). *Science* **186,** 365–366.
Strosberg, A. D., Foriers, A., Van Driessche, F., Mole, L. E., and Kanarek, L. (1976). *Arch. Int. Physiol. Biochim.* **84,** 660–661.
Takahashi, T., Ramachandramurthy, P., and Liener, I. E. (1967). *Biochim. Biophys. Acta* **133,** 123–133.
Takahashi, T., Yamada, N., Iwamoto, K., Shimabayashi, Y., and Izutsu, K. (1973). *Agric. Biol. Chem.* **37,** 29–36.
Thomas, M. W., Walborg, E. F., Jr., and Jirgensons, B. (1977). *Arch. Biochem. Biophys.* **178,** 625–630.
Tichá, M., Entlicher, G., Kostir, J. V., and Kocourek, J. (1970). *Biochim. Biophys. Acta* **221,** 282–289.
Tobiška, J. (1964). "Die Phytohemagglutinine." Akademie-Verlag, Berlin.
Tobiška, J., and Lhotecká-Brázdová, A. (1960). *Z. Immunitaetsforsch.* **19,** 225–230.
Toms, G. C., and Turner, T. D. (1965). *J. Pharm. Pharmacol., Suppl.* **17,** 118–125.
Toms, G. C., and Western, A. (1971). *In* "Chemotaxonomy of the Leguminosae" (J. B. Harborne, D. Boutler, and B. L. Turner, eds.), pp. 367–462. Academic Press, New York.
Trowbridge, I. S. (1974). *J. Biol. Chem.* **249,** 6004–6012.
Turner, R. H., and Liener, I. E. (1975). *J. Agric. Food Chem.* **23,** 484–487.
Uhlenbruck, G., and Steinhausen, G. (1972). *Blut* **25,** 335–337.
Uhlenbruck, G., Pardoe, G. I., and Bird, G. W. G. (1969). *Z. Immunitaetsforsch.* **138,** 423–433.
Van Driessche, E., Strosberg, A. D., and Kanarek, L. (1976). *Arch. Int. Physiol. Biochim.* **84,** 677–679.
Vretblad, P. (1976). *Biochim. Biophys. Acta* **434,** 169–176.
Wada, S., Pallansch, M. J., and Liener, I. E. (1958). *J. Biol. Chem.* **233,** 395–400.
Wagh, P. V., Klaustermeier, D. F., Waibel, P. E., and Liener, I. E. (1963). *J. Nutr.* **80,** 191–195.
Walborg, E. F., Jr., Davis, E. M., Gilliam, E. B., Smith, D. F., and Nery, G. (1975). *In* "Cellular Membranes and Tumor Cell Behavior," pp. 337–360. Williams & Wilkins, Baltimore, Maryland.
Wang, I. L., Becker, I. W., Reeke, G. N., Jr., and Edelman, G. M. (1974). *J. Mol. Biol.* **88,** 259–262.
Weber, T., Aro, H., and Nordman, C. T. (1971). *Biochim. Biophys. Acta* **263,** 94–105.
Wienhaus, O. (1909). *Biochem. Z.* **18,** 228–260.
Wolpert, J. S., and Albersheim, P. (1976). *Biochem. Biophys. Res. Commum.* **70,** 729–737.
Yachnin, S., and Svenson, R. H. (1972). *Immunology* **22,** 871–883.
Youle, R. J., and Huang, A. H. C. (1976). *Plant Physiol.* **58,** 703–709.

CHAPTER 4

Glucosinolates

H. L. Tookey, C. H. VanEtten, and M. E. Daxenbichler

I. INTRODUCTION

Glucosinolates were called thioglucosides in much of the literature before about 1970; many of the individual substances in this class are still known by such trivial names as sinigrin, sinalbin, and progoitrin. Glucosinolates occurring in cultivated plants are responsible for the pungent flavors of the condiments horseradish and mustard and contribute to the characteristic flavors of turnip,

TOXIC CONSTITUENTS OF PLANT FOODSTUFFS, SECOND EDITION
ISBN 0-12-449960-0

rutabaga, cabbage, and related vegetables. In certain cruciferous plants, some of these substances have also been associated with endemic goiter—hypothyroidism with an enlargement of the thyroid gland.

During the past two decades, chemical knowledge of the glucosinolates has grown enormously. Analytical methods for the hydrolytic products derived from glucosinolates have improved enough to render some of the older methods and compositional data obsolete.

Because most of the biological effects, as well as the flavors esteemed by man, are not caused by the glucosinolates per se but by the enzymatic products from these compounds, a discussion of the hydrolytic process is necessary in order to understand the role of glucosinolates. To the knowledge of specific kinds of glucosinolates present in various domestic crops must therefore be added the nature of the hydrolytic products to be expected under various conditions of hydrolysis during processing of food or feed.

Safety considerations in food and feed consisting of glucosinolate-containing plants are evaluated by a critical review of some of the feeding tests that have been made with crucifers. Finally, consideration is given to the possible effects of future plant breeding on the glucosinolate content of new varieties of plants used for food and feed.

II. CRUCIFERS OF ECONOMIC VALUE

The majority of cultivated plants containing glucosinolates belong to the Cruciferae, although several other plant families produce glucosinolates (Ettlinger and Kjaer, 1968; Kjaer, 1973). Many of the crucifers that serve as sources of food and condiment belong to the genus *Brassica*. Examples of such foods are cabbage and related vegetables, turnip, rutabaga, and mustard greens. Condiment mustard seed is also a member of *Brassica*. Other genera of economic value include *Armoracia* (horseradish), *Eruca* and *Lepidium* (salad greens), and *Raphanus* (radish). All of these plants contain glucosinolates, which upon hydrolysis provide much of the culinary flavor characteristic of these plants.

Of the horticultural plants from *Brassica,* cabbage is the main crop. World production for 1974 was 13.9 million metric tons, including white, red, Savoy, and Chinese cabbage, brussels sprouts, sprouting broccoli, and green kale (Food and Agriculture Organization of the United Nations, 1974). The estimate does not include vegetables from family gardens. Annual production of cabbage in the United States in 1974 was 1.0 million short tons for the fresh market and 0.3 million short tons for sauerkraut (U.S. Department of Agriculture, 1975).

The genus *Brassica* includes plants that are used in many parts of the world as pasture, forage, or silage for livestock (Morrison, 1959). Dwarf Essex kale (*B. napus*) and marrow-stem kale (*B. oleracea*) are grown for these purposes. In addition, weeds from the Cruciferae often grow in pastures and are consumed by livestock.

Rapeseed, derived from varieties of *B. napus* and *B. campestris,* is one of the major oilseeds of commerce. World production in 1974 was 7.2 million metric tons (Food and Agriculture Organization of the United Nations, 1974), which is an increase from the average annual production of 2.8 million metric tons over the years 1948–1953 (Food and Agriculture Organization of the United Nations, 1965). Estimated production for 1973 was for Asia 3.3, Europe 2.5, and Canada 1.2 million metric tons. World production of mustard seed (principally *B. hirta, B. nigra,* and *B. juncea*) was estimated at 0.3 million tons.

The seed meal remaining after oil extraction can be fed to livestock, but in limited amounts because it contains toxic substances. A more satisfactory seed meal for animal feeds from crucifer seed is needed by the feed industry. The problem is increasing in economic importance because of the growing production of rape and mustard seed. Processed *Brassica* seed meals give varied results when fed to animals. In addition to growth depression and enlarged thyroids, pathology in other body organs is often detected (see Section IV). The recent development of rapeseed varieties, such as Bronowski or Tower, which are low in glucosinolates has been a significant step toward making satisfactory feeds from rapeseed meal.

III. GLUCOSINOLATES AND THEIR CHEMISTRY

A. Historical Background

The pungent nature of horseradish and mustard seed has been known for many centuries. A relationship between the mustard oil allyl isothiocyanate (allyl-NCS)* and its precursor in mustard seed was established following the isolation of a glucosinolate (subsequently named sinigrin) by Bussy in 1840 (Kjaer, 1960). Formation of allyl-NCS appeared to be the result of the action of an enzyme system associated with sinigrin in the seed. Before World War I, only two additional glucosinolates were isolated, namely, sinalbin from white mustard seed and glucocheirolin from wallflower seed.

The trivial names sinigrin and sinalbin were used before much was known of glucosinolate chemistry. In a later naming system adopted by Kjaer (1960) and others, the prefix gluco was attached to a part of the Latin species name from which the compound was first isolated. Glucocheirolin is an example of this nomenclature. A more recent system that relates to the chemical structure was suggested by Ettlinger and Dateo (1961). In this system, all members of the class

*The abbreviation NCS will be used for an isothiocyanate; SCN for an organic thiocyanate; CN for a nitrile; OZT for an oxazolidine-2-thione; GS for a glucosinolate.

are called glucosinolates, to which is added a prefix that chemically describes the variable part of the molecule, e.g., methylglucosinolate (methyl-GS) or benzyl-GS.

The number of identified plant glucosinolates is now more than 70 (Kjaer, 1973). Most of these have been characterized by Kjaer and co-workers. According to Kjaer (1966), of about 1500 known species of Cruciferae, all 300 examined contain from one to seven glucosinolates. Nearly all species from the related families Capparidaceae, Moringaceae, Tovariaceae, and Resedaceae, as well as some species from unrelated plant families, also contain glucosinolates (Kjaer, 1973).

Glucosinolates occur throughout a plant—in root, stem, leaf, and seed. They are always accompanied by an enzyme system capable of hydrolyzing the glucosinolates, but there is ample evidence that the enzyme is separated from its glucosinolate substrate in the intact plant. This separation may be breached in the germinating seed, where the glucosinolates are readily hydrolyzed (Tookey and Wolff, 1970). In the early literature the enzyme was called myrosin or myrosinase. Ettlinger *et al.* (1961) coined the name glucosinolase to parallel the name glucosinolate. Nevertheless, the term used here is thioglucosidase (EC 3.2.3.1), as recommended by the International Union of Biochemistry (Florkin and Stotz, 1965).

B. Formulas of Glucosinolates and Aglucon Products

The basic structure for a glucosinolate was revised by Ettlinger and Lundeen (1956) and confirmed by the synthesis of benzyl-GS (glucotropaeolin) (Ettlinger and Lundeen 1957). All glucosinolates contain β-D-thioglucose as the sugar moiety (Ettlinger and Kjaer, 1968), and all probably have the same anti configuration between the sulfate and the R group around the carbon–nitrogen double bond (Waser and Watson, 1963; Marsh and Waser, 1970) (Scheme I). The glucosinolates are anions and occur in plants as salts. They are usually regarded as potassium salts, although the complex organic cation sinapine occurs widely

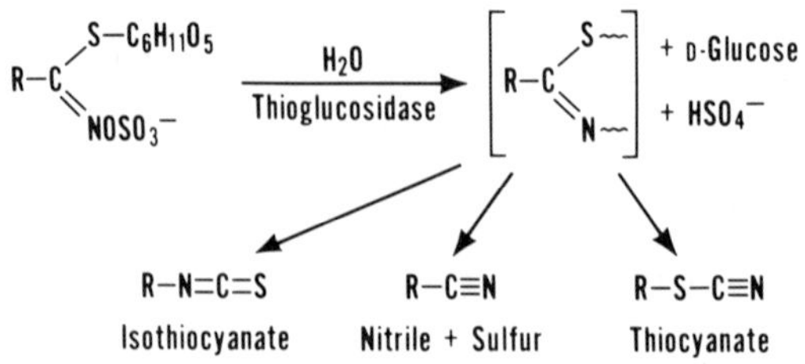

$$R-C(-S-C_6H_{11}O_5)(=NOSO_3^-) \xrightarrow[\text{Thioglucosidase}]{H_2O} \left[R-C(-S^{\sim})(=N^{\sim})\right] + \text{D-Glucose} + HSO_4^-$$

$$\longrightarrow \underset{\text{Isothiocyanate}}{R-N=C=S} \qquad \underset{\text{Nitrile + Sulfur}}{R-C\equiv N} \qquad \underset{\text{Thiocyanate}}{R-S-C\equiv N}$$

SCHEME I

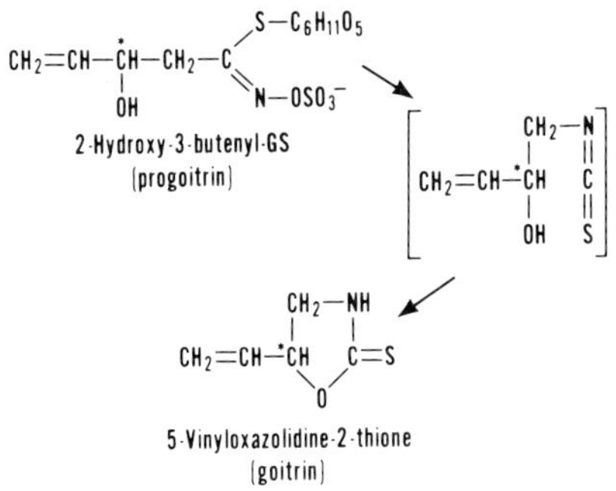

$CH_2{=}CH{-}\overset{\bullet}{C}H(OH){-}CH_2{-}C(S{-}C_6H_{11}O_5){=}N{-}OSO_3^-$

2-Hydroxy-3-butenyl-GS
(progoitrin)

$\rightarrow \left[CH_2{=}CH{-}\overset{*}{C}H(OH){-}CH_2{-}N{=}C{=}S \right] \rightarrow$

$CH_2{=}CH{-}\overset{*}{C}H{-}CH_2{-}NH{-}C({=}S){-}O$ (ring)

5-Vinyloxazolidine-2-thione
(goitrin)

SCHEME II

among crucifers (Schultz and Gmelin, 1953) and is the cation accompanying *p*-hydroxybenzyl-GS (sinalbin) as it is usually isolated (Kjaer, 1960).

Glucosinolates are hydrolyzed by an associated thioglucosidase enzyme system whenever wet, raw plant material is crushed. Glucose and acid sulfate ion are always released as products. The organic aglucon portion may undergo an intramolecular rearrangement following the hydrolysis to give an isothiocyanate. Without such a rearrangement, the aglucon forms a nitrile, often with the loss of sulfur. Alternatively, a rearrangement to an organic thiocyanate may occur (Scheme I). Conditions required for the formation of these various aglucon products are discussed under Section III,D. From some glucosinolates the expected isothiocyanates are not formed. They may cyclize to form oxazolidine-2-thiones. Hydrolysis of glucosinolates with a hydroxyl group at C-2 of the R group results in the formation of R′-OZT, as illustrated in Scheme II. Progoitrin [(*R*)-2-hydroxy-3-butenyl-GS] differs from *epi*-progoitrin [(*S*)-2-hydroxy-3-butenyl-GS] only by configuration at the carbon atom marked with an asterisk in Scheme II. The (*R*) and (*S*) designate stereochemistry by the sequence rule (Cahn, 1964). The absolute configuration of 5-vinyl-OZT from *Brassica* was established as (*S*) by Kjaer *et al.* (1959). Daxenbichler *et al.* (1965) established the configuration of 5-vinyl-OZT from *Crambe* as (*R*). Compounds in the same stereochemical series may have opposite (*R*) and (*S*) designations. Thus, (*R*)-2-hydroxy-3-butenyl-GS gives rise to (*S*)-5-vinyl-OZT, and (*S*)-2-hydroxy-3-butenyl-GS gives rise to (*R*)-5-vinyl-OZT. *epi*-Progoitrin was erroneously assigned the (*R*) configuration by Daxenbichler *et al.* (1965) but was later corrected (Daxenbichler *et al.*, 1966).

If the aglucon is unsaturated, under certain conditions (see Section III,D) diastereomeric epithionitriles can be formed from the aglucon at a newly created asymmetric center indicated by the asterisk in Scheme III.

Indolyl glucosinolates form nitriles or the presumably unstable isothiocyanates,

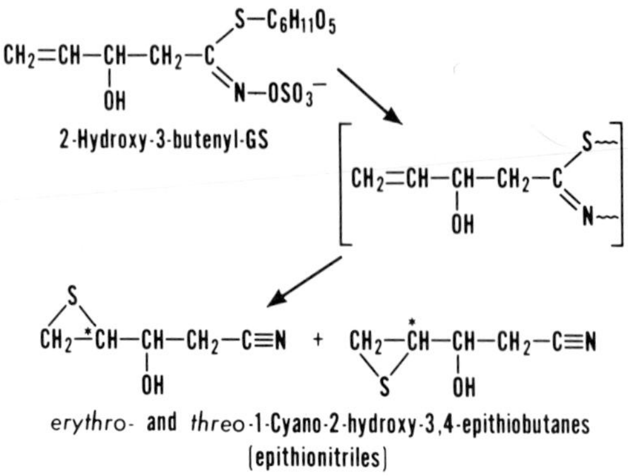

$CH_2{=}CH{-}CH(OH){-}CH_2{-}C(S{-}C_6H_{11}O_5){=}N{-}OSO_3^-$

2-Hydroxy-3-butenyl-GS

$[CH_2{=}CH{-}CH(OH){-}CH_2{-}C(S\sim){=}N\sim]$

$CH_2{-}{*}CH(S){-}CH(OH){-}CH_2{-}C{\equiv}N + CH_2{-}\overset{*}{C}H(S){-}CH(OH){-}CH_2{-}C{\equiv}N$

erythro- and *threo*-1-Cyano-2-hydroxy-3,4-epithiobutanes (epithionitriles)

SCHEME III

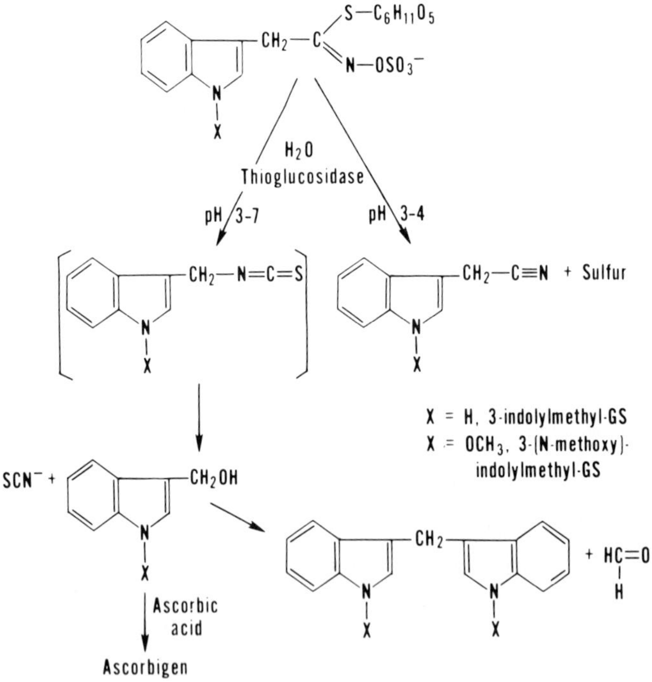

$S{-}C_6H_{11}O_5$; $CH_2{-}C$; $N{-}OSO_3^-$; N; X

H_2O
Thioglucosidase

pH 3-7 | pH 3-4

$[\ CH_2{-}N{=}C{=}S\ ]$ N X | $CH_2{-}C{\equiv}N$ + Sulfur N X

X = H, 3-indolylmethyl-GS
X = OCH_3, 3-(N-methoxy)-indolylmethyl-GS

SCN^- + CH_2OH N X → CH_2 N X N X + $HC{=}O$ H

Ascorbic acid

Ascorbigen

SCHEME IV

as shown in Scheme IV. The isothiocyanates degrade further to thiocyanate ion and indolyl alcohols, which may react with ascorbic acid (Gmelin and Virtanen, 1961, 1962). Enzymatic hydrolysis of *p*-hydroxybenzyl-GS may give *p*-hydroxybenzyl-NCS, which readily degrades to *p*-hydroxybenzyl alcohol and SCN^-, as well as other products in small amounts (Kawakishi *et al.*, 1967).

C. Occurrence in Cultivated Plants

Glucosinolates found in cultivated plants are listed in Table I. Although glucosinolates are present throughout the plant, their highest concentration is usually in the seed. Exceptions are 3-indolylmethyl-GS and 3-(*N*-methoxy)indolylmethyl-GS, which are rarely reported in seeds. Concentrations of glucosinolates, where known, are also presented in Table I. For brevity, only the major glucosinolates are shown when the particular plant contains many different glucosinolates.

Glucosinolates in which the R group is methyl, allyl, 3-butenyl, 4-pentenyl, 3-methylthiopropyl, 4-methylthiobutyl, 2-phenylethyl, benzyl, or *m*-methoxybenzyl give rise to pungent, steam-volatile isothiocyanates. These glucosinolates may hydrolyze instead to nitriles (Section III,D).

Progoitrin and *epi*-progoitrin (enantiomers of 2-hydroxy-3-butenyl-GS) are precursors of the antithyroid compounds called goitrins (5-vinyl-OZT), as well as precursors of nitriles and epithionitriles.

Indolyl glucosinolates are widespread among crucifers, being found in rutabaga, kale, cabbage, and radish (Table I). Several studies indicated that ascorbigen derived from these glucosinolates (Scheme IV) has vitamin C activity (Kiesvaara and Virtanen, 1963; Matano and Kato, 1967).

Both of the indolyl glucosinolates and *p*-hydroxybenzyl-GS may finally degrade to release thiocyanate ion. This degradation accounts for the presence of SCN^- in rutabaga, cabbage, radish, white mustard, and charlock (Michajlovskij and Langer, 1958).

D. Enzymatic Hydrolysis

Glucosinolates are hydrolyzed by the enzyme thioglucosidase to give glucose, acid sulfate ion, and one or more of the aglucon products indicated in Section III,B. Only one enzyme seems to be necessary for glucosinolate hydrolysis, but other factors are involved in determining which of the several types of aglucon products will predominate.

Early workers, e.g., von Euler and Erikson (1926) and Sandberg and Holly (1932), believed that two enzymes, a thioglucosidase and a sulfatase, were necessary for complete hydrolysis of glucosinolates. Nagashima and Uchiyama (1959a,b) provided convincing evidence that only a single enzyme is required.

TABLE I

GLUCOSINOLATES IN CULTIVATED PLANTS[a]

Plant name[b]	Plant part	R Group of glucosinolate	Common name of glucosinolate	Amount (μg/gm)[c]	References
Food				327	
Brassica campestris L. (turnip)	Root	(*R*)-2-Hydroxy-3-butenyl	Progoitrin	134	Mullin and Sahasrabudhe (1978)
		3-Indolylmethyl	Glucobrassicin	327	
		2-Phenylethyl	Gluconasturtiin	S	Lichtenstein *et al.* (1962)
	Leaf	(*R*)-2-Hydroxy-3-butenyl	Progoitrin	L	Tapper and MacGibbon (1967)
		(*R*)-2-Hydroxy-4-pentenyl	—	S	
Brassica campestris ssp. *Pekinensis* (Chinese cabbage)[d]	Leaf, stem (head)	3-Butenyl	Gluconapin	4–250	Daxenbichler *et al.* (1979)
		(*R*)-2-Hydroxy-3-butenyl	Progoitrin	3–195	
		4-Pentenyl	Glucobrassicanapin	13–275	
		2-Phenylethyl	Gluconasturtiin	17–258	
		3-Indolylmethyl and 3-(*N*-methoxy)-indolylmethyl	Glucobrassicin and neoglucobrassicin	92–533	
Brassica napus L. (rutabaga)	Root	(*R*)-2-Hydroxy-3-butenyl	Progoitrin	L	Astwood *et al.* (1949)
		3-Indolylmethyl	Glucobrassicin	S	Gmelin and Virtanen (1961)
		3-(*N*-Methoxy)indolyl-methyl	Neoglucobrassicin	S	Gmelin and Virtanen (1962)
Brassica oleracea L. (cabbage)[e]	Leaf (head)	Allyl	Sinigrin	17–584	VanEtten *et al.* (1976), Josefsson (1967)
		3-Methylsulfinylpropyl	Glucoiberin	57–464	
		(*R*)-2-Hydroxy-3-butenyl	Progoitrin	4–86	
		4-Methylsulfinylbutyl	Glucoraphanin	<1–319	
		2-Phenylethyl	Gluconasturtiin	2–17	
		3-Indolylmethyl and	Glucobrassicin and		

		3-(*N*-methoxy)-indolylmethyl	neoglucobrassicin	132–527	
(kohlrabi)	Stem	3-Methylsulfinylpropyl	Glucoiberin	97	Michajlovskij *et al.* (1969b)
		2-Butyl	—	115	
		3-Indolylmethyl 3-(*N*-Methoxy)indolyl-methyl	Glucobrassicin and Neoglucobrassicin	145	
(broccoli, brussels sprouts, cauliflower, collards, kale)	Leaf, flower bud, or stem		Similar to cabbage		Josefsson (1967)
Carica papaya L. (papaya)[e]	Fruit pulp (immature)	Benzyl	Glucotropaeolin	650	Tang (1971)
	Latex, dry	Benzyl	Glucotropaeolin	73,000–116,000	Tang (1973)
Lepidium sativum L. (garden cress)	Leaf	Benzyl	Glucotropaeolin	L	Gmelin and Virtanen (1959)
		3-Butenyl	Gluconapin	S	
		2-Phenylethyl	Gluconasturtiin	S	
		4-Pentenyl	Glucobrassicanapin	S	MacLeod and Islam (1976)
Nasturtium officinale R. Br. (water cress)	Leaf	2-Phenylethyl	Gluconasturtiin	L	Schultz and Gmelin (1952)
		7-Methylthioheptyl	—	S	MacLeod and Islam (1975)
		8-Methylthiooctyl	—	S	
Raphanus sativus L. (radish)	Root	4-Methylthio-3-butenyl	—	L	Friis and Kjaer (1966)
		3-Indolylmethyl	Glucobrassicin	S	Gmelin and Virtanen (1961)
Eruca sativa Mill. (rocket mustard)	Whole plant	4-Methylthiobutyl	Glucoerucin	24	Cole (1976)
Condiment					
Armoracia lapathifolia, *A. rusticana* (horseradish)	Root	Allyl	Sinigrin	5000	Stoll and Seebeck (1948), Hansen (1974)
		2-Phenylethyl	Gluconasturtiin	S	Stahmann *et al.* (1943)
Brassica carinata Braun. (Ethiopian rapeseed)	Seed	Allyl	Sinigrin	L	Ettlinger and Thompson (1962)

(*Continued*)

TABLE I (*Continued*)

Plant name[b]	Plant part	R Group of glucosinolate	Common name of glucosinolate	Amount (μg/gm)[c]	References
Brassica hirta Moench (*Sinapis alba* L.) (white mustard)	Seed	*p*-Hydroxybenzyl	Sinalbin	57,000–74,000	Kjaer and Rubinstein (1954), Josefsson (1970)
Brassica juncea (L.) Coss (brown or indian mustard)	Seed	Allyl	Sinigrin	L	Jensen *et al.* (1953)
		3-Butenyl	Gluconapin	S	Vaughan (1977)
Brassica kaber (D.C.) L.C. Wheeler, (*Sinapis arvensis* L.) (charlock)	Seed	*p*-Hydroxybenzyl	Sinalbin	L	Ettlinger and Thompson (1962)
Brassica nigra (L.) Koch (black mustard)	Seed	Allyl	Sinigrin	L	Jensen *et al.* (1953), Ettlinger and Thompson (1962)
Capparis spinosa L.[f] (caper)	Bud	Methyl	Glucocapparin		Ahmed and Rizk (1972)
		Allyl	Sinigrin		
		3-Methylsulfinylpropyl	Glucoiberin		
		2-Methyl-2-hydroxybutyl	Glucocleomin		
Feed					
Brassica campestris L. (rape, turnip rape, Polish rape, navette, rubsen)	Seed meal	3-Butenyl	Gluconapin	18,000–44,000	Josefsson and Appelqvist (1968)
		(*R*)-2-Hydroxy-3-butenyl	Progoitrin	2,000–20,000	
		4-Methylsulfinylbutyl	Glucoraphanin	S	Kjaer *et al.* (1953)
		4-Pentenyl	Glucobrassicanapin	S	Ettlinger and Thompson (1962)
		5-Methylsulfinylpentyl	Glucoalyssin	S	
Brassica napus L. (rape, winter rape, Argentine rape, colza)	Seed meal	3-Butenyl	Gluconapin	10,000–22,000	Josefsson and Appelqvist (1968)
		(*R*)-2-Hydroxy-3-butenyl	Progoitrin	18,000–50,000	
		3-Methylsulfinylpropyl	Glucoiberin	S	Schultz and Gmelin (1954)

		4-Pentenyl	Glucobrassicanapin	S	Kjaer *et al.* (1953)
		2-Phenylethyl	Gluconasturtiin	S	Astwood *et al.* (1949)
		p-Hydroxybenzyl	Sinalbin	S	Kjaer and Boe Jensen (1956)
(Bronowski variety)	Seed meal	3-Butenyl	Gluconapin	1,000–4,000	Josefsson and Appelqvist (1968)
		(*R*)-2-Hydroxy-3-butenyl	Progoitrin	1,000–23,000	
(dwarf Essex)	Leaf	(*R*)-2-Hydroxy-3-butenyl	Progoitrin	12	Tapper and MacGibbon (1967)
		2-Hydroxy-4-pentenyl	—	4	
Crambe abyssinica Hochst. ex R. E. Fries (crambe)	Seed meal (dehulled)	(*S*)-2-Hydroxy-3-butenyl	*epi*-Progoitrin	80,000–97,000	Daxenbichler *et al.* (1965), VanEtten *et al.* (1965)
		Allyl	Sinigrin	8,000–13,000	VanEtten *et al.* (1965), Daxenbichler *et al.* (1964)
		3-Butenyl	Gluconapin		
		2-Phenylethyl	Gluconasturtiin		
		4-Pentenyl or benzyl	—		
Limnanthes alba Benth.[f]	Seed meal	*m*-Methoxybenzyl	Glucolimnanthin	29,000–51,000	Daxenbichler and VanEtten (1974)
		2-Hydroxy-2-methylpropyl	Glucoconringiin	0–9,000	Miller *et al.* (1964), Ettlinger and Lundeen (1956)

[a]Reprinted with permission from C. H. VanEtten, M. E. Daxenbichler, and I. A. Wolff (1969), *J. Agric. Food Chem.* **17,** 483–491. Copyright by the American Chemical Society.

[b]The taxonomy of *Brassica* and related genera is reviewed by Downey (1965) and Vaughan (1977).

[c]Calculated as potassium glucosinolate in fresh weight for vegetative parts, in air dry weight for seed. If no numerical value is shown, the amount relative to total glucosinolate content is indicated by L (large) or S (small).

[d]Chinese cabbage also contains allyl-, 2-hydroxy-4-pentenyl-, 5-methylthiopentyl-, and 5-methylsulfinylpentylglucosinolates.

[e]Cabbage also contains (in lesser amounts) 3-methylthiopropyl-, butenyl-, 4-methylthiobutyl-, 4-methylsulfonylbutyl-, and benzylglucosinolates.

[f]Does not belong to Cruciferae.

Ettlinger *et al.* (1961) and Miller (1965) provided a theory consistent with complete glusocinolate hydrolysis being brought about by a single thioglucosidase.

Since the early 1960s, thioglucosidase from several plant sources in Cruciferae have been examined. In these studies, only a single enzyme was required for hydrolysis, but apparently a single plant may produce several isozymes, proteins that differ in minor ways but have the same enzymatic activity (Tsuruo and Hata, 1972; Vose, 1972; Henderson and McEwan, 1972).

Glucosinolate-hydrolyzing thioglucosidase appears not only in all the plants that contain glucosinolates, but in a few other organisms as well. Certain intestinal bacteria exhibit thioglucosidase activity (Oginsky *et al.*, 1965; Tani *et al.*, 1974). This activity assumes importance when intact glucosinolates are fed to animals. Although the nature and specificity of these bacterial enzymes have not been well studied, they may differ from the plant thioglucosidases; e.g., ascorbate is reported to inhibit the *Paracolobactrum* enzyme (Oginsky *et al.*, 1965) and the *Enterobacter* enzyme (Tani *et al.*, 1974), whereas it usually activates the plant enzymes. Some fungi also are able to hydrolyze glucosinolates (Reese *et al.*, 1958), but the enzyme from *Aspergillus niger* shows a larger K_m than that from mustard seed (Ohtsuru *et al.*, 1969) and is indifferent to ascorbate (Ohtsuru and Hata, 1973).

Ascorbic acid generally enhances the activity of plant thioglucosidase, sometimes more from one plant source than from another (Ettlinger *et al.*, 1961; Vose, 1972; Tookey, 1973b). The amount of activation needs to be interpreted with caution, since the ionic strength of the medium influences the degree of activation (Tsuruo and Hata, 1968). The enzymes that have been characterized are glycoproteins (Bjorkman and Janson, 1972; Lönnerdal and Janson, 1973) with sulfhydryl groups essential to their activity (Nagashima and Uchiyama, 1959a; Tsuruo and Hata, 1967; Tookey, 1973a).

Thioglucosidase from *Brassica hirta* (white mustard) seed has been widely used in analytical methods in which glucosinolates are converted to isothiocyanates or to oxazolidine-2-thiones (cyclized isothiocyanates). Miller (1965) showed that allyl-GS was hydrolyzed to form allyl-NCS at pH 5, but primarily to allyl-CN at pH 3. Similarly, *B. hirta* thioglucosidase acting at pH 5–7 on *p*-hydroxybenzyl-GS produces *p*-hydroxybenzyl-NCS, which on standing hydrolyzes to *p*-hydroxybenzyl alcohol and SCN^-; however, at pH 3–4 *p*-hydroxybenzyl-CN becomes the major aglucon product (Kawakishi *et al.*, 1967). *Brassica hirta* enzyme gives similar results when acting on another glucosinolate whose isothiocyanate cyclizes to oxazolidine-2-thione: 2-hydroxy-3-butenyl-GS is hydrolyzed quantitatively to 5-vinyl-OZT at pH near 7, but as the pH is lowered from 6 to 3, the aglucon product becomes nearly half 1-cyano-2-hydroxy-3-butene (Daxenbichler *et al.*, 1966).

Crushed plant materials may not act in the same way as enzyme from *B. hirta*. Meal from freshly harvested *Crambe abyssinica* seed incubated with water hy-

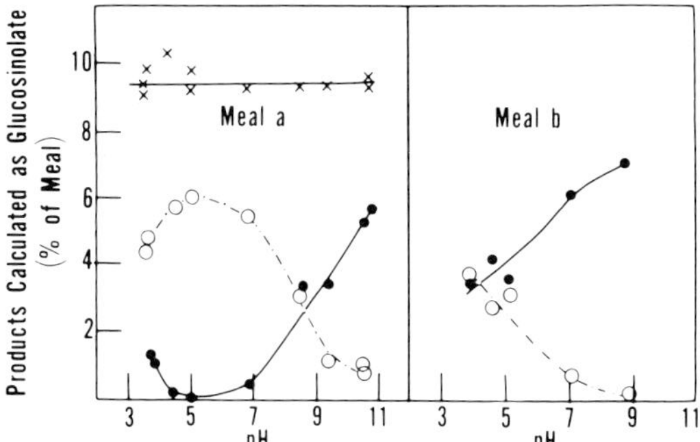

FIG. 1. Aglucon products from 2-hydroxy-3-butenyl-GS in *C. abyssinica* seed meal as a function of pH. Conditions: 1 gm of meal per 5 ml water at room temperature. Meal a gives no 5-vinyl-OZT and Meal b considerable 5-vinyl-OZT on autolysis in water at room temperature (x—x, total glucosinolates hydrolyzed; ●—●, 5-vinyl-OZT; ○---○, all nitriles calculated as 1-cyano-2-hydroxy-3-butene). Reproduced by permission from C. H. VanEtten, M. E. Daxenbichler, J. E. Peters, and H. L. Tookey (1966), *J. Agric. Food Chem.* **14**, 426–430. Copyright by the American Chemical Society.

drolyzes endogenous 2-hydroxy-3-butenyl-GS primarily to nitriles at pH 4–7; 5-vinyl-OZT does not predominate until the pH is raised above 9 (VanEtten *et al.*, 1966) (Fig. 1, Meal a). Aged crambe meal produces 5-vinyl-OZT at a lower pH, as shown in Fig. 1 (Meal b). However, the latter meal in the presence of an SH reagent (0.07 *M* HS—CH_2—CH_2—OH) responds essentially as did the fresh meal (Tookey and Wolff, 1970), thus indicating that some subtle oxidation has occurred in aged meal. Further evidence of a labile factor involved in the formation of nitriles is shown by the effect of increasing the temperature of autolysis (Fig. 2). Seed meal from *B. napus* and *C. abyssinica* both respond similarly to

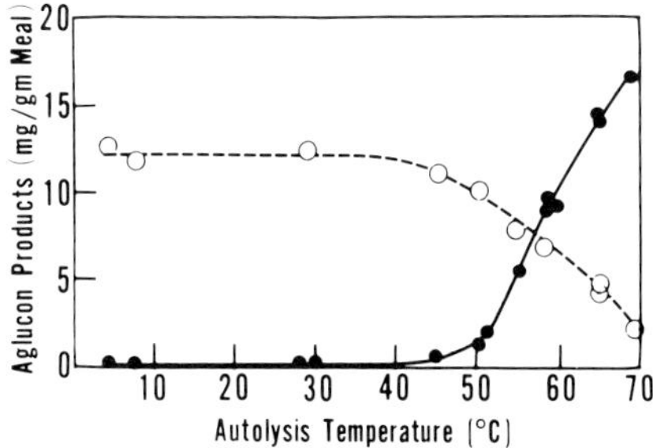

FIG. 2. Products from 2-hydroxy-3-butenyl-GS in *C. abyssinica* seed meal as a function of autolysis temperature. Conditions: 3 gm of meal per 15 ml of water, pH 4.0–5.0 (●—●, 5-vinyl-OZT; ○-○, all nitriles calculated as 1-cyano-2-hydroxy-3-butene). Total glucosinolate in the meal, 7.2%. Reproduced by permission from C. H. VanEtten, M. E. Daxenbichler, J. E. Peters, and H. L. Tookey (1966), *J. Agric. Food Chem.* **14**, 426–430. Copyright by the American Chemical Society.

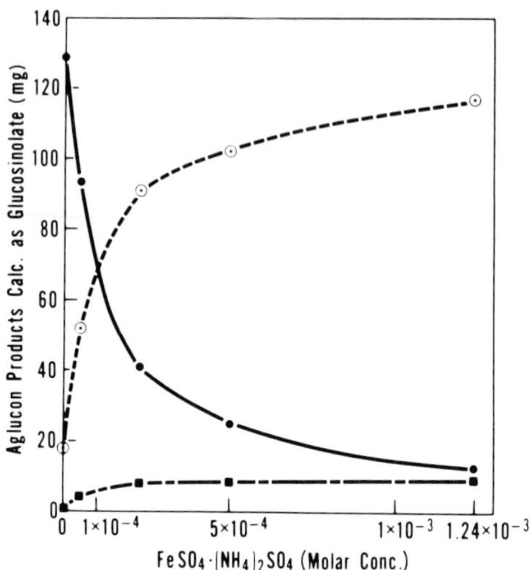

FIG. 3. Products from 2-hydroxy-3-butenyl-GS versus ferrous ion concentration. Thioglucosidase (100 mg) and glucosinolate (1.2×10^{-2} M) in 30 ml 0.07 M ascorbate, pH 5.3 [●—●, 5-vinyl-OZT; ⊙---⊙, all nitriles; ■---■, thionamide (3-hydroxy-4-pentenylthionamide)]. Thionamide is a nonenzymatic product from reaction of Fe^{2+} and glucosinolates (Austin *et al.*, 1968). Reproduced by permission of the National Research Council of Canada from Tookey and Wolff (1970).

dry heat: heat treatment of these meals at 100°–120°C favors 5-vinyl-OZT formation at the expense of nitriles (VanEtten *et al.*, 1966).

Several factors involved in the thioglucosidase hydrolysis of 2-hydroxy-3-butenyl-GS have been elucidated. Partially purified thioglucosidase from crambe hydrolyzes the 2-hydroxy-3-butenyl-GS to form 5-vinyl-OZT (Tookey, 1973a,b). The addition of ferrous ion to crambe thioglucosidase shifts the aglucon product from 5-vinyl-OZT to nitriles (Tookey and Wolff, 1970; Tookey, 1973a) (Fig. 3). This shift from oxazolidine-2-thiones to nitriles is not accounted for by a change in pH as noted earlier for *B. hirta* thioglucosidase but represents a direct involvement of ferrous ion in this enzymatic reaction (Tookey, 1973b). The ratio of nitrile to oxazolidine-2-thione is not proportional to acidity in the presence of Fe^{2+}: no oxazolidine-2-thione appears between pH 4.6 and 6.0 in a crude crambe meal extract (Fig. 4).

Autolysis of fresh crambe meal leads to the formation of both nitrile and epithionitriles (Daxenbichler *et al.*, 1968; Carlson *et al.*, 1970). This formation of epithionitriles in unsaturated aglucons is an added factor superimposed on the induced formation of nitrile products. Tookey (1973b) isolated an epithiospecifier protein (ESP) of 30,000–40,000 daltons that directs the inclusion of sulfur into a terminal unsaturation under conditions that would be expected to yield 1-cyano-3-hydroxy-3-butene (Fig. 5).

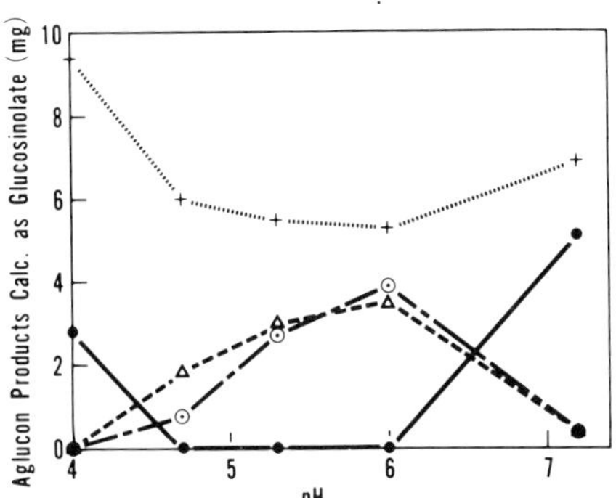

FIG. 4. Aglucon products from 2-hydroxy-3-butenyl-GS versus pH of hydrolysis. Crambe meal extract prepared in 0.2 *M* NaCl, 1.2×10^{-3} *M* $Fe(NH_4)_2(SO_4)_2$, and 10^{-3} *M* dithiothreitol and incubated with glucosinolate at 25°C in a nitrogen atmosphere. The pH was initially adjusted with NaOH or HCl (●——●, 5-vinyl-OZT; +---+, 1-cyano-2-hydroxy-3-butene; ○---○, *erythro*-1-cyano-2-hydroxy-3,4-epithiobutane; △---△, *threo*-1-cyano-2-hydroxy-3,4-epithiobutane). Reproduced by permission of the National Research Council of Canada from Tookey (1973b).

Both Fe^{2+} and ESP are required for the formation of epithionitriles (Scheme V). If 2-hydroxy-3-butenyl-GS and purified thioglucosidase are incubated at pH 5.8 with or without ESP, 5-vinyl-OZT is the principal aglucon product; only if both ESP and Fe^{2+} are present do the epithionitriles appear (Tookey, 1973b; H. L. Tookey, unpublished).

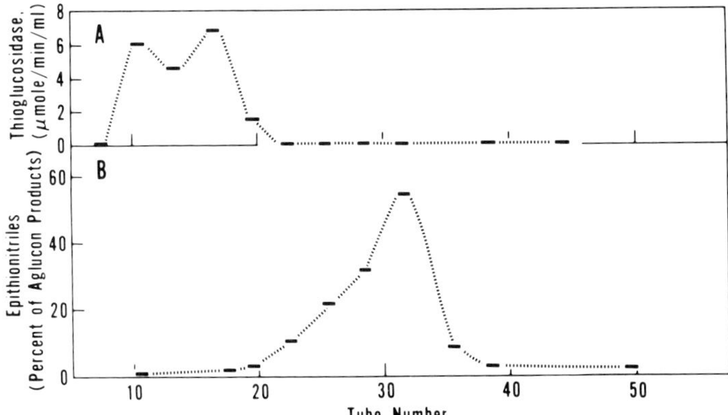

FIG. 5. Elution of an ammonium sulfate fraction of crambe meal extract from a Sephadex G-100 column: (A) thioglucosidase activity eluting from column, (B) percentage of 1-cyano-2-hydroxy-3,4-epithiobutanes from 2-hydroxy-3-butenyl-GS following incubation of eluate fraction with added thioglucosidase: tubes 21–38 contain ESP. Concentration of Fe^{2+} is 2.4×10^{-3} *M*; dithiothreitol is 6.5×10^{-4} *M*; pH is 5.8. Reproduced by permission of the National Research Council of Canada from Tookey (1973b).

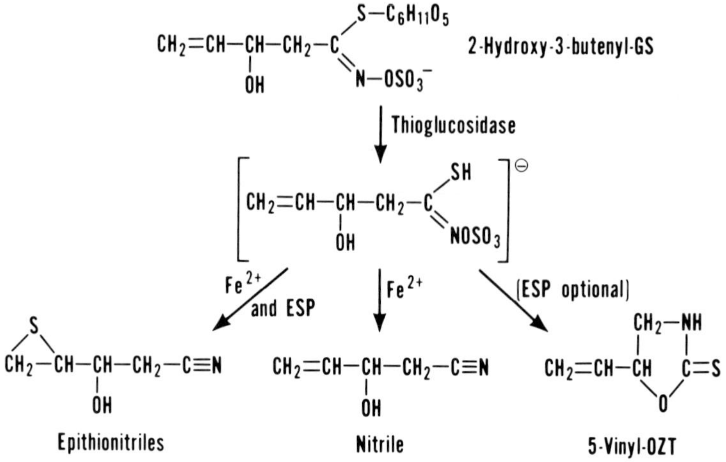

$CH_2{=}CH{-}CH(OH){-}CH_2{-}C(S{-}C_6H_{11}O_5){=}N{-}OSO_3^-$ 2-Hydroxy-3-butenyl-GS

↓ Thioglucosidase

$[CH_2{=}CH{-}CH(OH){-}CH_2{-}C(SH){=}NOSO_3]^{\ominus}$

Fe^{2+} and ESP ↙ | ↓ Fe^{2+} | ↘ (ESP optional)

$\overset{S}{CH_2{-}CH}{-}CH(OH){-}CH_2{-}C{\equiv}N$ | $CH_2{=}CH{-}CH(OH){-}CH_2{-}C{\equiv}N$ | $CH_2{=}CH{-}CH{-}O{-}C({=}S){-}NH{-}CH_2$ (ring)

Epithionitriles | Nitrile | 5-Vinyl-OZT

SCHEME V

This mode of formation of nitriles and epithionitriles may be of broad significance. Daxenbichler *et al.* (1967) demonstrated epithionitrile formation from progoitrin of autolyzed rapeseed (*B. napus*) meal in proportions similar to those of crambe. This result was confirmed at pH 7 in autolyzed *B. napus* seed by Srivastava and Hill (1974). Kirk and MacDonald (1974) autolyzed *B. campestris* meal containing 3-butenyl-GS and found 1-cyano-3,4-epithiobutanes. Their results demonstrate that a hydroxyl group is not required for epithionitrile formation. Several of these compounds were recently identified from autolyzates of fresh cabbage (Daxenbichler *et al.*, 1977) and from autolyzed leaves of both *C. abyssinica* and *B. napus* (VanEtten and Daxenbichler, 1971). Epithionitriles have also been noted in *Nasturtium* (MacLeod and Islam, 1975) and in *Alyssum, Cakile, Cardamine, Hirchfeldia, Lobularia,* and *Sisymbrium* (Cole, 1976).

Thus far, some explanation has been offered for the formation of isothicyanates (Scheme I) and their cyclization to oxazolidine-2-thiones (Scheme II). Formation of nitriles, and in some cases epithionitriles (Scheme III), has been discussed, but the formation of thiocyanates (Scheme I) has not yet been satisfactorily explained.

Virtanen and co-workers described the hydrolysis of benzyl-GS (glucotropaeolin) in *Lepidium* seed meal as yielding benzyl-SCN as well as benzyl-NCS and benzyl-CN (Gmelin and Virtanen, 1959). An isomerase converting R-NCS to R-SCN was postulated (Virtanen and Saarivirta, 1962), but could not be confirmed (Virtanen, 1965; Schluter and Gmelin, 1972). Saarivirta (1973) concluded that R-SCN arises directly from the glucosinolate. He pointed out interesting parallels between R-SCN formation in *Lepidium* and epithionitrile formation in crambe and *B. napus*. Similarly, R-SCN is formed in crushed seeds of *Lepidium* and *Thlaspi* (Gmelin and Virtanen, 1959). Miller (1965) studied *T. arvense* seed and found that allyl-SCN was favored in freshly ground seed by the addition of L-ascorbic acid and by low temperatures. This indicated that the factor(s) leading to R-SCN were labile and sensitive to oxidation—again a striking parallel to the formation of epithionitriles in crambe previously described. Cole (1976) found R-SCN in *Lepidium* and *Alyssum* by mass spectral analysis.

Glucobrassicin is a known source of thiocyanate ion. The glucosinolate contains an indole structure that, on hydrolysis through the postulated isothiocyanate intermediate at pH 7.0, forms SCN^- and other products (Scheme IV). At pH 3–4 3-indolylacetonitrile and sulfur are formed (Gmelin and Virtanen, 1961). Neoglucobrassicin [3-(*N*-methoxy)indolylmethyl-GS] from turnip yields hydrolytic products analogous to those of glucobrassicin (Gmelin and Virtanen, 1962).

p-Hydroxybenzyl-GS (sinalbin) from white mustard seed is also a source of thiocyanate ion (Ettlinger and Thompson, 1962). *p*-Hydroxybenzyl-NCS formed by thioglucosidase hydrolysis decomposes to *p*-hydroxybenzyl alcohol and thiocyanate ion or to di-(*p*-hydroxybenzyl) disulfide (Kawakishi and Muramatsu, 1966).

E. Analytical Methods

1. *Total Glucosinolates*

If the total glucosinolate content of a natural product is desired, estimation of either the sulfate ion or the glucose released by thioglucosidase may be used. Insofar as is known, sulfate and glucose formation are quantitative and independent of the conditions under which the glucosinolate is hydrolyzed by thioglucosidase. Such methods have been published by Sandberg and Holly (1932), Schultz and Gmelin (1954), Nagashima and Uchiyama (1959a), VanEtten *et al.* (1965), and McGhee *et al.* (1965).

More recently, micro procedures have been developed in which the amount of glucose from glucosinolates is determined by glucose oxidase (Lein, 1970; Bjorkman, 1972). Lein's method can be applied to a sample as small as a half-seed and so is useful in plant breeding. Lein (1972) modified the procedure to include estimation of isothiocyanates and oxazolidine-2-thiones on half-seeds. Similar procedures for larger samples of seed were developed by VanEtten *et al.* (1974) and by McGregor and Downey (1975). The latter procedures depend on charcoal for the removal of interfering substances and are not applicable to vegetable samples containing much free glucose. Free glucose can be separated from glucosinolates by retention of the glucosinolates on an anion-exchange column. While retained on the resin, the glucosinolates are enzymatically hydrolyzed in aqueous buffer with added methylene chloride (VanEtten and Daxenbichler, 1977). Separation of aqueous and methylene chloride phases allows simultaneous measurement of glucose and aglucon products.

Total glucosinolates may also be measured by titration of HSO_4^- released by thioglucosidase hydrolysis. This method is used for assay of the enzyme (Tookey and Wolff, 1970; Tookey, 1973a) but has restricted value in the analysis of glucosinolates because of buffers that are commonly present in mixtures separated from plants.

2. *Aglucon Products of Glucosinolate Hydrolysis*

Early methods were designed to measure steam-volatile isothiocyanates only. Because pH control during hydrolysis was lacking or the pH was unsuitable, the products included nitriles, but only isothiocyanates were measured. The methods for product measurement are still valid, but results cannot be interpreted in terms of all glucosinolates present. This deficiency was caused by the limited knowledge of the variable products that could be formed enzymatically under uncontrolled conditions. Volatile isothiocyanates may be distilled into ammonia and the resulting thioureas measured argentimetrically (Wetter, 1955). Kjaer and Rubinstein (1953) identified the thioureas by paper chromatography; Appelqvist and Josefsson (1967) estimated the thiourea derivatives by ultraviolet (uv) absorption. Nonvolatile isothiocyanates were also measured by extraction of the aglucon products into ether before conversion to thioureas (Ettlinger and Thompson, 1962). Astwood *et al.* (1949) and Kreula and Kiesvaara (1959) described a procedure for determining oxazolidine-2-thiones in which the compounds are extracted from the autolyzate with ether and measured by their uv absorption. Wetter (1957) also utilized uv measurement of oxazolidine-2-thione on the same sample used for isothiocyanate determination.

Enzymatic hydrolysis at pH 6–7 by added thioglucosidase derived from white mustard (*B. hirta*) can be expected to yield only isothiocyanates or oxazolidine-2-thiones from complete hydrolysis of glucosinolates. However, care should be taken to destroy the endogenous thioglucosidase system by using a hot-water extract of plant material for the analytical sample. Methods using these conditions have been used successfully by numerous workers (Schwimmer, 1961; Appelqvist and Josefsson, 1967; VanEtten *et al.*, 1965; Daxenbichler *et al.*, 1966; Youngs and Wetter, 1967; Wetter and Youngs, 1976).

The extent to which hydrolysis of glucosinolates in plant species other than *Brassica* or *Crambe* as affected by conditions during autolysis of the plant material has been studied in only a few plants. Any research requiring accurate analysis for a glucosinolate or a specific hydrolytic product must be preceded by assurance that the hydrolytic product measured is quantitatively formed from the glucosinolate.

Gas–liquid chromatography (GLC) was used to estimate volatile isothiocyanates in rapeseed by Youngs and Wetter (1967) and allyl-NCS in mustard seed by Anderson (1970). A GLC method for estimating nitriles, epithionitriles, and oxazolidine-2-thiones in crambe and rapeseed was developed by Daxenbichler *et al.* (1970) and later modified to determine these compounds at the parts per million level in body tissues of animals (VanEtten *et al.*, 1977). Further modifications provided a method of estimating each of the ten isothiocyanates and goitrin found in cabbage (Daxenbichler and VanEtten, 1977). The method also gives the amount of nitriles if formed instead of isothiocyanates and oxazolidine-2-thiones. The latter methods measure the aglucon hydrolytic products regardless of the conditions of hydrolysis of the glucosinolates and are thus useful in determining the fate of glucosinolates during processing of food or feed. A method for measuring nitriles in the infrared may still be useful as an adjunct or where GLC is not available (Daxenbichler *et al.*, 1966).

Organic thiocyanates have been extracted into ether and then measured by the color developed when ferric nitrate is added (Saarivirta and Virtanen, 1963). The method is based on the reaction of SCN^- to form a colored ferric complex. Alternatively, the organic thiocyanates may be extracted into chloroform and then measured at the thiocyanate band at 2160 cm^{-1} in the infrared (Schluter and Gmelin, 1972).

Thiocyanate ion must be measured by a separate procedure. Michajlovskij and Langer (1958, 1959) reported the SCN^- content of vegetables from *Brassica*. These authors used the reaction of thiocyanate ion with bromine to give cyanogen bromide, which forms a measurable color on reaction with benzidine in dilute pyridine. Later investigators have adapted the well-known reaction of SCN^- in acid solution with ferric iron to give a measurable color. This latter method is simple, and the use of benzidine, which is a carcinogen, is not required. With such procedures, Chong and Bible (1975) determined the SCN^- content of radish root and leaves, Johnston and Jones (1966) of kale leaves, Josefsson (1968) and Ettlinger and Thompson (1962) of white mustard seed, and VanEtten *et al.* (1976) of cabbage. The thiocyanate ion in white mustard seed is used as an estimate of the glucosinolate because the *p*-hydroxybenzyl-NCS formed by enzymatic hydrolysis decomposes in an alkaline medium to give SCN^- as one of the products. The indolyl-containing glucosinolates are seldom found in crucifer seeds. Indolyl glucosinolates were, however, reported in seed of *Isatis* (Elliot and Stowe, 1971). In other parts of the plant, if enzymatic hydrolysis occurs under conditions in which isothiocyanates are formed instead of nitriles, the amount of SCN^- that comes from decomposition of 3-indolylmethyl-NCS and 3-(*N*-methoxy)indolylmethyl-NCS is a measure of these two glucosinolates. No detailed study of the source of SCN^- has been made to make certain it derives from only the indolyl glucosinolates or that they always quantitatively form SCN^-. Interference due to color is a problem with some vegetables.

3. *Intact Glucosinolates*

GLC methods for determining trimethylsilyl derivatives of intact glucosinolates have recently been reported (Underhill and Kirkland, 1971; Thies, 1976).

F. Biosynthesis

The biosynthetic route to glucosinolates has been established for a number of glucosinolates, but details of some of the steps are still missing. The starting compounds are either common amino acids or "chain-elongated" amino acids. Because a detailed discussion of biosynthesis is outside the scope of this chapter, only one example of the proposed biosynthetic scheme is outlined here (Scheme VI).

Methionine (A) undergoes elongation by two carbon atoms to 2-amino-6-(methylthio)caproic acid (B). The amino nitrogen is oxidized stepwise to the aldoxime (C). Reaction of an activated intermediate with an unknown sulfur carrier leads to (D), the thiohydroximic acid. Uridine diphosphoglucose acts as a supplier of glucose and 3′-phosphoadenosine-5′-sulfatophosphate supplies the sulfate to form a glucosinolate (F).

The methylthio group may be retained in the glucosinolate or eliminated as shown. An extensive discussion of glucosinolate biosynthesis can be found in several reviews (Ettlinger and Kjaer, 1968; Kjaer and Larsen, 1973; Underhill *et al.*, 1973).

IV. TOXIC PRODUCTS FROM GLUCOSINOLATES

A. Goitrin and Other Oxazolidinethiones

Goitrin from rapeseed [(*S*)-5-vinyl-OZT] was shown to be thyroid-suppressing by measurement of reduced uptake of radioactive iodine (Astwood *et al.*, 1949) or by thyroid hyperplasia (Carroll, 1949) after feeding (*S*)-5-vinyl-OZT to rats. Greer (1962b) showed that racemic 5-vinyl-OZT has antithyroid activity equal to that of (*S*)-5-vinyl-OZT; hence, both (*R*) and (*S*) isomers are equivalent. They

$$CH_3{-}S{-}CH_2{-}CH_2{-}\underset{\displaystyle NH_2}{\underset{|}{CH}}{-}COOH \;\longrightarrow\; CH_3{-}S{-}(CH_2)_4{-}\underset{\displaystyle NH_2}{\underset{|}{CH}}{-}COOH$$

(A) (B)

$$CH_3{-}S{-}(CH_2)_4{-}C\begin{matrix}\diagup S{-}X\\ \diagdown\!\!\diagdown N{-}OH\end{matrix} \;\longleftarrow\; CH_3{-}S{-}(CH_2)_4{-}CH{=}N{-}OH$$

(D) (C)

$$CH_3{-}S{-}(CH_2)_4{-}C\begin{matrix}\diagup S{-}C_6H_{11}O_5\\ \diagdown\!\!\diagdown N{-}OH\end{matrix} \;\longrightarrow\; CH_2{=}CH{-}CH_2{-}CH_2{-}C\begin{matrix}\diagup S{-}C_6H_{11}O_5\\ \diagdown\!\!\diagdown N{-}OSO_3^-\end{matrix}$$

(E) (F)

SCHEME VI

have only 2% the activity of propylthiouracil (an antithyroid drug) when tested in rats, but 133% of propylthiouracil activity when tested in man (Greer, 1962). Racemic 5-vinyl-OZT fed to chicks at 0.15% of the ration caused depression of growth rate, hyperplasia, and hypertrophy of the thyroid. Uptake of radioiodine by the thyroid was depressed during initial feeding, but after longer feeding (25 days) thyroid function returned to normal (Clandinin *et al.*, 1966). The authors concluded that the chicken eventually reached physiological equilibrium at an increased ratio of thyroid to body weight. (*R*)-5-Vinyl-OZT from crambe fed to rats for 90 days as 0.23% of the ration caused mild hyperplasia of the thyroid and reduced body weights to 85% of the controls (VanEtten *et al.*, 1969c). Doses of 186 mg/kg (*R*)- and (*S*)-5-vinyl-OZT are not teratogenic in rats (Khera, 1977).

Oxazolidine-2-thiones other than those formed from progoitrin and *epi*-progoitrin are also goitrogenic. The 5,5-dimethyl-OZT isolated from hare's ear mustard, *Conringia orientalis* (Hopkins, 1938; Kjaer *et al.*, 1956), has about the same antithyroid activity as goitrin (Astwood *et al.*, 1945). Barbarin [(−)-5-phenyl-OZT] was isolated by Kjaer and Gmelin (1957) from various species of *Barbarea* and *Reseda luteola*. According to Greer and Whallon (1961), 5-phenyl-OZT has about one-half the antithyroid activity of goitrin. 5-Allyl-OZT occurs in turnip greens (Tapper and MacGibbon, 1967) and Chinese cabbage (Daxenbichler *et al.*, 1979). Other oxazolidine-2-thiones have been isolated and identified from the *Cleome* genus in the family Capparidaceae (Kjaer and Thomsen, 1962) and two from *Sisymbrium austriacum* (Kjaer and Christensen, 1959, 1962).

Oxazolidine-2-thiones presumably act by inhibiting the incorporation of iodine into precursors of thyroxine and interfering with secretion of thyroxine (Matsumoto *et al.*, 1968; Akiba and Matsumoto, 1973, 1976). In contrast to the activity of thiocyanate ion, the antithyroid effect of oxazolidine-2-thiones is not overcome by larger amounts of iodine in the diet (Greer *et al.*, 1964).

B. Thiocyanate Ion and Isothiocyanates

Thiocyanate ion appears in foods and feeds as a result of decomposition of isothiocyanates, e.g., from the indolyl glucosinolates and *p*-hydroxybenzyl-GS (see Section III,B). Barker (1936), in giving human patients SCN^- to treat hypertension, discovered that this ion causes thyroid enlargement. Because thiocyanate ion inhibits uptake of iodine by the thyroid (VanderLaan and VanderLaan, 1947), the antithyroid effect is most likely to be seen when the diet is low in iodine (Astwood, 1943). High levels of SCN^- inhibit iodine uptake even when the dietary iodine is adequate (Greer *et al.*, 1966).

Michajlovskij and Langer (1958) analyzed cabbage and related edible plants and found SCN^- ranging from 0.7 to 10.2 mg per 100 gm fresh material. Ingestion of these *Brassica* vegetables caused a rise in SCN^- in the blood followed by its appearance in the urine of animals and man (Michajlovskij and Langer, 1958).

A single feeding of cabbage to guinea pigs or rats maintained on a low-iodine diet inhibited iodine uptake by the thyroid more than expected from the SCN^- found in the cabbage (Langer, 1961; Langer and Stolc, 1964). This result indicated the presence of additional goitrogens in the cabbage. A mixture of SCN^-, allyl-NCS, and 5-vinyl-OZT fed to rats at levels equivalent to those in cabbage fed *ad libitum* led to inhibition of iodine uptake by the thyroid and histological changes in the thyroid similar to results expected from a cabbage diet (Langer, 1966).

Conversion of allyl-NCS and other isothiocyanates to SCN^- is part of a metabolic detoxification of isothiocyanate (Greer, 1950; Wood, 1975). Hence, the goitrogenic effect of these compounds may be accounted for by conversion to SCN^-. Doses of 2–4 mg allyl-NCS force fed to rats inhibited thyroid uptake of radioactive iodine and increased plasma SCN^- concentration (Langer, 1964). After 60 daily doses (equal to eating 40 gm cabbage daily), the rats' thyroids showed histological changes indicative of goiter (Langer and Stolc, 1965). Iodine uptake by the rat thyroid is also inhibited by 3-methylsulfonylpropyl-NCS (Bachelard and Trikojus, 1960) from a pasture weed of Australia, *Rapistrum rugosum*. These authors found that 3-methylsulfonylpropyl-NCS was converted by rumen fluid to the disubstituted thiourea, which was weakly goitrogenic.

Since isothiocyanates are irritating and act as vesicants at higher concentrations, it would be surprising if amounts large enough to cause malfunction of the thyroid could be eaten. However, if they were consumed as their precursor glucosinolates and were later slowly released in the intestinal tract, they might act as goitrogens, especially if the diet contained iodine below the optimal requirement.

C. Nitriles

The formation of toxic nitriles from glucosinolates has been more thoroughly studied in seed meal of *Crambe abyssinica* than elsewhere. Numerous workers, however, have reported toxic substances other than known goitrogens in crucifer seed meals (Bell, 1955, 1965; Bowland *et al.*, 1963; Holmes, 1965; Hesketh *et al.*, 1963).

Feeding of several enzyme-modified crambe meals to rats demonstrated that meals containing mixed nitriles were lethal to rats at 28% of the ration, whereas meals from which these nitriles were removed were essentially nontoxic. These meals supported growth at 88% of the controls. All body organs were histologically normal and of normal size except for possible thyroid enlargement (Tookey *et al.*, 1965). The nitriles were removed from the meals by extraction with aqueous acetone.

Mixed nitriles in such aqueous acetone extracts contained no goitrin, 43–65% nitrile, calculated as 1-cyano-2-hydroxy-3-butene, and 16–18% sulfur (VanEtten *et al.*, 1966). Later work (Daxenbichler *et al.*, 1968) showed that the sulfur must be present as 1-cyano-2-hydroxy-3,4-epithiobutane, corresponding to 57–65% of the 1-cyano-2-hydroxy-3,4-epithiobutanes and 0–15% 1-cyano-2-hydroxy-3-butene by weight. This agrees well with the expected products found in autolyzed crambe meal (Daxenbichler *et al.*, 1970; Tookey, 1973b). Rats fed a mixed-nitriles fraction at 0.2% of the ration died within 14 days. Rats fed at the 0.1% level for 106 days grew at only 17% of the weight gain of the controls and had major lesions in the liver and kidneys, but no evidence of tumor formation was reported (VanEtten *et al.*, 1969c). The livers exhibited bile duct hyperplasia and necrotic areas; the kidneys had megalocytosis of the tubular epithelium.

The above findings, indicating chronic toxicity of these mixed nitriles, are consistent with the acute toxicities of these compounds compared to 5-vinyl-OZT (VanEtten *et al.*, 1969b) as shown in the tabulation below:

Aglucon product	LD_{50} (mg/kg)
(*R*)-5-Vinyl-OZT	1260–1415
1-Cyano-2-hydroxy-3-butene	170
1-Cyano-2-hydroxy-3,4-epithiobutane (erythro)	178
1-Cyano-2-hydroxy-3,4-epithiobutane (threo)	240

Similar evidence for toxic nitriles from rapeseed continues to accumulate. Lo and Hill (1972a,b) suggested that the greater toxicity of raw seed meal (*B. napus*) compared to heated seed meal could be accounted for by the greater production of 1-cyano-2-hydroxy-3-butene and epithionitriles in the raw meal. *Brassica napus* seed meal was prepared to contain high levels of mixed nitriles or to contain 5-vinyl-OZT and was then feed to rats and chicks. Weight gains were inversely related to the amount of nitrile-rich meal in the ration (Srivastava *et al.*, 1975) in both species. Enlarged, pale kidneys were observed in the rats, but histologic studies were not done. Josefsson (1975b) showed that the toxic substances present in an autolyzed Bronowski (*B. napus*) seed meal could be removed by dialysis or by methylene chloride extraction and that the extract contained the mixed nitriles. Heat treatment before or after autolysis provided evidence that products formed during autolysis were responsible for the deleterious effects and that these products were the mixed nitriles. Furthermore, a meal freed of glucosinolates by cold aqueous acetone extraction was satisfactory as a sole protein source for mice, but if these glucosinolates were added back to the enzyme-active meal, the growth response was as poor as with the original meal (Josefsson and Uppström, 1976a). These authors concluded that protein factors in raw meal caused hydrolysis of glucosinolates to the more toxic nitriles rather than to isothiocyanates or 5-vinyl-OZT.

According to Greer (1950), previous work with rabbits indicated that acetonitrile produced thyroid hyperplasia that was overcome by iodine administration, but no later work connects nitriles to thyroid enlargement as their primary effect.

An organic polysulfide fraction from cabbage containing 1,2-dithiocyclopent-4-en-3-thione caused histological changes in the thyroid of rats (Jirousek, 1956; Jirousek and Starka, 1958). The relation of this polysulfide fraction to glucosinolates remains unclear.

D. Intact Glucosinolates

(*R*)-2-Hydroxy-3-butenyl-GS given to rats and man inhibits the uptake of iodine by the thyroid (Greer and Deeney, 1959). Following the feeding of the glucosinolate, 5-vinyl-OZT appeared in blood serum and urine. Greer (1962b) demonstrated that thioglucosidase is produced by bacteria in the intestinal contents of rats and man. Hydrolysis by the microflora is much slower than by plant thioglucosidase. In poultry, thioglucosidase activity of the intestinal flora increases when thioglucosidase-free rapeseed meal is fed (Marangos and Hill, 1974). Glucosinolates in such heat-treated meal are goitrogenic. The intestinal flora of rats can hydrolyze glucosinolates to a variety of aglucon products. VanEtten *et al.* (1969c) fed rats a ration containing 0.85% *epi*-progoitrin and observed pathology indicative of both 5-vinyl-OZT and mixed nitriles. Coprophagy by rats probably adds to the deleterious effects of glucosinolates fed to rats because these rodents reconsume some of the aglucon products produced in the gut.

V. RELATION OF GLUCOSINOLATES TO HUMAN GOITER

A. Direct Consumption of Crucifers

The literature contains many reports on endemic goiter and suggestions that the condition may be caused by eating cruciferous plants. However, goiter in human beings caused by eating these plants has not been proved. Goiter has been attributed to the consumption of large amounts of cabbage (Kelly and Snedden, 1960) or of kale (Michajlovskij *et al.*, 1969a). To keep a proper perspective, 96% of all human goiter (hypothyroid) is probably caused by iodine deficiency (Greer, 1962b). The only goiters related to glucosinolates are those associated with hypothyroidism.

Definite evidence of a goitrogen in food was first discovered by Chesney *et al.* (1928), who reported greatly enlarge thyroids in rabbits that were fed cabbage as a major part of their diet. Not all workers in other laboratories were able to confirm this report, including Hercus and Purves (1936). Unsuccessful with cabbage leaves, they fed rats cabbage and related *Brassica* seeds, which readily caused enlarged thyroids because of the high glucosinolate content of seeds relative to crucifer vegetables. Of many foods that were tested by measurement of thyroid uptake of radioactive iodine, turnip and rutabaga gave results consistent with goitrogenicity (Greer, 1962).

Given the present state of knowledge, it seems plausible that in regions of the world where iodine content is low, benign goiter (iodine deficiency) may be accentuated by the ingestion of excessive amounts of *Brassica* vegetables. Even this has not been proved to be a public health problem.

B. Glucosinolate Aglucon Products in Milk

The leaves of *Brassica* forage plants yield thiocyanate ion (Michajlovskij and Langer, 1958) and oxazolidine-2-thiones (Kreula and Kiesvaara, 1959) in small amounts. The plants have been eaten extensively as a forage by livestock, yet there are few reports that such practice causes goiter in animals. Examples are given by Wright and Sinclair (1959), Clements and Wishart (1956), and Peltola (1965). Severe poisoning of sheep and cattle may occur if the animals are allowed access to the seed or are pastured on the plant after the seed is formed (Holmes, 1965). Cattle are also often grazed on pastures consisting in part of wild plants that contain glucosinolates (Bachelard and Trikojus, 1960).

Transmittal of goitrogens to man by cow's milk was investigated in detail by Clements and Wishart (1956). They cited evidence that endemic goiter in the population of parts of Tasmania was due to goitrogens transmitted through the milk of cows fed large amounts of marrow-stem kale. They believed that the goitrogens were 5-vinyl-OZT and SCN^- or possibly the isothiocyanate from 3-methylsulfonylpropyl-GS (glucocheirolin), but no such compounds were isolated. There was also evidence of the suspected milk-inhibited thyroid uptake of radioactive iodine following ingestion of test samples. Similar experiments in England by Greene *et al.* (1958) and in Holland by Van der Veen and Hart (1955) did not confirm the results of the Australians.

Virtanen *et al.* (1958, 1963) estimated the amount of goitrin transferred to cow's milk after feeding moistened rapeseed, rape forage, 2-hydroxy-3-butenyl-GS, and 5-vinyl-OZT. These studies did not take other aglucon products into account. Thus, measurement of 5-vinyl-OZT only might underestimate the goitrogenicity of milk.

In attempts to confirm the work of Clements and Wishart (1956) directly, Vilkki *et al.* (1962) tested for the accumulation of radioiodine by the thyroids of men who had drunk large amounts of milk from cows on a very high ration of *Brassica* forage. Radioiodine uptake by the thyroid was not depressed. Virtanen *et al.* (1963) reported that the method of estimating radioiodine uptake by the thyroids in human beings was very sensitive to fluctuations in iodine supply and that test samples high in salt content caused inhibition of radioiodine uptake not correlated with known goitrogens. False readings caused by the ingestion of iodine in the test meal could occur (Greer and Astwood, 1948). These observations are offered as possible explanations of the positive radioiodine tests for goitrogens in milk found by Clements and Wishart (1956).

However, from the work of Peltola (1965) it appears that the radioiodine test may not be satisfactory for detecting the cause of endemic goiter if it is caused by a goitrogen. He reported that the milk from cattle in an area in Finland where goiter was endemic contained 50–100 μg/liter of 5-vinyl-OZT. He demonstrated that the thyroids of rats increased in weight significantly as the amount of 5-vinyl-OZT fed was increased over the range 0.0–2.0 μg/day. With the rat as test animal, Peltola found that at least 100 μg of 5-vinyl-OZT in a single dose was required to depress radioiodine uptake by the thyroid. From his experiments it is apparent that the radioiodine test applied to rats cannot be used as a criterion for endemic goiter, which may be caused by ingestion of small amounts of 5-vinyl-OZT over long periods of time. He also reported that rats fed the milk containing 50–100 μg/liter of 5-vinyl-OZT developed enlarged thyroids. His work supports the implications of Clements and Wishart (1956).

In studies on the transfer of thiocyanate ion to milk, Piironen and Virtanen (1963) found that milk from cows not fed cruciferous plants contained about 2 mg SCN^- per liter but that milk from those fed 30 kg of marrow kale or swedes per day contained 5–8 mg/liter. These studies also showed that feeding SCN^- equal to the amount in 30–60 kg of cruciferous forage produced levels of SCN^- in milk as high as 17.5 mg/liter. Vilkki *et al.* (1962) and Stanley and Astwood (1947) reported that a dose of 200–1000 mg of SCN^- ion is required to inhibit radioiodine uptake in the thyroid of man. They concluded that not enough SCN^- ion could be transferred to the milk to cause it to be goitrogenic.

In the course of their investigations, Piironen and Virtanen (1963) measured the amount of iodine in the milk during the feeding of SCN^- and after the SCN^- was no longer fed. They found that the iodine content of the milk was lowered by the thiocyanate feeding but rose to high levels as soon as the thiocyanate feeding was stopped. Apparently, SCN^- feeding inhibits iodine absorption by the mammary gland as well as by the thyroid. Such an effect on the mammary gland conserves the iodine for the lactating animal but lowers the iodine content of the milk for her young. Such an effect has been cited by others (Garner *et al.*, 1960; Miller *et al.*, 1965). It appears from this work that a lactating mammal ingesting SCN^- could possibly cause goiter in her young or in those drinking large amounts of her milk.

VI. FEEDING STUDIES OF SEED MEALS

A. Rodents

Goering *et al.* (1960) reported that thioglucosidase hydrolysis of allyl-GS in mustard seed meal (*Brassica juncea*), followed by removal of the allyl-NCS by volatization with steam, gave a satisfactory feed meal. Their conclusion was based on rat feeding with the mustard seed meal as the only protein source. Mustakas *et al.* (1965) and Mustakas and Kirk (1963) described a procedure for processing mustard seed in which they recovered the allyl-NCS. They reported that the processed mustard seed meal fed to rats as 20% of the ration gave optimal growth. As 30% of the ration, this meal afforded less than optimal growth. Histological examinations of the thyroids showed no abnormalities.

Early studies on rodents fed rapeseed meals are difficult to interpret because the species of rape and the meal processing were often not reported. Kennedy and Purves (1941) fed rats on a ration containing 45% rapeseed for up to 224 days. They noted a marked increase in the thyroid weight during the first 3 weeks, after which thyroid and rat weight increased at the same relative rate. Growth was depressed, hypertrophy of the adrenal cortex was noted, ovary development in the young females was delayed, and histological changes in the pituitary were observed. Daily doses of 1.3 mg of potassium iodide decreased thyroid enlargement, but hypertrophy and hyperplasia were still present.

Bell and Williams (1953), Bell (1957a,b), Bell and Baker (1957), and Belzile *et al.* (1963) discussed the effects of feeding rapeseed to mice. Expeller-processed rapeseed meal caused growth depression when the meal was fed as 18% of the ration. The residue meals after hot-water or aqueous alcohol extraction consistently gave marked improvement in weight gain. However, rapeseed meal that had been steam-cooked up to 20 min at 15 psi or subjected to acid hydrolysis gave little or no improvement. Additions of flavoring agents, iodinated casein, potassium iodide, aureomycin, skim milk powder, fish meal, and lysine were not beneficial. Feeding of *B. napus* and *B. campestris* types gave similar growth responses.

Manns and Bowland (1963) and Manns *et al.* (1963) fed commercial solvent-extracted rapeseed meal, which assayed 4.9–6.4 mg 5-vinyl-OZT per gram, to rats. Those fed meal at 15.6% of the ration gained only 75% as fast as the controls. Supplementation with lysine did not increase growth. Increase in thyroid weight, decrease in standard metabolic rate, but no change in serum protein-bound iodine occurred as the amount of rapeseed meal fed was increased.

Nakaya and Nakamura (1963a,b) and Nakaya (1964) fed rapeseed meal, hot-water-extracted meal, and the hot-water extract to rats and poultry. From their results they concluded that the meal contains, in addition to growth inhibitors, water-extractable materials that cause enlargement of thyroid, liver, adrenals, and kidneys. They suggested measurement of the reduced glutathione in the liver to estimate the effect on rats of feeding rapeseed meal, because the reduced glutathione increased as the amount of rapeseed fed was increased.

In order to eliminate toxic components from rapeseed, recent work has been directed toward feeding varieties of rape that have been selected by plant breeding to be low in glucosinolates or toward preparing protein fractions from the seed meals.

Seed meal from Bronowski cultivar of *B. napus* contains only about 0.5% glucosinolate, compared to 5–6% in most other cultivars (Josefsson and Appelqvist, 1968) and appears to be a better feed than the high-glucosinolate cultivars. Josefsson and co-workers demonstrated that some of the remaining deleterious effects of Bronowski meal are still caused by glucosinolates if an enzyme system that produces nitriles is present (Joseffson, 1975a,b; Joseffson and Uppström, 1976a,b) (see Section IV,C). Srivastava and Hill (1976) fed low-glucosinolate rapeseed meal to mice. They could find no toxic nitriles in autolyzates containing 1.3 mg glucosinolate per gram meal and concluded that the beneficial effect of heat treatment was not related to glucosinolates.

It should be noted that sinapine, a bitter substance in white mustard and in rapeseed (Schultz and Gmelin, 1952), is not a significant toxicant. Clandinin (1961) fed chicks sinapine for 4 weeks with no ill effects. Austin *et al.* (1968) fed sinapine to rats for 7 days with no effect, and Josefsson and Uppström (1976b) compared sinapine to *p*-hydroxybenzyl-GS and showed that sinapine has only a small effect. Protein-rich fractions of excellent quality can be obtained from rapeseed meal but not always in a fashion that is economically feasible. Water-washed rapeseed press cake (dehulled) gave protein efficiency ratio (PER) and productive protein values equal to those of methionine-enriched casein (Ågren and Eklund, 1972) and yielded satisfactory results when used as the sole source of protein for rats for 90 days (Eklund *et al.*, 1974).

Sarwar *et al.* (1973) found that a glucosinolate-free rapeseed protein isolate was superior to casein for PER and weight gain of mice, and Tape *et al.* (1970) produced protein flour of good quality for feeding rats. Shah *et al.* (1976) showed that the adverse effects of *Brassica* protein concentrates on rats were caused by phytates and could be overcome by zinc. Lo and Hill (1971) prepared rapeseed protein concentrate by aqueous extraction of Bronowski (low-glucosinolate) cultivar and obtained results similar to those of Ågren and Eklund. Lo and Hill used protein materials soluble in water (after dialysis to remove small molecules), whereas Ågren and Eklund rejected water-soluble fractions.

Working with *Crambe abyssinica,* VanEtten *et al.* (1965) found that rats fed dehulled and defatted seed as 5% of the ration gained weight 90% as much as the controls. When the meal was fed as 15% of the ration, however, all the animals lost weight and died within 90 days. Aqueous acetone extraction of the meal without prior hydrolysis of the glucosinolates gave a meal containing only traces of glucosinolates, isothiocyanates, or oxazolidine-2-thiones. In a short-term experiment in which this extracted meal was fed to rats as 23% of the ration, weight gain was 94% that of the controls (compare Section IV).

Many modified procedures have been proposed for the processing of crucifer oilseeds to give a high-quality oil and feed meal. They include treatment of the meal during processing with soda ash (Mustakas *et al.*, 1968), with ferrous iron (Kirk *et al.*, 1971), and with ammonia (Kirk *et al.*, 1966), water extraction of the defatted seed meal (Ballester *et al.*, 1973; Mustakas *et al.*, 1976), water extraction of the whole or cracked seed before extraction of the oil (Tape *et al.*, 1970; VanEtten *et al.*, 1969a; Sosulski *et al.*, 1972), and fermentation to remove glucosinolates and their aglucons (Staron, 1970). Most of these procedures gave a meal of apparently improved quality for feeding, but increased processing costs and the disposal of waste by-products have prevented their adoption by industry.

B. Poultry

Pettit *et al.* (1944) and Turner (1946) were the first to point out that the feeding of rapeseed to chickens causes enlarge thyroids and growth depression. Frolich (1952) found that Swedish rapeseed meal in rations fed to poultry caused thyroid enlargement; after water or alcohol extraction, the meal caused less thyroid enlargement.

A report by Renner *et al.* (1955), which was confirmed by Clandinin *et al.* (1959), indicated that the Argentine-type (*B. napus*) rapeseed meal is more goitrogenic to poultry than is the Polish type (*B. campestris*). Feeding increased amounts of either type gave less growth. According to Klain *et al.* (1956), addition of iodinated casein or potassium iodide to either type gave no better growth. The iodinated casein reduced the size of the thyroids, but the potassium iodide did not.

In the feeding of processed rapeseed meals at higher levels, in which adequate essential amino acids were not supplied from other protein sources, additional lysine was required for poultry. Excessive heat treatment in processing increased the lysine requirement. This complicating factor in the study of rapeseed toxic substances in poultry feeding was studied by Clandinin *et al.* (1959).

Hesketh *et al.* (1963) compared crambe seed meal with soybean meal as a source of protein for growing chicks. Rations containing 5–42% crambe meal depressed growth and enlarged thyroids, which increased with the greater amount of crambe meal fed. Kirk *et al.* (1966) reported that a crambe meal treated with ammonia under pressure gave an improved growth response when fed to chicks as 20% of the ration for 4 weeks. Weight gains were 82% of those of the controls. No abnormalities of body organs were observed. In a later paper, these workers reported a similar response of chicks fed crambe meal treated with soda ash (Mustakas *et al.*, 1968). By analysis, nearly all the glucosinolate and 5-vinyl-OZT was absent from both the ammonia- and soda ash-treated meals. The nitrile content of the meals was not reported.

Feeding of rapeseed meal to poultry has been linked to perosis, lowered egg production, off-flavors in eggs, and liver damage, as well as to poor growth response and thyroid enlargement as previously discussed. Perosis in chickens caused by feeding rapeseed meal may be partially alleviated by removing alcohol-soluble substances from the meal (Holmes and Roberts, 1963). On the other hand, Seth and Clandinin (1973) linked perosis to some interference with mineral nutrition. However, the evidence on this point is still contradictory.

Egg production is lowered by the addition of 10–15% raw rapeseed to the ration of laying hens (Leslie and Summers, 1972). A difference between breeds of hens was noted by Hawrysh *et al.* (1975): Rhode Island Red hens fed a diet containing 6.8% *B. napus* meal laid eggs with a "fishy" off-flavor, but white Plymouth Rock and Hyline White Leghorn hens laid eggs of normal flavor. Addition of a small amount of ground raw rapeseed (to provide thioglucosidase) increased the incidence of "fishiness" in eggs. Fishiness in eggs has been linked to trimethylamine. Feeding of rapeseed meals to certain breeds of hens caused trimethylamine to appear in eggs, but the source of this compound in rapeseed is still obscure (Hobson-Frohock *et al.*, 1973, 1975).

Liver hemorrhage and reticulolysis in poultry fed rapeseed meal lead to increased mortality. Jackson (1969) reported deaths from liver hemorrhage in Hyline hens fed rations of 8% or more Algerian rapeseed meal although heavy-weight hens did not have excess deaths when fed similar diets. Olomu *et al.* (1975) and Marangos *et al.* (1974) also reported liver hemorrhage associated with feeding of rapeseed meal. Rations containing 50% rapeseed meals caused extensive liver hemorrhage, but no significant difference was seen between Bronowski meal (0.49% glucosinolate) and Target meal (3.62%) (Smith and Campbell, 1976). The cause of liver hemorrhage is unknown (Hall, 1974), but by analogy to evidence shown in Section IV,C, the nitriles should be suspect.

In view of these data on the performance of laying hens, commercial rapeseed meal should be limited to 5% of the ration (Hawrysh *et al.*, 1975). In the feeding of broilers, prepress solvent-extracted rapeseed meals may be fed up to 15% of the ration (Rapeseed Association of Canada, 1972). However, these recommendations were most likely made for *B. campestris* meal, which is lower in progoitrin than the meal from *B. napus*.

C. Swine

Nordfeldt *et al.* (1954) found that feeding 10–20% rapeseed meal in the diet of growing pigs caused growth depression, hyperplasia of thyroids, and enlargement of livers and kidneys. When the meal was fed after it had been extracted with warm water, growth rate improved and the body organs were essentially normal. Supplementation of the untreated rapeseed meal with iodinated casein decreased thyroid weights, but the glands had a histological picture similar to colloid goiter. Feeding organically bound iodine had no effect on pig growth.

Manns and Bowland (1963) and Manns *et al.* (1963) observed that growing pigs responded to rapeseed meal feeding in the same way that rats responded when fed the identical rapeseed meals at the same percentage of the ration.

Protein-bound iodine in the blood serum was slightly decreased in pigs receiving rapeseed meal. Difficulty of conception of gilts occurred. Also, litter size and weight of the young at weaning were lowered. These sows on rapeseed meal, and the finishing pigs to a lesser extent, had larger livers than the controls (Bowland *et al.*, 1963).

Bell (1965) fed solvent-extracted *B. campestris* meal containing no detectable thioglucosidase to growing pigs at a starting level of 5 and 10%. In a second group, 0.022% iodinated casein was added. To a third group, a source of thioglucosidase (ground rapeseed screenings) was included. During the growth period, both levels of rapeseed meal depressed the rate of gain, which was not counteracted by the addition of iodinated casein. Addition of screenings as a source of thioglucosidase caused 26% slower gains.

Finishing rations for swine may contain up to 11% rapeseed meal (*B. campestris*) if supplemental lysine and methionine are included (Bayley *et al.*, 1969). Many of these feeding studies are difficult to interpret because the glucosinolate content of the meal was not known, and other factors such as lysine and methionine availability or mineral nutrition complicate the picture. Josefsson (1972) summarized the amounts of rapeseed that might be successfully fed to various classes of swine. Bowland and Hardin (1973) found that 6% of the ration of breeding sows could be *B. campestris* seed meal without detrimental effects occurring. Most recent work on swine feeding has centered on low-glucosinolate rapeseed meals such as Bronowski (*B. napus*) or Tower cultivars. These may be fed at higher levels than the standard, high-glucosinolate cultivars (Bell, 1975; Bowland, 1975).

D. Ruminants

Whiting (1965) concluded that commercial rapeseed meals could be fed to cattle at 10% of the total ration with no ill effects. Ruminants are less susceptible than monogastric animals to the toxic effects of rapeseed meals (Josefsson, 1972). However, less experimental work has been done on cattle than on other animal species.

Seed meal of crambe (which is similar to rape in glucosinolate content) has been fed in experimental finishing rations for beef cattle. Preliminary trials indicated that crambe could substitute for soy meal at up to 10% of the ration with satisfactory gains and carcass quality (Lambert *et al.*, 1970). In other experiments crambe meal at 8.4% was satisfactory in weight gains and feed efficiency in finishing rations (Perry *et al.*, 1979). The meals in these experiments contained 1.1–3.7% *epi*-progoitrin, 1.5–0.6% of derived nitriles, respectively, but little 5-vinyl-OZT and no enzyme.

Cattle were fed rations containing 10% crambe meal for 14–30 days in a study to determine if any of the known aglucon products from the glucosinolate appeared in meat, fat, liver, or kidney. None of these products were detected by methods sensitive to 1 ppm (VanEtten *et al.,* 1977).

The fate of glucosinolates in the rumen is not well understood. Bachelard and Trikojus (1963) showed conversion of 3-methylsufonylpropyl-NCS to the disubstituted thiourea by rumen fluid. Virtanen *et al.* (1963) indicated that progoitrin was not hydrolyzed in the rumen, but Lanzani *et al.* (1974) reported the formation of 5-vinyl-1,3-thiazolidine-2-one in small amounts instead of 5-vinyl-OZT from progoitrin when it was incubated in sheep rumen fluid. In work initiated at the Northern Regional Research Center (C. H. VanEtten *et al.,* unpublished), no 2-hydroxy-3-butenyl-GS or aglucon products were found in rumen fluid from a number of cattle that were fed processed crambe meal as 10% of the ration.

VII. OUTLOOK

In the literature before 1966, the major emphasis on possible toxic effects from glucosinolate-bearing plants centered on enlargement of the thyroid associated with SCN^-, isothiocyanates, and 5-vinyl-OZT. More recently, it has been shown that some of the nitrile aglucon products are liver or kidney toxins (Section IV,D). Glucosinolates need to be examined both as sources of goitrogens and as sources of toxicants to other body organs. The use of better analytical methods for data on feed composition makes it possible to relate the results of feeding experiments to the glucosinolate products in the ration. The effects of processing need to be evaluated because heat or other treatments given may determine which type of aglucon product is fed, whether the intact glucosinolate survives processing, or whether the products in question have been effectively detoxified.

Maximal feeding levels of commercial rapeseed meals have been established but are based only on the identity of the rapeseed and the method of processing (Rapeseed Association of Canada, 1972). Permissible feeding levels based on both thioglucosidase activity and content of glucosinolates and their hydrolytic products in the processed seed meal should be a more valid criterion.

The discovery in Canada and Sweden that the Bronowski rapeseed variety was low in total glucosinolates (Josefsson and Appelqvist, 1968; Lööf and Appelqvist, 1972) provided a better variety for animal feeding and a genetic source with which to breed plants for zero glucosinolate content. Thus, the development of better varieties may be expected in the future. Already a "double low" rapeseed (low erucic acid and low glucosinolate) has been developed (Downey, 1976; Stefansson, 1974).

Before judging possible deleterious effects from the use of cruciferous vegetables in the human diet, one should consider the results of animal feeding tests with rapeseed and related seed meals. The level of glucosinolates in these meals is usually several orders of magnitude greater than that in vegetables commonly used as human food. Condiments are, however, more nearly comparable to seed meals (Table I). Animal feeding trials with low-glucosinolate varieties of rapeseed should be evaluated carefully for signs of toxicity attributable to glucosinolates.

The exact composition of a food product coupled with animal feeding of this material will be required to establish levels of consumption at which harmful effects may appear. The method of food preparation also needs to be considered. For example, cole slaw—raw cabbage—contains an active enzyme system so that a mixture of aglucon products rich in nitriles and epithionitriles would be expected. Cooked or boiled cabbage would probably contain intact glucosinolates and no active enzyme. Since the hydrolysis in the gut is likely to be slower, as is the case for 2-hydroxy-3-butenyl-GS (Greer, 1962), it may be that a milder dose of aglucon products results from eating cooked cruciferous vegetables.

A third type of cabbage—kraut—is fermented, and its contents are still largely unknown. Daxenbichler *et al.* (1977) recently found 4–16 μg/gm of 1-cyano-3-methylsulfinylpropane in sauerkraut. Until appropriate animal experiments are run on cruciferous foods, caution should be used in postulating potential harm in the practical use in our diet of various horticultural crops from the Cruciferae. Isolation of nitriles, isothiocyanates, and 5-vinyl-OZT in sufficient amounts for animal testing is an alternate method of investigation.

Proposed regulations by the Food and Drug Administration of new varieties of vegetables that differ significantly from previous varieties in content of natural toxicants (Senti and Rizek, 1974) make it important to know exactly how the aglucon products affect animals.

More information is needed on the nature of the thioglucosidase enzyme system and its relation to glucosinolate hydrolysis in different *Brassica* species. The aglucon products formed by microbial enzymes in the digestive tract of monogastric animals have been studied somewhat, but more data are needed to define which aglucon products will be formed. In view of the apparent greater tolerance of ruminants to glucosinolates, the effect of rumen fluid on glucosinolates needs further investigation. Techniques involving the use of an artificial rumen should be useful in this area.

There have been reports on the nature of the volatile components from boiled cabbage, broccoli, and cauliflower (MacLeod and MacLeod, 1968) and from fresh watercress (MacLeod and Islam, 1975). In order to evaluate these horticultural crops as to flavor and potential hazards, both volatile and nonvolatile glucosinolate aglucon products must be studied. Identification of the flavor-producing substances should aid in developing better varieties. As the specific glucosinolate contents of various *Brassica* horticultural crops become known, those of low glucosinolate content may be used in the development of new varieties. However, disease and insect resistance of these plants may be associated with higher glucosinolate content. Glucosinolates may act as attractants to some insects or as repellents to others. Allyl-NCS was reported to be an attractant for cabbage maggots (Eckenrode and Arn, 1972) and for adult cabbage flea beetles (Hicks, 1974). On the other hand, 2-phenylethyl-NCS is toxic to several insects (Lichtenstein, 1966), and allyl-GS acts as a "barrier" against the black swallowtail butterfly (Erickson and Feeney, 1974).

Certain of the glucosinolate aglucon products may have potential as protectors against carcinogens. 3-Indolylacetonitrile from brussels sprouts increases rat liver oxidation of benzo[*a*]pyrene (Loub and Wattenberg, 1974). Low-level antibiotic activity has been attributed to certain isothiocyanates (Virtanen, 1965). The abnormal growth of *B. rapa* roots in club root disease is reported to be caused by the hydrolysis of 3-indolymethyl-GS to 3-indolylacetonitrile (Butcher *et al.*, 1974). This compound as well as its metabolite, indolylacetic acid, causes swelling of root cortical tissue.

The evaluation of the glucosinolate aglucon products for possible chronic and acute toxicity is most desirable. If some are potentially more hazardous than others, varieties may be developed that contain less of the more hazardous components. In view of the variation of glucosinolate content from plant to plant and in different parts of the plant, considerable analytical work will be required to establish a baseline for plant breeding.

References

Ågren, G., and Eklund, A. (1972). *J. Sci. Food Agric.* **23,** 1457–1462.

Ahmed, Z. F., and Rizk, A. M. (1972). *Phytochemistry* **11,** 251–256.

Akiba, Y., and Matsumoto, T. (1973). *Poult. Sci.* **52,** 562–567.

Akiba, Y., and Matsumoto, T. (1976). *Poult. Sci.* **55,** 716–719.

Anderson, D. L. (1970). *J. Assoc. Off. Anal. Chem.* **53,** 1–3.

Appelqvist, L. A., and Josefsson, E. (1967). *J. Sci. Food Agric.* **18,** 510.

Astwood, E. B. (1943). *J. Pharmacol. Exp. Ther.* **78,** 79–89.

Astwood, E. B., Bissell, A., and Hughes, A. M. (1945). *Endocrinology* **37,** 456–481.

Astwood, E. B., Greer, M. A., and Ettlinger, M. G. (1949). *J. Biol. Chem.* **181,** 121–130.

Austin, F. L., Gent, C., and Wolff, I. A. (1968). *J. Agric. Food Chem.* **16,** 752–755.

Bachelard, H. S., and Trikojus, V. M. (1960). *Nature (London)* **185,** 80–82.

Bachelard, H. S., and Trikojus, V. M. (1963). *Aust. J. Biol. Sci.* **16,** 166–176.

Ballester, D., Rodriguez, B., Rojas, M., Brunser, O., Reid, A., Yanez, E., and Mönckeberg, F. (1973). *J. Sci. Food Agric.* **24,** 127–138.

Barker, M. H. (1936). *J. Am. Med. Assoc.* **106,** 762–767.

Bayley, H. S., Cho, C. Y., and Summers, J. D. (1969). *Can. J. Anim. Sci.* **49,** 367–373.

Bell, J. M. (1955). *Can. J. Agric. Sci.* **35,** 242–251.

Bell, J. M. (1957a). *Can. J. Anim. Sci.* **37,** 31–42.

Bell, J. M. (1957b). *Can. J. Anim. Sci.* **37,** 43–49.

Bell, J. M. (1965). *J. Anim. Sci.* **24,** 1147–1151.

Bell, J. M. (1975). *Can. J. Anim. Sci.* **55,** 61–70.

Bell, J. M., and Baker, E. (1957). *Can. J. Anim. Sci.* **37,** 21–30.

Bell, J. M., and Williams, K. (1953). *Can. J. Agric. Sci.* **33,** 201-209.
Belzile, R., Bell, J. M., and Wetter, L. R. (1963). *Can. J. Anim. Sci.* **43,** 169-173.
Bjorkman, R. (1972). *Acta Chem. Scand.* **26,** 1111-1116.
Bjorkman, R., and Janson, J.-C. (1972). *Biochim. Biophys. Acta* **276,** 508-518.
Bowland, J. P. (1975). *Can. J. Anim. Sci.* **55,** 409-419.
Bowland, J. P., and Hardin, R. T. (1973). *Can. J. Anim. Sci.* **53,** 355-363.
Bowland, J. P., Zivković, S., and Manns, J. G. (1963). *Can. J. Anim. Sci.* **43,** 279-284.
Butcher, D. N., El-Tigani, S., and Ingram, D. S. (1974). *Physiol. Plant Pathol.* **4,** 127-140.
Cahn, R. S. (1964). *J. Chem. Educ.* **41,** 116-125.
Carlson, K. D., Weisleder, D., and Daxenbichler, M. E. (1970). *J. Am. Chem. Soc.* **92,** 6232-6238.
Carroll, K. K. (1949). *Proc. Soc. Exp. Biol. Med.* **71,** 622-624.
Chesney, A. M., Clawson, T. A., and Webster, B. (1928). *Bull. Johns Hopkins Hosp.* **43,** 261-277.
Chong, C., and Bible, B. (1975). *J. Sci. Food Agric.* **26,** 105-108.
Clandinin, D. R. (1961). *Poult. Sci.* **40,** 484-487.
Clandinin, D. R., Renner, R., and Robblee, A. R. (1959). *Poult. Sci.* **38,** 1367-1372.
Clandinin, D. R., Bayly, L., and Caballero, A. (1966). *Poult. Sci.* **45,** 833-838.
Clements, F. W., and Wishart, J. W. (1956). *Metab., Clin. Exp.* **5,** 623-639.
Cole, R. A. (1976). *Phytochemistry* **15,** 759-762.
Daxenbichler, M. E., and VanEtten, C. H. (1974). *J. Am. Oil Chem. Soc.* **51,** 449-450.
Daxenbichler, M. E., and VanEtten, C. H. (1977). *J. Assoc. Off. Anal. Chem.* **60,** 950-953.
Daxenbichler, M. E., VanEtten, C. H., Brown, F. S., and Jones, Q. (1964). *J. Agric. Food Chem.* **12,** 127-130.
Daxenbichler, M. E., VanEtten, C. H., and Wolff, I. A. (1965). *Biochemistry* **4,** 318-323.
Daxenbichler, M. E., VanEtten, C. H., and Wolff, I. A. (1966). *Biochemistry* **5,** 692-697.
Daxenbichler, M. E., VanEtten, C. H., Tallent, W. H., and Wolff, I. A. (1967). *Can J. Chem.* **45,** 1971-1974.
Daxenbichler, M. E., VanEtten, C. H., and Wolff, I. A. (1968). *Phytochemistry* **7,** 989-996.
Daxenbichler, M. E., Spencer, G. F., Kleiman, R., VanEtten, C. H., and Wolff, I. A. (1970). *Anal. Biochem.* **38,** 374-382.
Daxenbichler, M. E., VanEtten, C. H., and Spencer, G. F. (1977). *J. Agric. Food Chem.* **25,** 121-124.
Daxenbichler, M. E., VanEtten, C. H., and Williams, P. H. (1979). *J. Agric. Food Chem.* **27,** 34-37.
Downey, R. K. (1965). *Can., Dep. Agric., Publ.* **1257,** 7-23.
Downey, R. K. (1976). *Chem. Ind. (London)* pp. 401-406.
Eckenrode, C. J., and Arn, H. (1972). *J. Econ. Entomol.* **65,** 1343-1345.
Eklund, A., Ågren, G., Nordgren, H., and Stenram, U. (1974). *J. Sci. Food Agric.* **25,** 343-356.
Elliot, M. C., and Stowe, B. B. (1971). *Plant Physiol.* **48,** 498-503.
Erickson, J. M., and Feeney, P. (1974). *Ecology* **55,** 103-111.
Ettlinger, M. G., and Dateo, J. P., Jr. (1961). *In* "Studies of Mustard Oil Glucosides," Contract No. DA19-129 QM-1059, Proj. No. 7-84-06-032, Simplified Food Logistics, Final Rep., pp. 1-96. Department of Chemistry, Rice Institute, Houston, Texas.
Ettlinger, M. G., and Kjaer, A. (1968). *Recent Adv. Phytochem.* **1,** 59-144.
Ettlinger, M. G., and Lundeen, A. J. (1956). *J. Am. Chem. Soc.* **78,** 4172-4173.
Ettlinger, M. G., and Lundeen, A. J. (1957). *J. Am. Chem. Soc.* **79,** 1764-1765.
Ettlinger, M. G., and Thompson, C. P. (1962). *In* "Studies of Mustard Oil Glucosides (II)," Contract No. DA19-129 QM-1689, Proj. No. 7-99-01-001, Simplified Food Logistics, Final Rep., pp. 1-106. Department of Chemistry, Rice Institute, Houston, Texas.
Ettlinger, M. G., Dateo, G. P., Jr., Harrison, B. W., Mabry, T. J., and Thompson, C. P. (1961). *Proc. Natl. Acad. Sci. U.S.A.* **47,** 1875-1880.

Florkin, M., and Stotz, E. H. (1965). *Compr. Biochem.* **13,** 142.
Food and Agriculture Organization of the United Nations (1965). "Production Yearbook," Vol. 19, pp. 123–136. FAO, Rome, Italy.
Food and Agriculture Organization of the United Nations (1974). "Production Yearbook," Vol. 28-1, pp. 101–103, and 133. FAO, Rome, Italy.
Friis, P., and Kjaer, A. (1966). *Acta Chem. Scand.* **20,** 698–705.
Frolich, A. (1952). *Lantbrukshoegsk. Ann.* **19,** 205–208.
Garner, R. J., Sansom, B. F., and Jones, H. G. (1960). *J. Agric. Sci.* **55,** 283–286.
Gmelin, R., and Virtanen, A. I. (1959). *Acta Chem. Scand.* **13,** 1474–1475.
Gmelin, R., and Virtanen, A. I. (1961). *Ann. Acad. Sci. Fenn., Ser. A2* **107,** 1–25.
Gmelin, R., and Virtanen, A. I. (1962). *Acta Chem. Scand.* **16,** 1378–1384.
Goering, K. J., Thomas, O. O., Beardsley, D. R., and Curran, W. A., Jr. (1960). *J. Nutr.* **72,** 210–216.
Greene, R., Farran, H., and Glascock, R. F. (1958). *J. Endocrinol.* **17,** 272–279.
Greer, M. A. (1950). *Physiol. Rev.* **30,** 513–548.
Greer, M. A. (1962). *Recent Prog. Horm. Res.* **18,** 187–219.
Greer, M. A., and Astwood, E. B. (1948). *Endocrinology* **43,** 105–119.
Greer, M. A., and Deeney, J. M. (1959). *J. Clin. Invest.* **38,** 1465–1474.
Greer, M. A., and Whallon, J. (1961). *Proc. Soc. Exp. Biol. Med.* **107,** 802–804.
Greer, M. A., Kendall, J. W., and Smith, M. (1964). *In* "The Thyroid Gland" (R. Pitt-Rivers and W. R. Trotter, eds.), pp. 357–389. Butterworth, London.
Greer, M. A., Stott, A. K., and Milne, K. A. (1966). *Endocrinology* **79,** 237–247.
Hall, S. A. (1974). *Vet. Rec.* **94,** 42–44.
Hansen, H. (1974). *Tidsskr. Planteavl* **78**(3), 408–410; *Chem. Abstr.* **82,** 70257f (1975).
Hawrysh, Z. J., Clandinin, D. R., Robblee, A. R., Hardin, R. T., and Darlington, K. (1975). *Can. Inst. Food Sci. Technol. J.* **8,** 51–54.
Henderson, H. M., and McEwan, T. J. (1972). *Phytochemistry* **11,** 3127–3133.
Hercus, C. E., and Purves, H. D. (1936). *J. Hyg.* **36,** 182–203.
Hesketh, H. R., Creger, C. R., and Couch, J. R. (1963). *Poult. Sci.* **42,** 1276.
Hicks, K. L. (1974). *Ann. Entomol. Soc. Am.* **67,** 261–264.
Hobson-Frohock, A., Land, D. G., Griffiths, N. M., and Curtis, R. F. (1973). *Nature (London)* **243,** 304–305.
Hobson-Frohock, A., Fenwick, R. G., Land, D. G., Curtis, R. F., and Gulliver, A. L. (1975). *Br. Poult. Sci.* **16,** 219–222.
Holmes, R. G. (1965). *Vet. Rec.* **77,** 480–481.
Holmes, W. B., and Roberts, R. (1963). *Poult. Sci.* **42,** 803–809.
Hopkins, C. Y. (1938). *Can. J. Res., Sect. B* **16,** 84–93.
Jackson, N. (1969). *J. Sci. Food Agric.* **20,** 734–740.
Jensen, K. A., Conti, J., and Kjaer, A. (1953). *Acta Chem. Scand.* **7,** 1267–1270.
Jirousek, L. (1956). *Naturwissenschaften* **43,** 328–329.
Jirousek, L., and Starka, L. (1958). *Naturwissenschaften* **45,** 386–387.
Johnston, T. D., and Jones, D. I. H. (1966). *J. Sci. Food Agric.* **17,** 70–71.
Josefsson, E. (1967). *Phytochemistry* **6,** 1617–1627.
Josefsson, E. (1968). *J. Sci. Food Agric.* **19,** 192–194.
Josefsson, E. (1970). *J. Sci. Food Agric.* **21,** 94–97.
Josefsson, E. (1972). *In* "Rapeseed: Cultivation, Composition, Processing, and Utilization" (L. A. Appelqvist and R. Ohlson, eds.), pp. 354–378. Elsevier, Amsterdam.
Josefsson, E. (1975a). *J. Sci. Food Agric.* **26,** 157–164.

Josefsson, E. (1975b). *J. Sci. Food Agric.* **26,** 1299–1310.
Josefsson, E., and Appelqvist, L. A. (1968). *J. Sci. Food Agric.* **19,** 564–570.
Josefsson, E., and Uppström, B. (1976a). *J. Sci. Food Agric.* **27,** 433–437.
Josefsson, E., and Uppström, B. (1976b). *J. Sci. Food Agric.* **27,** 438–442.
Kawakishi, S., and Muramatsu, K. (1966). *Agric. Biol. Chem.* **30,** 688–692.
Kawakishi, S., Namiki, M., and Watanabe, H. (1967). *Agric. Biol. Chem.* **31,** 823–830.
Kelly, F. C., and Snedden, W. W. (1960). *W.H.O., Monogr. Ser.* **44,** 91.
Kennedy, T. H., and Purves, H. D. (1941). *Br. J. Exp. Pathol.* **22,** 241–244.
Khera, K. S. (1977). *Food Cosmet. Toxicol.* **15,** 61–62.
Kiesvaara, M., and Virtanen, A. I. (1963). *Acta Chem. Scand.* **17,** 849–853.
Kirk, J. T. O., and MacDonald, C. G. (1974). *Phytochemistry* **13,** 2611–2615.
Kirk, L. D., Mustakas, G. C., and Griffin, E. L., Jr. (1966). *J. Am. Oil Chem. Soc.* **43,** 550–555.
Kirk, L. D., Mustakas, G. C., and Griffin, E. L., Jr. (1971). *J. Am. Oil Chem. Soc.* **48,** 845–850.
Kjaer, A. (1960). *In* "The Chemistry of Organic Natural Products" (L. Zechmeister, ed.), pp. 122–176. Springer-Verlag, Berlin and New York.
Kjaer, A. (1966). *In* "Comparative Phytochemistry" (T. Swain, ed.), pp. 187–194. Academic Press, New York.
Kjaer, A. (1973). *Chem. Bot. Classification, Proc. Nobel Symp., 25th, 1973* pp. 229–234.
Kjaer, A., and Boe Jensen, R. (1956). *Acta Chem. Scand.* **10,** 1365–1371.
Kjaer, A., and Christensen, B. W. (1959). *Acta Chem. Scand.* **13,** 1575–1584.
Kjaer, A., and Christensen, B. W. (1962). *Acta Chem. Scand.* **16,** 71–77.
Kjaer, A., and Gmelin, R. (1957). *Acta Chem. Scand.* **11,** 906–907.
Kjaer, A., and Larsen, P. (1973). *Biosynthesis* **2,** 71–105.
Kjaer, A., and Rubinstein, K. (1953). *Acta Chem. Scand.* **7,** 528–536.
Kjaer, A., and Rubinstein, K. (1954). *Acta Chem. Scand.* **8,** 598–602.
Kjaer, A., and Thomsen, H. (1962). *Acta Chem. Scand.* **16,** 591–598.
Kjaer, A., Conti, J., and Jensen, K. A. (1953). *Acta Chem. Scand.* **7,** 1271–1275.
Kjaer, A., Gmelin, R., and Boe Jensen, R. (1956). *Acta. Chem. Scand.* **10,** 432–438.
Kjaer, A., Christensen, B. W., and Hansen, S. E. (1959). *Acta Chem. Scand.* **13,** 144–150.
Klain, G. J., Hill, D. C., Branion, H. D., and Gray, J. A. (1956). *Poult. Sci.* **35,** 1315–1326.
Kreula, M., and Kiesvaara, M. (1959). *Acta Chem. Scand.* **13,** 1375–1382.
Lambert, J. L., Clanton, D. C., Wolff, I. A., and Mustakas, G. C. (1970). *J. Anim. Sci.* **31,** 601–607.
Langer, P. (1961). *Hoppe-Seyler's Z. Physiol. Chem.* **323,** 194–198.
Langer, P. (1964). *Hoppe-Seyler's Z. Physiol. Chem.* **339,** 33–35.
Langer, P. (1966). *Endocrinology* **79,** 1117–1122.
Langer, P., and Stolc, V. (1964). *Hoppe-Seyler's Z. Physiol. Chem.* **335,** 216–220.
Langer, P., and Stolc, V. (1965). *Endocrinology* **76,** 151–155.
Lanzani, A., Piana, G., Piva, G., Cardillo, M., Rastelli, A., and Jacini, G. (1974). *J. Am. Oil Chem. Soc.* **51,** 517–518.
Lein, K.-A. (1970). *Z. Pflanzenzuecht.* **63,** 137–154.
Lein, K.-A. (1972). *Angew. Bot.* **46,** 263–284.
Leslie, A. J., and Summers, J. D. (1972). *Can. J. Anim. Sci.* **52,** 563–566.
Lichtenstein, E. P. (1966). *Adv. Chem. Ser.* **53,** 34–38.
Lichtenstein, E. P., Strong, F. M., and Morgan, D. G. (1962). *J. Agric. Food Chem.* **10,** 30–33.
Lo, M. T., and Hill, D. C. (1971). *J. Sci. Food Agric.* **22,** 128–130.
Lo, M. T., and Hill, D. C. (1972a). *Can. J. Physiol. Pharmacol.* **50,** 373–377.
Lo, M. T., and Hill, D. C. (1972b). *Can. J. Physiol. Pharmacol.* **50,** 962–966.

Lönnerdal, B., and Janson, J.-C. (1973). *Biochim. Biophys. Acta* **315,** 421–429.
Lööf, B., and Appelqvist, L. A. (1972). *In* "Rapeseed: Cultivation, Composition, Processing, and Utilization" (L. A. Appelqvist and R. Ohlson, eds.), p. 120. Elsevier, Amsterdam.
Loub, W. D., and Wattenburg, L. W. (1974). *Lloydia* **36,** 436 (abstr.).
McGhee, J. E., Kirk, L. D., and Mustakas, G. C. (1965). *J. Am. Oil Chem. Soc.* **42,** 889–891.
McGregor, D. I., and Downey, R. K. (1975). *Can. J. Plant Sci.* **55,** 191–196.
MacLeod, A. J., and Islam, R. (1975). *J. Sci. Food Agric.* **26,** 1545–1550.
MacLeod, A. J., and Islam, R. (1976). *J. Sci. Food Agric.* **27,** 909–912.
MacLeod, A. J., and MacLeod, G. (1968). *J. Sci. Food Agric.* **19,** 273–276.
Manns, J. G., and Bowland, J. P. (1963). *Can. J. Anim. Sci.* **43,** 252–263.
Manns, J. G., Bowland, J. P., Mendel, V. E., and Zivković, S. (1963). *Can. J. Anim. Sci.* **43,** 271–278.
Marangos, A., and Hill, R. (1974). *Proc. Nutr. Soc.* **33,** 90A.
Marangos, A., Hill, R., Laws, B. M., and Muschamp, D. (1974). *Br. Poult. Sci.* **15,** 405–414; *Chem. Abstr.* **82,** 56442f (1975).
Marsh, R. E., and Waser, J. (1970). *Acta Crystallogr., Sect. B* **26,** 1030–1037.
Matano, K., and Kato, N. (1967). *Acta Chem. Scand.* **21,** 2886–2887.
Matsumoto, T., Itoh, H., and Akiba, Y. (1968). *Poult. Sci.* **47,** 1323–1330.
Michajlovskij, N., and Langer, P. (1958). *Hoppe-Seyler's Z. Physiol. Chem.* **312,** 26–30.
Michajlovskij, N., and Langer, P. (1959). *Hoppe-Seyler's Z. Physiol. Chem.* **317,** 30–33.
Michajlovskij, N., Sedlak, J., Jusic, M., and Buzina, R. (1969a). *Endocrinol. Exp.* **3,** 65 (*Chem. Abstr.* **73,** 118299q.).
Michajilovskij, N., Sedlack, J., and Kostekova, O. (1969b). *Rev. Czech. Med.* **15,** 132–144.
Miller, H. E. (1965). Master's Thesis, pp. 1–78. Rice University, Houston, Texas.
Miller, J. K., Swanson, E. W., and Cragle, R. G. (1965). *J. Dairy Sci.* **48,** 1118–1121.
Miller, R. W., Daxenbichler, M. E., Earle, F. R., and Gentry, H. S. (1964). *J. Am. Oil Chem. Soc.* **41,** 167–169.
Morrison, F. B. (1959). *In* "Feeds and Feeding", 22nd ed., pp. 394–396. Morrison, Clinton, Iowa.
Mullin, W. J., and Sahasrabudhe, M. R. (1978). *Can. Inst. Food Sci. Technol. J.* **11,** 50–52.
Mustakas, G. C., and Kirk, L. D. (1963). U.S. Patent 3,106,649.
Mustakas, G. C., Kirk, L. D., Sohns, V. E., and Griffin, E. L., Jr. (1965). *J. Am. Oil Chem. Soc.* **42,** 33–37.
Mustakas, G. C., Kirk, L. D., Griffin, E. L., Jr., and Clanton, D. C. (1968). *J. Am. Oil Chem. Soc.* **45,** 53–57.
Mustakas, G. C., Kirk, L. D., and Griffin, E. L., Jr. (1976). *J. Am. Oil Chem. Soc.* **53,** 12–16.
Nagashima, Z., and Uchiyama, M. (1959a). *Nippon Nogei Kagaku Kaishi* **33,** 478, 484, 881, 980, 1068, and 1144.
Nagashima, Z., and Uchiyama, M. (1959b). *Bull. Agric. Chem. Soc. Jpn.* **23,** 555–556.
Nakaya, T. (1964). *Jpn. J. Zootech. Sci.* **35,** 107–112.
Nakaya, T., and Nakamura, R. (1963a). *Jpn. J. Zootech. Sci.* **34,** 253–261, 263–268.
Nakaya, T., and Nakamura, R. (1963b). *Jpn. J. Zootech. Sci.* **34,** 319–322, 323–327.
Nordfeldt, S., Gellerstedt, N., and Falkmer, S. (1954). *Acta Pathol. Microbiol. Scand.* **35,** 217–236.
Oginsky, E. L., Stein, A. E., and Greer, M. A. (1965). *Proc. Soc. Exp. Biol. Med.* **119,** 360–364.
Ohtsuru, M., and Hata, T. (1973). *Agric. Biol. Chem.* **37,** 2543–2548.
Ohtsuru, M., Tsuruo, I., and Hata, T. (1969). *Agric. Biol. Chem.* **33,** 1309, 1315, and 1320–1325.
Olomu, J. M., Robblee, A. R., Clandinin, D. R., and Hardin, R. T. (1975). *Can. J. Anim. Sci.* **55,** 219–222.
Peltola, P. (1965). *Thyroid Res., Proc. Int. Thyroid Conf., 5th, 1965* pp. 872–876. Academic Press, New York.

Perry, T. W., Kwolek, W. F., Tookey, H. L., Princen, L. H., Beeson W. M., and Mohler, M. T. (1979). *J. Anim. Sci.* **48,** 758–763.

Pettit, J. H., Slinger, S. J., Evans, E. V., and Marcellus, F. N. (1944). *Sci. Agric.* **24,** 201–213.

Piironen, E., and Virtanen, A. I. (1963). *Z. Ernaehrungswiss.* **3,** 140–147.

Rapeseed Association of Canada (1972). "Canadian Rapeseed Meal in Poultry and Animal Feeding," Publ. No. 16. Rapeseed Assoc. Can., Winnipeg, Manitoba, Canada.

Reese, E. T., Clapp, R. C., and Mandels, M. (1958). *Arch. Biochem. Biophys.* **75,** 228–242.

Renner, R., Clandinin, D. R., and Robblee, A. R. (1955). *Poult. Sci.* **34,** 1233.

Saarivirta, M. (1973). *Farm. Aikak.* **82,** 11–15.

Saarivirta, M., and Virtanen, A. I. (1963). *Acta Chem. Scand.* **17,** S74–S78.

Sandberg, M., and Holly, O. M. (1932). *J. Biol. Chem.* **96,** 443–447.

Sarwar, G., Sosulski, F. W., and Bell, J. M. (1973). *Can. Inst. Food Sci. Technol. J.* **6,** 17–21.

Schluter, M., and Gmelin, R. (1972). *Phytochemistry* **11,** 3427–3432.

Schultz, O. E., and Gmelin, R. (1952). *Z. Naturforsch., Teil B* **7,** 500–508.

Schultz, O. E., and Gmelin, R. (1953). *Z. Naturforsch., Teil B* **8,** 151–156.

Schultz, O. E., and Gmelin, R. (1954). *Z. Naturforsch., Teil B* **9,** 27–29.

Schwimmer, S. (1961). *Acta Chem. Scand.* **15,** 535–544.

Senti, F. R., and Rizek, R. L. (1974). *In* "The Effect of FDA Regulations (GRAS) on Plant Breeding and Processing" (C. H. Hanson, ed.), pp. 7–20. Crop Sci. Soc. Am., Madison, Wisconsin. Spec. Publ. No. 5.

Seth, P. C., and Clandinin, D. R. (1973). *Poult. Sci.* **52,** 1158–1160.

Shah, B. G., Jones, J. D., McLaughlin, J. M., and Bear-Rogers, J. L. (1976). *Nutr. Rep. Int.* **13,** 1–8.

Smith, T. K., and Campbell, L. D. (1976). *Poult. Sci.* **55,** 861–867.

Sosulski, F. W., Soliman, F. S., and Bhatty, R. S. (1972). *Can. Inst. Food Sci. Technol. J.* **5**(2), 101–104.

Srivastava, V. K., and Hill, D. C. (1974). *Phytochemistry* **13,** 1043–1046.

Srivastava, V. K., and Hill, D. C. (1976). *J. Sci. Food Agric.* **27,** 953–958.

Srivastava, V. K., Philbrick, D. J., and Hill, D. C. (1975). *Can. J. Anim. Sci.* **55,** 331–335.

Stahmann, M. A., Link, K. P., and Walker, J. C. (1943). *J. Agric. Res.* **67,** 49–63.

Stanley, M. M., and Astwood, E. B. (1947). *Endocrinology* **41,** 66–84.

Staron, T. (1970). *Proc. Int. Conf. Sci. Technol., Marketing of Rapeseed Rapessed Prod., 1970* pp. 321–337.

Stefansson, B. R. (1974). *In* "The Story of Rapeseed in Western Canada" (A. D. MacLeod, ed.), pp. 22–25. Saskatchewan Wheat Pool, 50th Anniv. Publ., Regina, Sask., Canada.

Stoll, A., and Seebeck, E. (1948). *Helv. Chim. Acta* **31,** 1432–1434.

Tang, C.-S. (1971). *Phytochemistry* **10,** 117–121.

Tang, C.-S. (1973). *Phytochemistry* **12,** 769–773.

Tani, N., Ohtsuru, M., and Hata, T. (1974). *Agric. Biol. Chem.* **38,** 1623–1630.

Tape, N. W., Sabry, Z. I., and Eapen, K. E. (1970). *Can. Inst. Food Technol. J.* **3,** 78–81.

Tapper, B. A., and MacGibbon, D. B. (1967). *Phytochemistry* **6,** 749–753.

Thies, W. (1976). *Fette, Seifen, Anstrichm.* **78,** 231–234.

Tookey, H. L. (1973a). *Can. J. Biochem.* **51,** 1305–1310.

Tookey, H. L. (1973b). *Can. J. Biochem.* **51,** 1654–1660.

Tookey, H. L., and Wolff, I. A. (1970). *Can. J. Biochem.* **48,** 1024–1028.

Tookey, H. L., VanEtten, C. H., Peters, J. E., and Wolff, I. A. (1965). *Cereal Chem.* **42,** 507–514.

Tsuruo, I., and Hata, T. (1967). *Agric. Biol. Chem.* **31,** 27–32.

Tsuruo, I., and Hata, T. (1968). *Agric. Biol. Chem.* **32,** 479–483.

Tsuruo, I., and Hata, T. (1972). *Agric. Biol. Chem.* **36,** 2495–2503.

Turner, C. W. (1946). *Poult. Sci.* **25,** 186–187.

Underhill, E. W., and Kirkland, D. F. (1971). *J. Chromatogr.* **57,** 47–54.

Underhill, E. W., Wetter, L. R., and Chisholm, M. D. (1973). *Biochem. Soc. Symp.* **38,** 303–326.

U.S. Department of Agriculture. (1975). "Agricultural Statistics," p. 157. USDA, Washington, D.C.

VanderLaan, J. E., and VanderLaan, W. P. (1947). *Endocrinology* **40,** 403–416.

Van der Veen, H. E., and Hart, P. C. (1955). *Voeding* **16,** 12–22.

VanEtten, C. H., and Daxenbichler, M. E. (1971). *J. Agric. Food Chem.* **19,** 194–195.

VanEtten, C. H., and Daxenbichler, M. E. (1977). *J. Assoc. Off. Anal. Chem.* **60,** 946–949.

VanEtten, C. H., Daxenbichler, M. E., Peters, J. E., Wolff, I. A., and Booth, A. N. (1965). *J. Agric. Food Chem.* **13,** 24–27.

VanEtten, C. H., Daxenbichler, M. E., Peters, J. E., and Tookey, H. L. (1966). *J. Agric. Food Chem.* **14,** 426–430.

VanEtten, C. H., Daxenbichler, M. E., Booth, A. N., Robbins, D. J., and Wolff, I. A. (1969a). *158th Natl. Meet., Am. Chem. Soc.* Abstract No. 53, AFCD.

VanEtten, C. H., Daxenbichler, M. E., and Wolff, I. A. (1969b). *J. Agric. Food Chem.* **17,** 483–491.

VanEtten, C. H., Gagne, W. E., Robbins, D. J., Booth, A. N., Daxenbichler, M. E., and Wolff, I. A. (1969c). *Cereal Chem.* **46,** 145–155.

VanEtten, C. H., McGrew, C. E., and Daxenbichler, M. E. (1974). *J. Agric. Food Chem.* **22,** 483–487.

VanEtten, C. H., Daxenbichler, M. E., Williams, P. H., and Kwolek, W. F. (1976). *J. Agric. Food Chem.* **24,** 452–455.

VanEtten, C. H., Daxenbichler, M. E., Schroeder, W., Princen, L. H., and Perry, T. W. (1977). *Can. J. Anim. Sci.* **57,** 75–80.

Vaughan, J. G. (1977). *BioScience* **27,** 35–40.

Vilkki, P., Kreula, M., and Piironen, E. (1962). *Ann. Acad. Sci. Fenn., Ser. A2* **110,** 1–13.

Virtanen, A. I. (1965). *Phytochemistry* **4,** 207–228.

Virtanen, A. I., and Saarivirta, M. S. (1962). *Suom. Kemistil. B* **35,** 102–104, 248–249.

Virtanen, A. I., Kreula, M., and Kiesvaara, M. (1958). *Acta Chem. Scand.* **12,** 580–581.

Virtanen, A. I., Kreula, M., and Kiesvaara, M. (1963). *Z. Ernaehrungswiss., Suppl.* **3,** 23–37.

von Euler, H., and Erikson, S. E. (1926). *Fermentforschung* **8,** 518.

Vose, J. R. (1972). *Phytochemistry* **11,** 1649–1653.

Waser, J., and Watson, W. H. (1963). *Nature (London)* **198,** 1297–1298.

Wetter, L. R. (1955). *Can. J. Biochem. Physiol.* **33,** 980–984.

Wetter, L. R. (1957). *Can. J. Biochem. Physiol.* **35,** 293–297.

Wetter, L. R., and Youngs, C. G. (1976). *J. Am. Oil Chem. Soc.* **53,** 162–164.

Whiting, F. (1965). *Can., Dep. Agric., Publ.* **1257,** 61–68.

Wood, J. L. (1975). *In* "Chemistry and Biochemistry of Thiocyanic Acid and its Derivatives" (A. A. Newman, ed.), pp. 156–221. Academic Press, New York.

Wright, E., and Sinclair, D. P. (1959). *N.Z.J. Agric. Res.* **2,** 933–937.

Youngs, C. G., and Wetter, L. R. (1967). *J. Am. Oil Chem. Soc.* **44,** 551–554.

Chapter 5

Cyanogens

R. D. MONTGOMERY

I. DIETARY SOURCES OF CYANOGEN

Cyanide in trace amounts is widespread in the plant kingdom and occurs mainly in the form of cyanogenetic glucosides. Relatively high concentrations are found in certain grasses, pulses, root crops, and fruit kernels. Many of these are consumed by animals, and a few are known to be of practical importance in human nutrition. Although certain aspects of animal poisoning are relevant, our main concern in this chapter is with the toxicity of human foodstuffs. Although this principally involves cyanogenic glucosides of plant origin, it should be borne in mind that glucosides may not be the only source of cyanide (Paris, 1963), and that cyanide may be released not only from plants, but from fungi, bacteria, and even the animal kingdom. Cyanogenesis has been shown to be an important defensive mechanism in moths (Jones *et al.*, 1962) and millipedes (Eisner *et al.*, 1963).

Lists of cyanogenic glucosides are recorded by Gibbs (1963) and Kingsbury

TOXIC CONSTITUENTS OF PLANT FOODSTUFFS, SECOND EDITION

ISBN 0-12-449960-0

(1964), but it seems that only four have been of practical importance with regard to human toxicity in edible plants: amygdalin, dhurrin, linamarin, and lotaustralin. Amygdalin was first identified in the bitter almond and is also present in the kernel of other fruits. Dhurrin, a closely related compound, occurs in sorghum and other grasses. Linamarin (phaseolunatin) and methyllinamarin (lotaustralin) are the glucosides of pulses, linseed (flax), and cassava. The following is a list of the sources of these glucosides, which are habitually consumed by man: cassava, sweet potato, and yam; maize and millets (especially *Sorghum vulgare*); bamboo and sugarcane; peas and beans (especially *Phaseolus lunatus*—lima or butter bean); kernel of almond, lemon, lime, apple, pear, cherry, apricot, prune, and plum.

The maximal yields reported from varieties of certain plants eaten by man are recorded in Table I. In addition to these plants, various leaves, particularly of *Sorbus* and *Prunus* species, may cause animal poisoning, as may linseed, which is an important cattle food. Other grasses are listed by Couch (1934) as having caused poisoning in cattle in the United States. Linseed may yield up to 53 mg of HCN per 100 gm (Auld, 1912–1913) and arrow grass up to 77 mg per 100 gm

TABLE I

HIGHEST REPORTED FIGURES OF CYANIDE YIELD IN VARIOUS PLANT FOODSTUFFS

Plant	HCN yield (mg per 100 gm)	Reference
Bitter almonds	250	Polson and Tattersall (1959)
Bitter cassava		
Dried root cortex	245	Collens (1915)
Whole root	53	Clark (1936)
Fresh root bark	89	de Bruijn (1973)
Fresh stem bark	133	de Bruijn (1973)
Leaves	104	de Bruijn (1973)
Sorghum		
Whole plant, immature	250	Anonymous (1903)
Bamboo		
Tip of immature shoots	800	Baggchi and Ganguli (1943)
Stem, immature	300	Baggchi and Ganguli (1943)
Lima bean (*Phaseolus lunatus*) varieties		
Java, colored	312	Guignard (1907)
Puerto Rico, black	300	Viehoever (1940)
Burma, white	210	Kohn-Abrest (1906)
Arizona, colored	17	Montgomery (1964)
America, white	10	Montgomery (1964)

(Couch, 1934). The poisoning of General Kitchener's transport animals in the Sudan by *Lotus arabicus* (Egyptian vetch, Sudan grass) led to the identification of phaseolunatin (Dunstan and Henry, 1903). This grass occurs in the United States. A significant yield of cyanogen may also be obtained from certain varieties of *Vicia sativa* (common vetch) (Bertrand, 1907; Montgomery, 1964). The structure of the glucoside vicianin derived from these plants is uncertain.

In grasses and cane, the highest yield is obtained from the tip of the immature plant, particularly if grown in rich soil (Couch, 1934; Baggchi and Ganguli, 1943). In *Phaseolus lunatus* and in linseed, the cyanide occurs in all parts of the plant throughout life, as well as in the seed. The highest yield of cyanide in *Phaseolus lunatus* is in the black varieties. Systematic cultivation of white varieties has greatly reduced the cyanide content (Anonymous, 1920), but probably no variety can be produced that is entirely free of cyanogen. Native American strains tend to contain less than those originating in Africa and the East.

Concentrations of cyanogen of the order of 2 mg per 100 gm may be found also in *Phaseolus vulgaris* (Kidney, haricot, or navy bean, red pea), *Vigna sinensis* (black-eye pea), and *Pisum sativum* (garden pea) (Jaffé, 1950; Montgomery, 1964).

Within individual plant genera, there are striking species differences with regard to the presence or absence of cyanogen, and this fact, together with the presence of identical biosynthetic pathways in different plant families (Abrol and Conn, 1966), is of considerable genetic interest. The genetic and ecological aspects of cyanogenesis have been reviewed by Jones (1972, 1973). Although the importance of the protective role of cyanogen in plants has been questioned, there is no doubt that in many cases it conveys a biological advantage by offering a defense against insects, as, for example, in the kernels of fruit and in the cortex of cassava (Conn and Butler, 1969).

In cassava, the highest quantities of cyanogen can be found in the bitter variety, but there is in fact no clear differentiation between bitter and sweet strains (Oyenuga and Amazigo, 1957; Nartey, 1973).

II. CHEMISTRY OF CYANOGENETIC GLUCOSIDES

A. Structure

Amygdalin is a glucoside of benzaldehyde cyanohydrin (mandelonitrile), which on complete hydrolysis yields glucose, benzaldehyde, and hydrogen cyanide.

With regulated enzymatic hydrolysis, the glucose is removed in two stages. With alkali or concentrated acid, amygdalinic acid is produced. Figure 1 shows the structure of amygdalin and its hydrolytic products.

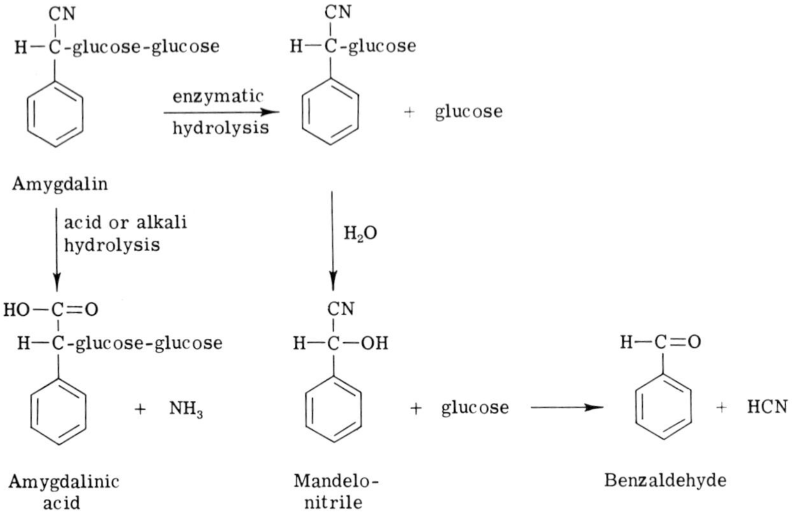

FIG. 1. The structure of amygdalin and its hydrolysis products.

Dhurrin is the glucoside of a closely related compound, *p*-hydroxybenzaldehyde cyanohydrin, and is hydrolyzed similarly to *p*-hydroxybenzaldehyde (Fig. 2a), glucose, and HCN.

Linamarin is the glucoside of acetone cyanohydrin, yielding acetone and HCN on hydrolysis (Fig. 2b). It is now known that most plants containing linamarin also harbor a methyl analogue, lotaustralin, which hydrolyzes to methyl ethyl ketone (Butler, 1965; Nartey, 1968; Bissett *et al.*, 1969).

All of these compounds are β-glucosides, which are poorly soluble in water. Because of this property, they are well suited as vehicles for the storage of noxious substances such as cyanide until such time as the latter are required to perform a biological function (Henry, 1929).

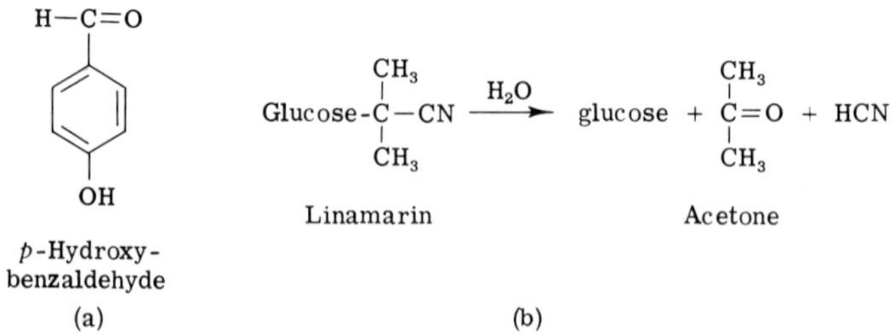

FIG. 2. (a) *p*-Hydroxybenzaldehyde, a hydrolysis products of dhurrin. (b) Linamarin and its hydrolysis products.

B. Autolysis

Spontaneous release of hydrogen cyanide (HCN) from the plant depends on the presence of a specific glucosidase and water. The enzymes are extracellular and gain access to the glucoside after physical disruption of the cell. They will act in the cold and are readily destroyed by heat. Autolysis is enhanced if the plant is soaked in water after crushing. Bruising without soaking, however, will lead to slow release of HCN, and it is well recognized that bruised cassava root is unfit for consumption. The boiling point of HCN is 26°C; storage in hot, humid conditions therefore leads to gradual loss of HCN from the plant as a result of minor trauma. However, with either dry or cold storage the potential HCN yield may be maintained over a period of years (Viehoever, 1940).

In growing grasses, autolysis may be initiated by frost, severe drought, or trampling of the grass.

Thus, the rate of HCN release depends greatly on the physical conditions. Viehoever (1940) found that the HCN yield from *Phaseolus lunatus* was maximal after 45 min if the bean was soaked in water at 37°C after crushing. Winkler (1951) considered that the yield was greater if the crushed bean was soaked for 24 hr at room temperature. In the case of linseed cake used for cattle feeding, Auld (1912) found that one-half of the available HCN might be released after a 15-min soaking in the cold.

The degradation of cyanogenetic glucosides is discussed in more detail by Wood (1966), Butler *et al.*, (1973), and Zitnak (1973).

C. Methods of Extraction and Assay

Because of the volatility of HCN, all extraction procedures must be carried out in a closed apparatus. To ensure full recovery it is desirable that the extraction be carried out fairly rapidly, and methods involving heating and acid hydrolysis are, in general, preferable to enzymatic autolysis. According to Wood (1966), however, acid hydrolysis is inconsistent in the case of linamarin and may produce artifacts, and methods using the specific enzyme β-linamarase are preferable.

In contrast to the glucosidase, the intact glucoside is remarkably stable to heat. In *Phaseolus lunatus,* according to Viehoever (1940), full recovery of HCN can be achieved (by acid hydrolysis) after the intact bean has been boiled for several hours or subjected to autoclaving at 160°C. Acid hydrolysis is ineffective at temperatures less than 60°C and is incomplete with hydrochloric or sulfuric acid stronger than 5% due to oxidation to ammonia. A satisfactory method of extraction, therefore, is to hydrolyze in a closed steam distillation apparatus with 5% hydrochloric acid at 100°C for at least 3 hr (Montgomery, 1964). The distillate is collected in 2.5% NaOH, and HCN is then assayed by a modification of Leibig's iodine titration (Horwitz, 1960). The end point of this titration depends on the detection of a silver iodide precipitate, and the method is therefore inaccurate at very low concentrations.

The Guignard test (Guignard, 1907) may be used qualitatively to detect cyanogenesis. Sodium picrate paper is prepared by dipping strips of filter paper in 1% picric acid, drying, and then dipping into 10% sodium carbonate and drying. The test material is chopped finely and moistened in a test tube. Moist sodium picrate paper is inserted without touching the material, a few drops of chloroform are added, and the tube is tightly stoppered. The paper slowly changes color to orange or brick red. This test is positive with material yielding approximately 50 μg HCN per gram.

Quantitative colorimetry derived from this method can estimate HCN in the range of 5–50 μg/gm extract (Gilchrist *et al.*, 1967; Indira and Sinha, 1969).

Sensitive colorimetric micromethods of HCN assay were also described by Boxer and Rickards (1950) and Wokes and Willimott (1951). Others were reviewed by Zitnak (1973), who stressed that the principal source of error is not in the assay itself, but in the variability of extraction, isolation, and sampling procedures.

III. HUMAN METABOLISM OF INORGANIC CYANIDE

A. Metabolic Pathways

Ingested cyanide is rapidly absorbed from the upper gastrointestinal tract. It also passes readily through the skin, and HCN gas is rapidly absorbed from the lungs. The most widely studied metabolic pathway is by combination with thiosulfate to form thiocyanate and sulfite, the reaction being catalyzed by rhodanese, an enzyme that may be described as a sulfurtransferase (Lang, 1933). This enzyme is widespread in living tissues, reaching its highest concentrations in the liver, kidney, thyroid, adrenal, and pancreas (Rosenthal, 1948). The resulting thiocyanate is excreted in the urine and also in the saliva, from which it may be either reabsorbed or metabolized in the gastrointestinal tract. Thiocyanate is slowly oxidized to sulfate, but this does not appear to be an important factor in body tissues (Clemedson *et al.*, 1960). In a second pathway to thiocyanate, which has been demonstrated in experimental animals, cyanide reacts with 3-mercaptopyruvate, aided by another sulfurtransferase (Fiedler and Wood, 1956). These reactions require the presence of adequate amounts of cystine as a sulfur donor. The metabolic disposal of inorganic cyanide is summarized in Fig. 3.

Cystine may react more directly with cyanide to produce 2-amino-4-thiozoline-carboxylic acid (Wood and Colley, 1956). Isotope studies indicate that some of

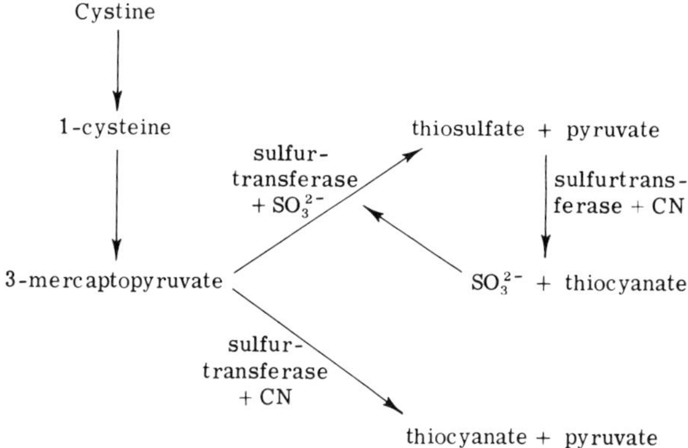

FIG. 3. The metabolic disposal of inorganic cyanide.

the carbon of ingested cyanide is lost as carbon dioxide via a 1-carbon pathway (Boter and Rickards, 1952). It is now believed that this involves hydroxocobalamin (vitamin B_{12}) as a co-enzyme.

Man is continually exposed to small doses of cyanide—in his diet, in atmospheric pollution, and particularly in cigarette smoke (Osborne *et al.*, 1956; Anonymous, 1964; Densen *et al.*, 1967) and also as a result of urinary infection with cyanogenic bacteria, such as *Pseudomonas pyocyanea* (Adams *et al.*, 1966). The presence of trace amounts in the body can possibly be regarded as physiological and may act as a brake in cellular oxidative processes. The thiocyanate pathway is not all in one direction; Goldstein and Rieders (1953) demonstrated an endogenous oxidase in red blood cells which converts thiocyanate to free cyanide.

It is well known that hydroxocobalamin takes up cyanide as cyanocobalamin and readily liberates it on exposure to light (Wokes *et al.*, 1953). This could be merely a secondary effect of the presence of cyanide in the body, but there is evidence that hydroxocobalamin plays a more active role in cyanide detoxication. The administration of hydroxocobalamin has been shown to protect mice against cyanide poisoning to a remarkable extent (Mushett *et al.*, 1952). Urinary thiocyanate excretion was found to be increased not only by the ingestion of cyanide (Hartmann and Wagner, 1949) but also by vitamin B_{12} deficiency (Wokes *et al.*, 1955). The latter finding could be interpreted as evidence of an interference with cyanide detoxication. Wilson and Langman (1966) proposed that the linkage of cyanide with cobalamin may be the usual method of disposal of minute traces of HCN in the body. Cima *et al.* (1967) showed that the enzymatic process of decyanation of cobalamin calls for adequate supplies of nicotinic acid and riboflavin. Clark in 1936 proposed that both HCN and sulfite (derived from thiosulfate) inactivated thiamine. Hence, a vicious circle could be established by a state of deficiency of the various components of the vitamin B complex in the presence of HCN.

While these considerations weigh heavily on the question of chronic toxicity, there are few biochemical problems involved in severe acute poisoning. Cyanide has a strong affinity for cytochrome oxidase, forming a link that results in immediate inhibition of cellular respiration. Death ensues from generalized cytotoxic anoxia, the brain probably being the most susceptible organ. The link is reversible, and with sublethal doses the cyanide may be cleared from the cell, leaving no permanent enzymatic derangement. There is no experimental evidence for a cumulative effect of HCN on this enzyme system.

B. Significance of Thiocyanate

Although, as stated, cyanogen may to some extent be handled in other ways, a clear rise in serum thiocyanate after ingestion of organic cyanogen compounds is good evidence that some degradation to HCN must have taken place (Støa, 1957). Because of its highly reactive nature, very little free HCN can be found in body fluids, and it is probable that in the past certain estimations of cyanide in the tissues also included thiocyanate. The colorimetric method of Aldridge (1945) depends on the bromination of either cyanide or thiocyanate and does not distinguish between the two. Free cyanide in small quantities in body fluid can be estimated by the method of Boxer and Rickards (1950), but in extreme dilutions it is doubtful accuracy, and it requires deproteination, so that the important factor of protein binding cannot be assessed. Osuntokun *et al.* (1970) described a modification for the estimation of "free" HCN; this, however, may also artifactually release HCN from thiocyanate (Vaizey and Wilson, 1977).

The estimation of thiocyanate either in blood saliva or in urine is not a simple index of the amount of cyanide ingested for a number of reasons. Firstly, there is the question of other metabolic pathways, although this factor may be small. Second, there is a large thiocyanate metabolic pool in the body, which will fluctuate widely with the variations in the dietary intake of preformed thiocyanate. Støa (1957) demonstrated that thiocyanate is present in considerable but variable amounts in green vegetables, milk, and beer. He also found that the urinary excretion of thiocyanate varied directly with urine volume and with chloride excretion. Plasma levels of thiocyanate also varied with the degree of hydration of the subject and could be lowered by debility and repeated vomiting. It may therefore be difficult to detect, or assess the importance of, fluctuations in the thiocyanate pool following the ingestion of cyanogenetic glucosides in food.

Thiocyanate is well recognized as a goitrogen. Further medical interest has been aroused by the observation that it is a powerful catalyst of nitrosation (Boyland and Walker, 1974). The nitrate content of gastric juice may be high in hypochlorhydric stomachs (Ruddell *et al.*, 1977). In the presence of increased thiocyanate in the saliva and hence in the gastric juice, the production of carcinogenic nitrosamine is facilitated. This may be relevant to gastric cancer and its relationship to diet, smoking, and hypochlorhydria.

C. Features of Acute HCN Toxicity

The minimal lethal dose of HCN taken by mouth has been estimated to be between 0.5 and 3.5 mg/kg body weight (Chen *et al.*, 1934; Gettler and Baine, 1938; Halstrom and Moller, 1945). The lethal dose of the alkaline cyanides is approximately twice that of HCN. A relatively large dose of HCN will cause death within a few minutes, but survival for up to 3 hr has been reported after smaller doses. Survivors have described initial symptoms of peripheral numbness and light-headedness. This is followed by mental confusion and stupor, cyanosis, twitching and convulsion, with terminal coma. Smaller, nonfatal doses may produce headache, a sensation of tightness in the throat and chest, palpitations, and muscle weakness. The same symptoms may occur from exposure to low concentrations of HCN gas (Polson and Tattersall, 1959). No other symptoms directly attributable to HCN have been described.

In experimental animals the principal manifestations of toxicity are stupor and convulsions. With regard to the poisoning of cattle, it should be noted that well-fed animals can show a considerable degree of tolerance to a steady intake of cyanogenetic grasses with a potential cyanide yield of as much as 50 mg/kg HCN per day; starved cattle, on the other hand, taking acute doses daily may succumb to a very much lower intake (Rose, 1941). This may be due not only to the speed of digestion and absorption, but probably to impairment of detoxification in the malnourished state. Such findings are relevant to the human situation.

IV. TOXICOLOGY OF CYANOGENETIC PLANTS

A. Mechanism of Human Poisoning

Probably the earliest and one of the best descriptions of human poisoning by cyanogen of plant origin is that of Davidson and Stevenson (1884). This occurred in Mauritius because of the ingestion of *Phaseolus lunatus,* known locally as "pois d'Achery." The clinical features were those of mental confusion, generalized muscle paresis, and respiratory distress, as in acute poisoning with inorganic cyanide. These symptoms, however, were delayed and were preceded by severe abdominal pains and vomiting.

Human poisoning by *Phaseolus lunatus* was reported briefly by Kohn-Abrest (1906), Guignard (1907), Lang (1907), Anonymous (1912), and Viehoever (1940). The latter report dealt with a series of family outbreaks in Puerto Rico between 1917 and 1925 with seven deaths. These were all due to black or dark varieties of the species, known variously as Achery, Java, or Kratok beans.

The poisonous properties of cassava were first recorded by Clusius in 1605, but there are more recent references from the West Indies (Carmody, 1900) and West Africa (Clark, 1936). Even today, deaths after heavy meals of cassava continue to be reported in the Nigerian press (Osuntokun, 1973). Clark (1936) also referred to poisoning by unripe millet in Nigeria, and Baggchi and Ganguli (1943) reported poisoning of cattle by sorghum in India. According to the same authors, cattle are frequently killed by young bamboo shoots, and human poisoning is also known.

With regard to amygdalin, bitter almonds have caused accidental deaths (Blyth and Blyth, 1920), and oil of bitter almonds used to be a not uncommon agent of suicide or homicide. Accidental poisoning has occasionally followed the ingestion of peach kernels (Falck, 1894), Kalabey apricot (McTaggart, 1936), and choke cherries (Pijoian, 1942).

If these studies are taken in conjunction with a much larger literature on animal poisoning, both accidental and experimental, there can be little doubt that the acute toxicity of these various plants is related to the release of HCN. The poisonous nature of cassava in particular has been recognized for hundreds of years, and it is significant that in various parts of the tropics the long-established local traditions relating to its preparation as food serve, in effect, to reduce the HCN yield.

As already indicated, hydrolysis occurs when the plant is macerated in water. The free HCN is then readily volatilized on boiling. If, however, the cooking utensil is covered with a lid, condensation occurs and the cooking water may become heavily contaminated. Particularly in the case of cassava, bruising of the root may lead to considerable autolysis before it ever reaches the cooking pot. It is tradition, therefore, to wash cassava before cooking, and this should be done in running water. If the cassava is washed in still water, the cyanogen may be inadequately removed.

Cassava is a foodstuff of enormous economic importance and potential, both for human beings and animals (Nestel, 1973), and it meets some 10% of the world's caloric requirements (Anonymous, 1973). A few of the many ways in which it is processed for food in different parts of the tropics are illustrated in Fig. 4.

Immature bamboo shoots are a popular delicacy in many Eastern countries. Their collection and handling may readily lead to damage, or they may be trampled by man or cattle. The shoots are commonly cooked, and the same conditions apply as for cassava. In addition, however, they may be incorporated uncooked in pickles and chutney, from which HCN may subsequently be liberated experimentally (Baggchi and Ganguli, 1943).

Not only may autolysis occur before the glucosidase has been destroyed, but

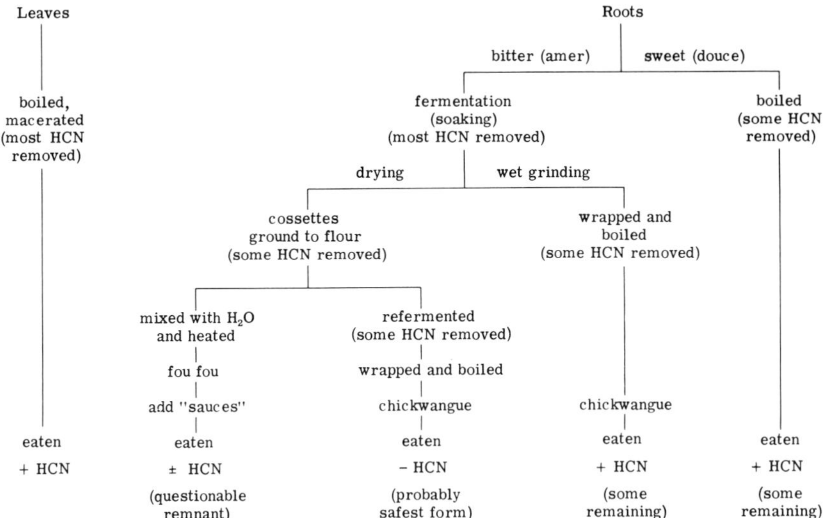

FIG. 4. Diagram showing varied usage of cassava in Zaire. Reproduced from Rogers *et al.* (1973) with the permission of the International Development Research Centre, Ottawa, Canada.

after it has been inactivated by heating, contact with fresh glucosidase may still lead to the decomposition of the glucoside, which is heat stable. Thus, cooked bamboo, beans, or cassava that are left for any time and then mixed with fresh, uncooked vegetables may become subject to autolysis. It is common practice in the East for a mixture of the evening's hot meal to be left overnight and eaten the following morning.

We have so far considered only the possibility of hydrolysis before the food is eaten. The question remains whether the whole intact plant or the intact glucoside itself may be toxic after ingestion. Auld (1912–1913) showed that the glucosidase was inactive at the pH of human saliva or of gastric juice and was also inactivated by the presence of cellulose or glucose. It is theoretically possible, however, that if a large vegetable meal were eaten hurriedly, particularly by poorly nourished people with low gastric acidity, autolysis could continue in the stomach for an appreciable length of time, until the contents were fully permeated by gastric juice. With regard to the intact glucoside, it has been shown experimentally that, in the absence of effective glucosidase, neither saliva nor dilute hydrochloric acid at body temperature can release significant amounts of HCN from *Phaseolus lunatus* (Montgomery, 1964). Auld (1912–1913) fed phaseolunatin and amygdalin to guinea pigs in amounts that would theoretically yield up to 12 lethal doses of HCN per day, without ill effect.

In spite of these negative findings, the ultimate fate of the ingested cyanogenetic glucoside remains unknown, and it may quite possibly lead to an appreciable increase in the body's cyanide or thiocyanate pool. In certain circumstances, the microbial flora of the intestine might play a part in the decomposition of the glucoside (Winkler, 1958). It is not always clear why outbreaks of human poisoning have occurred, In the instance reported from Mauritius (Davidson and Stevenson, 1884), the *Phaseolus* beans had been boiled, separated from their cooking water, and reboiled. Gabel and Druger (1920) reported that vomiting was produced experimentally in human subjects by cyanogenetic lima beans that had been boiled for 2½ hr, and cyanide was detected in the urine. The absolute safety of cooked food containing such glucosides must still be questioined (de Bruijn, 1973; Sadik and Ki Hahn, 1973).

B. Prevention of Poisoning

Immediate requirements to prevent poisoning are clear from the preceding discussion. Precautions must be taken to ensure that any HCN that has been formed is thoroughly removed from the food in the process of cooking and to ensure that the cooking water is discarded. Cassava should be eaten only if it is fresh, should be rejected if it is bruised, should be freshly peeled, and should be washed in running water after peeling. Ideally, cassava known to be bitter should not be eaten at all. Other cyanogenetic plants should not be stored for long periods and again should not be damaged in harvesting, marketing, or preparation for eating. Cooked plants should not be mixed with fresh vegetables.

Another aspect of prevention is to limit the availability of cyanogenetic plants for consumption. This has obvious applications in animal husbandry. In the case of human food, attention in the past was centered on *Phaseolus lunatus*. Over a period of many years, the Imperial Institute in London fostered the selective growing of varieties with low cyanogen yield and advocated the selective importation of these species (Anonymous, 1912, 1920). This has also been accepted policy in the United States.

Such crops as sorghum and maize tend to be consumed in the immature state only when there is a local food shortage, and prevention of this aspect of poisoning involves the question of even distribution of resources in tropical countries, as well as local education policies.

C. Chronic Toxicity and the Nervous System

Because of the powerful cytotoxic effect of cyanogen and the dramatic picture of acute poisoning, there has inevitably been speculation whether chronic pathological changes might arise from prolonged exposure to low concentrations, either in the atmosphere or from other sources. Not only is man exposed to HCN in polluted air and cigarette smoke, but infection with certain bacteria may be accompanied by considerable production of cyanide (Adams *et al.*, 1966). Many millions of people consume cassava as a staple item of diet, and particularly in tropical Africa the consumption of pulses may be as high as 300 gm per day (Anonymous, 1959).

The possible goitrogenic and carcinogenic roles of thiocyanate have already been mentioned. In addition, chronic cassava toxicity has been incriminated (so far inconclusively) in the high incidence of chronic pancreatic disease in south India and tropical Africa (Thomas and Pitchumoni, 1973). In animals under natural conditions, chronic ill effects of ingested cyanogen are difficult to demonstrate, even though the blood thiocyanate levels may be raised. There is a slight tendency for goiter to occur, although it is preventable by an increased iodine intake, and there may be some failure of growth; the latter may be related to a relative lack of essential sulfur-containing amino acids.

Apart from this, the attention of research workers has been on the nervous system. Collins and Martland (1908) described a case somewhat resembling poliomyelitis in a worker chronically exposed to KCN. Wicke (1935) reported demyelination of the pallidal zone of the brain in dogs and man as a result of cyanide poisoning. He reported the case of a man showing evidence of damage to this area of the brain, who had worked for 8 years in a brass foundry, where he was exposed to KCN or HCN. Neither of these findings has ever been substantiated by further case reports.

Large numbers of experiments in various species have been performed using repeated injections of sublethal doses of cyanide. Wilson (1965) gives 13 references. In all these experiments, pathological changes occurred in only a minority, and in nearly all of them the animals had suffered from repeated episodes of convulsions and asphyxia, which could induce brain damage. Smith and Duckett (1965) reported possibly more specific changes accompanied by a rise in blood levels of thiocyanate, following the administration of smaller doses of cyanide to rats over a period of several months. Myelin pallor and cell degeneration appear to be preceded and accompanied by proliferation of the oligodendroglia (Ibrahim *et al.*, 1963; Bass, 1968).

Optic atrophy as a result of repeated injections of cyanide was demonstrated by Ferraro (1933) and Hurst (1940) and more recently by Lessell (1971). These findings are significant in view of the extensive investigations that have been made of blindness occurring in smokers (tobacco amblyopia) and of the relation of this lesion to a disturbance of vitamin B_{12} metabolism. Optic nerve damage similar to tobacco amblyopia occurs in vitamin B_{12} deficiency. Heaton *et al.* (1958) first showed that the amblyopia of pernicious anemia occurred predominantly in males and perhaps exclusively in smokers. Wokes (1958) proposed that the causal factor was cyanide, accentuated by failure to detoxicate via hydroxocobalamin. Amblyopia of tobacco smokers was shown to improve with cobalamin therapy even if smoking was continued (Freeman and Heaton, 1961), and hydroxocobalamin was more effective in this respect that cyanocobalamin

(Chisholm *et al.*, 1967). Wilson *et al.* (1971) showed that there was a significant increase in the proportion of cyanocobalamin in the blood of these patients.

Against this background, we may consider the theory first advanced by Moore (1934), that amblyopia in West Africa is a manifestation of chronic cassava poisoning. Clark in 1936 proposed that the toxic effect of cassava was due to cyanide and stressed the importance of adequate sulfur-containing amino acids for its detoxication, which could be of critical importance in a protein-depleted population. Moore also emphasized the importance of adequate supplies of the vitamin B complex in maintaining the integrity of the nervous system when exposed to toxins. This condition, now known as tropical amblyopia, is common throughout the tropics among malnourished communities, and it also occurred in Far Eastern prison camps during the Second World War. In spite of an extensive literature, little has been added in 50 years to the original theories of Moore and Clark. The pattern of visual field defects is variable, irregular, and bizarre. It is consistent with the effect of a toxin diffusing into the retina and affecting those areas that are least protected from light by the overlying nerve fibers. This could be due to the release of free HCN from cyanocobalamin by photolysis. In more severe cases direct toxic demyelination of the optic nerve can be postulated. Although this form of amblyopia is commonly an isolated lesion, it may also be a feature of a more generalized neurological syndrome that includes sensory ataxia, nerve deafness, mental changes, and spastic weakness. In all communities in which this syndrome occurs the diet tends to be low in animal protein and relatively rich in vegetables, including cassava and pulses. The earlier literature on this subject was reviewed by Denny-Brown (1947) and Spillane (1947), and more recent articles are those of Cruickshank (1956), Money (1959), Montgomery *et al.* (1964), Collomb *et al.* (1967), Ashcroft *et al.* (1967), and Osuntokun (1968).

Wilson (1963, 1965) in a study of Leber's disease, a rare hereditary form of optic atrophy, observed other manifestations of the disease rather similar to those found in the tropical syndrome; he also found indirect evidence of an inherent defect in cyanide metabolism in these subjects. Smokers with this condition failed to show the expected rise in plasma thiocyanate concentration, suggesting a failure to handle cyanide in this way. Differential studies showed that there was an abnormal increase in plasma cyanocobalamin regardless of the subjects' smoking habits. Similar observations were also made in some of their relatives (Wilson *et al.*, 1971).

Studies in patients with the tropical ataxic syndrome in a cassava-eating area of Nigeria showed a striking elevation of plasma thiocyanate levels compared to healthy controls (Monekosso and Wilson, 1966). This was the first positive biochemical lead to have been discovered in this condition. Detailed field studies by Osuntokun (1968) furnished further confirmation of exposure to cyanide, and it was significant that severe depletion of sulfur-containing amino acids was also demonstrated (Osuntokun *et al.*, 1968), suggesting that the subjects' body cyanide content could be even greater than was revealed by the serum thiocyanate level due to a relative lack of sulfurtransferase. More recent information on this subject was summarized by Osuntokun in 1973. In some villages in endemic areas with a heavy consumption of cassava the prevalence of neurological disease was as high as 3%. Significantly, there was an increased incidence of goiter among the neurological patients. Levels of plasma thiocyanate, plasma cyanide, and urinary thiocyanate were all significantly raised and varied with changes in the cassava diet inside and outside the hospital. There was no evidence of a deficiency of total vitamin B_{12}, and in many tropical areas plasma levels of B_{12} tend to be high. In addition to these extensive and sophisticated studies in Nigeria, a rise in the serum thiocyanate levels was found in a small group of patients suffering from tropical sensory ataxia in Tanzania (Makene and Wilson, 1972), again in an area where cassava is an important food item.

The evidence is now strong that cyanogen ingestion can give rise to chronic neurological disease in man. This evidence is so far confined to tropical ataxic neuropathy and tropical amblyopia seen in Nigeria and possibly in Tanzania, with cassava as the source. The role of cyanide in other cassava areas and of other cyanogenetic plants remains to be defined, but it seems logical that under certain circumstances chronic toxicity could occur. The importance of the individual's nutritional status, particularly with regard to sulfur-containing amino acids and the vitamin B complex, cannot be overemphasized, and these factors are all interrelated. Future preventive work in this field must involve not only the reduction of exposure to dietary cyanide but also improvement in the nutrition of the exposed population.

References

Abrol, Y.P., and Conn, E. E. (1966). *Phytochemistry* **5,** 237–242.

Adams, J. H., Blackwood, W., and Wilson, J. (1966). *Brain* **89,** 15–26.

Aldridge, W. N. (1945). *Analyst* **70,** 474–475.

Anonymous (1903). *Bull. Imp. Inst., London* **1,** 12–18.

Anonymous (1912). *Bull. Imp. Inst., London* **10,** 653–655.

Anonymous (1920). *Bull. Imp. Inst., London* **18,** 471–476.

Anonymous (1959). Report of the FAO/CCTA Technical Meeting on Legumes in Agriculture and Human Nutrition in Africa, Rome, Italy.

Anonymous (1964). "Smoking and Health: Report of the Advisory Committee to the Surgeon General of the Health Service," Public Health Serv. Publ. No. 1103, p.60. U.S. Dept. of Health, Education and Welfare, Washington, D.C.

Anonymous (1973). *Lancet* **2,** 245–246.

Ashcroft, M. T., Cruickshank, E. K., Hinchcliffe, R., Jones, W. I., Miall, W. E., and Wallace, J. (1967). *West Indian Med. J.* **16,** 233–245.

Auld, S. J. M. (1912). *J. Board Agric.* **19,** 446–460.

Auld, S. J. M. (1912-1913). *J. Agric. Sci.* **5,** 409–417.

Baggchi, K. N., and Ganguli, H. D. (1943). *Med. Gaz.* **78,** 40–42.

Bass, N. H. (1968). *Neurology* **18,** 167–177.

Bertrand, G. (1907). *Bull Sci. Pharmacol.* **4,** 65.

Bissett, F. H., Clapp, R. C., Coburn, R. A., Ettingler, M. G., and Long, L. (1969). *Phytochemistry* **8,** 2235–2247.

Blyth, A. W., and Blyth, M. W. (1920). "Poisons," 5th ed., pp. 215 and 223. Griffiths, London.

Boxer, G. E., and Rickards, J. C. (1950). *Arch. Biochem. Biophys.* **30,** 372–381.

Boxer, G. E., and Rickards, J. C. (1952). *Arch. Biochem. Biophys.* **37,** 7–26.

Boyland, E., and Walker, S. A. (1974). *Nature (London)* **248,** 601–602.

Butler, G. W. (1965). *Phytochemistry* **4,** 127–131.

Butler, G. W., Reay, P. F., and Tapper, B. A. (1973). *In* "Chronic Cassava Toxicity" (B. Nestel and R. MacIntyre, eds.), Mongr. IDRC-010e, pp. 65–71. Int. Dev. Res. Cent., London.

Carmody. (1900). *Lancet* **2,** 736–737.

Chen, K. K., Rose, C. L., and Clowes, G. H. A. (1934). *Am. J. Med. Sci.* **188,** 767–781.

Chisholm, I. A., Bronte-Stewart, J., and Foulds, W. S. (1967). *Lancet* **2,** 450–451.

Cima, L., Levorato, C., and Mantovan, R. (1967). *J. Pharm. Pharmacol.* **19,** 32–40.

Clark, A. (1936). *J. Trop. Med.* **39,** 269–276 and 285–295.

Clemedson, C. J., Sorbo, B., and Ullberg, S. (1960). *Acta Physiol. Scand.* **48,** 382–389.

Clusius, C. (1605). "Liber Exoticorum." Leyden [cited in *Bull. Imp. Inst., London* **4,** 334 (1906)].

Collens, A. E. (1915). *Bull. Dep. Agric., Trinidad Tobago* **14,** 54.

Collins, J., and Martland, H. S. (1908). *J. Nerv. Ment. Dis.* **35,** 417–426.

Collomb, M., Quere, M. A., Cross, J., and Giordano, C. (1967). *J. Neurol. Sci.* **5,** 159–179.

Conn, E. E., and Butler, G. W. (1969). *In* "Perspectives in Phytochemistry" (J. B. Harborne and T. Swain, eds.), pp. 47–74. Academic Press, New York.

Couch, J. F. (1934). *U.S., Dep. Agric. Leafl.* **88,** 2–4.

Cruickshank, E. K. (1956). *West Indian Med. J.* **5,** 147–158.

Davidson, A., and Stevenson, T. (1884). *Practitioner* **32,** 435–439.

de Bruijn, G. H. (1973). *In* "Chronic Cassava Toxicity" (B. Nestel and R. MacIntyre, eds.), Monogr. IDRC-010e, pp. 43–48. Int. Dev. Res. Cent., London.

Denny-Brown, D. (1947). *Medicine (Baltimore)* **26,** 41–113.

Densen, P. M., Davidow, B., Bass, H. E., and Jones, E. W. (1967). *Arch. Environ. Health* **14,** 865–874.

Dunstan, W. R., and Henry, T. A. (1903). *Proc. R. Soc. London, Ser. B* **72,** 85–88.

Eisner, T., Eisner, H. E., Hurst, J. J., Kafatos, F. D., and Meinwald, J. (1963). *Science* **139,** 1218–1220.

Falck. (1894). *Med. Rec.* **45,** 633 (Cited in Schrady).

Ferraro, A. (1933). *Arch. Neurol. Psychiatry* **29,** 1364–1367.

Fiedler, H., and Wood, J. L. (1956). *J. Biol. Chem* **222,** 387–397.

Freeman, A. G., and Heaton, J. M. (1961). *Lancet* **1,** 908–911.

Gabel, W., and Kruger, W. (1920). *Muensch. Med. Wochenschr.* **67,** 214–215.

Gettler, A. O., and Baine, J. O. (1938). *Am. J. Med. Sci.* **195,** 182–198.

Gibbs, R. D. (1963). *In* "Chemical Plant Taxonomy" (T. Swain, ed.), pp. 6082. Academic Press, New York.

Gilchrist, D. G., Leuschen, W. E., and Hittle, C. N. (1967). *Trop. Sci.* **7,** 267–268.

Goldstein, F., and Rieders, F. (1953). *Am. J. Physiol.* **173,** 287–290.

Guignard, L. (1907). *Bull. Sci. Pharmacol.* **14,** 556–557.

Halstrom, F., and Moller, K. D. (1945). *Acta Pharmacol. Toxicol.* **1,** 18–28.

Hartmann, F., and Wagner, K. H. (1949). *Dtsch. Arch. Klin. Med.* **196,** 432–434.

Heaton, J. M., McCormick, J. A., and Freeman, A. G. (1958). *Lancet* **2,** 286–290.

Henry, T. A. (1969). *In* "Encyclopedia Britannica," 14th ed., Vol. 10, pp. 444–445.
Horwitz, W., ed (1960). "Official Methods of Analysis of the Association of Official Agricultural Chemists," 9th ed., p. 293. AOAC, Washington, D.C.
Hurst, E. W. (1940). *Aust. J. Exp. Biol. Med. Sci.* **18,** 201–223.
Ibrahim, M. Z. M., Briscoe, P. B., Bayliss, O. B., and Adams, C. W. M. (1963). *Neurol., Neurosurg. Psychiatry* **26,** 479–486.
Indira, P., and Sinha, J. K. (1969). *Indian J. Agric. Sci.* **39,** 1021–1023.
Jaffé, W. G. (1950). *Acta Cient. Venez.* **1,** 62–64.
Jones, D. A. (1972). *In* "Phytochemical Ecology" (J. B. Harborne, ed.), pp. 103–124. Academic Press, New York.
Jones, D. A. (1973). *In* "Taxonomy and Ecology" (V. H. Heywood, ed.), pp. 213–242. Academic Press, New York.
Jones, D. A., Parsons, J., and Rothschild, M. (1962). *Nature (London)* **193,** 52–53.
Kingsbury, J. M. (1964). "Poisonous Plants of the United States and Canada." Prentice-Hall, Englewood Cliffs, New Jersey.
Kohn-Abrest, E. (1906). *C.R. Hebd. Seances Acad. Sci.* **142,** 586.
Lang, K. (1933). *Biochem. Z.* **259,** 243–256.
Lange, W. (1907). *Arb. Kaiserl. Gesundh.* **25,** 478.
Lessell, S. (1971). *Arch. Ophthalmol.* **84,** 194–204.
McTaggart, C. M. (1936). *Br. Med. J.* **2,** 100.
Makene, W. J., and Wilson, J. (1972). *J. Neurol., Neurosurg. Psychiatry* **35,** 31–33.
Monekosso, G. L., and Wilson, J. (1966). *Lancet* **1,** 1062–1064.
Money, G. L. (1959). *West Afr. Med. J.* **8,** 3–17.
Montgomery, R. D. (1964). *West Indian Med. J.* **13,** 1–11.
Montgomery, R. D., Cruickshank, E. K., Robertson, W. B., and McMenemey, W. H. (1964). *Brain* **87,** 425–462.
Moore, D. G. F. (1934). *Ann. Trop. Med.* **28,** 295–303.
Mushett, C. W., Kelley, K. L., Boxer, G. E., and Rickards, J. C. (1952). *Proc. Soc. Exp. Biol. Med.* **81,** 234–237.
Nartey, F. (1968). *Phytochemistry* **7,** 1307–1312.
Nartey, F. (1973). *In* "Chronic Cassava Toxicity" (B. Nestel and R. MacIntyre, eds.), Monogr. IDRC-010e, pp. 73–87. Int. Dev. Res. Cent., London.
Nestel, B. (1973). *In* "Chronic Cassava Toxicity" (B. Nestel and R. MacIntyre, eds.), Monogr. IDRC-010e, pp. 5–26. Int. Dev. Res. Cent., London.
Osborne, J. S., Adamek, S., and Hobbs, M. E. (1956). *Anal. Chem.* **28,** 211–215.
Osuntokun, B. O. (1968). *Brain* **91,** 215–247.
Osuntokun, B. O. (1973). *In* "Chronic Cassava Toxicity" (B. Nestel and R. MacIntyre. eds.), Monogr. IDRC-010e, pp. 127–138. Int. Dev. Res. Cent., London.
Osuntokun, B. O., Durowoju, J. E., McFarlane, H., and Wilson, J. (1968). *Br. Med. J.* **2,** 647–649.
Osuntokun, B. O., Aladetoyinbo, A., and Adeuja, A. O. G. (1970). *Lancet* **2,** 372–373.
Oyenuga, V. A., and Amzigo, E. O. (1957). *West Afr. J. Biol. Chem.* **1,** 39–43.
Paris, R. (1963). *In* "Chemical Plant Taxonomy" (T. Swain, ed.), pp. 337–358. Academic Press, New York.
Pijoian, M. (1942). *Am. J. Med. Sci.* **204,** 550–553.
Polson, C. J., and Tattersall, R. N. (1959). *In* "Clinical Toxicology," p. 114. English Univ. Press, London.
Rogers, D., Slater, C., and Hersch, G. (1973). *In* "Chronic Cassava Toxicity" (B. Nestel and R. MacIntyre, eds.), Monogr. IDRC-010e, p. 18. Int. Dev. Res. Cent., London.
Rose, A. L. (1941). *Aust. Vet. J.* **17,** 211–219.
Rosenthal, O. (1948). *Fed. Proc., Fed. Am. Soc. Exp. Biol.* **7,** 181–182.

Ruddell, W. S. J., Blendis, L. M., and Walters, C. L. (1977). *Gut* **18,** 73–77.

Sadik, S., and Ki Hahn, S. (1973). *In* "Chronic Cassava Toxicity" (B. Nestel and R. MacIntyre, eds.), Monogr. IDRC-010e, pp. 41–42. Int. Dev. Res. Cent., London.

Smith, A. D. M. (1961). *Lancet* **1,** 1001–1002.

Smith, A. D. M., and Duckett, S. (1965). *Br. J. Exp. Pathol.* **46,** 615–622.

Spillane, J. D. (1947). "Nutritional Disorders of the Nervous System." Livingstone, Edinburgh.

Støa, K. F. (1957). *Arbok Univ. Bergen, Med. Ser.* **2,** 1–165.

Thomas, E., and Pitchumoni, C. S. (1973). *Lancet* **2,** 1397–1398.

Vaizey, C., and Wilson, J. (1977). *Br. J. Pharmacol.* (in press).

Viehoever, A. (1940). *Thai. Sci. Bull.* **2,** 1–99.

Wicke, R. (1935). *Med. Welt* **9,** 1216–1217.

Wilson, J. (1963). *Brain* **86,** 347–362.

Wilson, J. (1965). *Clin. Sci.* **29,** 505–515.

Wilson, J., and Langman, M. J. S. (1966). *Nature (London)* **212,** 787–789.

Wilson, J., Linnell, J. C., and Matthews, D. M. (1971). *Lancet* **1,** 259–261.

Winkler, W. O. (1951). *J. Assoc. Off. Agric. Chem.* **34,** 541.

Winkler, W. O. (1958). *J. Assoc. Off. Agric. Chem.* **41,** 282–287.

Wokes, F. (1958). *Lancet* **2,** 526–527.

Wokes, F., and Willimott, S. C. (1951). *J. Pharm. Pharmacol.* **3,** 905–917.

Wokes, F., Baxter, N., Horsford, J., and Preston, B. (1953). *Biochem. J.* **53,** xix–xx.

Wokes, F., Badenoch, J., and Sinclair, H. M. (1955). *Am. J. Clin. Nutr.* **3,** 375–382.

Wood, T. (1966). *J. Sci. Food Agric.* **17,** 85–90.

Wood, J. L., and Cooley, S. L. (1956). *J. Biol. Chem.* **218,** 449–457.

Zitnak, A. (1973). *In* "Chronic Cassava Toxicity" (B. Nestel and R. MacIntyre, eds.), Monogr. IDRC-010e, pp. 89–96. Int. Dev. Res. Cent., London.

CHAPTER 6

Saponins*

YEHUDITH BIRK AND IRENA PERI

I. INTRODUCTION

The saponins are glycosides that occur in a wide variety of plants. They are generally characterized by their bitter taste, foaming in aqueous solutions, and their ability to hemolyze red blood cells. The saponins are highly toxic to cold-blooded animals, their toxicity being related to their activity in lowering surface tension. They are commonly isolated by extraction of the plant material with hot water or ethanol. Upon complete hydrolysis they yield sapogenins, which are

*This chapter is dedicated to the memory of our late colleagues Benjamin Gestetner, Yaacov Assa, and Gad Reshef. Their valuable contribution to the structure–function relationships of legume saponins is fully documented in this chapter.

TOXIC CONSTITUENTS OF PLANT FOODSTUFFS, SECOND EDITION

ISBN 0-12-449960-0

either steriods (C_{27}) or triterpenoids (C_{30}), and sugars (hexoses, pentoses, and saccharic acids). The saponins in foods and feeds have been studied very little and different properties that were attributed to them have not always been verified.

Earlier knowledge on saponins was reviewed by Birk (1969) in the preceding edition and also by Bondi *et al.* (1973). This chapter will deal primarily with the isolation and characterization of triterpenoid saponins, particularly from alfalfa, and with their mode of action on the molecular level.

It is hoped that further elucidation of the mechanism of interaction of saponins with cellular and membranal components will lead to a better understanding of the role of saponins in nutrition and medicine.

II. DISTRIBUTION AND CHEMICAL COMPOSITION OF ALFALFA SAPONINS

Gestetner (1971) isolated and fully characterized the unique, biologically active alfalfa saponin medicagenic acid 3-β-*O*-triglucoside (Fig. 1). In addition to the known components of alfalfa saponins, the structure of the unidentified soyasapogenol U has been established as hederagenin (Shany *et al.,* 1972). Hederagenin differs from medicagenic acid at C-23, where it bears a CH_2OH group instead of COOH (Fig. 2). Exact analyses of saponins present in such alfalfa varieties as DuPuits and Lahontan were carried out by Hanson *et al.* (1973). Four other unidentified soybean sapogenins were classified by mass spectrometric analyses, as related to the pentacyclic triterpenoids medicagenic acid and soybean sapogenins (Table I).

The saponins of alfalfa varieties (*Medicago sativa* L.) differ in composition as well as in quantity (Pedersen *et al.,* 1966; Hanson *et al.,* 1973). DuPuits and Lahontan alfalfa varieties contain 25 different saponins, but the chemical composition of the saponins in the two varieties is not the same (Berrang *et al.,* 1974). The aglycones of DuPuits saponins are mainly of the medicagenic acid

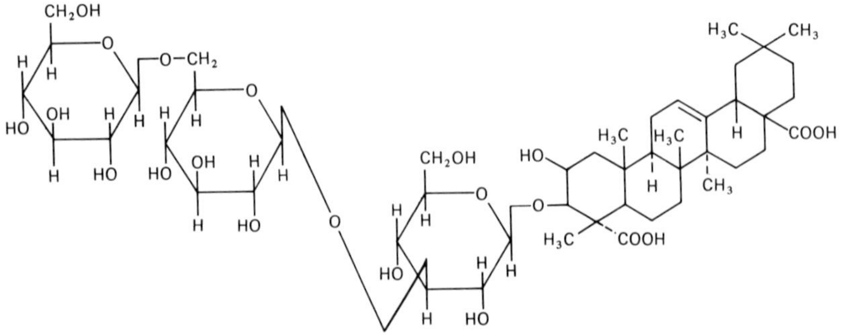

FIG. 1. Structural formula of the alfalfa saponin medicagenic acid 3-β-*O*-triglucoside.

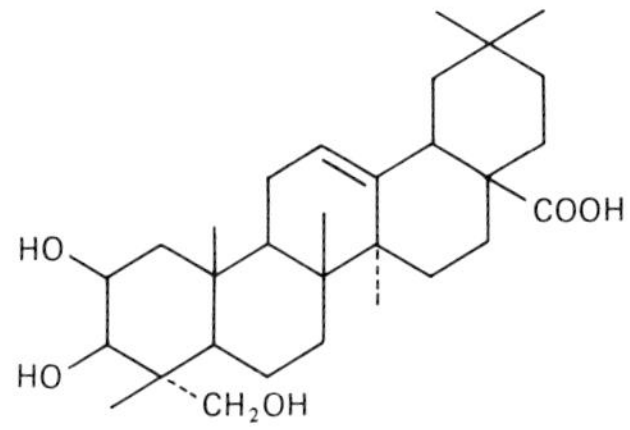

FIG. 2. Structural formula of hederagenin.

type, whereas the saponins of Lahontan contain soybean sapogenins and monocarboxylic acids as aglycones. Also, DuPuits saponins lack D-galactose, which is present in lahontan saponins. Polish cultivated varieties of alfalfa (*Medicago media* Pers.) have about the same saponin content and chemical composition (Jurzysta, 1975).

The first analysis of saponins in a single alfalfa seed was carried out by Jurzysta (1973). Soyasapogenols B, C, and E were found, whereas soya-

TABLE I

AGLYCONES OF DUPUITS AND LAHONTAN ALFALFA SAPONINS[a]

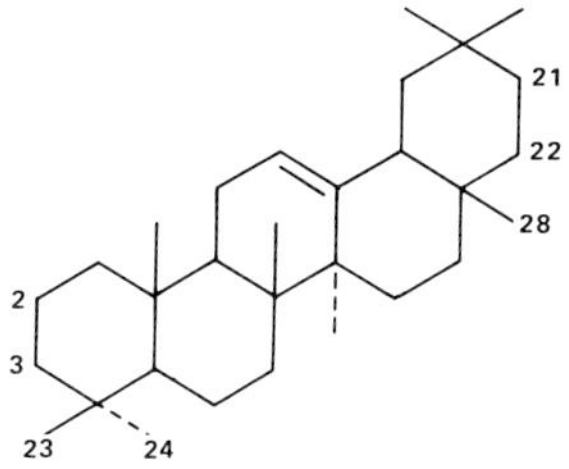

Name	Number of substituents		Structural Features
	OH	COOH	
Medicagenic acid	2	2	2,3-OH; 23,28-COOH
Unknown	3	2	OH in ring D or E; 28-COOH
Soyasapogenol A	4	—	3,21,22,23-OH
Unknown	3	1	OH in ring D or E; 28-COOH
Lucernic acid	3	1	Lactone
Soyasapogenol B	3	—	3,21,23-OH
Unknown	5	—	Two OH in ring D or E
Unknown	4	1	Two OH in ring D or E; 28-COOH

[a]From Hanson *et al.* (1973).

sapogenols A and D and medicagenic acid were not. Thus, the composition of alfalfa seed saponins seems to be similar to that of soybean seeds.

Hanson *et al.* (1963) studied the effect of varieties, cuttings, and locations on saponin content in whole alfalfa plants. The relatively high proportion of the variety variance component indicated the possibility of obtaining plants of low saponin content by selection. Later, Jones (1969) found high heritability for saponin content.

Attempts were made to change saponin concentration by selection and breeding (Pedersen and Wang, 1971). As a result of the selection, the average saponin content of the "low" lines was reduced by 28.6% and that of the "high" lines was increased by 83%, as compared to the unselected parent variety. Progress in selection for high saponin content was more rapid than selection for low saponin content. Selection for saponin content also significantly influenced protein yield and forage yield, but it did not affect the yield of fiber, fat, ash, nitrogen-free extract, or seed yield. Later, Pedersen *et al.* (1973) reported that selecting for high saponin concentration increased the relative concentration of carboxylic acid-type aglycones, whereas selecting for low saponin concentration increased the relative concentration of the neutral aglycones. This tendency has been demonstrated by the high medicagenic acid content of the "high-saponin" line DuPuits and the increased content of soybean sapogenins of Lahontan. Similar results were obtained by Quazi (1975), who cultivated a variety called Wairau, which, like DuPuits, contains mainly the toxic saponin with the carboxylic acid type of aglycones, and the Washoe variety, which, like Lahontan, contains the neutral saponins. Seasonal influence was observed in the Wairau variety on the ratio of total content of saponins to medicagenic acid-containing saponins. Examination of root extracts from various tissues of two alfalfa varieties indicated that the saponins are concentrated more in the outer portion of the alfalfa root, i.e., in the bark, rather than in the xylem (Quazi, 1976).

III. BIOLOGICAL ACTIVITIES OF ALFALFA SAPONINS

A. Hemolysis

Hemolysis, one of the characteristic properties of saponins, was used as early as 1919 to identify extracted compounds from alfalfa as saponins (Jacobson, 1919). Since then, numerous assays and quantitative determinations of saponins have been based on their hemolytic properties. Hemolysis of erythrocytes has also been used to demonstrate the presence of saponins throughout the alfalfa plant. Jones (1969) reported differences in degree of hemolysis between alfalfa leaf and stem saponin extracts. Shany *et al.* (1970a) showed that alfalfa root extracts have higher surface activity and are more hemolytic than those from alfalfa tops.

A series of studies was carried out to relate the hemolytic activity to the structure and composition of the saponin molecule. It was shown that the higher hemolytic activity of the root extract, as compared to that of the top, results from the relatively higher content of medicagenic acid, as well as from the higher ratio of sapogenin to sugar in the alfalfa root saponin extract (Gestetner *et al.*, 1971a). It was also shown that the saponins containing medicagenic acid and hederagenin as their aglycones are precipitable by cholesterol and have a strong hemolytic activity (Gestetner *et al.*, 1971a). The main difference between medicagenic acid and the soybean sapogenins, which comprise the other aglycones of alfalfa saponins, is in the presence of two carboxylic groups. Blocking of the carboxylic groups in medicagenic acid results in complete loss of the hemolytic activity of the saponin (Gestetner *et al.*, 1971b). The alfalfa saponins that contain soyasapogenols A, B, C, D, and E as their sapogenins and do not lyse red blood cells differ from soybean saponins, which contain the same sapogenins but do have considerable hemolytic activity (Gestetner *et al.*, 1971a). This discrepancy is explained by the lower sapogenin/carbohydrate ratio (1:5) in alfalfa saponins as compared to the 1:1 ratio in soybean saponins, thus relating the hemolytic activity to the overall hydrophobicity of the saponin molecule.

Horber *et al.* (1974) supported the conclusion of Gestetner *et al.* (1971a) that medicagenic acid contributes to the hemolytic activity of saponins. Their data indicate that DuPuits saponins as a group contain the aglycone medicagenic acid and are by far more hemolytic than Lahontan extracts, which lack medicagenic acid. However, the presence or absence of medicagenic acid explains only part of the difference in hemolytic activity between DuPuits and Lahontan varieties. Medicagenic acid is not the only acid sapogenin that is strongly active in hemolysis. Hanson *et al.* (1973) described an unidentified acid sapogenin chemically related to, but distinctly different from, medicagenic acid. This sapogenin is responsible for the relatively high hemolytic activity of three out of ten fractions of Lahontan saponin, as compared to the three corresponding DuPuits fractions that contain no medicagenic acid (Horber *et al.*, 1974).

The hemolytic activity of saponins is generally attributed to their interaction with cholesterol in the erythrocyte membrane (Dourmashkin *et al.*, 1962; Glauret *et al.*, 1962). In the case of alfalfa saponins, this activity could be abolished by the presence of extraneous cholesterol (Shany, 1971). Shany *et al.* (1974) supported the concept that alfalfa saponins and certain hemolytic proteins (e.g., streptolysin O, creolysin) share a common binding site on the erythrocyte membrane, and this binding site is cholesterol. Treatment of membranes with one of these agents prevents the binding of the other. Since the binding site of saponin is membranal cholesterol (Schroeder *et al.*, 1971) it seems that this is also the binding site for the hemolytic proteins. Addition of alfalfa saponins to membranes prepared by osmosis of rabbit erythrocytes and to cholesterol-containing liposomes resulted in formation of pits or holes surrounded by rings, as revealed by negative staining followed by electron microscopic examination.

Assa *et al.* (1973) used the unique and well-defined alfalfa saponin, medicagenic acid 3-*β*-*O*-triglucoside, to investigate the membranal elements that might be involved in the mechanism of hemolysis of erythrocytes by saponins.

This saponin formed interaction products of different stability with membranal cholesterol, proteins, and phospholipids. No interaction was found with fatty acids of the membrane. The structural changes resulting from the interaction of membranal components with the saponin affected the activity of membranal enzymes only slightly. It was suggested that breakdown of the structure of the membrane stems from a combination of nonspecific interactions of saponins with membrane proteins, phospholipids, and cholesterol leading consequently to hemolysis.

B. Effect on Fungi and Other Microorganisms

Selective toxicity of saponins to fungi has been reported by Focke (1970), Tscheche and Wulff (1965), Wolters (1968), Zimmer *et al.* (1967), Olsen (1971a), and Leath *et al.* (1972). The last group tested the influence of alfalfa saponin and non-saponin fractions from DuPuits and Lahontan varieties on 15 alfalfa pathogens. Some of the test organisms were inhibited by the saponin fractions, some were stimulated, and others were unaffected. Saponin fractions from DuPuits exerted greater growth-inhibiting activity than similar fractions from Lahontan, suggesting that individual saponins might be responsible for the difference in activity. These findings were later confirmed by Horber *et al.* (1974). The non-saponin fractions of alfalfa also affected the growth of the organisms, but varietal differences were not observed. Recently, breeding for low saponin content has been proposed as the more effective method for reducing the saponin concentrations in alfalfa feed products. Pedersen *et al.* (1976) reported that it was possible to select for low saponin content without generally increasing the susceptibility of the plant to alfalfa pathogens.

The effect of alfalfa saponins on *Trichoderma viride* was studied by several groups of investigators. According to Horber *et al.* (1974) the presence or absence of medicagenic acid seemed to explain most of the differences in biological activity of DuPuits and Lahontans varieties. However, medicagenic acid is not the only acid sapogenin that is strongly active toward pathogens. In a manner similar to the hemolysis experiments, these investigators demonstrated the growth inhibition of *Trichoderma viride* by three saponin fractions from Lahontan. These fractions lack medicagenic acid but contain an unknown acid sapogenin that is chemically related to, but distinctly different from, medicagenic acid. It should be pointed out that the bioassays with *Trichoderma viride* have been successfully used for the estimation of alfalfa saponin activity and concentration. Leath *et al.* (1972) showed that, of 15 microorganisms tested, *Trichoderma viride* was the most sensitive to alfalfa saponins.

The role of saponins in disease resistance was discussed by Leath *et al.* (1972). No evidence was provided for the interaction of saponins and pathogens within the plant, but it is known that the concentration of saponins in the alfalfa plant is higher than the concentration that is active against pathogens *in vitro* (Leath *et al.*, 1972), and therefore the interaction seems possible. Pedersen

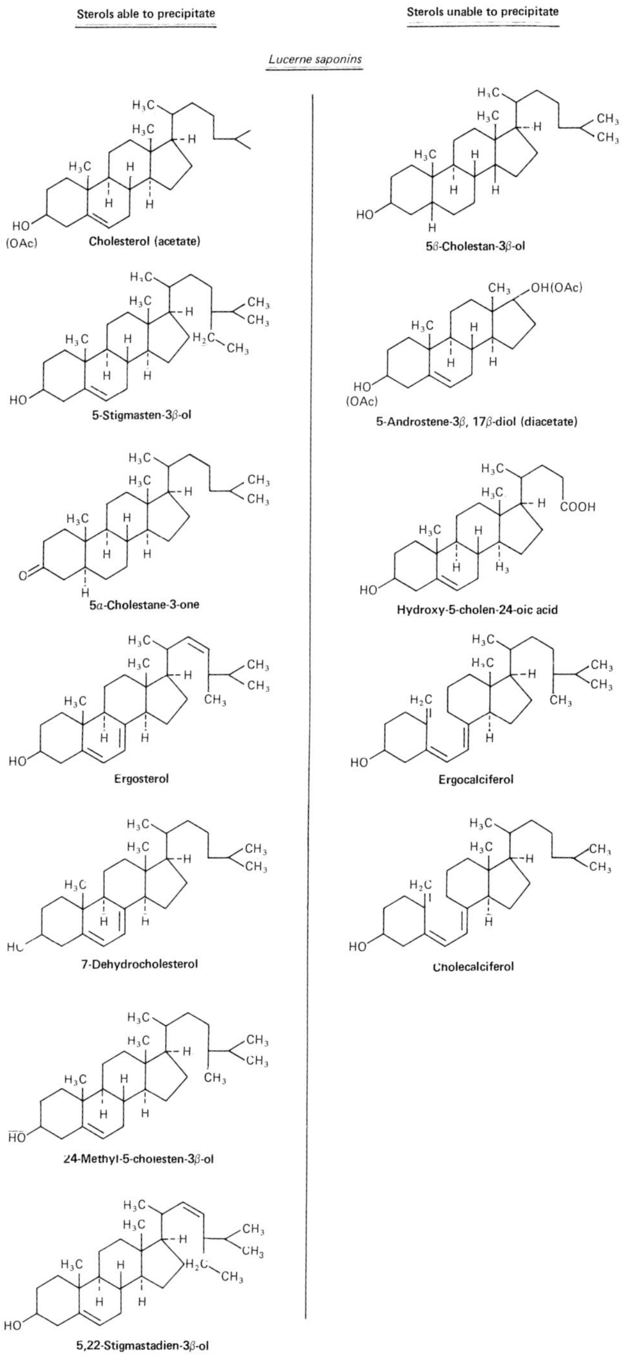

Sterols able to precipitate

Sterols unable to precipitate

Lucerne saponins

Cholesterol (acetate)

5-Stigmasten-3β-ol

5α-Cholestane-3-one

Ergosterol

7-Dehydrocholesterol

24-Methyl-5-cholesten-3β-ol

5,22-Stigmastadien-3β-ol

5β-Cholestan-3β-ol

5-Androstene-3β, 17β-diol (diacetate)

Hydroxy-5-cholen-24-oic acid

Ergocalciferol

Cholecalciferol

FIG. 3. Structural formulas of sterols that do or do not interact with alfalfa saponins.

(1975) found that the saponin was concentrated mostly in the cortex of the mature roots; the concentration decreased toward the center, and there was no evidence of saponin in the stele. The distribution and concentration of saponins in the cortex suggest that the saponin may protect the root from saponin-sensitive organisms. An interaction of alfalfa saponins with soil organisms, either in the rhizosphere of living plants or as crop residues after the crop is plowed under, has also been suggested.

Knowledge on the mode of action of saponins on fungi is still limited. Several investigators have suggested that saponins exert their fungicidal effect by precipitating sterols in the cell membrane. Gestetner *et al.* (1972) showed that the presence of an intact steroid ring structure with the conformation of cholesterol, to which a side chain characteristic of cholesterol or phytosterols is attached, is essential for the formation of a sterol–saponin addition product with alfalfa saponins (Fig. 3). The effect of alfalfa saponins from tops and roots, and of their constituents, on the mycelial growth of *Sclerotium rolfsii,* on various species and strains of *Pythium,* and on *Mycoplasma* was studied by Gestetner *et al.* (1971a) and Assa *et al.* (1972). They showed that only the sterol-precipitable saponins are able to arrest mycelial growth. The antifungal activity of alfalfa saponins can be counteracted by the addition of cholesterol or 7-dehydrosterol, but not by ergosterol, to the culture medium (Fig. 4). The finding that ergosterol, which is a

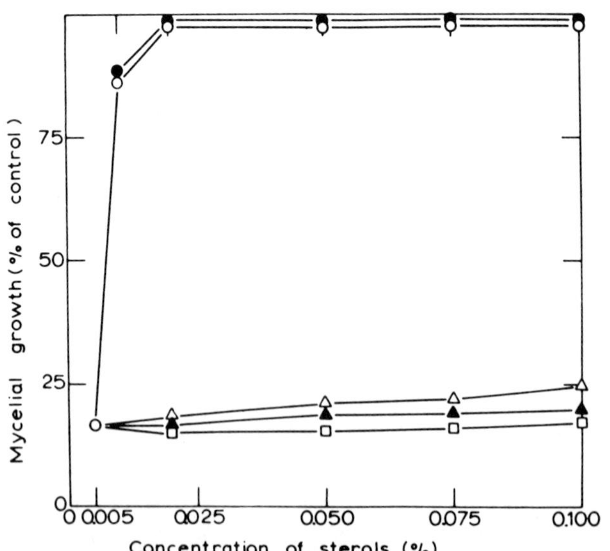

FIG. 4. Effect of different concentrations of various sterols on the growth impairment of *S. rolfsii* sacc. caused by 0.005% sterol-precipitable alfalfa saponins (○, cholesterol; △, 5-stigmastene-3β-ol; ●, 7-dehydrocholesterol; ▲, ergosterol; □, 5α-cholestan-3-one). (From Assa *et al.*, reproduced by permission.)

plant sterol, is ineffective in counteracting fungal growth impairment caused by alfalfa saponins may be of practical significance in the use of alfalfa roots in the soil for prevention of root rots caused by fungi.

Gestetner *et al.* (1971a) showed that the growth of *Sclerotium rolfsii* is very strongly inhibited not only be saponins but also by the aglycone medicagenic acid, whereas the soybean sapogenins have no effect. The two carboxylic groups at C-23 and C-28 and the hydroxyl of medicagenic acid are essential for fungal growth depression. The finding that the toxicity of medicagenic acid is reduced, but not completely abolished, by methylation of the carboxylic groups indicates that other structural features of this sapogenin are essential for fungal growth inhibition. Assa *et al.* (1972) studied the comparative effect of alfalfa saponins on *Pythium,* a fungus that contains cholesterol in the membrane, and on *Mycoplasma,* which lacks cholesterol in its membrane, and showed that the toxic effect of alfalfa saponins is exerted only on those fungi that contain cholesterol in their membrane. Furthermore, the fungistatic effect of alfalfa saponins is related to the amount of cholesterol in the fungus. The findings that all saponin-sensitive fungi do contain sterols, point toward an interaction between saponins and membranal sterol, which may bring about a drastic change in membrane permeability and may lead to an immediate and total loss of the cell contents (Fig. 5). Assa *et*

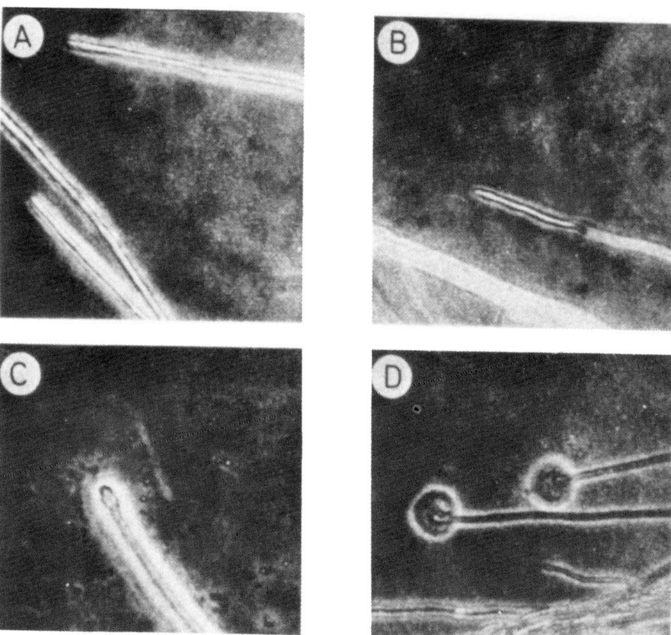

FIG. 5. Morphological changes, as a function of time, at the mycelial tips of *S. rolfsii* affected by alfalfa saponins. (A) control; (B–D) 30-sec time intervals from start of saponin spraying. (From Assa *et al.*, 1972, reproduced by permission.)

al. (1975) also studied the interaction of alfalfa saponins with membranal components and the consequent changes in permeability as prelytic events. Using *Physarum polycephalum* for interaction with saponins, they found that alfalfa saponins form interaction products with membranal sterols, proteins, and phospholipids. Although the interaction of the saponin with sterols is much more specific than that with proteins or phospholipids, the growth inhibition or lysis of *Physarum polycephalum* plasmodia should not be attributed solely to the interaction of saponin with sterols but rather to a concerted attack on the various membranal constituents. This results in changes of permeability of plasmodial membrane, which are further expressed by an increased influx of inorganic sodium ions and water, as observed by lysis, while no efflux of ions occurs. The ability of saponins to complex with membrane sterols seems to be the major mode of interaction in mycelial growth inhibition, since alfalfa saponins also inhibit growth of the *Rhizobium meliloti* (Leath *et al.*, 1972), which lacks sterols in its cell membrane.

C. Effect on Insects

The working hypothesis that saponins constitute a specific metabolic defense mechanism evolved in the plant against insects (Birk, 1969; Applebaum *et al.*, 1969; Applebaum and Birk, 1972) was also examined by other investigators. It is a well-known fact that third instar larvae of *Costelytra zealandica* (White) feed voraciously, and often with very severe effect, on roots of a wide range of pasture plants. Kain and Atkinson (1970) and Farrell and Sweney (1972) demonstrated resistance to grass grub attack in alfalfa (*Medicago sativa* L.) both in the field and in the laboratory. Alfalfa is also resistant to attack by larvae of the European scarabaeid *Melolontha vulgaris* F. Working with grubs of this species, Horber (1965) found antibiosis and nonpreference related to high saponin content in roots of resistant strains. This observation suggested a possible relationship between saponin and resistance to phytophagous insects. In view of the findings that alfalfa root saponins are particularly active as insect repellents and in antibiosis, it may be expected that resistant alfalfa plants could be effective against various other soil insects, e.g. white grubs, wireworms, and rootworms. The factors responsible for the resistance of alfalfa to attack by grubs have not yet been elucidated. Kain and Atkinson (1970) found that alfalfa seedlings were not as resistant to attack as were mature stands of the plant. They assumed that larvae were unable to consume sufficient amounts of the woody taproot of the older plants, and they consequently died of starvation. On the other hand, Farrell and Sweney (1972) suggested that the poor survival and the reduced weight gain of grubs reared on alfalfa resulted from antibiosis. Sutherland *et al.* (1975) studied the effect of saponins on the feeding response of *Costelytra zealandica* larvae. A crude extract of alfalfa root (*Medicago sativa* "Wairau") was found to contain a strong feeding deterrent for the larvae at concentrations within the range expected for saponins in fresh alfalfa root. Saponins purified from the crude extract were more effective, confirming the saponin nature of the deterrent.

The study of the effect of saponins on stored product pests (Birk, 1969) was further extended by Shany *et al.* (1970a) and Gestetner *et al.* (1970), who demonstrated that alfalfa root saponins are more toxic to *Tribolium castaneum* larvae than alfalfa top saponins. Larval growth can be restored to its normal level by addition of cholesterol to the diet. The negative effects of alfalfa saponins on *Tribolium castaneum* larvae seem to be exerted by their aglycone moiety, since partly degraded saponins at the carbohydrate moiety are more toxic than the intact extracts. Isolated alfalfa sapogenins, in particular medicagenic acid and hederagenin, are most toxic (Shany *et al.,* 1970b). On the other hand, the soyasapogenins have been found to be harmless. The detrimental influence of alfalfa sapogenins could also be fully prevented by incorporation of cholesterol or plant sterols into the diet. The extent of the preventive effect of the sterols depends on the saponin/sterol or sapogenin/sterol ratios in the diet (Shany *et al.,* 1970b). This ratio serves as a clue as to whether plant sterols act as an intrinsic defense mechanism against the antibiological effects of saponins.

Thorp and Briggs (1972) found mortality of immature alfalfa leaf cutter bees (*Megachile rotundata* F.) in alfalfa seed fields. Bohart (1971) reported a similar observation with *Megachile pacifica* (panzer). These bees are very efficient pollinators of alfalfa in the United States. As a result of mortality of young larvae, alfalfa seed growers usually have to renew bee stocks annually instead of obtaining natural field increases. Thorp and Briggs (1972) found that crude extracts and purified saponins from alfalfa, injected into the pollen food mass, caused increased mortality among immature bees. Since these bees use pieces of alfalfa leaves to line their brood cells—and in this sense they are unique among alfalfa pollinators—the investigators suggested that saponins were responsible for the high mortality. Eves (1973), however, found no evidence that alfalfa leaves caused any higher mortality of bees than did similar leaf pieces from other plant species. Parker and Pedersen (1974) studied the effect of leaves from two strains of alfalfa—one with a high and one with a low concentration of saponin—on the mortality of immature *Megachile pacifica* (panzer). The average difference between strains was insignificant. Thus, the claim that saponins in alfalfa leaf pieces used in lining the brood cell are responsible for high mortality of leaf-cutting bees foraging on alfalfa is still controversial.

In view of the evidence that alfalfa saponins interfere with normal insect activity (Horber, 1965, 1972), a technique was devised to use the potato leafhopper, *Empasca fabae* (Harvis), and pea aphid *Acyrthosiphon pisum,* to assay alfalfa saponins (Roof *et al.,* 1972; Harbor *et al.,* 1974). DuPuits and Lahontan saponins were separated by Horber *et al.* (1974) by thin-layer chromatography into 10 fractions and then assayed with potato leafhoppers and pea aphids. It was shown that DuPuits saponins, which contain medicagenic acid, were highly toxic to these insects, whereas Lahontan saponins, which lack medicagenic acid, had little or no toxicity. Medicagenic acid showed an antibiotic action on phytophagous insects. Pedersen *et al.* (1975, 1976) examined the resistance of insects to various varieties and selections of alfalfa. No correlation was found between resistance and saponin content of foliage. It appears that the resistance was dependent on saponin content of the roots, which was not directly related to that of the leaves and of other plant compounds. In a preference test with cockroaches (*Blatta orienlalis* L.) the aqueous ethanol extract of resistant selections showed pronounced repellent action (Horber, 1972). *Drosophila funebris* L., maintained on a diet supplemented with alfalfa ground roots of higly resistant selection, produced no larvae. On the other hand, growth of the highly specialized alfalfa weevil (*Hypera postica* Gyll.) was not inhibited by any of 16 saponin fractions isolated from alfalfa leaves incorporated into artificial diets (Hsiao, 1969).

Research on the mode of action of saponins on insects was conducted by N. Levin, S. W. Applebaum, and Y. Birk (unpublished results, 1977), using the African migratory locust. A significant decrease was noted in the weight of adult locust reared on alfalfa in comparison to those reared on grass. Saponin solution applied to grass caused a growth depression of adult locusts and a smaller effect on the first larval instars. Growth depression may be explained by the morphological changes observed in the gut of saponin-treated locusts. The mode of action of saponins is complex, and it includes a disturbance in water resorption in the rectum and possibly also inhibition of enzyme activity in the digestive system.

In conclusion, the mode of action of saponins in relation to insect growth and survival has not yet been elucidated. Thorp and Briggs (1972) and Parker and Pedersen (1974) found a relationship between alfalfa saponins and mortality of immature leaf cutter bees, but they did not speculate on the mechanism causing their death. Horber (1965) ascribed the antibiosis he observed in white grubs (*Melolontha vulgaris*) reared on resistant varieties of alfalfa to the high saponin content of the plants, but later, in a reappraisal of the same experiments, the possibility of starvation caused by nonpreference simulating an ''antibiotic'' effect was raised (Horber, 1972). A similar doubt about the mode of action of saponins was expressed by Roof *et al.* (1972). They did not resolve the question whether the mortality of *Empoasca fabae* larvae reared on artificial diets containing 0.1–5.0% commercial saponin resulted from *antibiosis* or from the saponin acting as *feeding deterrent,* thereby causing death through dehydration or starvation. Thus, alfalfa saponins may be significant in host plant resistance to polyphagous insects, such as white grubs, pea aphids, and leafhoppers. However, oligophagous insects, narrowly specialized on alfalfa, sometime during coadaptive evolution may have found a way to overcome saponins as a barrier.

D. Effect on Chicks and Other Animals

The studies on the effects of alfalfa saponins on farm animals (Ishaaya *et al.*, 1969; Birk, 1969) were extended by Cheeke (1971), who showed species differences in the response of monogastric animals to dietary saponins, poultry being much more sensitive than the other monogastrics. Levels of approximately 20% alfalfa meal in chick rations (equivalent to about 0.3% saponin) resulted in growth depression, which was attributed entirely to the saponin content. The same levels of alfalfa meal were incorporated into swine grower rations with no adverse effects on growth. Cheeke (1971) observed that mice are much less tolerant to alfalfa saponins than rats. At a dietary level of 2% alfalfa saponin, feed intake of mice was markedly depressed, resulting in weight loss, whereas the rats were not affected. On the other hand, Reshef *et al.* (1976) reported that the growth of mice and quails was not affected by 2% alfalfa top saponins in the diet—a concentration equivalent to saponin content of a diet composed solely of alfalfa meal. Growth depression was observed only when the diet was supplemented with 0.5% alfalfa root saponins. The extent of growth inhibition was related to the content of medicagenic acid in the saponin preparation used.

Saponins have been considered as a factor limiting the level of alfalfa that can be incorporated in the diet. The recent development of alfalfa varieties low in saponin is of agricultural significance because of their higher feeding value for poultry. Results of Pedersen *et al.* (1972) illustrated the improved growth of broilers fed low-saponin alfalfa meal at a dietary level of 10%. Similar results were obtained with rats. The adverse saponin effect was much greater with DuPuits alfalfa than with Lahontan, owing to the predominance of the medicagenic acid aglycone in DuPuits. These findings are consistent with the conclusions of Reshef *et al.* (1976) and with previous data, which show that various biological activities of alfalfa saponins are attributable mainly to the medicagenic acid content of their saponins (Shany *et al.*, 1970b; Gestetner *et al.*, 1971a,b).

Very little is known about the mode of action of alfalfa saponins on monogastric animals. Cheeke (1976) suggested that the main factor limiting the direct use of alfalfa by monogastrics is the fact that they reject the feed because of the bitter taste of the saponins. The better growth response of chicks and rats to low-saponin alfalfa than to high-saponin alfalfa (Pedersen *et al.*, 1972) may be due simply to greater palatability of the low-saponin meal. A similar opinion was expressed by Kendall and Leath (1976), who used a meadow voles (*Microtus pennsylvanicus*) and showed that meals of low-saponin lines of alfalfa were generally more palatable than their corresponding high saponin lines, but only very high concentrations of saponin in the diet (4%) affected palatability to voles. In view of the predominant assumption that saponins are a major component of the antipalatability factor, selection of alfalfa plants for low saponin concentration should greatly increase the potential for the use of this crop as a source of protein for monogastrics (Cheeke, 1971). Such low-saponin varieties certainly have superior nutritional value. Although the original interest in the development of low-saponin alfalfa was attributed to the postulated involvement of saponins in ruminant bloat, the major significance of the plant breeding development is presently in the area of monogastric animal nutrition. From an agronomic viewpoint, it could be expected that such selection may increase the plant's susceptibility to insects and diseases, but this has actually been found in only a few instances.

E. Effect on Seed Germination

The inhibitory effect of alfalfa saponins on the germination of cottonseeds, reported by Mishustin and Naumova (1955) and by Pedersen (1965) (for review see Birk, 1969), was studied by Shany *et al.* (1970a) with different alfalfa saponin preparations. The mechanism by which alfalfa root saponins inhibit the germination of cottonseeds was elucidated in a series of investigations by Marchaim *et al.* (1970, 1972, 1974, 1975). It was shown that the alfalfa saponins inhibit germination by a process associated with the cottonseed coat but not with the seed embryo. Although the rate of water uptake during imbibition was similar for intact seeds immersed in either water or in 0.5% saponin solution, there was a distinct decrease in oxygen diffusion through the seed coats and membranes of seeds preimmersed in saponins (Marchaim *et al.,* 1972) (Fig. 6). The diffusion decrease in oxygen consumption as indicated by studies with $^{18}O_2$ (Mayevsky and previously immersed in saponin solution showed a lower rate of respiration than those preimmersed in water (Marchaim *et al.,* 1972, 1975), this being a true decrease in oxygen consumption as indicated by studies with $^{18}0_2$ (Mayevsky and Marchaim, 1972). Yet excised embryos of seeds preimmersed either in water or in saponin solution had identical respiration rates (Marchaim *et al.,* 1972; Mayevsky and Marchaim, 1972). It was concluded that saponins, as potent surfactants, interact with the seed coats and/or seed membranes, without penetrating the embryo (Marchaim *et al.,* 1975). This interaction results in a reduction in oxygen consumption by the embryo and is expressed in inhibition of germination, although the respiratory ability of the embryo has not been affected. The extent of inhibition of germination depends on the period of preimmersion in saponin solution. Inhibition of germination was irreversible after preimmersion in saponins for 6 hr or more (Marchaim *et al.,* 1975).

As to the mode of action of the saponins—being surface-active glycosides with hydrophobic as well as hydrophilic properties—they may cause changes in the microstructure of natural cell membranes, as was shown by Bangham and Horne (1962) for membranal lipids. This hypothesis was further supported by experiments with synthetic surfactants of different chemical composition and/or

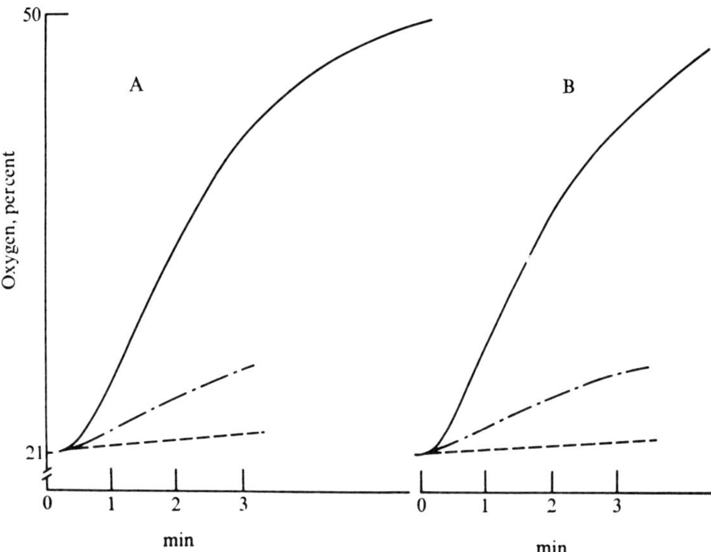

FIG. 6. Effect of preimmersion of seeds in 0.5% alfalfa root saponins (---) or in water (—·—) on the diffusion of O_2 through the chalazal regions of the seed coat (A) or the membrane (B) as compared to the O_2 diffusion in absence of the separating tissue in the apparatus (—). (From Marchaim *et al.*, 1972, reproduced by permission.)

charge, which inhibited cottonseed germination and oxygen consumption, much as did alfalfa saponins (Marchaim *et al.*, 1974, 1975). Observations with light and scanning electron microscopes showed that preimmersion of cottonseeds in alfalfa saponins and other surfactant solutions caused structural changes in the membranes that might have affected their permeability to oxygen (Marchaim *et al.*, 1974).

Jurzysta (1970) studied the effect of saponins isolated from seeds of alfalfa (Polish varieties) on germination of cereal seeds and showed that 0.1 and 0.5% saponin solution of *Medicago lupulina* significantly inhibited the growth of four species of cereals (barley, oats, wheat, and rye). The same concentrations of saponins of *Medicago media* delayed considerably the growth of seedlings of barley and oats, did not affect wheat, and inhibited only slightly the growth of rye.

F. Effect on Blood and Tissue Cholesterol Levels

It has been shown that dietary alfalfa saponins form complexes with cholesterol and help to lower the plasma cholesterol levels of chicks (for review, see Cheeke, 1971). The feeding of dietary cholesterol also reduces the growth inhibition attributed to saponins. Since many saponins form insoluble complexes with cholesterol, it has been assumed that dietary saponins in the guts of monogastrics can combine with endogenous cholesterol excreted via the bile. This prevents the cholesterol reabsorption and results in a reduction in serum cholesterol (Cheeke, 1971). In ruminants, neither growth depression nor reduced tissue cholesterol levels, due to saponin ingestion, have been demonstrated, possibly because of bacterial dissimilation of the saponin in the rumen.

The cholesterol-complexing property of alfalfa saponins, which was demonstrated *in vitro* (Gestetner *et al.*, 1972), indicated that saponins may be involved in certain aspects of lipid metabolism. Reshef *et al.* (1976) studied the effect of alfalfa saponins on lipid metabolism in mice and quails and showed that saponin–cholesterol complexes are formed in the gut of mice but not in the gut of quails. Proof of the formation of such a complex *in vivo* following ingestion of alfalfa saponins by mice was provided by the presence of a lower level of cholesterol in the liver, accompanied by appearance of cholesterol in larger amounts in the feces, and followed by an increased rate of cholesterol synthesis in the liver. It should be noted that in quails only an increase in the rate of cholesterol biosynthesis was observed. However, this additional amount of cholesterol was not utilized by quails for "neutralizing" the effect of dietary alfalfa saponins. Presumably, it was metabolized for other purposes, such as formation of bile acids, which in turn may be the cause of the slightly increased absorption of lipids.

Malinow *et al.* (1977a,b) demonstrated that the inclusion of alfalfa saponins in high-cholesterol diets prevents hypercholesterolemia in monkeys. These observations suggest that the effect of the saponins is due to interference with intestinal absorption of cholesterol, probably through the formation of insoluble complexes of the saponins with cholesterol. Since alfalfa saponins are not adsorbed from the gastrointestinal tract in mammals (Birk, 1969) and since they prevent dietary-induced elevation of cholesterolemia in monkeys, Malinow *et al.* (1977a,b) suggested that alfalfa saponins may constitute an important new therapy for human hypercholesterolemia. This suggestion is supported by the findings that alfalfa saponins are tolerated by rats and monkeys. Since saponins are present in small amounts in many plants commonly ingested by man, one may assume that they are also tolerated to a certain extent by man.

G. Effect on Enzymes

An interference by saponins with enzymes concerned with energy metabolism could account for toxicity symptoms resulting from ingestion of legume saponins by monogastric animals. Of particular importance are the enzymes concerned with the citric acid cycle, since inhibition of this major metabolic pathway could have pronounced effects on the efficiency of nutrient utilization and on animal growth. Cheeke and Oldfield (1970) studies the effect of alfalfa saponins on the *in vitro* oxidation of succinate, a key citric acid cycle intermediate. Alfalfa saponins were found to inhibit *in vitro* succinate oxidation by rat liver enzymes. Ethanolic extracts of alfalfa also ihibited *in vitro* succinoxidase activity. Besides apparent effects on cellular metabolism, saponins might affect the activity of digestive enzymes, as reported by Ishaaya and Birk (1965).

IV. SAPONINS IN OTHER FOODS AND FEEDS

A. Saponins from French Beans (*Phaseolus vulgaris*)

Pantic *et al.* (1974) extracted saponins from French beans (*Phaseolus vulgaris*). Acid hydrolyzates of these saponin preparations contained galactose, glucose, and three other unidentified fractions, as well as four sapogenins that, according to their R_f values on paper chromatography, corresponded to those of soybean sapogenins. The French bean saponin preparation possessed only marginal hemolytic activity and also lacked the pronounced effects shown by alfalfa saponins on the mitotic rate, revealing a decrease of cell population in the small intestine, bone marrow, spleen, and testes.

B. Saponins from *Pisum arvense* Seeds

Saponins from *P. arvense* L. seeds were fractionated according to their solubility in the presence of excess CaO (Gaither, 1972). The insoluble saponin fraction could be separated into seven spots by paper chromatography and had a very bitter taste and high surface activity. The aglycone moiety of this preparation was composed of soyasapogenols B, C, and E plus an unidentified sapogenin, whereas arabinose, glucuronic acid, and xylose were identified in the carbohydrate moiety. The soluble saponin fraction showed four spots by paper chromatography, exhibited only slight bitterness and surface activity, and contained no identifiable sapogenins or monosaccharides. Thus, most, if not all *P. arvense* saponins appear to be precipitable with excess CaO. Neither saponin preparation showed significant *in vitro* hemolysis.

C. Saponins from Bird's Foot Trefoil Seeds (*Lotus corniculatus* L.)

Jurzysta (1973) isolated saponins from bird's foot trefoil seeds in the form of a Ca–Mg salt and claimed that this salt is the natural form of this saponin in the seed. Acid hydrolysis of this preparation yielded rhamnose, xylose, glucose, galactose, and glucuronic acid as well as the soyasapogenols B, C, and E (traces) and an unidentified aglycone.

V. BIOSYNTHESIS OF SAPONINS IN ALFALFA AND SOYA PLANTS

Arigoni (1958) showed that when germinating soybeans are grown on a substrate containing ^{14}C-labeled acetic acid, the specific activity is equally distributed between the soybean sapogenins, which are pentacyclic terpenoids, and the steroid sitosterols and is also found in squalene. The role of squalene as the progenitor of polycyclic triterpenoids is now well established. The intermediacy of 2,3-epoxysqualene in the formation of C-3-oxygenated triterpenes has been well substantiated. However, in the triterpenoids found in primitive plants, which have no oxygen function at C-3, it was shown that their biogenesis may involve a direct proton-initiated cyclization of squalene (Corey and Ortiz de Montellano, 1967).

The biosynthesis of sapogenins and especially medicagenic acid from [2-^{14}C]acetate was investigated in germinating seeds of alfalfa (*Medicago media*) by Nowacki *et al.* (1976). During the first 10 days, the highest radioactivity was found in medicagenic acid and not in the sapogenins. It was suggested that medicagenic acid is the first synthesized sapogenin in germinating seeds, and the other sapogenins are formed from medicagenic acid.

The biosynthesis of sapogenins from [2-^{14}C]mevalonic acid was studied in alfalfa and soya plants in our laboratory (I. Peri, U. Mor, and A. Bondi, unpublished results, 1978). In order to follow the biosynthesis of saponins *in vivo,* alfalfa and soybean seeds were incubated with radioactive mevalonic acid. Labeled saponins and respective sapogenins could be detected by cochromatography, scanning for labeled sapogenins, and autoradiography. In the soya plant the radioactive mevalonic acid was readily incorporated into the saponin fraction of 5- and 10-day old soya seedlings. After hydrolysis the sapogenins were identified by thin-layer chromatography (TLC). The formation of labeled sapogenins again supports the validity of the scheme whereby mevalonic acid serves as the source of five carbon atoms to form the isoprenoid units and, via squalene, the sterol or triterpenoid structure.

VI. METHODS FOR IDENTIFICATION AND QUANTITATIVE DETERMINATION OF SAPONINS

A. Determination of the Chemical Composition of Saponins

The determination of the chemical composition of a saponin by analyzing the carbohydrate and aglycone moieties in the acid hydrolyzate still remains the most reliable method for identification and definition. Techniques employed for the sterochemical and structural studies of saponins were described by Kulshreshta *et al.* (1972).

The structure of the sugar moiety is generally determined by mild acidic or alkaline hydrolysis of the premethylated glycosides. The individual methylated sugars are then identified either by direct chromatographic comparison or by gas chromatography on columns of 5% neopenthylglycol adipate polyester on Chromosorb w (operating temperature, 150°C). Mass spectrometry is also emerging as a valuable tool in these determinations.

Sapogenin fractions obtained by acid hydrolysis are identified by high-resolution mass spectrometry of the corresponding acetates and methyl esters. In addition to mass spectrometry, sapogenins are also subjected to TLC analysis and identified by comparison with the respective authentic compounds. Each aglycone yields a characteristic color upon spraying with concentrated H_2SO_4. Sapogenins can also be identified by using gas chromatography. For gas chromatographic analyses the sapogenins are silylated by rapid treatment with TRISIL/BSA in order to produce volatile trimethyl silyl (TMS) ethers. The best resolution of silylated sapogenins is observed by 1% coating of SE-30 (or OV-1) on Gas-Chrom Q and employing a temperature of 280°C.

B. Methods for Quantitative Determination

1. Spectrophotometric Methods

A reliable quantitative procedure for the analysis of saponins in saponin extracts and in leguminous plants was developed by Gestetner *et al.* (1966) for soybean saponins. This method was based on the spectrophotometric determination, with a modified Lieberman–Burchard reagent, of the sapogenin content in the saponin hydrolyzate, using a constant sapogenin/sugar ratio as a conversion factor of the sapogenin to saponin content. It should be pointed out that different saponin fractions (e.g., soybean and alfalfa) have different sapogenin/sugar ratios.

Since the modified Lieberman–Burchard reagent was not suitable for the determination of medicagenic acid, a titrimetric method for a quantitative determination of the content of this aglycone constituent was developed (Tencer *et al.*, 1972). The method is based on the titration of the two carboxylic groups (C-23, C-28), and it can be employed in the presence of the other sapogenins that interfere in the colorimetric assays of medicagenic acid.

2. Bioassays

In addition to the common bioassay techniques, such as the numerous methods based on blood cell hemolysis and *Trichoderma viride* inhibition (for review, see Birk, 1969), a technique was devised to use potato leafhopper and pea aphid survival to assay activities of alfalfa saponins (Horber *et al.*, 1974). This technique was developed because of previous evidence that legume seed saponins (Applebaum *et al.*, 1969) and alfalfa saponins (Horber, 1965, 1972) interfere with normal insect activity. The bioassay technique clearly differentiated between saponins from different origins (e.g., yucca sp., *Medicago sativa*) and different varieties (e.g., Lahontan and DuPuits). In order to lower the saponin level in alfalfa by plant breeding programs, a bioassay with the fungus *Trichoderma viride* Pers. ex Fr. was used (Zimmer *et al.*, 1967). Recently, a microbiological assay has been described for determining the level of medicagenic acid-type saponins in dried alfalfa, leaf protein concentrates, and alfalfa sprouts (Livingston *et al.*, 1977). The use of minnow and erythrocyte assays for isolating plants of low saponin content was described by Jones and Elliott (1969).

Acknowledgments

This work was supported by Grant Nos. 161 and 751 from the United States-Israel Binational Science Foundation (B.S.F.).

References

Applebaum, S. W., and Birk, Y. (1972). *In* "Insect and Mite Nutrition" (J. G. Rodriguez, ed.), pp. 629–636. North-Holland Publ., Amsterdam.

Applebaum, S. W., Marco, S., and Birk, Y. (1969). *J. Agric. Food Chem.* **17**(3), 618–622.

Arigoni, D. (1958). *Experientia* **14,** 153–155.

Assa, Y., Gestetner, B., Chet, I., and Henis, Y. (1972). *Life Sci.* **11,** Part 2, 637–647.

Assa, Y., Shany, S., Gestetner, B., Tencer, Y., Birk, Y., and Bondi, A. (1973). *Biochim. Biophys. Acta* **307,** 83–91.

Assa, Y., Chet, I., Gestetner, B., Govrin, R., Birk, Y., and Bondi, A. (1975). *Arch. Microbiol.* **103,** 77–81.

Bangham, A. D., and Horne, R. W. (1962). *Nature (London)* **196,** 952–953.

Berrang, B., Davis, K. H., Jr., Wall, M. E., Hanson, C. H., and Pedersen, M. W. (1974). *Phytochemistry* **13,** 2253–2260.

Birk, Y. (1969). *In* "Toxic Constituents of Plant Foodstuffs" (I. E. Liener, ed.), 1st ed., pp. 169–210. Academic Press, New York.

Bohart, G. E. (1971). *Proc. Tall Timbers Conf. Ecol. Anim. Control Habitat Manage.* pp. 253–266.

Bondi, A., Birk, Y., and Gestetner, B. (1973). *In* "Chemistry and Biochemistry of Herbage" (G. W. Butler and R. W. Bailey, eds.), Vol. 1, pp. 511–528. Academic Press, New York.

Cheeke, P. R. (1971). *Can. J. Anim. Sci.* **51,** 621–632.

Cheeke, P. R. (1976). *Nutr. Rep. Int.* **13**(3), 315–324.

Cheeke, P. R., and Oldfield, J. E. (1970). *Can. J. Anim. Sci.* **50,** 107–112.

Corey, E. J., and Ortiz de Montellano, P. R. (1967). *J. Am. Chem. Soc.* **89,** 3362–3363.

Dourmashkin, R. R., Dougherty, R. M., and Harris, R. J. C. (1962). *Nature (London)* **194,** 1116–1119.

Eves, J. D. (1973). Ph.D. Dissertation, Department of Entomology, Washington State University, Pullman.

Farrell, J. A. K., and Sweney, W. J. (1972). *N.Z. J. Agric. Res.* **15,** 904-908.

Focke, I. (1970). *Arch. Pflanzenschutz* **6,** 119-124.

Gaither, D. H. (1972). *J. Tenn. Acad. Sci.* **48**(3), 91-93.

Gestetner, B. (1971). *Phytochemistry* **10,** 2221-2223.

Gestetner, B., Birk, Y., and Bondi, A. (1966). *Phytochemistry* **5,** 799-806.

Gestetner, B., Shany, S., Tencer, Y., Birk, Y., and Bondi, A. (1970). *J. Sci. Food Agric.* **21,** 502-507.

Gestetner, B., Assa, Y., Henis, Y., Birk, Y., and Bondi, A. (1971a). *J. Sci. Food Agric.* **22,** 168-172.

Gestetner, B., Shany, S., and Assa, Y. (1971b). *Experientia* **27,** 40-41.

Gestetner, B., Assa, Y., Henis, Y., Tencer, Y., Rotman, M., Birk, Y., and Bondi, A. (1972). *Biochim. Biophys. Acta* **270,** 181-187.

Glauert, A. M., Dingle, J. T., and Lucy, J. A. (1962). *Nature (London)* **196,** 953-955.

Hanson, C. H., Kohler, G. O., Dudley, J. W., Sorensen, E. L., Van Atta, G. R., Taylor, K. W., Pedersen, M. W., Carnahn, H. L., Wilsie, C. P., Kehr, W. R., Lowe, C. C., Standford, E. H., and Yungen, J. A. (1963). *U.S., Dep. Agric. Res. Serv., Bull.* pp. 33-44.

Hanson, C. H., Pedersen, M. W., Berrang, B., Wall, M. E., and Davis, K. H., Jr. (1973). *In* "Anti-Quality Components of Forages" (A. G. Matches, ed.), Spec. Publ. No. 4, p. 33. Crop Sci. Soc. Am., Madison, Wisconsin.

Horber, E. (1965). *Physiol. Insect Cent. Nerv. Syst., Pap. Int. Congr. Entomol., 12th, 1964* pp. 540-541.

Horber, E. (1972). *In* "Insect and Mite Nutrition" (J. G. Rodriguez, ed.), pp. 611-627. North-Holland Publ., Amsterdam.

Horber, E., Leath, K. T., Berrang, B., Marcarian, V., and Hanson, C. H. (1974). *Entomol. Exp. Appl.* **17,** 410-424.

Hsiao, T. H. (1969). *Proc. Int. Symp. "Insect and Hostplant," 2nd,* pp. 777-778.

Ishaaya, I., and Birk, Y. (1965). *J. Food Sci.* **30,** 118-120.

Ishaaya, I., Birk, Y., Bondi, A., and Tencer, Y. (1969). *J. Sci. Food Agric.* **20,** 433-436.

Jacobson, D. A. (1919). *J. Am. Chem. Soc.* **41,** 640-648.

Jones, M. (1969). Ph.D. Dissertation, Michigan State University, East Lansing.

Jones, M., and Elliott, F. C. (1969). *Crop. Sci.* **9,** 688-691.

Jurzysta, M. (1970). *Zesz. Nauk. Univ. Mikolaja Kopernika Toruniu, Nanki Mat.-Przyr.* **13, 23,** 253-256.

Jurzysta, M. (1973). *Acta Soc. Bot. Pol.* **42**(2), 201-207.

Jurzysta, M. (1975). *Pamiet. Pulawski* **62,** 99-107.

Kain, W. M., and Atkinson, D. S. (1970). *Proc. 23rd N. Z. Weed Pest Control Conf.* **23,** 180-183.

Kendall, W. A., and Leath, K. T. (1976). *Agron. J.* **68,** 473-476.

Kulshreshtha, M. J., Kulshreshta, D. K., and Rastogi, R. P. (1972). *Phytochemistry* **11,** 2369-2381.

Leath, K. T., Davis, K. H., Jr., Wall, M. E., and Hanson, C. H. (1972). *Crop Sci.* **12,** 851-856.

Livingston, A. L., Whitehand, L. C., and Kohler, G. O. (1977). *J. Assoc. Off. Anal. Chem.* **60**(4), 957-960.

Malinow, M. R., McLaughlin, P., Kohler, G. O., and Livingston, A. L. (1977a). *Steroids* **29**(1), 105-110.

Malinow, M. R., McLaughlin, P., Kohler, G. O., and Livingston, A. L. (1977b). *Clin. Res.* **25**(2), 97A.

Marchaim, U., Birk, Y., Dovrat, A., and Berman, T. (1970). *Plant Cell Physiol.* **11,** 511-514.

Marchaim, U., Birk, Y., Dovrat, A., and Berman, T. (1972). *J. Exp. Bot.* **23**(75), 302-309.

Marchaim, U., Werker, E., and Thomas, W. D. E. (1974). *Bot. Gaz. (Chicago)* **135**(2), 139–146.
Marchaim, U., Birk, Y., Dovrat, A., and Berman, T. (1975). *Plant Cell Physiol.* **16,** 857–864.
Mayevsky, A., and Marchaim, U. (1972). *Plant Cell Physiol.* **13,** 927–930.
Mishustin, B. N., and Naumova, A. N. (1955). *Izv. Akad. Nauk SSSR, Ser. Biol.* **6,** 3–9.
Nowacki, E., Jurzysta, M., and Dietrych-Szostak, D. (1976). *Biochem. Physiol. Pflanz.* **169,** 183–186.
Olsen, R. A. (1971a). *Physiol. Plant.* **24,** 534–543.
Olsen, R. A. (1971b). *Physiol. Plant* **25,** 204–212 and 503–508.
Pantic, V. R., Vucurevic, N., and Sekulic, M. (1974). *Arch. Biol. Nauka (Beograd)* **25,** 119–124.
Parker, F. D., and Pedersen, M. W. (1974). *Environ. Entomol.* **4**(1), 103–104.
Pedersen, M. W. (1965). *Argon. J.* **57,** 516–517.
Pedersen, M. W. (1975). *Crop Sci.* **15,** 541–543.
Pedersen, M. W., and Wang, L.-C. (1971). *Crop Sci.* **11,** 833–835.
Pedersen, M. W., Zimmer, D. E., Anderson, J. O., and McGuire, C. F. (1966). *Proc. Int. Grassl. Congr., 10th, 1966* pp. 693–698.
Pedersen, M. W., Anderson, J. O., Street, J. C., Wang, L.-C., and Baker, R. (1972). *Poult. Sci.* **51**(2), 458–463.
Pedersen, M. W., Berrang, B., Wall, M. E., and Davis, K. H., Jr. (1973). *Crop Sci.* **13,** 731–735.
Pedersen, M. W., Sorensen, E. L., and Anderson, M. J. (1975). *Crop Sci.* **15,** 254–256.
Pedersen, M. W., Barnes, D. K., Sorensen, E. L., Griffin, G. D., Nielson, M. W., Hill, R. R., Jr., Frosheiser, F. I., Sonoda, R. M., Hanson, C. H., Hunt, O. J., Peaden, R. N., Elgin, J. H., Jr., Devine, T. E., Anderson, M. J., Goplen, B. P., Elling, L. J., and Howarth, R. E. (1976). *Crop Sci.* **16**(2), 193–199.
Quazi, H. M. (1975). *N.Z.J. Agric. Res.* **18,** 227–232.
Quazi, H. M. (1976). *N.Z.J. Agric. Res.* **19,** 347–348.
Reshef, G., Gestetner, B., Birk, Y., and Bondi, A. (1976). *J. Sci. Food Agric.* **27,** 63–72.
Roof, M., Horber, E., and Sorensen, E. L. (1972). *Proc. North Cent. Branch Entomol. Soc. Am.* **27,** 140–143.
Schroeder, F., Holland, J. F., and Bieber, L. I. (1971). *J. Antibiot.* **24,** 846–849.
Shany, S. (1971). Ph.D. Dissertation, Hebrew University of Jerusalem.
Shany, S., Birk, Y., Gestetner, G., and Bondi, A. (1970a). *J. Sci. Food Agric.* **21,** 131–135.
Shany, S., Gestetner, B., Birk, Y., and Bondi, A. (1970b). *J. Sci. Food Agric.* **21,** 508–510.
Shany, S., Gestetner, B., Birk, Y., and Bondi, A. (1972). *Isr. J. Chem.* **10,** 881–884.
Shany, S., Bernheimer, A. W., Grushoff, P. S., and Kim, K.-S. (1974). *Mol. Cell. Biochem.* **3**(3), 179–186.
Sutherland, O. R. W., Hood, N. D., and Hillier, J. R. (1975). *N.Z.J. Zool.* **2**(1), 93–100.
Tencer, Y., Shany, S., Gestetner, B., Birk, Y., and Bondi, A. (1972). *J. Agric. Good Chem.* **20**(6), 1149–1151.
Thorp, R. W., and Briggs, D. L. (1972). *Environ. Entomol.* **1**(4), 399–401.
Tschesche, R., and Wulff, G. (1965). *Z. Naturforsch., Teil B* **20,** 543–546.
Wolters, B. (1968). *Planta* **79,** 77–83.
Zimmer, D. E., Pedersen, M. W., and McGuire, C. F. (1967). *Crop Sci.* **7,** 223–224.

Chapter 7

Gossypol

LEAH C. BERARDI AND LEO A. GOLDBLATT

TOXIC CONSTITUENTS OF PLANT FOODSTUFFS, SECOND EDITION

ISBN 0-12-449960-0

I. INTRODUCTION

The polyphenolic gossypol pigments (Boatner, 1948; Altschul *et al.*, 1958; Adams *et al.*, 1960; Markman and Rzhekhin, 1965; Bell and Stipanovic, 1977) are indigenous in the genus *Gossypium* and in certain other members of the order Malvales. In the cotton plant, they are contained almost exclusively within discrete bodies commonly called pigment glands (Boatner, 1948), which are found in the leaves, stems, roots, and seed of cotton plants. The occurrence of the gossypol pigments in cottonseed is of considerable economic importance. Although cottonseed is a by-product of the cotton fiber, the processing of cottonseed is a major industry in the cotton-producing areas of the world. World production of cottonseed in 1976 was estimated at 25.6 million tons, of which 4.0 million tons were produced in the United States. In the 14 cotton-producing states of the United States the seed represents a multimillion dollar agricultural crop. Oils and meals obtained from cottonseed are recognized as highly desirable or preferred products in the markets in which they compete. The principal quality deficiencies of these cottonseed products are due to their contents of gossypol pigments (Harper, 1963).

Of the various products (oil, meal, linters, and hulls) obtained from cottonseed, oil is the most valuable. Almost all of the cottonseed oil produced in the United States is used in the manufacture of edible products, such as salad oil, margarine, and shortening. Crude oils must be processed and bleached to acceptable color standards. Some color present in crude cottonseed oil, especially "fixed" color attributed to gossypol pigments, is often quite resistant to removal and presents a special problem. It is estimated that about 25% of crude cottonseed oil cannot be refined and bleached to colors acceptable for the manufacture of high-quality shortening by usual commercial procedures (Smith, 1962). Consequently, many processors operate their mills under conditions that confine most of the gossypol pigments to the cottonseed meal.

Until rather recently, cottonseed meal was used principally as a protein supplement for ruminant livestock. The naturally occurring gossypol pigments of the cottonseed are recognized toxicants to monogastric animals, and this has limited the use of cottonseed meals in feeds for swine and poultry. One of the objectives of cottonseed processing is the binding of the gossypol pigments in the processed meal. The phenomenon of binding allows for detoxification of the gossypol pigments but, at the same time, causes a lowering of the protein and biological value of the meal. However, a concentrated research effort demonstrated that cottonseed meal is a valuable source of protein for the feeding of swine and poultry (Harper, 1964), and an increasing proportion is now used in swine and poultry rations.

Cottonseed also represents a potentially excellent source of low-cost, high-quality protein products for food (Martinez *et al.*, 1970). The use of cottonseed protein concentrates for feeding to human beings was established at a conference on cottonseed protein for animal and man (Anonymous, 1962). Utilization of cottonseed meals as protein supplements for human beings in world areas where shortages of animal proteins exist was investigated under a worldwide program sponsored jointly by agencies such as the United Nations Children's Fund (UNICEF), the Agency for International Development (AID), and the U.S. Department of Agriculture. Two commercial products containing cottonseed flour have been used successfully for supplemental human feeding in Latin American countries (Lambou *et al.*, 1966).

Cottonseed meal was the leading source of domestic protein concentrates until 1942 when soybean meal took leadership, but cottonseed meal still represents the second largest U. S. source of vegetable protein concentrate. Unfortunately, cottonseed meal is not a product of uniform quality, and there is a wide variability in the gossypol contents of cottonseed meals (Altschul *et al.*, 1958). Because of the economic influence of the gossypol pigments on cottonseed protein products and oil values, the pigments have been the subject of numerous studies. To obtain maximal usage of cottonseed meal in feeds for nonruminants and to make appropriate use of the meal in foods for human beings, upgrading of cottonseed and cottonseed meal quality is the subject of much current research. This chapter reviews the more important results of the studies conducted on gossypol and its derivatives and on gossypol pigmentation of cottonseed and its products.

II. GOSSYPOL PIGMENTS

Boatner (1948) reported the presence of at least 15 gossypol pigments or derivatives in extracts of cottonseed or cottonseed oils and meals, but only about eight have been isolated in more or less purified form and characterized. They include gossypol (yellow), diaminogossypol (yellow), 6-methoxygossypol (yellow), 6,6′-dimethoxygossypol (yellow), gossypurpurin (purple), gossyfulvin (orange), gossycaerulin (blue), and gossyverdurin (green). Gossypol occurs in greater amount in raw cottonseed than in cottonseed that has been subjected to moist heat treatment (cooked) during processing, whereas more gossypurpurin and gossyfulvin occur in the cooked seed. Gossycaerulin occurs almost exclusively in cooked cottonseed. Gossypol is converted to gossypurpurin during maturation and prolonged storage of the seed (Rzhekhin *et al.*, 1968).

A. Gossypol

The predominant naturally occurring gossypol pigment, and the one that has been investigated far more thoroughly than any of the others, is the yellow pigment gossypol, $C_{30}H_{30}O_8$. Named by Marchlewski in 1899, gossypol is 1,1′,6,6′,7,7′-hexahydroxy-5,5′-diisopropyl-3,3′-dimethyl[2,2′-binaphthalene]-8,8′-dicarboxaldehyde (Adams *et al.*, 1960). The structure of gossypol, derived by Adams *et al.* (1938b) on the basis of classic studies of the reactions, properties, and degradation products, is shown in Fig. 1. The postulation of three tautomeric structures was necessary to explain many of the reactions of gossypol. Of the three tautomeric modifications of gossypol shown in Fig. 1, structure (1a) represents the hydroxyaldehyde tautomer, (1b) the lactol tautomer, and (1c) the cyclic carbonyl tautomeric form. The structure of gossypol proposed by Adams *et al.* was corroborated in studies by Shirley and Dean (1955, 1957) and confirmed with the total synthesis of gossypol reported by Edwards (1958, 1970). Isolation of an optically active form of gossypol ($[\alpha]_D^{19}$ + 445 ± 10°, chloroform) from the bark and flowers of *Thespesia populnea* was reported by King and De Silva (1968). The optical activity was ascribed to restricted rotation of the substituted binaphthyl groups. Dechary and Pradel (1971) later reported the isolation of both (±)-gossypol and (+)-gossypol from cottonseed.

Gossypol is markedly reactive and exhibits strongly acidic properties. It is capable of acting as a phenolic and as an aldehydic compound. Gossypol reacts

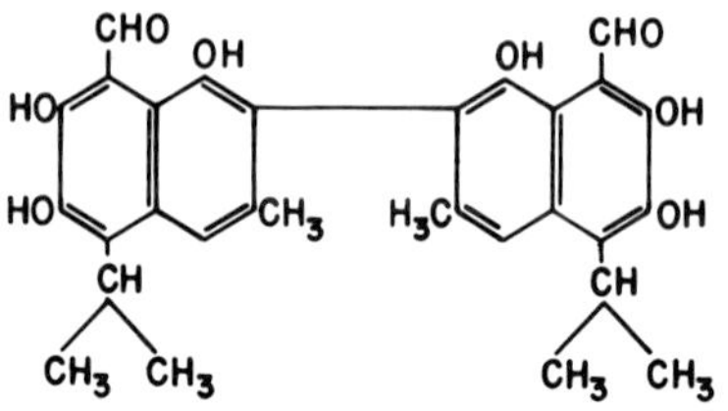

(Ia)

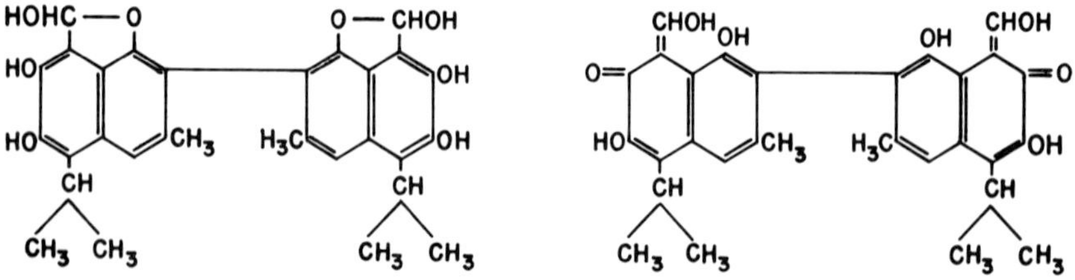

(Ib)

(Ic)

FIG. 1. Structures of the various tautomeric forms of gossypol.

as a strong dibasic acid to form neutral salts when dissolved in dilute aqueous alkali. In alcoholic solutions, it is extremely sensitive to oxidation. It forms brightly colored compounds when reacted with metallic ions. The phenolic groups of gossypol react readily to form esters and ethers. The aldehyde groups react with amines to form Schiff bases and with organic acids to form heat-labile compounds (Adams *et al.*, 1960). The reaction with aromatic amines, as with aniline to form dianilinogossypol, is significant for analysis.

Gossypol, molecular weight 518.5, is soluble in many organic solvents, and it in insoluble in low-boiling petroleum ether (bp 30°–60°C) and in water. Crystalline gossypol and most of its solutions in organic solvents are photosensitive (Boatner, 1948; Pominski *et al.*, 1951a). Gossypol of mp 184°C is obtained upon crystallization from ether, of mp 199°C from chloroform, and of mp 214°C from ligroin. Campbell *et al.* (1937) attributed this wide range of melting temperatures to polymorphism of gossypol. Shirley (1966) reported obtaining gossypol of mp 195°C from benzene. His analysis of the crystalline gossypol showed 0.5 mole of benzene per mole of gossypol, and he concluded that this was a molecular compound. The other substances comprising the naturally occurring gossypol pigments occur in much smaller quantities than gossypol and have not been studied as extensively.

B. Gossycaerulin

Kuhlmann (1861), in the first published report on the pigments of cottonseed, noted the formation of a blue-colored material when acidified cottonseed oil soapstock was steam-distilled to recover free fatty acids. Gossycaerulin is the blue coloring matter found in acidulated cottonseed oil soapstocks and in cooked cottonseed (Boatner, 1948; Boatner *et al.*, 1947a). Gossycaerulin has also been detected in cottonseed meals and crude cottonseed oils. It acts as an indicator and thus changes color with changes in pH. It is blue under acidic conditions but changes to green and then to yellow under alkaline conditions. Gossycaerulin, $C_{30}H_{30}O_8$ and therefore an isomer of gossypol, melts at 169°C but can be sublimed or distilled *in vacuo*. It is soluble in alcohols, diethyl ether, chloroform, acetic acid, and acetic anhydride. It is relatively insoluble in petroleum ether, benzene, toluene, and water. It gives reactions which indicate that it has an aldehyde group adjacent to a hydroxy group; it has been postulated that the carbonyl groups are quinonoid in nature (Boatner *et al.*, 1947a), but the complete structure has not been established.

Gossycaerulin can be isolated from water extracts of acidulated cottonseed oil soapstock, particularly soapstock of hydraulic-press oil (Kuhlmann, 1861; Boatner, 1948). It can be purified by recrystallization from a mixture of ethanol and petroleum ether. Gossycaerulin can be prepared by heating gossypol dissolved in sulfuric acid.

C. Diaminogossypol

The presence of diaminogossypol in cottonseed that has been stored at high temperatures was reported by Castillon *et al.* (1949). Miller and Adams (1937) reported the synthesis of diaminogossypol by reaction of gossypol with liquid ammonia and assigned the formula $C_{30}H_{32}O_6N_2$ to the yellow solid product that melted with decomposition at 228°–230°C. However, Pominski *et al.* (1951b) reported the synthesis of a bright yellow solid melting at 219°–221°C, also designated as diaminogossypol but formulated as $C_{30}H_{34}O_7N_2$, by reaction of gaseous ammonia with a warm solution of gossypol in chloroform. The product was reported to react with aniline and with *p*-anisidine to form dianilinogossypol and di-*p*-anisidinogossypol, respectively.

D. Gossypurpurin

Another of the naturally occurring, nitrogen-containing pigments found in small amounts in cottonseed is gossypurpurin. This gossypol derivative is purple and melts at 200°–204°C. Von der Haar and Pominski (1952) found the molecular weight of gossypurpurin to be approximately 1200, based on cryoscopic determinations in benzene solution. They proposed $C_{60}H_{64}N_2O_{14}$ as a tentative molecular formula for gossypurpurin. Later, Manevich *et al.* (1964a) reported the molecular weight of gossypurpurin, determined by thermistor ebulliometry, to be between 427 and 608, and they proposed $C_{30}H_{32}NO_7$ as the formula for gossypurpurin. More recently, Rzhekhin *et al.* (1969) synthesized gossypurpurin by heating an intimate mixture of gossypol with various amino acids, e.g., methionine, phenylalanine, and leucine in a molar ratio of 3:1. They reported that solutions of gossypurpurin in hydrophobic solvents such as $CHCl_3$ have a purple color with characteristic absorption at 530 and 565 nm, but solutions in aliphatic alcohols are yellow in color with no characteristic absorption in the visible region. They explained the transition by photochemical oxidation–reduction–solvent interactions induced by visible light radiation and assigned to gossypurpurin the structure shown in Fig. 2.

Gossypurpurin is soluble in dioxane, acetone, pyridine, chloroform, and benzene. It is slightly soluble in petroleum ether, methanol, and ethanol and is insoluble in water. Solutions of gossypurpurin are very unstable to heat and light, and the gossypurpurin is thereby converted to a yellow product that is not gossypol (Boatner, 1948). Gossypurpurin is hydrolyzed by acid to yield gossypol. Like diaminogossypol, it reacts with aniline and with *p*-anisidine to form dianilinogossypol and di-*p*-anisidinogossypol, respectively (Pominski and Von der Haar, 1951).

The gossypurpurin content of cottonseed increases during prolonged storage, the amount formed depending on the temperature and period of storage. The

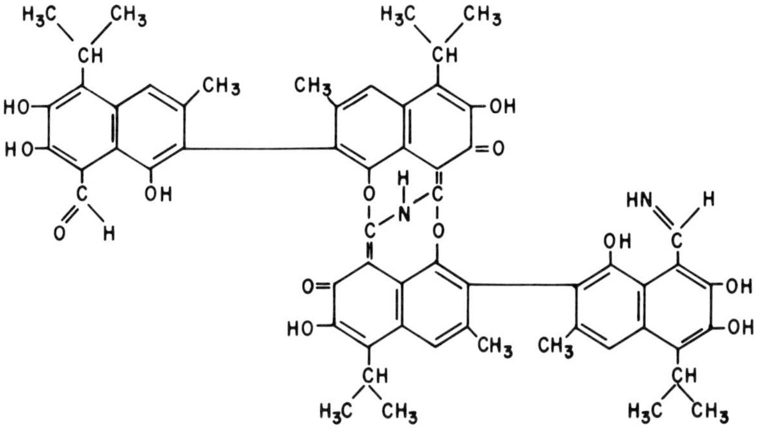

FIG. 2. Structure of gossypurpurin.

isolation of gossypurpurin from cottonseed and from pigment glands was described by Pominski *et al.* (1951b), El-Nockrashy *et al.* (1963), and Manevich *et al.* (1946b).

The unstable "red gossypol" pigment reported by Podolskaya (1936) to exist in cottonseed was shown to be a mixture of gossypol and gossypurpurin (Boatner, 1948).

E. Gossyfulvin

Gossyfulvin is an orange compound and is occasionally found in raw cottonseed. It is formed during the cooking of cottonseed and has been detected in hydraulic-press cottonseed oil and meal and in some samples of crude cottonseed oil prepared by solvent extraction. Crystalline gossyfulvin melts with decomposition at 238°–239°C. Acid hydrolysis of gossyfulvin yields gossypol in amounts of 82–86% of the weight of gossyfulvin treated. Unlike gossypol, gossyfulvin does not react with aniline, Fehling's solution, or fuchsin–aldehyde reagent and is insoluble in aqueous alkali. The formula $C_{34}H_{34}N_2O_8$ was proposed by Boatner *et al.* (1947b) for gossyfulvin on the basis of elemental analysis and the yield of gossypol obtained on acid hydrolysis. However, the analytical values reported are in better agreement with the formula $C_{35}H_{34}N_2O_8$. Gossyfulvin in chloroform solution exhibits an absorption spectrum in the ultraviolet and visible regions identical with that of a chloroform solution of dianilinogossypol (Boatner, 1948). Gossyfulvin can be prepared from ethereal extracts of cottonseed after separation of gossypol, fatty acids, and other materials from the extracts (Boatner *et al.*, 1944). It can be purified by recrystallization from cold mixtures of chloroform and diethyl ether.

F. Gossyverdurin

Isolation of the unstable green pigment gossyverdurin in partially purified form from acetone extracts of pigment glands was reported by Lyman *et al.* (1963). Gossyverdurin is soluble in chloroform, methanol, ethanol, acetone, and diethyl ether and is insoluble in petroleum ether. Gossyverdurin becomes brown at 210°C but does not melt when heated to temperatures as high as 310°C. Analysis of gossyverdurin showed the following composition: C, 62.92%; H, 6.19%; N, 1.90%; O, 21.09%, ash, 8.20% (Lyman *et al.*, 1963). They reported a structural relationship between gossyverdurin and gossypol. Gossyverdurin, when analyzed by the usual procedures, gave values for 25% apparent free gossypol* and 32.5% apparent total gossypol.

G. Other Gossypol Pigments

Stipanovic *et al.* (1975) reported the isolation of 6-methoxygossypol and 6,6′-dimethoxygossypol from the seed of *G. barbadense* L. The former compound was obtained as yellow crystals of mp 146°–149°C upon recrystallization from benzene–hexane and the latter as bright yellow crystals of mp 181°–184°C. Mass, NMR, and IR spectral data of the crystalline compounds agreed with those of synthesized similar compounds, establishing their chemical structures and the fact that 6-methoxygossypol is an asymmetric molecule of methoxylated gossypol.

Dechary (1957) reported the isolation from crude cottonseed oil of a nitrogen-containing gossypol pigment that accounted for 4% of the gossypol pigments of the oil and contained 0.43% nitrogen. He showed the presence of several amino acids combined with gossypol in this pigment.

Martin (1959) and Mattson *et al.* (1960) reported the separation of a "soluble bound gossypol" pigment from the free gossypol pigments contained in an aqueous acetone extract of cottonseed that had previously been extracted with petroleum ether. They reported fractionation of this pigment into two groups: phospholipid-bound gossypol and hydrophilic "soluble bound gossypol," although both would normally be included in the free gossypol category as determined by the standard analytical procedures (see Section IV). The phospholipid–gossypol was approximately equal to gossypol in physiological activity, as judged by the effect on egg yolk discoloration, but the hydrophilic soluble bound gossypol, if active at all, was appreciably less so.

The bound gossypol pigments, which are formed during processing of cottonseed for recovery of meal and oil, never have been characterized as definite chemical entities. It is assumed that they are formed during seed processing by

*See Section IV,C, for the significance of the terms "free," "bound," and "total" gossypol.

reaction of free gossypol pigments with extraglandular seed components (Altschul *et al.*, 1958).

Several gossypol compounds containing amino acids and sugars were separated from pigment glands of cottonseed by Rzhekhin *et al.* (1971). Gossypol-related pigments have been isolated from the roots, stems, leaves, and flower buds of cotton plants. In addition to gossypol and the O-methylated gossypol pigments, they include anhydrogossypol, dianhydrogossypol, hemigossypol. 6-methoxyhemigossypol, desoxyhemigossypol, desoxy-6-methoxyhemigossypol, hemigossypolone, and others (Bell and Stipanovic, 1977).

H. Spectral Characteristics

The ultraviolet and visible absorption characteristics of gossypol and some of the gossypol-derived pigments were discussed in detail by Boatner (1948). The spectrum of gossypol is affected by the mode of crystallization, solvent, temperature used for dissolving it, and the age of the solution. Most of the observed differences may be ascribed to the reactivity of gossypol and the tautomeric modifications. Boatner reported that ultraviolet absorption spectra of chloroform solutions of gossypol of different melting points (184°, 199°, and 214°C) differed only in the intensities of the absorption maxima. A decrease in intensity and a bathochromic shift were noted with increasing polarity of solvent.

In Table I, reported ultraviolet and visible absorption maxima of chloroform or

TABLE I

ULTRAVIOLET AND VISIBLE ABSORPTION MAXIMA OF GOSSYPOL PIGMENTS

Pigment	Solvent	Maxima (nm)
Gossypol[a]	$CHCl_3$	276–279, 288–289, 362–365
Gossycaerulin[b]	$CHCl_3$	605
Diaminogossypol[c]	$CHCl_3$	250, 378
Gossypurpurin[c,d]	$CHCl_3$	326–327, 370, 530–532, 565–568
Gossyfulvin[e]	$CHCl_3$	250–251, 312–313, 439–440
Gossyverdurin[f]	$CHCl_3$	250, 370, 560
6-Methoxygossypol[g]	ETOH	235, 288, 369
6,6′-Dimethoxygossypol[g]	ETOH	231, 253, 287, 360, 390 (sh)

[a]Boatner (1948).
[b]Boatner *et al.* (1947a); spectrum of blue (acid) form.
[c]Pominski *et al.* (1951b).
[d]Rzhekhin *et al.* (1969).
[e]Boatner *et al.* (1947b).
[f]Lyman *et al.* (1963).
[g]Stipanovic *et al.* (1975).

ethanol solutions of gossypol pigments of cottonseed are listed. Dechary (1957) reported that chloroform solutions of a nitrogen-containing gossypol pigment isolated from crude cottonseed oil exhibited a broad maximum at 370–380 nm. Martin (1959) reported that gossypol–phospholipid in methanol–chloroform (80:20) also exhibited a broad maximum between 370 and 400 nm.

O'Connor *et al.* (1954) reported the results of a detailed study of the spectra of chloroform solutions of gossypol, diaminogossypol, and 12 other derivatives of gossypol in the infrared region from 2 to 12 nm. This study supported the tautomeric structures of gossypol.

Shirley (1966) reported that the NMR spectra of gossypol in deuterochloroform and in dioxane indicated that gossypol is in the "aldehyde form" (Fig. 1a) in these solvents. The aldehyde tautomer predominated in neutral and acidic solvents, whereas the cyclic carbonyl form (Fig. 1c) was reported (Stipanovic *et al.*, 1973) to be the major tautomer of gossypol in basic solvent systems.

III. OCCURRENCE IN COTTONSEED

Of the more than 20 recognized species of the genus *Gossypium* (subtribe Hibisceae, order Malvales), only four are cultivated for fiber production. Two amphidiploid species, *G. barbadense* L. and *G. hirsutum* L. are cultivated mainly in North, South, and Central America. *Gossypium barbadense* L. and two diploid Asiatic species, *G. arboreum* L. and *G. herbaceum* L., are cultivated in other world areas (Stephens, 1958). After about 5–6 months of growth of the cotton plants, the cotton bolls (fruit) are harvested. Each mature boll is divided into three, four, or five locks; each lock contains several seeds with their masses of long (staple) fibers. Ginning removes the staple fibers from the seed. For every bale of 500 pounds of staple fiber there are produced about 825–875 pounds of cottonseed. After ginning, the North American seed varieties are densely covered with short fibers (linters). Ginned Egyptian or Sea Island seed, *G. barbadense* L., usually has no adhering short fibers and appears "bald" or "naked" and black in color.

Gossypol is the predominant pigment in the pigment glands of the seed. The other gossypol-related pigments, except the methoxylated gossypol compounds in seed of *G. barbadense* L., occur to a much smaller extent. Stipanovic *et al.* (1975) reported that 6-methoxygossypol and 6,6′-dimethoxygossypol constitute about 30% of the gossypol pigments in one variety of *G. barbadense* L. seed. On the other hand, only traces of the Ō-methylated pigments have been detected in seed of *G. hirsutum* L. Ranges of 0.68–2.36% gossypol in kernels of various cottonseed varieties grown in India and of 0.33–2.40% gossypol in kernels of seed grown in the USSR were reported (Markman and Rzhekhin, 1965; Murti and Achaya, 1975). Carter *et al.* (1966) reported that the gossypol contents of seed from 11 (cultivated and noncultivated) species of the genus *Gossypium* varied from 0% for the glandless variety (see Section III,D) of *G. hirsutum* L. to more than 9% for *G. klotzschianum* var. *davidsoni* (Kellogg).

A. Development of Pigments

The gossypol pigments occur in greater quantity in cotton roots than in the seed and in a much smaller amount in other parts of the cotton plant (Rao and Sarma, 1945). Most, or all, of the deeply colored gossypol pigments of the seed are concentrated in the pigment glands, sometimes also referred to as gum or resin glands.

After incubation of excised root tips from glanded cottonseed for 4 weeks, Smith (1961) reported that their gossypol contents increased from 0.08 to 6.32%. Heinstein *et al.* (1962) reported that labeled gossypol is formed metabolically via the isoprenoid pathway. An enzyme system that permits biosynthesis of gossypol from ^{14}C-labeled acetate and mevalonate via a specific cyclization of *cis,cis*-farnesyl pyrophosphate was isolated from cotton roots (Heinstein *et al.*, 1970). In the process, six molecules of [2-^{14}C]mevalonate were utilized for one molecule of gossypol. Metabolic inhibitors, pathogens, and certain metallic ions may stimulate gossypol formation in all tissues of the cotton plant (Bell *et al.*, 1974).

Development of the cottonseed embryo begins with flower fertilization, which is completed within from 24 to 30 hr after opening of the cotton flower (Gore, 1932). The precusor pigments first appear in the glands of the 18-day-old cotton embryo, and seed maturation begins about the 32nd day (Reeves and Beasley, 1935). The formation of pigments takes place rapidly at first and then more slowly during maturation of the embryo. The orange and yellow precursors of gossypol in the immature seed have not been characterized, but they are converted to dianilinogossypol upon treatment with aniline (Boatner, 1948).

B. Pigment Glands

The mature, anatropous cottonseed is a pointed ovoid about 8–12 mm in length and consists of two major components: hull or spermaderm, and embryo or kernel. The seed is composed almost equally of kernels and of hulls with their linters (Altschul *et al.*, 1958). Longitudinal and transverse sections of the cottonseed are illustrated in Fig. 3. The long (staple) fibers and short fibers (linters) arise from the hull. A thin membrane separates the embryo from the hull.

The pigment glands are visible as small dark specks scattered throughout the tissue of the kernel. When observed with a microscope, the colors of the glands range from light yellow to orange, red, and purple. All of the pigment glands in a single kernal are not the same color, and the colors may vary with growth and

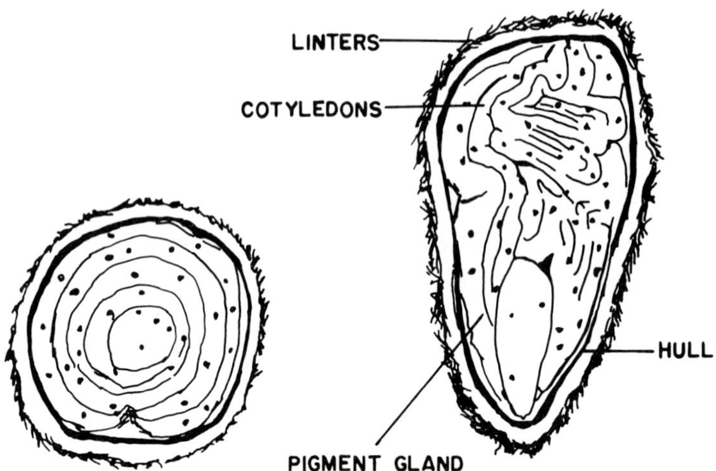

FIG. 3. Transverse and longitudinal sections of cottonseed, *Gossypium hirsutum*.

environmental conditions. The pigment glands are spherical or ovoid bodies, measuring from 100 to 400 μm on the long axis, and the average size may differ from seed to seed. The pigment glands comprise from 2.4 to 4.8% of the weight of the kernel (Boatner, 1948). The colored glandular contents are contained in discrete packages that are from 1 to less than 2 μm in diameter and are held by an inner membranous meshlike network within the glands (Moore and Rollins, 1961).

Gossypol constitutes from 39 to 50% of the weight of the pigment glands (Boatner, 1948), gossypurpurin from 0.612 to 1.73% (Boatner, 1948), and gossyverdurin about 2.0% (Lyman *et al.*, 1963). The glands also contain gossypol compounds containing amino acids and sugar residues (Rzhekhin *et al.*, 1971). Free sugars and amino acids were also found in the glands.

Each pigment gland is surrounded by a water-sensitive wall. Boatner (1948) concluded that the wall was not made of cells but of thick, curved "plates" containing celluloses, pectin, and uronic acid derivatives. Electron miscroscopy studies indicate that the glands are surrounded by a layer of tangentially flattened cells (Yatsu *et al.*, 1974).

The existence of the highly resistant wall of the cottonseed pigment glands can explain the anomaly of the relative stability of the highly reactive polyphenolic gossypol pigments even though the pigments are in close proximity to the other seed components. The behavior of the pigment gland wall during commercial cottonseed processing greatly influences the distribution of gossypol pigments in the final cottonseed meals and oils produced.

The pigment gland wall is resistant to rupture under pressure, such as that applied for flaking of cottonseed kernels during processing or during agitation in nonpolar solvents. Some organic solvents, such as chloroform or diethyl ether, completely extract the pigments from the pigment glands on prolonged contact (Boatner and Hall, 1946). Contact of the pigment gland with water brings about an immediate discharge of the glandular contents in the form of a rapidly moving stream of finely divided particles. Dilute aqueous solutions of organic or inorganic compounds cause a similar rapid rupture of the gland and discharge of the contents. However, saturated aqueous solutions of ammonium sulfate and sulfates of Al, Cd, Cu, Mg, Ni, Zn, and chlorides of Ca, Fe, and Mg do not readily rupture the glands (Muller *et al.,* 1976).

C. Influence of Genetics and Environment

The various gossypol pigments found in cottonseed have been described in Section II. The effects of inherent, genetic factors on the contents of gossypol pigments in cottonseed are about equal to the effects of environmental conditions during seed development. Smirnova (1936) observed that the average content of gossypol pigments of all varieties of seed of a given species is characteristic of the species, even though different varieties of the same species may differ with respect to the content of gossypol pigments when the seed is grown under different conditions. Boatner *et al.* (1949a) reported that *G. barbadense* seed contains more gossypol and gossypurpurin than seed of the species *G. hirsutum.* Within the species *G. barbadense,* Sea Island seed contains more gossypol and less gossypurpurin than Egyptian seed.

On the basis of a study of eight varieties of seed grown at 13 different locations for three successive years, Pons *et al.* (1953a) reported that the gossypol contents of the moisture-free kernels ranged from 0.39 to 1.70% and that both variety of seed and environment influenced the content of pigments. They reported that the content of gossypol pigments is negatively correlated with temperature of environment and positively correlated with rainfall. Cherry *et al.* (1978a) have shown that both free and total gossypol content in cottonseed were affected at a statistically significant level by the variables of cultivar, growing location, and their interaction term, cultivar × location. In studies of the relationships between contents of oil and other cottonseed constituents, Stansbury *et al.* (1956) reported finding a significant positive correlation between gossypol pigments and oil and a significant negative correlation between gossypol pigments and nitrogen. Carter *et al.* (1966) reported a significant positive correlation between total gossypol pigments in cottonseed and their lysine contents. Cherry *et al.* (1978b) showed negative correlation coefficients for protein, oil, and total gossypol in cottonseed, which suggested that the potential exists for specifically improving percentages of each constitutent.

The pigments of cottonseed may undergo changes during storage of the seed. As the ginned seed must often be stored before processing, such changes play an important role in determining the types and kinds of gossypol pigments present in the finished meals and oils. Goldovskii (1936) reported binding of the gossypol pigments in seed that had deteriorated during storage. Podolskaya (1939) reported that gossypol pigments decreased from 1.15 to 0.75% in seed stored for 4 months. Boatner *et al.* (1949a) reported that no consistent pattern of change was observed in the content of gossypol of different varieties of seed stored for several months at a temperature of 27° ± 3°C. Studies with three varieties of *G. hirsutum* seed stored at 3°, 25°, and 30°C revealed that the content of gossypol decreased, but the content of gossypurpurin increased during storage. The increase in gossypurpurin was reported to be dependent on the temperature and period of seed storage. Rzhekhin *et al.* (1968) obtained direct evidence that gossypol is converted to gossypurpurin during maturation and lengthy storage of cottonseed. They also reported that, with the lowering of seed quality, the contents of gossypurpurin are increased and those of gossypol decreased. The presence of diaminogossypol in stored seed was also reported (Castillon *et al.*, 1949).

D. Glandless Cottonseed

The commercial development of glandless cottonseed has been the objective of many plant breeders and the hope of many processors and users of cottonseed meals and oils as a means of overcoming the manifold difficulties resulting from the presence of the gossypol pigments (Anonymous, 1978a).

As indicated in Section III,C, Smirnova (1936) concluded that both environmental and genetic factors affect the gossypol content of cottonseed. Boatner *et al.* (1949a) confirmed Smirnova's findings but also inferred that "all of the gossypol and gossypurpurin of cottonseed occurs in isolated pockets or glands." An exciting development was the report by McMichael (1954) of the discovery of a completely glandless cotton boll. Although the first proposed use for the glandless boll characteristic was as a genetic marker for measuring the amount of natural crossing in cotton, the obvious benefits of glandless (and hence gossypol-free) seed to the cottonseed processing industry soon led to attempts to breed for removal of pigment glands from cottonseed (McMichael, 1959).

The first glandless strain derived from a cross between Acala and Hopi Moencopi, a noncultivated primitive cotton from Arizona, had glandless seed but it also had small bolls, low lint yield, inferior fiber properties, late maturity, and other undesirable characteristics (Lewis, 1965). The advent of commercial glandless varieties rests not only on the success of the breeder but also on successful integration into the economic system involving the seedsman, farmer, oil mill, and ultimately the consumer of cotton and cottonseed products. An extensive breeding program resulted in the combination of the glandless seed characteristic with the total complex of desirable traits required for a successful commercial variety. By 1969, the gland-free or glandless characteristic had been introduced into all major varieties of cotton (Miravalle, 1969). Ironically, the first large-scale commercial planting of glandless seed (approximately 15,000 acres of Gregg 25-V cottonseed in the Texas Plains area in 1966) resulted not primarily because of this variety's glandless character but because of its resistance to fusarium wilt. However, commercial plantings of glandless seed have not yet achieved early expectations, and less than 1% of U.S. cotton acreage is currently (1979) planted with glandless varieties. Breeding of glandless varieties has also been reported in Chad, India, Nigeria, and other countries.

Another approach to the development of glandless cottonseed is by way of radiation-induced mutation. Afifi *et al.* (1965) announced a new strain of glandless cottonseed, Bahtim-110, obtained by treatment of Egyptian cotton variety Giza with radioactive phosphorus. The new strain was reported to possess the desirable merits of the commercial, glanded variety. Although the Giza glanded seed contained two gossypol pigments, Bahtim-110 cottonseed and its seedlings were said to be completely free of gossypol. The glandless character was dominant and simply inherited (Nassar, 1969). In contrast, although American glandless cottonseed is free of gossypol, the glandless character is recessive, and it produces seedlings the roots, bark, and leaves of which contain gossypol (Wilson and Shaver, 1973). However, Bell and Stipanovic (1976) reported they had never found even a trace of gossypol in healthy glandless cotyledons, leaves, buds, petioles, or young stems and considered that previous measurements of gossypol might be greatly in error. They concluded that aniline and other amine derivatives should not be used for the determination of gossypol in leaves, flower buds, stem, bolls, and other foliar parts because carotenoids and other terpenoid aldehydes besides gossypol may be present and may form adducts that are indistinguishable from the gossypol–aniline complex by the customary measurement of absorbance at 440 nm (see Section IV,C). They recommended reaction with phloroglucinol as more sensitive and specific for collectively measuring the terpenoid aldehydes of *Gossypium*.

In the last decade, it became apparent that the pigment gland contents may provide resistance to certain insects and may decrease the need for insecticides. In discussing the dilemma of the pigment glands of cotton, Bell and Stipanovic (1977) suggested that one solution to the problem may be to breed cotton with the minimal pigment glands required for the insect population of specific localities. They further suggested that glandless cottons may be suitable for growth on the Texas High Plains and certain areas of California. For other areas, a program has been initiated to increase cotton's resistance to insects by genetically increasing the number of pigment glands in flower buds of the cotton plant. The resulting cotton plants, called "highly glanded" or "high-gossypol" cottons, should provide more resistance to most insect pests of cotton, even though seed are glandless (Lukefahr and Houghtaling, 1975).

Laboratory and pilot plant studies indicate that oil from glandless seed has all the desirable qualities of oil prepared commerically from glanded seed but without the undesirable color problem associated with some oils from glanded seed (Thaung *et al.*, 1961). Meals and other protein products prepared from glandless seed were shown to have high nutritive value (Jonston and Watts, 1964; Martinez and Hopkins, 1975), to be essentially free of gossypol, and to exhibit certain desirable functional characteristics in food formulations (Martinez *et al.*, 1970).

According to Hess (1977), the release of improved glandless cotton varieties together with a price premium to the cotton grower should encourage the production of enough acreage of glandless varieties so that glandless seed products are available within the next few years. Commercial lots of glandless cottonseed are now traded in the United States according to three classes: Class A contains not more than 400 parts per million (ppm) of total gossypol; Class AA not more than 100 ppm; and Class AAA not more than 10 ppm (Anonymous, 1978b). Although gossypol-free cottonseed may be an attainable goal, until such time as the commercial cotton crops in the world market are derived from glandless seed, the cottonseed-processing industry must continue to deal with the problem of the gossypol pigments.

IV. ISOLATION, REACTIONS, AND ANALYSIS OF GOSSYPOL

A. Isolation

The first yellow pigment of cottonseed was isolated by Longmore (1886) from soapstock obtained on refining hydraulic-press cottonseed oil. Marchlewski (1899) purified the pigment obtained from the same source, demonstrated its polyphenolic character, and named it gossypol, from gossyp(ium phen)ol, to indicate both its origin and chemical nature. Carruth (1918) described three procedures for the isolation and purification of gossypol from ethereal extracts of cottonseed. These three procedures, which have served as the basis for most subsequent published procedures, are based on (1) formation of an ether-insoluble gossypol–acetic acid complex, (2) formation of a water-soluble sodium salt and purification by way of the acetic acid complex, and (3) treatment with aniline to form ether-insoluble dianilinogossypol, followed by hydrolysis and purification as the acetic acid complex. Boatner (1948) reviewed in detail these procedures and numerous modifications.

All methods thus far proposed for the preparation of gossypol are tedious and time-consuming. The method of choice for the preparation of gossypol in quantity is determined primarily by the availability of raw material. The first source was soapstock from refining crude hydraulic-press cottonseed oil, but such soapstocks are now rarely available in the United States, and soapstocks from cottonseed oils obtained by the newer methods of processing are generally unsuitable. Raw cottonseed, preferably after defatting, is a commonly used source. Separated cottonseed pigment glands, if available, serve as an ideal source, and Castillon *et al.* (1948) described a procedure whereby about 50% of the glandular pigments can be recovered as gossypol. Royce *et al.* (1941) and others recommended the use of cotton root bark by reason of its high gossypol content and freedom from oil.

Pons *et al.* (1959a) developed a procedure for the isolation of gossypol from cottonseed gums by water washing of crude, direct solvent-extracted cottonseed oil. The gums were refluxed with methyl ethyl ketone (2-butanone) containing oxalic or phosphoric acid to hydrolyze bound gossypol pigments. Upon cooling the mixture, two phases separated with the gossypol, chiefly in the upper (ketone) phase. After reextraction of the aqueous phase with additional butanone and concentration, gossypol was precipitated with acetic acid as the acetic acid complex. The crude complex, after washing with hexane, decolorization with carbon, and recrystallization from butanone–acetic acid, afforded the bright yellow gossypol–acetic acid complex in better than 98% purity. One kilogram of gums containing about 50 gm of total gossypol pigments can yield up to 25 gm of purified gossypol–acetic acid. This can readily be converted to gossypol by solution in diethyl ether and evaporation over water to remove the ether and hydrolyze the complex (Boatner, 1948) or by solution in dilute sodium carbonate under anaerobic conditions and acidification with mineral acid (Pons *et al.*, 1959a). Such cottonseed gums were at one time commercially available, and many pounds of gossypol have been prepared by the above described procedure. Recently, Shukurov *et al.* (1978) reported the production of technical gossypol by acid hydrolysis of gossypol anthranilate obtained as a by-product of cottonseed processing.

Smith (1963) described a method for the isolation of free and bound gossypol in the form of dianilinogossypol from porcine livers.

B. Reactions

1. Structure of Gossypol

Most of the early studies of the chemical reactions of gossypol centered around the elucidation of the structure of gossypol. Clark (1927) established the molecular formula as $C_{30}H_{30}O_8$ and reported procedures for the preparation of several important degradation products. Of particular importance was the compound he designated as apogossypol, formed, together with 2 equivalents of formic acid, by the action of strong alkali on gossypol. Other important derivatives proposed by Clark included the condensation product of gossypol with two molecules of aniline, which he called dianilinogossypol ($C_{42}H_{40}N_2O_6$), a dioxime, and a crystalline hexaacetate. He concluded that, of the eight oxygen atoms of gossypol, two are present in carbonyl groups and six in hydroxy groups and that, of these, two are much more acidic than are the other four.

At the start of an intensive investigation of the structure of gossypol by Adams and his associates (1937–1941), the following structural characteristics were accepted: empirical formula of $C_{30}H_{30}O_8$, two carbonyl groups, six hydroxy groups including two ortho phenolic hydroxy groups and a hydroxy group peri or ortho to a carbonyl group, an aliphatic side chain, probably isopropyl, isobutyl, or isoamyl, and possibly a naphthalene nucleus. The results of this classic investigation are too extensive to be dealt with in detail here but are described in an excellent review by Adams *et al.* (1960). To summarize, however, the conversion of gossypol to apogossypol was formulated as the loss of two aldehyde residues, characteristic of many aromatic hydroxyaldehydes on treatment with hot aqueous alkali, and the change from apogossypol to desapogossypol as a consequence of the removal of two isopropyl groups. A binaphthyl structure was postulated on the basis of ultraviolet absorption spectra of gossypol and many of its derivatives. The structure 2,2′-bi[1,6,7-trihydroxy-5-isopropyl-3-methylnaphthyl] was thus formulated for apogossypol. The two carbonyl groups of gossypol were designated as formyl groups in the 8,8′ positions on the basis of acidic oxidation products obtained. Postulation of the three tautomeric forms of gossypol shown in Fig. 1 was necessary to account for the numerous reaction products encountered in the establishment of the structure of gossypol and their characteristics. Thus, the hydroxyaldehyde tautomer (Fig. 1a) is responsible for most of the normal aldehyde reactions of gossypol, and this is the predominant form in several organic solvents (O'Connor *et al.*, 1954; Shirley, 1966). The hexamethyl ether was formulated from the lactol tautomer (Fig. 1b) to account for the unusual stability to alkali, as contrasted to the ease of hydrolysis under acidic conditions with the loss of two methoxyl groups. The cyclic carbonyl tautomer (Fig. 1c) accounts for the ready formation of anhydrogossypol and of identical Diels–Alder-type adducts of both gossypol and anhydrogossypol with dienes such as butadiene.

The total synthesis of gossypol was accomplished by Edwards (1958). The synthesis of apogossypol hexamethyl ether had previously been reported (Edwards and Cashaw, 1956). Demethylation by the use of boron trifluroide afforded apogossypol which was identical with apogossypol prepared from gossypol. To accomplish the total synthesis of gossypol it was then necessary only to introduce two formyl groups in the 8,8′ positions of apogossypol. Reaction of apogossypol with *N*,*N*′-diphenylformamidine gave dianilinogossypol identical with that derived from gossypol. Since (1) reaction of *N,N'*-diphenylformamidine with phenols is reported always to introduce the entering group (—$CH{=}NC_6H_5$) ortho to a hydroxyl group and (2) hydrolysis of dianilinogossypol has previously been reported to produce gossypol, this comprised a total synthesis of gossypol and conclusively confirmed the structure formulated by Adams *et al.* (1938b).

Preparation of radioactive gossypol labeled in the formyl position with ^{14}C was described by Lyman *et al.* (1969).

2. *Ethers*

Since gossypol is very sensitive to numerous reagents and the ethers are more stable and suitable for stepwise degradation, numerous ethers of gossypol and its derivatives, especially apogossypol and desapogossypol, were prepared and characterized in investigations of the structure of gossypol. Indeed, the ethers served as the starting point for the preparation of various derivatives, from which important conclusions as to the structure of gossypol were deduced. As indicated previously, the behavior of the hexamethyl ether toward hydrolysis played an important role in the postulation of the lactol tautomer. The dimethyl and tetramethyl ethers also played important roles in the elucidation of the structure of gossypol. The hexamethyl ether exists in several forms, but all react in the same way chemically. The reader is referred to the previously mentioned review by Adams *et al.* (1960) for much additional information concerning the ethers of gossypol, apogossypol, desapogossypol, and their oxidation products. Preparation of a trimethylsilyl ether by reaction of gossypol with bis-(trimethylsilyl) acetamide was reported by Raju and Cater (1967). Abou-Donia *et al.* (1970) determined the mass spectra of gossypol hexamethyl ether and the trimethylsilyl ether of gossypol. The spectra supported the structure assigned to gossypol by Adams and associates.

3. *Esters*

Gossypol reacts with organic acids and the esterfication of gossypol has also been investigated extensively, although the esters were not as useful as the ethers in the studies on structure. In particular, the hexaacetate, which exists in a white form melting at ca. 280°C and a yellow form melting at ca. 185°C, has been the subject of much study (Adams *et al.*, 1960). NMR spectra indicate that the white form has the structure derived from the lactol tautomer (Fig. 1b) (Shirley, 1966). The hexaacetate is obtained by reaction of gossypol under rather severe conditins, e.g., reaction with acetic anhydride and sodium acetate under reflux. A quite different type of reaction product is readily formed by reaction of gossypol with acetic acid at room temperature. Under these conditions, acetic acid combines with gossypol in a 1:1 molar ratio to produce a gossypol–acetic acid complex. This complex, which is much more stable than is gossypol, has been found very useful in the isolation of gossypol, its preservation for use as a reagent, and as a reference standard for analysis. Shirley (1966) postulated that this complex, called gossypol–acetic acid, is a clathrate. Hexabenzoates and hexapalmitate have been described (Adams *et al.*, 1960). Correa *et al.* (1966) reported the synthesis of the gossypol hexaesters of the eight normal fatty acids of propionic through decanoic acid. Berardi and Frampton (1957) concluded that gossypol undergoes an ester exchange reaction with triglycerides of crude cottonseed oil and thus accounted for some of the difficulties encountered in refining certain highly colored crude oils.

4. *Anils*

The reaction of gossypol with amines has also been studied extensively. Gossypol with liquid ammonia, or in chloroform solution treated with gaseous ammonia, reacts to form diaminogossypol (Miller and Adams, 1937; Pominski *et al.*, 1951b). Gossypol reacts readily with 2 moles of aniline to eliminate two molecules of water and yields a condensation product commonly called dianilinogossypol. Although Adams *et al.* (1938a) concluded that dianilinogossypol is not a simple Schiff base and postulated the existence of a phenylaminomethylene structure, the anils derived from gossypol are generally formulated as Schiff-type bases derived from the hydroxyaldehyde tautomer (1a) in Fig. 1. Generally, gossypol condenses with two molecules of primary amines with elimination of two molecules of water, but *p,p'*-methylenedianiline was reported by Dechary and Brown (1956) to react in a 1:1 ratio. Similar reaction with only one aldehyde group of gossypol was reported by Alley and Shirley (1959) with the methyl ester of lysine and of glycylglycine, possibly because of the low solubility of the reaction product. Dechary and Brown (1956) reported the reaction of gossypol with various aromatic amines including α- and β-naphthylamine, *p*-aminobenzoic acid, *o*-, *m*-, and *p*-hydroxyaniline, *o*-, *m*-, and *p*-nitroaniline, *o*-phenylaniline, *o*-chloroaniline, and *o*-mercaptoaniline. The scope of the reaction between gossypol and primary amines was investigated by Alley and Shirley (1959). They succeeded in preparing anils of widely varying types of representative aliphatic and arylaliphatic amines including allylamine, *n*-octadecylamine, diethylenetriamine, 3-dimethylaminopropylamine, *p*-nitrobenzylamine, *p*-bromobenzylamine, β-phenylethylamine, and the methyl ester of glycine. Correa *et al.* (1966) reported the preparation and characterization of 20 new N-substituted gossypol amines of probable industrial or biological interest. The reaction with aniline and with *p*-anisidine has been extensively utilized for the quantitative determination of gossypol and gossypol-related pigments (see Section IV,C).

5. *Oxidation*

Gossypol readily reduces Fehling's solution and ammoniacal silver nitrate and is extremely sensitive to oxidation. Crystalline gossypol oxidizes in air at room temperature but can be stored without undergoing oxidative destruction, provided it is protected from light (Pominski *et al.*, 1951a). Hove and Hove (1944) reported that gossypol is an effective antioxidant, and Wachs (1957) determined that the four hydroxy groups in the 6,6′ and 7,7′ positions are responsible for the inhibition of autoxidation by gossypol.

Clark (1928) observed that gossypol is converted by even the mildest oxidizing agents to small and, on the whole, unrecognizable fragments. Accordingly, as indicated previously, most oxidative degradation studies, so necessary for the elucidation of the structure of complex molecules, were carried out on more stable derivatives such as the ethers. Fundamental to the practical problem associated with the presence of gossypol in cottonseed products is the fate of gossypol under oxidizing conditions. Thus, crude cottonseed oils containing gossypol must be refined promptly to prevent the formation of highly colored oxidation products that are not removed by the usual refining and bleaching procedures. It has long been known that gossypol is rapidly attacked in alkaline solution (used in refining crude oils) by atmospheric oxygen, yielding intensely colored, labile compounds, but few definitive reports are to be found as to the nature of the oxidation products. Scheiffele and Shirley (1964) reported rapid uptake of oxygen by gossypol in alkaline solution at 5°–10°C and that absorption of 1.0–1.7 moles of oxygen per mole of gossypol corresponded to maximal production of a deep red–brown (neutral) or deep purple (alkaline) compound, which was formulated as the *o*-binaphthoquinone (I), 1,1′,6,6-tetrahydroxy-5,5′-diisopropyl-3,3′-dimethyl(2,2′-binaphthalene)-7,7′,8,8′-tetrone (**I**):

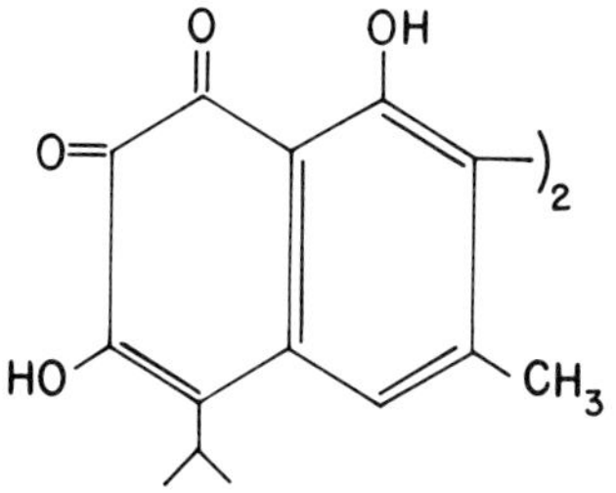

O OH O=)₂ HO CH₃

(I)

Mild oxidation with hydrogen peroxide resulted in rupture of the naphthalene ring, and several quinones and hydroxyquinones were obtained (Farmer, 1963). One of the oxidation products isolated from the early stages of oxidation was a compound formulated as (**II**):

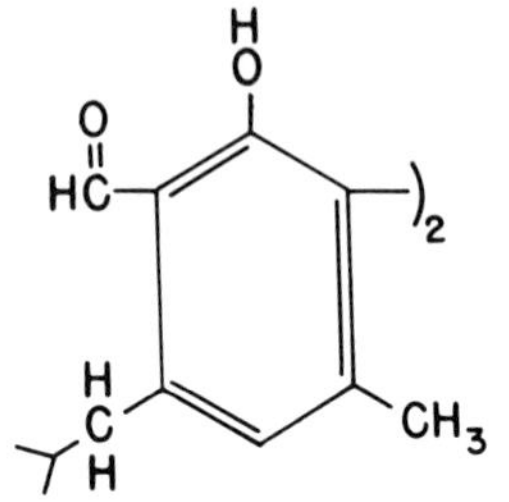

(II)

Another oxidation product isolated from the later stages of oxidation was formulated as (**III**):

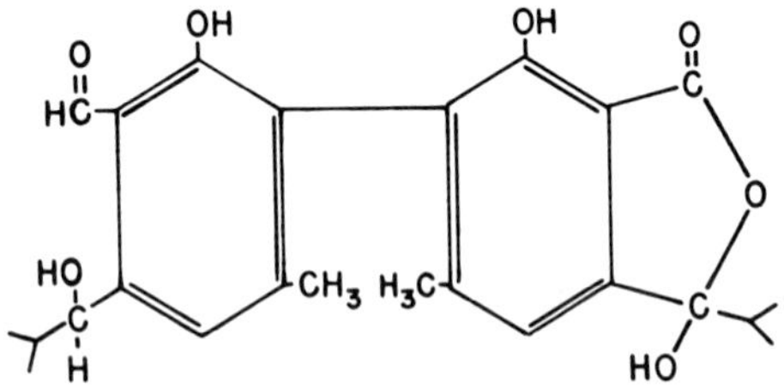

(III)

Later, Haas and Shirley (1965) reported the formation of gossypolone (**IV**) by the

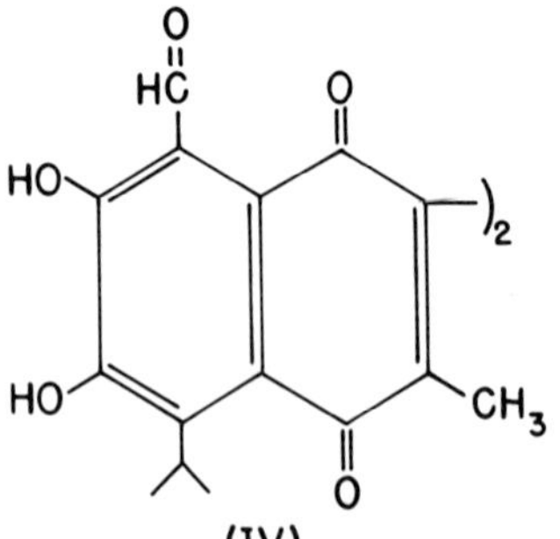

(IV)

oxidation of gossypol with ferric chloride in acetic acid–acetone. Formation of this compound was considered to be due to complexation of the *o*-hydroxyaldehyde moiety by the ferric chloride, which thus allowed oxidation of the other and normally less reactive phenolic ring and steric blocking of the normal oxidative coupling reaction of phenols.

6. *Metal Salts*

Metal salts of gossypol have aroused considerable interest, particularly because of their potential utility in analysis and the possibility that such salts might counteract toxicity due to gossypol. The proceedings of a conference devoted to inactivation of gossypol with mineral salts have been published (Anonymous, 1966). Gossypol reacts as a strong dibasic acid and forms neutral disodium and dipotassium salts. The sodium, potassium, calcium, and barium salts are soluble in water; the lead and iron salts are insoluble (Boatner, 1948). Jonassen and Demint (1955) and Ramaswamy and O'Connor (1969) reported that sodium gossypolate reacts with ferrous ions in a 1:1 molar ratio. Values of pK for dissociation of the combination averaged 7.3. It was postulated that the *peri*-hydroxyls (1-hydroxyls) of gossypol are the site of reaction with ferrous ion. The same site was believed to be the combination point when gossypol reacts with ferric ion in a 1:1 molar ratio. The pK value for the gossypol–ferric ion reaction product was calculated to be 6.75 (Muzaffarudin and Saxena, 1966). Addition of calcium ions to the soluble ferrous–gossypol chelate renders it insoluble. This suggests that a similar occurrence may take place during the biological synergism of calcium and iron to inactivate gossypol (Shieh *et al.*, 1968).

Gossypol is also known to react with such other metals as tin, antimony, and molybdenum (Adams *et al.*, 1960), and it has been proposed for use as a reagent in the qualitative or quantitative analysis for molybdenum, niobium, uranium, and vanadium (see Section VIII).

C. Analysis

Numerous procedures have been published for the determination of gossypol and gossypol pigments. Most procedures have involved extraction with a suitable solvent, treatment with an aromatic amine (generally aniline or *p*-anisidine), and determination of the reaction product gravimetrically or spectrophotometrically. Improved spectrophotometric procedures have generally replaced the earlier gravimetric methods. Other reagents that have been used for the development of colors suitable for spectrophotometric methods include antimony trichloride (Hall *et al.*, 1948), phloroglucinol (Storherr and Holley, 1954), and borax (Sadykov *et al.*, 1959). A polarographic method was reported by Markman and Kolesov (1956), and Vil'kova and Markman (1958) described a method based on the quenching by gossypol of luminescence exhibited by a solution of chloroform and acetone. Paper chromatographic methods have been reported by

Rakhmanov and Yakubov (1957) and Schramm and Benedict (1958). Thin-layer chromatographic methods have been reported by Mace *et al.* (1974, 1976), who also reported histochemical methods for the identification and measurement of gossypol. Raju and Cater (1967) and McClure (1971) reported the use of a trimethylsilyl ether for the determination of gossypol by gas–liquid chromatography.

Nearly all the early analytical methods for the determination of gossypol in cottonseed and cottonseed products were based on exhaustive extraction with diethyl ether. Pons and Guthrie (1949) and Pons and Hoffpauir (1955) developed procedures based on the use of aqueous acetone as an extractant, and these have served as the basis for procedures recommended by the American Oil Chemists' Society (1977).

Gossypol as it occurs in the cottonseed kernel is readily extracted by 70% aqueous acetone [acetone–water, 70:30 (v/v)]. When cottonseed is processed, the amount of gossypol that is extracted by this solvent decreases, and, in general, the more severe the processing conditions the less gossypol is extracted. However, all, or most, of the residual gossypol is rendered soluble if the sample is treated (hydrolyzed) with oxalic acid. That gossypol which is extracted with 70% acqueous acetone (without prior hydrolysis) is defined as *free* gossypol. Gossypol is not determined directly but as the colored reaction product with an aromatic amine. Compounds related to gossypol and which are soluble in 70% aqueous acetone, such as diaminogossypol and gossypurpurin, give reaction products with the same color. Accordingly, the value for free gossypol, as determined, includes these pigments, if they are present, as well as gossypol. (See also Section III,D, for other possible inclusions.) *Bound* gossypol, which is defined as that portion not extracted directly by 70% acetone but extracted after hydrolysis with oxalic acid under prescribed conditions, is not determined as such. Instead, *total* gossypol is determined as the total amount of gossypol and gossypol-like pigments extracted after hydrolysis. Bound gossypol is calculated as the difference between the values obtained for total and free gossypol. The amount of gossypol extracted from a cottonseed meal after hydrolysis frequently falls short of that to be expected from that found in the raw seed, particularly if the seed has been exposed to especially severe conditions during processing. There are undoubtedly many forms of bound gossypol, exhibiting a broad spectrum of degrees of binding, and some of these may not be hydrolyzed under the prescribed conditions. This may account for the well-known phenomenon of apparent disappearance of gossypol during processing of cottonseed, i.e., decrease in *total* gossypol content. In turn, some of the most loosely bound gossypol at the other end of the spectrum may account for the slight physiological activity bound gossypol appears to possess.

The AOCS has adopted as official a method based on extraction with aqueous acetone for the determination of free gossypol in cottonseed and cottonseed meals (Anonymous, 1976a). The sample is extracted with acetone–water (70:30 by volume) at room temperature, an aliquot is diluted with alcohol, treated with aniline, and the gossypol content is estimated from the absorbance at 440 nm. Total gossypol is determined in an analogous manner (Anonymous, 1976b) but after the sample has been hydrolyzed with 0.1 *M* oxalic acid in methyl ethyl ketone–water azeotrope. The total gossypol content of cottonseed oils may be determined by AOCS Tentative Method CA-13-56 (Anonymous, 1976c). In this method, the sample is dissolved in isopropanol–hexane (60:40 by volume), an aliquot is treated with 2% *p*-anisidine in isopropanol containing 4% glacial acetic acid, and the total gossypol content is estimated from the absorbance at 460 nm.

Analytical methods designed primarily for the determination of gossypol in cottonseed meals have been unsatisfactory when applied to mixed feeds by reason of incomplete removal of free gossypol and extraction of feed constituents that interfere in the subsequent spectrophotometric determination. Pons and Hoffpauir (1957) described procedures that are used extensively for the determination of free and total gossypol in mixed feeds. The methods are based on the use of 3-amino-1-propanol as a complexing agent. Free gossypol is extracted with a mixture of isopropanol–hexane–water containing a small amount of 3-amino-1-propanol neutralized with acetic acid. An aliquot is treated with aniline, and the gossypol content is estimated from the absorbance at 440 nm. For determination of total gossypol, the sample is first treated at 100°C for 30 min with a solution of neutralized 3-amino-1-propanol in *N,N*-dimethylformamide. After cooling to room temperature, a mixture of isopropanol and hexane (60:40 by volume) is added, an aliquot is treated with anline, and gossypol content is estimated from the absorbance at 440 nm. Although designed primarily for mixed feeds, these methods are also applicable to cottonseed meals and are less time-consuming than are the AOCS methods.

Spectrophotometric methods have been reported for the determination of free and bound gossypol in swine tissues (Smith, 1965) and for the determination of the gossypol–cephalin content in eggs from laying hens (Grau *et al.*, 1954).

The spectrophotometric procedures require the use of a reference standard for instrument calibration and the preparation of working curves. As gossypol has not been available commerically, the Southern Regional Research Center of the U.S. Department of Agriculture, Agricultural Research Service, has for some years supplied purified gossypol (as gossypol–acetic acid) suitable for this purpose to laboratories requiring such a reference standard.

Gossypol-like pigments are generally not determined as such, but Boatner (1948) determined the relative concentrations of gossycaerulin, gossyfulvin, and gossypurpurin in chloroform extracts of cottonseed, of cottonseed meals, and oils on the basis of spectrophotometric analysis at certain principal wavelengths before and after acid treatment of the extracts. Yusupova *et al.* (1963) reported a spectrophotometric micromethod for the determination of gossypurpurin as well as of free and bound gossypol in cottonseed pigment glands.

Pons (1977) has reviewed past and present methods for gossypol analysis in cottonseed and its products.

V. GOSSYPOL IN PROTEIN PRODUCTS, OILS, AND SOAPSTOCK

A. Processing of Cottonseed

Commercial processing of cottonseed is generally carried out by one of four basic methods: hydraulic pressing, screw pressing, prepress solvent extraction, or direct solvent extraction. The proportion of cottonseed processed in the United States by these four methods has changed drastically in recent years. As recently as 1945, approximately 95% of the seed was processed by hydraulic pressing, but 20 years later this had decreased to about 3%. In 1965 about 51% was processed by screw pressing, 25% by prepress solvent extraction, and 22% by direct solvent extraction (Harper, 1966). A ton of cottonseed generally yields 320–350 pounds of oil and 930–960 pounds of meal.

As mentioned in Section I, one of the many objectives during processing cottonseed is the binding of the free gossypol pigments in the meal while preventing the pigments from being extruded into the oil. At the same time, damage to the meal protein has to be prevented, or at least minimized. Much research has been directed toward determination of the effects of variations in each stage of processing, including preparation of meats and flakes, cooking, conditions of pressing or solvent extraction, and meal desolventization on the gossypol pigments. Variations in the content of free and total gossypol in meals are due to differences in gossypol content of the seed as well as the conditions used in processing. However, processing conditions influence the content of free gossypol in cottonseed meal to a greater extent than they influence the content of total gossypol.

The processing of seed follows similar lines with all processes, although each mill operates under specially selected conditions to obtain maximal efficiency and highest quality products with the seed available. The seed brought to the mills is cleaned and then freed of linters by delinting machines, which are similar to the gins used to remove lint cotton, except that the saw blades are closer together. After removal of the linters, the seed is transferred to the huller machines, which are designed to cut or crack the seed so that the hulls are loosened from the kernels (meats). The cracked seed is passed through a series of beaters, shakers, and separators that separate the kernels from the loosened hulls. The kernels are cracked or rolled, flattened into flakes that are about 0.008–0.012 inches in thickness to expose as much surface as possible, and cooked. Flakes of controlled moisture content are cooked and mixed for 60–90 min at temperatures in the range 93°–135°C. Rolling is essential for effective cooking and binds some of the free gossypol. Cooking aids in the removal of oil and binds additional free gossypol.

Oil is removed from the cooked flakes by one of the four above-mentioned methods. The maximal pressure obtained during hydraulic pressing is about 2000 psi. The residual press cake has an oil content of 4.5–7.5% and a free gossypol content of 0.04–0.10%. In screw pressing, pressures up to about 20,000 psi are obtained. The oil content of the press cake ranges from about 2.5 to 5% and the free gossypol from 0.02 to 0.05%. For prepress solvent extraction about two-thirds of the oil is pressed out mechanically, and the press cake is reflaked and extracted with a petroleum hydrocarbon solvent, generally commercial hexane. After desolventization, the extracted cake contains from 0.4 to 1% oil and 0.02 to 0.07% free gossypol. Alternatively, the cooked meats may be extracted directly with solvent without prepressing. Residual oil in meal obtained by direct solvent extraction (hexane) averages slightly higher than that resulting from prepress solvent extraction but is close to 1%. The free gossypol content is also distinctly higher and generally ranges from 0.1 to 0.5%. The total gossypol content of the meal obtained depends primarily on the seed used and the conditions used in the preparation of meats prior to extraction of oil and is in the range 0.5–1.2%. The bound gossypol, the difference between the total and free gossypol, consequently varies accordingly. In some direct solvent extraction plants, the kernels are not cooked before extraction.

Removal of gossypol by solvents other than the widely used commercial hexane has been investigated extensively. Solvents studied include acetone, butanone, dioxane, ethanol, isopropanol, mixtures of water with acetone, ethanol, or isopropanol, ethanol–hydrocarbon mixtures, and various ternary solvent mixtures including acetone–hexane–water, acetone–cyclohexane–water, and methanol–hexane–water. Although such solvents have generally been successful in reducing the gossypol content of cottonseed meals to quite low levels, oil mills in the United States have not yet adopted any of these solvents because of technical or economic problems or both. One process using acetone (the Vaccarino process), which has apparently had limited commercial success in Sicily, has been described in detail by Vaccarino (1961, 1965).

A patented process (Rice, 1952) whereby gossypol was firmly bound in finished meal by the addition of aniline was in commercial use in the 1950s, and the product was marketed as degossypolized cottonseed meal (Curtin, 1953). In this modification of a direct extraction process, the oil-free meal was treated with aniline, which combined with the gossypol. Excess aniline was removed by steam distillation, and the gossypol–aniline reaction product was retained in the meal, although it could be removed by further extraction if desired. The use of aliphatic amines added to cottonseed kernels or to cottonseed meals for the purpose of forming soluble amine complexes with the gossypol pigments which could be removed by solvent extraction has also been suggested for production of cottonseed meals with low free and total gossypol contents (Clark *et al.*, 1965). Degossypolized cottonseed meal was formerly defined by the Trading Rules of the National Cottonseed Products Association as "cottonseed meal in which the gossypol has been deactivated, so as to contain not more than 0.04% free gossypol." (Anonymous, 1954). More recently, the term "low-gossypol" cottonseed meal has been used to designate cottonseed meal containing not more than 0.04% free gossypol (Anonymous, 1978c).

Various mechanical procedures for the removal of gossypol have been investigated on a laboratory or pilot-plant scale from time to time but have not as yet attained commercial status. Procedures investigated include separation of intact pigment glands by gland flotation (Boatner *et al.*, 1949b; Rzhekhin *et al.*, 1963), by differential settling in commercial hexane (Vix *et al.*, 1949), and by air classification (dynamic gaseous classification) of solvent-extracted, ground cottonseed flour (Meinke and Reiser, 1962; Martinez, 1969; Kadan *et al.*, 1977). Meinke and Reiser also reported that treatment with chlorine of cottonseed flour fractions thus obtained or of the solvent-wet flakes before desolventization, grinding, and air classification not only reduced still further the free and total gossypol contents of the finished meal, but also produced a lighter colored product. Swain and O'Connor (1975, 1976) reported that a combination of air classification and washing with water at pH 4–6 or aqueous ethanol provides light-colored protein concentrates from cottonseed flour.

A process that simultaneously removes pigment glands and lipids and concentrates protein from ground cottonseed kernels by classification in liquid (hexane), the so-called Liquid Cyclone Process (LCP) developed at the Southern Regional Research Center, is currently undergoing commercial development. In this process, dried kernels from prime cottonseed are dry-milled in a wide-chamber, sieveless, impact stud mill, suspended in hexane, and the slurry is fractionated by liquid centrifugation in a cyclone into a protein-rich, low-gossypol overflow fraction and an underflow fraction containing the pigment glands. The gland-free overflow, after filtration, washing, and desolventization, yields an edible-grade flour typically containing less than 0.04% free and 0.12% total gossypol and ca. 68% protein (dry weight basis) in a yield of ca. 45% based on weight of oil-free solids used (Ridlehuber and Gardner, 1974). The U.S. Food and Drug Administration (FDA) has granted approval for use of this LCP cottonseed flour, free gossypol less than 0.045%, as a food additive (Fine, 1972). The underflow affords a high-gossypol, coarse meal that may be used as a feed for ruminants. If glandless kernels are used the underflow product containing 50–54% protein can also be used as a food additive or as a starting material for production of cottonseed protein isolates (see Section V,B). Baugher and Campbell (1969) reported that treatment of cottonseed meal with a suspension of spores from a fungus (a strain of *Diplodia*) reduced the amount of free gossypol by 90% and increased the bound gossypol to a marked extent.

B. Gossypol Pigments in Cottonseed Protein Products

The initial content of gossypol pigments of the seed, the conditions selected for the preparation of the seed prior to oil extraction, and the conditions used for extraction all determine the gossypol pigments in the defatted meals produced.

The effect of conditioning prior to extraction on the pigment gland wall is an important consideration. This influences the extent of rupture of the glands. The rupturing of glands allows the pigments to diffuse and react with extraglandular constituents during oil extraction and plays an important role in the final distribution of the gossypol pigments in the meals and oils. Additional effects may be noted upon storage of the meals.

Cottonseed has long been recognized as an excellent source of protein for feeds. Increased attention is now being focused on its protein products for foods. Color is an important criterion in the utilization and consumer acceptance of foods, and the gossypol pigments will play a role in the utilization of cottonseed protein products to impart nutrients and functionality into food formulations. Edible soybean products are marketed at three levels of protein: flours at a minimum of 50% protein, concentrates at 70% protein, and isolates at 90% protein. All three types of products containing less than 0.045% free gossypol, the limit permitted by the FDA, can be prepared from cottonseed. The preparation of such products and their utilization were reviewed by Martinez *et al.* (1970). Roasted full-fat glandless kernels containing less than 0.045% gossypol are also permitted (Fine, 1976).

As mentioned in Section I, one of the many objectives during the usual processing of cottonseed is the binding of the free gossypol pigments in the meal while preventing the pigments from being extracted into the oil. This results from the concept that developed over the years that free gossypol is physiologically active and bound gossypol is not. This concept has had great practical significance, although recent research indicates that a small portion of the bound gossypol is biologically available to nonruminants and a small proportion of the free gossypol is not (see Section VII,C). A relatively small amount of total gossypol is lost or destroyed (i.e., not accounted for in a material balance of gossypol in cottonseed meats versus gossypol in the meal and oil) in processing cottonseed by either hydraulic- or screw-pressing methods. Most of this loss occurs during preparation of the meats (Pons *et al.*, 1953b). Although cottonseed meals produced by any of the commonly used methods of processing are not uniform in composition, in general the meals prepared by direct solvent extraction are highest in free gossypol (0.1–0.5%), those prepared by screw pressing are lowest (0.02–0.05%), and those prepared by prepress solvent extraction are intermediate (0.02–0.07%).

The products obtained from binding of gossypol pigments in cottonseed meals have not been isolated and characterized as definite chemical entities. It was shown that lysine in cottonseed proteins can be destroyed when the meal is heated (Martinez and Frampton, 1958), that pure gossypol reacts with the free ϵ-amino groups of lysine (Conkerton and Frampton, 1959), and that the biological availability of the lysine is thereby reduced (Baliga *et al.*, 1959). However, the decrease in gossypol on cooking cottonseed meal cannot be entirely accounted for by the decrease in ϵ-amino groups. Cooking of cottonseed meal

(Incaparina) with water was reported to cause a decrease in free gossypol with no change in total gossypol or in available lysine (Bressani *et al.*, 1964a). It was suggested that nonproteinaceous constituents of the meals may bind with the gossypol pigments (Martinez, 1959; Martinez *et al.*, 1967). Studies by Damaty and Hudson (1975a,b) indicate that the gossypol–protein interaction is initiated through reaction of the formyl groups of gossypol with the ϵ-amino groups of lysine and arginine and that gossypol may react with the thiol groups of cysteine. The resulting reaction products may undergo intramolecular change and ultimately form insoluble, indigestible polymerization products.

Utilization of cottonseed protein products in food depends on many factors, and color is one of the most important. The gossypol and nongossypol pigments and their secondary reaction products, which contribute undesirable color bodies during the preparation, processing, and storage of the protein products, will have to be controlled before their full potential in foods can be realized. With available glandless cottonseed protein products, the problem of undesirable color from the nongossypol pigments must still be resolved before their unrestricted use in foods will be permitted.

C. Gossypol Pigments in Cottonseed Oil

Cottonseed oil is the most valuable economic product of the cottonseed. The oil is used chiefly for the preparation of salad or cooking oils, shortening, and margarine. For these purposes, oils of the lightest possible color are generally desired. Such oils are essentially completely free of gossypol pigments. However, crude cottonseed oil as it is normally obtained by any of the methods used commercially is deeply colored, ranging from brownish yellow to reddish brown in appearance. Gossypol pigments are believed to be the chief cause of the dark color, although other nongossypol pigments are also known to be present. These include chlorophyll or pheophytin, carotenoid pigments, anthocyanins, and flavonol glycosides (Boatner, 1948; Wender, 1959). Crude cottonseed oil also contains other unwanted materials, such as gums, phosphatides, and free fatty acids, as do most other crude vegetable oils. These components are generally removed in the process of refining, that is, treatment with aqueous alkali.

Removal of color is one of the more important problems of the refiner of cottonseed oil, and many innovations have been introduced, including miscella refining, i.e., refining in solvent. With most vegetable oils, removal of color is not a major problem, as those pigments not separated during the refining stage are easily removed by bleaching with adsorbents such as bleaching clays or earths. Cottonseed oil presents a special problem because the fixed colors derived from gossypol pigments are often quite resistant to removal during normal refining and bleaching. Gossypol as such, since it is readily soluble in alkali, is readily removed on refining but the fixed colors are not. Fixed colors in cottonseed oil may be present initially in the crude cottonseed oil, or they may develop during storage of the oil. The content of gossypol pigments of crude oils can vary from 0.1 to 0.75%, depending on the mode of processing. The rate of development of the red, fixed color in cottonseed oils stored under anaerobic conditions has been found to be proportional to the square of the gossypol concentration in the oil and to be temperature dependent, increasing 22-fold with a temperature increase from 40° to 80°C (Pons *et al.*, 1959b). Also, the increase in bleach color (the color after treatment with bleaching earth) is directly related to the initial gossypol content of the crude oil (Thurber *et al.*, 1954). Accordingly, release from the seed into the oil of as little gossypol as possible and refining of crude cottonseed oil as promptly as possible after production are recommended. Increased color can also develop upon storage of refined and bleached oils. This phenomenon is known as color reversion, but the cause is not known.

Although considerable empirical information has been obtained as to the conditions favoring color fixation, little is known as to the mechanism by which color fixation arises or as to the products formed. Verberg *et al.* (1961) fractionated the pigments of crude cottonseed oils by means of molecular sieves, countercurrent distribution, and low-temperature crystallization. All three methods of separation appeared to afford two fractions of the same type: a hydrophilic fraction and a hydrophobic fraction.

Boatner (1948) and Castillon *et al.* (1949) suggested that gossypurpurin, or a decomposition product of gossypurpurin, is responsible for the dark, fixed colors of oils prepared from stored cottonseed. Dechary (1957) suggested that the fixed color bodies in the oils might arise from the interaction of gossypol with amino acids from the seed protein. El-Nockrashy *et al.* (1976) reported the presence of gossypol, gossypurpurin, and several color-fixed pigments in crude cottonseed oil. They attributed the color fixation of the oil to anhydrogossypol, gossyfulvin, anthocyanins, and carotenoids. Carotenoids were the only pigments eliminated by bleaching.

On the basis of a study of the reaction of pure gossypol dissolved in refined, bleached, and deodorized cottonseed oil, Berardi and Frampton (1957) concluded that gossypol undergoes an ester exchange reaction with triglycerides of the oil and that secondary reactions of an unexplained nature result in the formation of a red color that is not removable from the oils by standard refining and bleaching.

Markman and Zalesov (1961) found that heating cottonseed oil with added gossypol and phosphatide resulted in the alteration of some of the gossypol so that it could no longer be extracted with aqueous alkali and formation of bound gossypol–phosphatide pigment. Popova *et al.* (1966) found that about 90% of the gossypol pigments of the crude oil occurred in the cephalin fraction of the crude oil phosphatides.

Pons *et al.* (1961) found that the adsorption of red, fixed pigments from refined, off-colored oils onto several adsorbents can be described by the empirical Freundlich isotherm and that activated alumina is an excellent adsorbent for these pigments. Numerous chemicals have been suggested for the treatment of crude oils or oil-solvent miscellas to form oil-insoluble or alkali-soluble gossypol reaction products. Among the chemicals investigated are sodium hypochlorite (Zharskii, 1953), *p*-aminobenzoic acid (Dechary *et al.*, 1954), diethylenetriamine (Frampton *et al.*, 1958), hydrogen peroxide (Norris, 1959), anthranilic acid (Rzhekhin *et al.*, 1961), borax (Seshadri and Chander, 1962), *p*-aminosalicylic acid (Patwari and Rao, 1963), and ferric chloride (Yatsu *et al.*, 1970.

D. Gossypol Pigments in Cottonseed Oil Soapstock

Cottonseed oil soapstock, or foots, is the principal by-product of the refining of cottonseed oils. Approximately 100 million pounds of cottonseed oil soapstock are produced annually in the United States. This material contains from 20 to 50% fatty acids (as soaps and neutral oil, phosphatides, etc.) together with concentrates of the minor nonglyceridic constituents of the oil. Minor constituents of the soapstock may include up to 12% phosphatides and from less than 1 to up to 10% gossypol pigments (Curtin and Raper, 1956). Ebenezer and Achaya (1964) reported as much as 20% gossypol in foots obtained from refined oils in India. In recent years, the quality of cottonseed soapstock has tended to become poorer, that is, it contains a lesser proportion of fatty acids and a larger proportion of nonglyceridic components, in part at least because of improvements realized in refining the oil.

Cottonseed oil foots are used in the manufacture of alkaline washing powders, household powdered soap, yellow bar laundry soap and in the manufacture of fatty acids and pitch. In recent years, increasing amounts of soapstock have been used in animal feeds. The soapstock, as such or after acidulation, or after conversion to methyl esters, is added back to the extent of 1–3% to solvent-extracted meal to improve appearance, palatability, and physical properties (e.g., as a lubricant in pelleting and to reduce dustiness) and to add energy to the feed.

The major portion of the gossypol pigments in raw soapstock is in the free form (Pack and Goldblatt, 1955). Stansbury *et al.* (1957) analyzed 99 representative samples of commercial acidulated cottonseed oil soapstocks and reported values ranging from 0.013 to 10.78% for gossypol pigments. Kuhlmann (1861), in the first recorded investigation of the pigments of cottonseed, described the isolation of gossycaerulin, the blue gossypol derivative, from acidified cottonseed oil soapstock, which had been subjected to steam distillation for the recovery of fatty acids. Zamyshlyaeva *et al.* (1963) reported that when the free fatty acids are removed from raw soapstock there remains a black pitch that retains 36% of gossypol conversion products.

Because of the increasing use of raw and acidulated soapstocks in feed, the detoxification of the contained gossypol derivatives has been investigated. Pack and Goldblatt (1955) reported that the gossypol pigments of raw soapstock can be inactivated by treatment with ferrous or ferric salts. Simple heating of raw soapstock in a closed vessel for from 1 to 3 min at a temperature of 210°–220°C sufficed to reduce the content of gossypol pigments to less than 0.02%, but the gossypol pigments in acidulated soapstocks were somewhat more resistant to destruction by similar heat treatment (Pack and Goldblatt, 1956). Gossypol content of acidulated soapstock can be reduced to less than 0.1% by saponification with concentrated sodium hydroxide prior to acidulation (Lipstein and Bornstein, 1964b). Such acidulated soapstocks can be fed safely to broilers at any dietary level (Lipstein and Bornstein, 1964a) and up to 3% in rations for laying hens (Lipstein and Bornstein, 1966). (See also Section VI,B.)

VI. ROLE OF GOSSYPOL IN UTILIZATION OF COTTONSEED MEAL

At first the demand for cottonseed meal was entirely as a fertilizer, but by the turn of this century only about one-third of domestic production was used for this purpose, and, by 1944–1945, this had decreased to about 5%, including that used as a "fertilizer" for fish ponds, where such usage was recommended as particularly effective.

The presence of gossypol presents two problems that are not easily separated. The presence of high levels of free gossypol causes unfavorable physiological effects, and the reaction between gossypol and protein during processing reduces protein quality, especially by a reduction in the biological availability of lysine. The widely different processing methods used result in quite different protein quality. Also, although a great many studies have been published concerning the role of gossypol as it affects the nutritive value of cottonseed meal, most are studies of gossypol plus gossypol-like pigments, and it is frequently difficult to relate the results of such studies to those conducted with purified gossypol. Additionally, the residual oil content of the meal, frequently ignored in the studies, may have an important effect.

Five conferences on processing as related to nutritive value of cottonseed meal were held at the Southern Regional Research Laboratory of the U.S. Department of Agriculture between 1950 and 1959 with emphasis on the nutritional value of cottonseed meal for animals. The scope of another conference, in 1960, was broadened to cottonseed protein for animal and man. A conference in 1964, on cottonseed protein concentrates (Anonymous, 1965), was concerned almost entirely with human nutrition. This reflects the increasing concern for world protein deficits and the potential offered by cottonseed meal for the development of protein-rich foods to combat human malnutrition and starvation. Throughout all these conferences, attention was focused on the role of gossypol. The latest (1972), "Conference on Oilseed Proteins—Characteristics and Food Industry Requirements," dealt exclusively with foods but included other oilseeds in addition to cottonseed.

A. Ruminants

Cottonseed meal has been highly valued as a protein concentrate for livestock (cattle and sheep) for many years and has been used extensively in rations for horses and mules throughout the Cotton Belt. At one time, it was believed that cottonseed meal caused blindness, swelling of the joints, and loss of appetite in cattle (''cottonseed meal poisoning''), and these effects were attributed to gossypol. However, it has now been established that these symptoms were due to vitamin A deficiency in the diets and not to anything contained in the meal (Hale and Lyman, 1948). The compositions of rations suitable for calves, beef cattle, dairy cattle, lambs, sheep, and goats at various stages of growth and under various conditions are described in detail by Hale and Lyman and in bulletins in the ''Feeding Practices'' series issued by the National Cottonseed Products Association (Anonymous, 1967). The combinations used depend not only on the type of cottonseed meal and quality of the particular protein supplement but on the availability, economics, and other ration components.

B. Poultry

The use of cottonseed meals for feeding ruminants is meeting increasing competition, especially from soybean meal and from urea and other nonprotein nitrogen materials, such as ammoniated agricultural residues, that can be converted to protein by microorganisms in the rumen. The economics of the feed industry are such that it is desirable to utilize cottonseed meal as a protein concentrate for nonruminants, especially swine and poultry (turkeys, broilers, and laying hens). A concentrated research effort has made possible increased utilization in these areas. As reported by Jones (1976), the estimated 17% of U.S. cottonseed meal consumed in poultry feeds in 1960 had increased to 29% in 1974; similarly, sales to feed manufacturers in feed formulations increased from 25% in 1967 to 52% in 1974. Smith (1970) reviewed the practical significance of gossypol in feeds for nonruminants.

That significant gossypol–mineral dietary interrelationships exist was recognized more than 50 years ago. Withers and Brewster (1913) found that they could protect rabbits, an especially susceptible animal species, from gossypol toxicity by oral administration of iron as ferric ammonium citrate. From time to time other investigators reported beneficial effects in overcoming adverse manifestations of gossypol, by the use not only of iron but also of other mineral salts including calcium, sodium, and potassium. Jonassen and Demint (1955) established that gossypol reacts with ferrous iron in a 1:1 molar ratio and concluded that a complex of the formula Fe^{2+}–gossypolate was formed. Since 1960 considerable research effort has been devoted to the study of iron to counteract gossypol, and in 1966 a conference on the inactivation of gossypol with mineral salts was held. At this conference it was concluded that ferrous sulfate, used as a ration additive, makes it possible to include cottonseed meal in significant amounts in swine and poultry rations without fear of undesirable effects from gossypol, and, as a result, specific recommendations were formulated. The proceedings of this conference were published and will be referred to when relevant (Anonymous, 1966).

It is generally assumed that bound gossypol is inactive physiologically, although reports indicate that some bound gossypol may have an effect (Phelps, 1966a; Smith, 1972). Free gossypol is considered the major biologically active form (see Section VII,C), and, in the following sections, the gossypol contents of rations refer to free gossypol, unless indicated otherwise.

1. Broilers

The effect of gossypol in diets of growing poultry has been the subject of many reports. As may be expected, many are contradictory, and some are incorrect or misleading. An excellent summary with numerous references was provided by Phelps (1966a). Various reports have shown that high levels of purified or pigment gland gossypol in poultry rations are associated with depressed weight gains. On the other hand, Heywang and Bird indicated that low levels of gossypol stimulated weight gains in some experiments (1954) but not in others (1955). Lipstein and Bornstein (1964a) and others showed a poor correlation between gossypol levels in the ration and growth of poultry. This apparent paradox was explained on the basis of a difference in ration components as well as levels of gossypol ingested (Phelps, 1966a). High levels of gossypol have also been shown to depress feed intake. Heywang and Bird (1954) indicated that low levels of gossypol stimulated feed efficiency. Lipstein and Bornstein (1964a) noted that gossypol usually affected feed efficiency adversely.

Increasing levels of gossypol have generally been correlated with increased mortality. Tolerance levels of gossypol reported in the literature vary widely, from as much as 1000 ppm (0.1%) to as little as 160 ppm (0.016%). Also, levels of gossypol in excess of those likely to be found in practical-type rations have a detrimental effect on metabolizable energy. A committee on the status of cottonseed meal in poultry rations formulated the following statement at the third conference on cottonseed processing and the nutritive value of the meal in 1953 and reaffirmed its validity at the fourth conference in 1957:

> Results presented thus far indicate that chick and broiler rations containing cottonseed meal and soybean meal in equal amounts on a nitrogen basis are equal or superior to rations based on cottonseed meal or soybean meal alone, when the cottonseed meal used has 0.04% or less free gossypol and 75% or more of nitrogen solubility in 0.02 *N* NaOH solution. (Anonymous, 1953, 1957)

A ration formulated with 15% cottonseed meal containing 0.04% free gossypol affords a ration level of only 60 ppm (0.006%) free gossypol. At the 1966 conference on the inactivation of gossypol with mineral salts, previously mentioned, a panel of nutritionists concluded that, for broilers, a ratio of 2 parts of iron to 1 part of gossypol by weight effectively prevents deleterious effects on growth by meals high in free gossypol content and recommended a supplemental level of iron up to 600 ppm in the presence of free gossypol levels up to 300 ppm. At the same conference, a panel of feed manufacturers recommended the use of iron at a ratio of 1:1 on a weight basis up to 600 ppm and concluded that this should be done primarily because of the tissue residue problem. The reader should consult the proceedings of the conference for additional details, discussions, and reports of other panels (Anonymous, 1966). Lipstein and Bornstein (1972a) reported that addition of hydrated ferrous sulfate to chick rations was not as reliable as the hot alkaline saponification of acidulated soapstocks for eliminating the deleterious effects of gossypol.

2. *Laying Hens*

There are numerous reports that incorporation of cottonseed meal in the rations of laying hens results in discoloration of the yolks and whites of many stored and of fresh eggs. Many of the reports suggest that gossypol is the cause of a dark (olive) discoloration of the yolks, cyclopropenoid fatty acids (normally present in cottonseed oils and meals) are the cause of a pink discoloration of the whites. Although some reports appeared to indicate a relationship between the degree of olive yolk discoloration and free gossypol, others did not, and it has been suggested that bound gossypol can contribute to egg yolk discoloration (Heywang and Vavich, 1965; Heywang *et al.*, 1965). It now appears well established that the primary chemical entity responsible for pink whites is a cyclopropenoid and that gossypol deposition in the yolk is responsible for a major part of yolk discoloration. Only recently has consideration been given to the possible interaction of the two factors.

The olive yolk discoloration in eggs from hens ingesting gossypol appears to be due to a reaction between gossypol and yolk iron. Ingestion of cyclopropenoids can enhance yolk discoloration by accelerating the increase in yolk pH during cold storage of eggs. Thus, the two factors, gossypol and cyclopropenoids, are interrelated, and both must be considered in attempting to predict the possibility of yolk discoloration as a result of ingestion of cottonseed meals. A chemical procedure termed the available gossypol units, or AGU, method for predicting olive yolk discoloration in fresh or stored eggs was proposed by Grau *et al.* (1954). This method is based on spectrophotometric measurement of the "gossypol–cephalin" fraction of egg yolks from hens fed cottonseed meals. Halloran and Cavanagh (1960) reported that cottonseed meals with egg-tested AGU values of less than 0.30 were fed successfully at the 5 and 10% level, and Phelps (1966a) stated that "thousands of tons of egg-tested meals have been included in laying hen rations without any problems from pink whites or discolored yolks."

At the previously mentioned conference on the inactivation of gossypol with mineral salts, the panel of nutritionists noted that the effect of gossypol on egg damage is the most sensitive biological indicator of gossypol activity but concluded that iron in a suitable salt form can be completely effective in preventing egg damage caused by gossypol. They stated, "The states and forms of gossypol in cottonseed meals are not sufficiently well understood to provide a quantitative basis for establishing the amount of iron needed for gossypol inactivation. At the present time the basis of the use of iron must be largely empirical." At the same time, the panel of feed manufacturers recommended for laying hens "4 parts iron to 1 part gossypol in the total ration up to a maximum of 1,600 ppm iron to 400 ppm free gossypol, with not more than 0.1% cottonseed lipids in the total rations" (Anonymous, 1966). Lipstein and Bornstein (1972b) reported that dietary iron in laying hen rations containing from 0.024 to 0.056% gossypol (introduced into the rations as acidulated soapstock) reduced but did not prevent yolk defects in cold storage eggs.

Increasing levels of gossypol have generally been correlated with decreased hatchability of eggs. The effect on hatchability is complicated by the possible presence of cyclopropenoids, but it appears that less than about 120 ppm of purified free gossypol in the ration does not depress hatchability. Although the evidence is clear that large amounts of gossypol adversely affect hatchability, the amounts that are required are substantially higher than the amounts that cause egg discoloration.

C. Swine

Studies of cottonseed meals for swine rations were reviewed by Robison (1934), Phelps (1962), and more recently by Smith (1970). The adverse physiological effects of cottonseed meals when fed to swine are to a large extent due to the presence of gossypol. However, protein quality is an important factor, and the adverse effects can be ameliorated or in some cases eliminated by increasing the level of good-quality protein. Several experiments conducted in recent years have shown that the addition of iron to the ration is beneficial in counteracting gossypol toxicity. The panel of swine nutritionists at the conference on the inactivation of gossypol with mineral salts recommended that when cottonseed meal is the sole source of supplemental protein the amount of free gossypol in the ration should be no greater than 0.01%. The panel also noted that ferrous sulfate is the most effective iron salt; the addition of 1 unit weight of iron as ferrous sulfate per unit weight of free gossypol increases weight gain, feed efficiency, and in some instances hemoglobin and hematocrit values; addition of calcium compounds may have a synergistic effect with iron in the inactivation of gossypol. The panel of feed manufacturers recommended for swine (weanling pigs and older) the addition of a 1:1 ratio of iron to free gossypol on a weight basis up to 400 ppm of free gossypol in the total ration. Again, the proceedings of the conference should be consulted for additional details and relevant discussion.

D. Man

Despite the difficulties imposed by the presence of gossypol, increasing attention is being given to cottonseed for the production of protein concentrates to alleviate the world protein shortage. Cottonseed flour as a protein supplement in human foods is being investigated in various parts of the world where cottonseed is available. Cottonseed is the most widely distributed oilseed in tropical and subtropical areas, with Brazil, Egypt, India, Mexico, and Pakistan as major producers (Milner, 1965). Approximately one-quarter of the flour potentially available from world cottonseed production could alleviate the edible protein shortages of hungry nations (Gillham, 1969).

A commercial operation to produce cottonseed protein concentrate for human food has existed in the United States for many years. This product, which was sold under the title of partially defatted, cooked cottonseed flour, or similar products derived from cottonseed intended for human consumption, specified a "free gossypol content not to exceed 450 parts per million," i.e., 0.045%. In addition, toasted, partially defatted, cooked cottonseed flour was cleared as a color additive (Larrick, 1964) and was used mainly in bakery products. Incaparina, a mixture containing 18–38% cottonseed flour (with 58% corn flour) developed by the Institute for Nutrition for Central America and Panama (INCAP), was reported by Shaw in 1967 to be a commercial success in Guatemala, approaching the break-even point in Colombia, Salvador, Honduras, Nicaragua, Panama, and Venezuela. Bressani (1969) reported excellent growth improvement of children fed Incaparina mixtures. The cottonseed flour used in Incaparina met the tentative quality and processing guidelines adopted for cottonseed flour for human consumption at the 1960 conference on cottonseed protein for animal and man (Anonymous, 1962), as amended by the WHO/FAO/UNICEF Protein Advisory Group in 1964 (M.M., 1965). The composition specifications of these guidelines called for a maximum of 0.06% free gossypol and 1.2% total gossypol. The Protein Advisory Group also noted that there is no evidence that iron is necessary for human beings in products meeting these quality guidelines. Because of the critical influence of processing in improving and/or damaging the protein concentrates, it was deemed necessary to develop specifications both on the processes and on the products to ensure useful and reproducible materials. The proceedings of the 1960 conference should be consulted for other recommendations with respect to composition, processing, and quality control. At the conclusion of the conference, Altschul (1962) summarized the situation with respect to gossypol as follows:

> The meaning of gossypol content is becoming more and more difficult to assess. There is no question that gossypol is a toxic material . . . but the question remains as to the relative

> importance of free and total gossypol . . . this [reports that the protein level in the diet has an influence on the toxicity of gossypol] suggests an interrelationship between protein content, and probably protein quality, and gossypol effects.

More recently, two developments in the United States have aroused increased interest in cottonseed as a potential source of low-cost, edible-grade protein products. The developments are the breeding of glandless cottonseed and development of the Liquid Cyclone Process (LCP) for preparation of gland-free cottonseed flour (see Section V,A). Preparation of various kinds of high-quality foods using LCP flour or its protein derivatives has been demonstrated on an experimental scale (Harden and Yang, 1973; Olson, 1973; Alford *et al.,* 1977).

VII. PHYSIOLOGICAL EFFECTS

The first published report of injury to livestock from ingestion of cottonseed meal was that of Voelcker (1858). Subsequently, cottonseed poisoning was variously attributed by early workers to the presence of choline, betaine, or nitrogenous bases of the ptomaine type and to pyrophosphates (Withers and Carruth, 1915). With the establishment by Withers and Carruth and others that purified gossypol has an effect similar to that of cottonseed when ingested by rabbits, rats, guinea pigs, and swine, the tendency arose, despite much disagreement, to attribute all of the toxicity of cottonseed to gossypol. During this time, the concept of bound gossypol developed. According to this theory, during processing of cottonseed more or less of the native, or free, gossypol combines with either free amino or free carboxyl groups of cottonseed protein to form compounds that are stable and physiologically inert, and only the uncombined or "free" gossypol is physiologically active. The lack of agreement among early investigators is understandable in view of the relatively crude analytical methods sometimes used and differences in the physical condition of test animals, in the amount and quality of dietary protein and other ration components, and in the mode of administration of gossypol, all of which are now known to affect tolerance levels for gossypol.

A new era, and more disagreement, began in the mid-1940s with the development of a relatively convenient procedure for the isolation of cottonseed pigment glands, the recognition of the potential role of other gossypol-like pigments, and the increased availability of purified gossypol and gossypol-related pigments. However, the published results as to the physiological activity of gossypol must still be evaluated and interpreted with great care because of the numerous factors that may affect the observations yet may not have been given due consideration.

General symptoms of gossypol toxicity are depressed appetite and loss of body weight. Pathological symptoms are many and varied depending on animal species. The mechanism by which gossypol causes tissue damage is not yet known. Accumulation of fluid in body cavities suggests that membrane permeability may be involved (Altschul *et al.*, 1958). The toxicity is greatly increased when gossypol is administered intravenously or intraperitoneally (Menaul, 1923; Ferguson *et al.*, 1959). Cardiac irregularity is the most common toxic effect of gossypol, and death is generally ascribed to circulatory failure. Subacute gossypol poisoning results in death from pulmonary edema; chronic poisoning results in pronounced cachexia and inanition (Alsberg and Schwartze, 1919). In nonruminants, death from gossypol effects is attributed to reduced oxygen-carrying capacity of the blood and hemolytic effects on erythrocytes (Danke and Tillman, 1965). Gossypol can reduce succinic dehydrogenase and cytochrome oxidase activity in chick liver (Ferguson *et al.*, 1959). Abou-Donia and Dieckert (1974) considered gossypol to exert its toxic effects in animals and poultry by uncoupling respiratory-linked phosphorylation. Swine, guinea pigs, and rabbits are more sensitive than are dogs, rats, and poultry. Even though the laying hen is less sensitive to gossypol toxicity than broilers, the avian egg shows a very sensitive threshold response to dietary gossypol that is manifested as egg yolk discoloration, and this is the most sensitive indicator of physiological activity of gossypol (see Section VI,B,2). The physiological effects of gossypol have been reviewed by Adams *et al.* (1960), Eagle (1960), Lyman and Widmer (1966), and more recently by Abou-Donia (1976).

A. Acute Toxicity

The acute oral toxicity of gossypol is relatively low, although LD_{50} values for gossypol and related pigments have been reported for only a few animal species. Eagle and Bialek (1950) reported LD_{50} values of gossypol for rats ranging from 2600 to 3340 mg/kg of body weight and El-Nockrashy *et al.* (1963) reported a value of 2.57 ± 0.25 gm/kg. Castillon *et al.* (1953) reported the LD_{50} value of gossypol for mice as 4.8 ± 0.6 gm/kg of body weight. Lyman *et al.* (1963) reported the oral LD_{50} of gossypol for swine as 0.55 gm/kg. El-Nockrashy *et al.* (1963) reported the LD_{50} for rats of diaminogossypol as 3.27 ± 0.22 gm/kg and of gossypurpurin as 6.68 ± 0.11 gm/kg. Lyman *et al.* (1963) reported the LD_{50} of gossyverdurin for rats as 0.66 gm/kg, and, therefore, this is the most toxic of the gossypol pigments thus far examined. Eagle *et al.* (1948) reported the LD_{50} value of different samples of pigment glands for several animal species. Expressed as milligrams per kilogram body weight, they ranged from 925–1350 for rats, to 500–950 for mice, 350–600 for rabbits, and 280–300 for guinea pigs.

B. Chronic Toxicity

The acute oral toxicity of gossypol is relatively low, but if gossypol is ingested even in small amounts for prolonged periods it can cause death. Ruminants are not adversely affected by gossypol, but young calves, before the rumen is fully

TABLE II

ANTEMORTEM AND POSTMORTEM FINDINGS IN SOME NONRUMINANTS ATTRIBUTED TO CHRONIC TOXICITY OF GOSSYPOL OR FREE GOSSYPOL PIGMENTS IN COTTONSEED MEALS

Animal species	Antemortem symptoms	Postmortem findings
Rats[a]	Loss of appetite: growth rate depression; diarrhea; anorexia; hair loss; anemia; reduced erythrocytes, hemoglobin, and packed cell volume; prevention of sperm mobility and production; reduced coital behavior	Intestinal dilation and impaction: hemorrhagic congestion of stomach and intestines; congestion in lungs and kidneys; duodenal enteritis; degeneration of seminiberous tubules; dilated mitochondria in middle of spermatids
Cats[b]	Spastic paralysis, usually of hind legs; rapid pulse; dyspnea; cardiac irregularity	Edema of lungs and heart; heart enlargement; degeneration of the sciatic nerve
Dogs[c]	Posterior incoordination; stupor; lethargy; diarrhea; anorexia; weight loss; vomiting	Lung edema; hypertrophy and edema of the heart; congestion and hemorrhages of the liver, small intestine, and stomach; fibrosis of the spleen and gallbladder; congestion of the splanchnic organs
Rabbits[d]	Stupor; lethargy; loss of appetite; diarrhea; hypoprothrombinemia; spastic paralysis; decrease in litter weights	Hemorrhages in small intestine, lungs, brain, and leg bones; enlarged gallbladder; edema and impaction of the large intestine
Poultry[e]	Loss of weight; decreased appetite; leg weakness; lowering of hemoglobin and red blood cell count; lowering of protein and albumin/globulin ratio of serum; decreased egg size; egg yolk discoloration; decreased egg hatchability	Fluid in body cavities; enlargement of gallbladder and pancreas; liver discoloration; many vacuoles and foamy spaces in the liver; ceroid-like pigment deposition in the liver, spleen, and intestinal mucosa
Swine[f]	"Thumps" or labored breathing; dyspnea; weakness; emaciation; increase in glutamic oxaloacetic transaminase; weight loss; hair discoloration; altered electrocardiogram; diarrhea; reduced hemoglobin and hematocrit; deficiency of lymphatic cells	Widespread congestion and edema of many organs; fluid in body cavities; edematous bladder and thyroid gland; flabby, dilated heart with microscopic lesions; renal lipidosis; atrophied spleens; myocardial injury

[a] Eagle and Bialek (1950), Clawson *et al.* (1962), Tone and Jensen (1973), Rong-xi and Rong-hua (1978); National Coordinating Group on Male Antifertility Agents (1978).

[b] Schwartze and Alsberg (1924).

[c] West (1940), Eagle (1949, 1950).

[d] Schwartze and Alsberg (1924), Holley *et al.* (1955).

[e] Rigdon *et al.* (1958), Narain *et al.* (1961), Waldroup and Goodner (1973).

[f] Smith (1957), Hale and Lyman (1957), Skutches *et al.* (1974).

functioning, are susceptible. Hollon *et al.* (1958) reported that calves suffering from gossypol toxicity showed an erratic appetite and dyspnea. Postmortem findings included fatty degeneration of liver, hydroperitoneum, and decreased blood-clotting time. Horses are relatively unaffected by gossypol.

The biological effects of gossypol in nonruminants are cumulative. The effects on many animals species have been investigated, and the pathological findings for rats, cats, dogs, rabbits, poultry, and swine associated with toxicity due to gossypol are summarized in Table II. Swine, guinea pigs, and rabbits are the animal species most sensitive to gossypol, cats and dogs are intermediate, and rats and poultry are least sensitive. Eagle (1960) stated that three dogs given 19 doses of 50, 100, and 200 mg of gossypol per kilogram of body weight over a period of 37 days all died. Repeated daily doses of 10–35 mg/kg (for up to 32 days) were also fatal. Ambrose and Robbins (1951) found a sex effect in rats; males suffered the greater depression in body weight. Rong-xi and Rong-hua (1978) reported an antifertility effect of gossypol in male rats. They found that gossypol administered by stomach tube inhibits the production and mobility of sperm, exerting its action during the late period of spermatogenesis in the testis. The effect of gossypol on the testis was due to the higher vulnerability of the testis and not due to selective concentration of gossypol within the testis (National Coordinating Group on Male Antifertility Agents, 1978). Clawson *et al.* (1961) reported that a dietary level of 0.02% gossypol is the borderline between toxic and nontoxic levels in swine.

C. Fate of Ingested Gossypol

The fate of ingested gossypol has been the subject of a number of investigations. It is generally assumed that ingested bound gossypol is stable, is eliminated essentially unchanged by nonruminants, and is therefore not deleterious. However, Bressani *et al.* (1964b) obtained evidence that some bound gossypol is liberated during passage through the gastrointestinal tract of dogs. The total gossypol content in the feces was essentially the same as the total intake of dietary gossypol, but fecal free gossypol was about 3.5 times greater than intake, suggesting that some of the bound gossypol was liberated during passage through the gastrointestinal tract. That bound gossypol in cottonseed meal can contribute to yolk discoloration is indicated in reports by Heywang and Vavich (1965) and Heywang *et al.* (1965). Lyman and Widmer (1966) reported that digestion with proteolytic enzymes of model gossypol–protein preparations containing no detectable amount of free gossypol resulted in conversion of up to 50% of the bound gossypol to free gossypol.

Heywang *et al.* (1955) reported that as little as 0.008% gossypol in the rations of laying hens resulted in severe green yolk discoloration of their eggs on storage. According to Woronick and Grau (1955), a portion of the gossypol in the egg yolk is in the form of a gossypol–cephalin complex, and another portion is bound to the yolk proteins. Spectrophotometric measurement of the "gossypol–cephalin" fraction of egg yolks from hens fed cottonseed meal provided the basis for selecting cottonseed meals suitable for use in rations for laying hens.

The biological half-life of [^{14}C]gossypol for hens is 30 hr (Abou-Donia and Lyman, 1970), for swine is 78 hr (Abou-Donia and Dieckert, 1975), and for rats is 48 hr but is 23 hr with added dietary iron (Abou-Donia *et al.*, 1970). Lyman and Widmer (1966) reported on the recovery of radioactive gossypol (or gossypol metabolites) in the feces and tissues of chicks fed gossypol labeled in the formyl (aldehyde) group with ^{14}C. An average of 89.3% of the ^{14}C activity was found in the feces and only 8.9% in the tissues. Of this, approximately one-half (4.4%) was found in the liver, almost as much (3.1%) in the muscle, 1.6% in the blood, and lesser amounts in the kidney, lung, heart, brain, and spleen. Abou-Donia and Lyman (1970) found that [^{14}C]gossypol ingested by laying hens was largely excreted in the feces and lesser amounts in the urine and expired CO_2. Much of the absorbed gossypol was transmitted from the bloodstream to the eggs, and a small portion was deposited in the tissues.

Lyman and Widmer (1966) also reported on the recovery of radioactive gossypol in the feces and tissues of pigs fed radioactive gossypol labeled with ^{14}C both in the formyl group and in the ring structure. In this case, a much larger proportion (25.1%) of the ^{14}C activity was recovered in the tissues and 61.6% in the feces. They concluded that the considerably higher proportion of gossypol absorbed from the digestive tract by the pig than the chick is probably the explanation for the greater sensitivity of the pig. Again, much the largest concentration of the activity was found in the liver. These results are in good agreement with those reported by Smith and Clawson (1966), who determined by chemical analyses the gossypol content of various tissues from pigs fed unlabeled gossypol. They found the highest concentration in the liver, decreasing amounts in the kidney, spleen, and lymph nodes, and a small amount in the blood serum. Abou-Donia and Dieckert (1975) found that in swine 95% of the radioactivity of a single oral dose of [^{14}C]gossypol was excreted in the feces. Of the absorbed gossypol, radioactivity was higher in muscle than in liver, adipose tissue, and blood. They suggested a series of reactions in the sequence of biotransformations of gossypol involving decarbonylation, oxidation, and conjugation.

D. Dietary Factors

It has long been recognized that the effects of gossypol may vary considerably depending on the composition of the ration, particularly as it relates to the content of various minerals and the quality and quantity of protein.

The benefit of dietary iron in counteracting the toxic effects now recognized as due to gossypol were reported as long ago as 1913 (Withers and Brewster, 1913).

Phelps (1966b) noted that several iron salts have been either completely or partially successful in counteracting undesirable effects of gossypol in at least nine series of experiments conducted between 1912 and 1953 with rabbits, rats, swine, and laying hens. The effect of iron salts was reviewed by Phelps (1966b), and the whole subject of the use of mineral salts is considered at length in the proceedings of the conference on the inactivation of gossypol with mineral salts (Anonymous, 1966). It should suffice to note here that it is well established that dietary iron, and sometimes other mineral salts, can be used successfully to counteract undesirable effects due to gossypol in properly balanced rations for gossypol-sensitive species, such as swine and laying hens. For example, yolk discoloration in stored shell eggs caused by gossypol was prevented to a large extent by addition of ferrous sulfate to the rations of laying hens. Also, Smith and Clawson (1966) reported that the gossypol accumulated in the liver of pigs varies inversely with the level of dietary iron. Rincon *et al.* (1978) found that added iron, in the ratios of 0.5:1 to 1:1 iron to free gossypol, in corn–raw cottonseed diets reduced the free gossypol content in livers of rats enough to alleviate symptoms of toxicity. Raw cottonseed treated with $FeSO_4$, added in a 1:1 weight ratio of iron to free gossypol, could be substituted for soybean meal (in the range of 20–80% on an equal portion basis) without effect on the performance of growing–finishing pigs.

The importance of quantity and quality of the protein in the rations in counteracting the effects of gossypol was reported by Cabell and Earle (1956), Hale and Lyman (1957), and Lyman (1966). It is noteworthy that Lyman (1966) found that addition of lysine does not exert as great a protective effect as does addition of an equivalent amount of protein of high biological value.

Kemmerer *et al.* (1961, 1963) found that cyclopropenoid fatty acids, normally present in cottonseed oil and in the residual lipids in cottonseed meal, have an enhancing effect on the capacity of gossypol to cause discoloration of egg yolks. Other dietary factors, such as carbohydrates, phosphatides, and flavones, have also been shown to affect the biological activity of gossypol (Martinez *et al.*, 1961, 1967; Cabell and Earle, 1961; Berardi and Martinez, 1966).

E. Effects in Man

There is currently much interest in the physiological effects and metabolic fate of gossypol in human beings, especially because of the substantial contribution cottonseed meal could make in alleviating protein shortages in many of the developing countries. It is of interest that in 1947, following the presentation of two papers (Zucker and Zucker, 1974a,b) in which it was stated that gossypol is an appetite depressant and that intestinal irritation and other toxic manifestations previously ascribed to gossypol are not found with a pure preparation in reasonable doses, there was widespread publicity as to the possible use of gossypol for the treatment of obesity in human beings. This prompted an investigation of the effect of daily doses of purified gossypol on the body weight and food consumption of dogs, and, following the sudden unexpected death of several dogs that had received relatively low levels of purified gossypol, a warning note was published (Eagle, 1949). Nevertheless, there are no recorded instances in the technical literature of gossypol toxicity in human beings who have consumed gossypol-containing products, and, on the other hand, there are various reports of the absence of deleterious effects when cottonseed products containing relatively low levels of gossypol are ingested in moderate amounts.

Adamova and Lebedeva (1947) reported that daily ingestion for 4.5 months of 60 gm of cottonseed cake containing 0.11–0.20% free gossypol produced no harmful effects. Nor were any deleterious effects observed after consumption for a year of bread containing 10% cottonseed cake. Summers *et al.* (1953) reported that bakery goods containing cottonseed flour were sold for 18 months in the Oklahoma A. & M. College food store and cafeteria with no reported instance of any dietary or other disturbance due to gossypol. Scrimshaw (1965) reported that the amount of gossypol excreted by children consuming cottonseed meal was essentially identical with that ingested and that nitrogen balance was not affected. Bressani *et al.* (1966) found no observable effects from gossypol in children fed Incaparina mixtures containing 38% cottonseed flour with up to 0.057% free gossypol and 0.88% total gossypol. Liver biopsy in children fed cottonseed protein concentrates showed no damage. M.M. (1965), in a summary report of a meeting of the Protein Advisory Group of WHO/FAO/UNICEF on cottonseed protein concentrates, noted that close clinical surveillance extending over 2 years in Guatemala and six months in Peru failed to disclose any evidence of toxicity in children fed foods containing cottonseed protein concentrates (flour). Some families in Guatemala have used Incaparina for 4 years without any indication of difficulty. No evidence of any toxicity was reported on feeding children various foods containing LCP cottonseed flour (Cater *et al.*, 1977) or on women given liquid diets containing the flour (Alford *et al.*, 1977) over extended periods.

The use in China of gossypol as an antifertility agent (pill) for men has been reported (National Coordinating Group on Male Antifertility Agents, 1978). The antifertility efficacy, evaluated by semen examination of more than 4000 healthy men who had received gossypol for more than 6 months, was 99.89%. Side effects from gossypol were said to be mild and of low incidence. Gossypol apparently acts by blocking sperm formation.

VIII. UTILIZATION

From an economic standpoint, gossypol has been, and is, a liability. Nevertheless, the large amounts of gossypol potentially available and the numerous derivatives that can be prepared have led to recurring interest in its utilization.

Koltun *et al.* (1959) estimated that the 5 million tons of cottonseed produced annually in the United States contain approximately 60 million pounds of gossypol. They also estimated that "gums," produced commercially at one cottonseed processing plant by wet washing the crude oil, could produce about 81,000 pounds of gossypol annually at a cost of about $3.80 per pound of purified gossypol–acetic acid. Decossas *et al.* (1968) estimated that the total world annual production of 25 million tons of cottonseed contained approximately 78,000 tons of gossypol.

The earliest investigators who first isolated the pigments gossypol and gossycaerulin from cottonseed oil soapstock were interested primarily in the utilization of these pigments as dyestuffs. However, their instability to light and other factors prevented their utilization as dyes (Kuhlmann, 1861; Longmore, 1886; Marchlewski *et al.*, 1896, 1897). Nazarova *et al.* (1976) reported on coupling gossypol with diazotized aromatic amines to produce dyes for cotton. It has been claimed that the gossypol resin from soapstock can be used as a promoter of setting within mineral bonding materials (Dreling, 1957), as a binder for foundry cores (Vilenskaya *et al.*, 1963), and as a modifier of bituminous lacquers (Osovetskii *et al.*, 1966). Gossypol can be used in lubricating compositions (Moiseev *et al.*, 1973).

Several investigators have considered the utility of gossypol as an antioxidant and as a stabilizer for vinyl compounds against polymerization. Royce (1933) showed that gossypol has marked antioxidant strength at low concentrations and noted the probable utility in nonedible products, such as those of the rubber and petroleum industries. Hove and Hove (1944) reported that gossypol is an effective antioxidant when used in concentration lower than that reported to be physiologically active. Katsui and Kato (1954) reported that gossypol could be employed as a stabilizer for vitamin A in solution. The use of gossypol as a stabilizer for vinyl compounds against polymerization has been patented (Pack *et al.*, 1954) Braun *et al.* (1976) patented the use of gossypol as an iron deactivator, pot life extender, and processing aid for solid rocket propellants. The use of gossypol anthranilate as a stabilizer for synthetic rubbers was reported by Piotrovskii *et al.*, (1975).

Gossypol has been suggested as a reagent for the qualitative or quantitative analysis for certain metals, including molybdenum (Vioque-Pizzaro and Malissa, 1953), uranium (Asamov *et al.*, 1963), niobium (Talipox and Khadeeva, 1964), vanadium (Inoyatov *et al.*, 1974a), and titanium (Inoyatov *et al.*, 1974b).

The application of gossypol and its derivatives in pharmaceuticals has been suggested. An alcoholic extract of cotton root bark (rich in gossypol pigments) has long been accepted as an emmenagogue and antihemorrhagic agent (Boatner, 1948). Gossypol exhibits insecticidal activity and imparts resistance to attack by bollworm, tobacco budworm, pink bollworm, and other insects. An inhibitory effect of gossypol on the growth of various microorganisms was reported by

Margalith (1967) and more recently by Aizikov *et al.* (1977). Sadykov (1965) reported that the thiosemicarbazone of gossypol (amizon) and the isonicotinoylhydrazone of gossypol (ftivazide) have attained importance as antitubercular drugs. Vermel and Kruglyok (1963) and Jolad *et al.* (1975) reported antitumor activity of gossypol, and Dorsett *et al.*, (1975) reported on antiviral activity of gossypol. Use, in China, of gossypol as an antifertility agent for men has been reported, as noted in Section VII,E.

IX. SUMMARY

Gossypol, 1,1′,6,6′,7,7′-hexahydroxy-5,5′-diisopropyl-3,3′-dimethyl(2,2′-binaphthalene)-8,8′-dicarboxaldehyde, the major pigment of cottonseed, and several closely related pigments occur in pigment glands in cottonseed. The gossypol pigments content of cottonseed varies with both variety of seed and environment, and domestic commercial glanded cottonseed kernels may contain from 0.4 to 1.7%. The occurrence of gossypol pigments in cottonseed is of commercial importance and is a major obstacle to the worldwide utilization of cottonseed protein as food for human beings. Glandless, gossypol-free cottonseed and a process for the removal of glands during conventional processing of glanded seed by solvent exraction are under development.

Gossypol is highly reactive chemically and has undesirable physiological effects. Gossypol reacts with other seed constituents during the processing of cottonseed and is responsible, in part, for the reduction in nutritive quality of the protein of cottonseed meal, for other adverse physiological effects in nonruminants, and for the production of dark-colored pigments in the oil that are not removed in conventional refining and bleaching operations. Current analytical methods for cottonseed products distinguish between gossypol in a free or readily extractable state and bound gossypol, extracted only after hydrolysis. Most free gossypol and a small amount of bound gossypol appear to be biologically active after ingestion. Adverse physiological effects may be counteracted by binding of free gossypol during processing and by the use of certain minerals, especially iron salts.

REFERENCES

Abou-Donia, M. B. (1976). *Res. Rev.* **61,** 125–160.

Abou-Donia, M. B., and Dieckert, J. W. (1974). *Life Sci.* **14,** 1955–1963.

Abou-Donia, M. B., and Dieckert, J. W. (1975). *Toxicol. Appl. Pharmacol.* **31,** 32–46.

Adou-Donia, M. B., and Lyman, C. M. (1970). *Toxicol. Appl. Pharmacol.* **17,** 4473–4486.

Abou-Donia, M.B., Lyman, C. M., and Dieckert, J. W. (1970). *Lipids* **5,** 938–946.

Adamova, A. A., and Lebedeva, M. A. (1947). *Gig. Sanit.* **12,** 33–35.

Adams, R., Price, C. C., and Dial, W. R. (1938a). *J. Am. Chem. Soc.* **60,** 2158–2160.

Adams, R., Morris, R. C., Geissman, T. A., Butterbaugh, D. J., and Kirkpatrick, E. C. (1938b). *J. Am. Chem. Soc.* **60,** 2193–2204.

Adams, R., Geissman, T. A., and Edwards, J. D. (1960). *Chem. Rev.* **60,** 555–574.

Afifi, A., Abdel-Bari, A., Kamel, S. A., and Heickal, I. (1965). *Bahtim. Exp. Stn., Tech. Bull.* **80,** pp. 1–35.

Aizikov, M. I., Kurmukov, A. G., and Isamukhamedov, I. (1977). *Dokl. Akad. Nauk Uzb. SSR* (6), pp. 41–42. *Chem Abstr.* **88,** 164324b (1978).

Alford, B. B., Kim, S., and Onley, K. (1977). *J. Am. Oil Chem. Soc.* **54,** 71–74.

Alley, P. W., and Shirley, D. A. (1959). *J. Org. Chem.* **24,** 1534–1536.

Alsberg, C. L., and Schwartze, E. W. (1919). *J. Pharmacol. Exp. Ther.* **13,** 504.

Altschul, A. M. (1962). *Proc. Conf. Cottonseed Protein Anim. Man, 1960* ARS 72–74, pp. 68–71.

Altschul, A. M., Lyman, C. M., and Thurber, F. H. (1958). *In* "Processed Plant Protein Feedstuffs" (A. M. Altschul, ed.). Chapter 17. Academic Press, New York.

Ambrose, A. M., and Robbins, D. J. (1951). *J. Nutr.* **43,** 357–370.

American Oil Chemists' Society (1977). *In* "Official and Tentative Methods of the American Oil Chemists' Society" (W. E. Link, ed.), 3rd ed., AOCS, Champaign, Illinois.

Anonymous (1953). *Proc. Conf. Process. as Related to Nutr. Value Cottonseed Meal, 3rd, 1953* pp. 1–59.

Anonymous (1954). *In* "Trading Rules 1954–1955," Rule 263. Natl. Cottonseed Prod. Memphis, Tennessee.

Anonymous (1957). *Proc. Nutr. Conf. Process. as Related to Value Cottonseed Meal, 4th, 1957,* pp. 1–110.

Anonymous (1962). *Proc. Conf. Process. as Related to Cottonseed Protein Animal Man, 6th, 1953,* pp. 1–77. ARS 72–24.

Anonymous (1965). *Proc. Conf. Cottonseed Protein Concentrates, 7th, 1964,* pp. 1–243. ARS 72–38.

Anonymous (1966). *Proc. Conf. Inactiv. Gossypol Miner. Salts, 1966* pp. 1–205.

Anonymous (1967). *Natl. Cottonseed Prod. Assoc. Bull.* **34,** 1–48.

Anonymous (1976a). "Official and Tentative Methods of the American Oil Chemists' Society," AOCS Off. Method Ba 7-58. AOCS, Champaign, Illinois.

Anonymous (1976b). "Official and Tentative Methods of the American Oil Chemists' Society," AOCS Off. Method Ba 8-55. AOCS, Champaign, Illinois.

Anonymous (1976c). "Offical and Tentative Methods of the American Oil Chemists' Society," AOCS Tentative Method Ca 13-56. AOCS, Champaign, Illinois.

Anonymous (1978a). *Proc. Conf. Glandless Cotton: Its Significance, Status, Prospects, 1977* pp. 1–184.

Anonymous (1978b). Rule 112. *In* "Trading Rules 1978–1979." Natl. Cottonseed Prod. Assoc., Memphis, Tennessee.

Anonymous (1978c). *In* "Trading Rules 1978–1979," Rule 263. Natl. Cottonseed Prod. Assoc., Memphis, Tennessee.

Asamov, K. A., Talipov, S. T., and Dzhiyanbaeva, R. K. (1963). *Dokl. Akad. Nauk Uzb., SSR* **20,** 32–36.

Baliga, B. P., Bayliss, M. E., and Lyman, C. M. (1959). *Arch. Biochem. Biophys.* **84,** 1–6.

Baugher, W. L., and Campbell, T. C. (1969). *Science* **164,** 1526–1527.

Bell, A. A., and Stipanovic, R. D. (1976). *Proc. Beltwide Cotton Prod. Res. Conf., 1976,* pp. 52–54.

Bell, A. A., and Stipanovic, R. D. (1977). *Proc. Beltwide Cotton Prod. Res. Conf., 1977,* pp. 244–258.

Bell, A. A., Stipanovic, R. D., Howell, C. R., and Mace, M. E. (1974). *Proc. Beltwide Cotton Prod. Res. Conf., 1974,* pp. 40–41.

Berardi, L. C., and Frampton, V. L. (1957). *J. Am. Oil Chem. Soc.* **34,** 399–401.

Berardi, L. C., and Martinez, W. H. (1966). *Proc. Conf. Inactiv. Gossypol Miner. Salts, 1966* pp. 167–173.

Boatner, C. H. (1948). *In* "Cottonseed and Cottonseed Products: Their Chemistry and Chemical Technology" (A. E. Bailey, ed.), Chapter 6. Wiley (Interscience), New York.

Boatner, C. H., and Hall, C. M. (1946). *Oil Soap* **23,** 123–128.

Boatner, C. H., Caravella, M., and Samuels, C. S. (1944). *J. Am. Chem. Soc.* **66,** 838–839.

Boatner, C. H., Samuels, C. S., Hall, C. M., and Curet, M. C. (1947a). *J. Am. Chem. Soc.* **69,** 668–672.

Boatner, C. H., O'Connor, R. T., Curet, M. C., and Samuels, C. S. (1947b). *J. Am. Chem. Soc.* **69,** 1268–1273.

Boatner, C. H., Castillon, L. E., Hall, C. M., and Neely, J. W. (1949a). *J. Am. Oil Chem. Soc.* **26,** 19–25.

Boatner, C. H., Hall, C. M., and Merrifield, A. L. (1949b). U.S. Patent 2,482,141.

Braun, J. D., Pickett, M. F., Gerrish, H. W., Jr., and Jonassen, H. B. (1976). U.S. Patent 3,953,260.

Bressani, R. (1969). *In* "Protein-Enriched Cereal Foods for World Needs" (M. Milner, ed.), pp. 49–66. Am. Assoc. Cereal Chem., St. Paul, Minnesota.

Bressani, R., Elias, L. G., Jarquin, R., and Braham, J. E. (1964a). *Food Technol.* **18,** 1599–1603.

Bressani, R., Elias, L. G., and Braham, J. E. (1964b). *J. Nutr.* **83,** 209–217.

Bressani, R., Elias, L. G., deZaghi, S., Mosovich, L., and Viteri, F. (1966). *J. Agric. Food Chem.* **14,** 493–496.

Cabell, C. A., and Earle, I. P. (1956). *J. Am. Oil Chem. Soc.* **33,** 416–419.

Cabell, C. A., and Earle, I. P. (1961). *Feedstuffs* **33,** No. 8, 10 (abstr.).

Campbell, K. N., Morris, R. C., and Adams, R. (1937). *J. Am. Chem. Soc.* **59,** 1723–1728.

Carruth, F. E. (1918). *J. Am. Chem. Soc.* **40,** 647–663.

Carter, F. L., Castillo, A. E., Frampton, V. L., and Kerr, T. (1966). *Phytochemistry* **5,** 1103–1112.

Castillon, L. E., Hall, C. M., and Boatner, C. H. (1948). *J. Am. Oil Chem. Soc.* **25,** 233–236.

Castillon, L. E., Hall, C. M., O'Connor, R. T., and Miller, C. B. (1949). *J. Am. Oil Chem. Soc.* **26,** 655–659.

Castillon, L. E., Karon, M., Altschul, A. M., and Martin, F. N. (1953). *Arch. Biochem. Biophys.* **44,** 181–188.

Cater, C. M., Mattil, K. F., Meinke, W. W., Taranto, M. V., Lawhon, J. T., and Alford, B. B. (1977). *J. Am. Oil Chem. Soc.* **54,** 90A–93A.

Cherry, J. P., Simmons, J. G., and Kohel, R. J. (1978a). *Proc. Beltwide Cotton Prod. Res. Conf., Dallas,* pp. 47–50.

Cherry, J. P., Simmons, J. G., and Kohel, R. J. (1978b). *In* "Nutritional Improvement of Food and Feed Proteins" (M. Friedman, ed.), pp. 343–363. Plenum, New York.

Clark, E. P. (1927). *J. Biol. Chem.* **75,** 725–739.

Clark, E. P. (1928). *J. Biol. Chem.* **77,** 81–87.

Clark, S. P., Deacon, B. D., and Lawhon, T. J. (1965). *Oil Mill Gazet.* **69,** 16–21.

Clawson, A. J., Smith, F. H., Osborne, J. C., and Barrick, E. R. (1961). *J. Anim. Sci.* **20,** 547–552.

Clawson, A. J., Smith, F. H., and Barrick, E. R. (1962). *J. Anim. Sci.* **21,** 911–915.

Conkerton, E. J., and Frampton, V. L. (1959). *Arch. Biochem. Biophys.* **81,** 130–134.

Correa, O. G., Cappi, H. M., Salem, M., and Staffa, C. (1966). *J. Am. Oil Chem. Soc.* **43,** 678–680.

Curtin, L. V. (1953). *East. Feed Merchant* **4,** 24–26 and 70–71.

Curtin, L. V., and Raper, J. T. (1956). *Poult. Sci.* **35,** 273–278.

Damaty, S., and Hudson, B. J. F. (1975a). *J. Sci. Food Agric.* **26,** 1667–1672.

Damaty, S., and Hudson, B. J. F. (1975b). *Proc. Nutr. Soc.* **34,** 48A.

Danke, R. J., and Tillman, A. D. (1965). *J. Nutr.* **87,** 493–498.

Dechary, J. M. (1957). *J. Am. Oil Chem. Soc.* **34,** 597-600.

Dechary, J. M., and Brown, L. E. (1956). *J. Am. Oil Chem. Soc.* **33,** 76–78.

Dechary, J. M., and Pradel, P. (1971). *J. Am. Oil Chem. Soc.* **48,** 563-564.

Dechary, J. M., Kupperman, R. P., Thurber, F. H., and O'Connor, R. T. (1954). *J. Am. Oil Chem. Soc.* **31,** 420–424.

Decossas, K. M., Molaison, L. J., Kleppinger, A. deB., and Laporte, V. L. (1968). *J. Am. Oil Chem. Soc.* **45,** 52A, 54A, and 83A–85A.

Dorsett, P. H., Kerstine, E. E., and Powers, L. J. (1975). *J. Pharm. Soc.* **64,** 1073–1075.

Dreling, P. E. (1957). U.S.S.R. Patent 104, 670.

Eagle, E. (1949). *Science* **109,** 361.

Eagle, E. (1950). *Arch. Biochem. Biophys.* **26,** 68–71.

Eagle, E. (1960). *J. Am. Oil Chem. Soc.* **37,** 40–43.

Eagle, E., and Bialek, H. F. (1950). *Food Res.* **15,** 232–236.

Eagle, E., Castillon, L. E., Hall, C. M., and Boatner, C. H. (1948). *Arch. Biochem. Biophys.* **18,** 271–277.

Ebenezer, S., and Achaya, K. T. (1964). *Indian Oilseeds J.* **9,** 20–21.

Edwards, J. D., Jr. (1958). *J. Am. Chem. Soc.* **80,** 3798-3799.

Edwards, J. D., Jr. (1970). *J. Am. Oil Chem. Soc.* **47,** 441–442.

Edwards, J. D., Jr., and Cashaw, J. L. (1956). *J. Am. Chem. Soc.* **78,** 3224-3225.

El-Nockrashy, A. S., Lyman, C. M., and Dollahite, J. W. (1963). *J. Am. Oil Chem. Soc.* **40,** 14–17.

El-Nockrashy, A. S., Zaher, F. A., and Osman, F. (1976). *Nahrung* **20,** 117–124.

Farmer, L. B. (1963). *Diss. Abstr.* **24,** 510-511.

Ferguson, T. M., Couch, J. R., and Rigdon, R. H. (1959). *Proc. Conf. Chem. Struct. React. Gossypol and Nongossypol Pigments Cottonseed, 1959* pp. 131–141.

Fine, S. D. (1972). *Fed. Regist.* **37,** 13713.

Fine, S. D. (1976). *Fed. Regist.* **41,** 19933.

Frampton, V. L., Kuck, J. C., Dechary, J. M., and Altschul, A. M. (1958). *J. Am. Oil Chem. Soc.* **35,** 18–20.

Gillham, F. E. M. (1969). *S. Afr. J. Sci.* **65,** 173–179.

Goldovskii, A. M. (1936). *Tr. Vses. Nauchno-Issled. Inst. Zhirov* pp. 5–31 (English summary, pp. 28–31).

Gore, U. R. (1932). *Am. J. Bot.* **19,** 795–807.

Grau, C. R., Allen, E., Nagumo, M., Woronick, C. L., and Zweigart, P. A. (1954). *J. Agric. Food Chem.* **2,** 982–986.

Haas, R. H., and Shirley, D. A. (1965). *J. Org. Chem.* **30,** 4111–4113.

Hale, F., and Lyman, C. M. (1948). *In* "Cottonseed and Cottonseed Products: Their Chemistry and Chemical Technology" (A. E. Bailey, ed.), Chapter 21. Wiley (Interscience), New York.

Hale, F., and Lyman, C. M. (1957). *J. Anim. Sci.* **16,** 364-369.

Hall, C. M., Castillon, L. E., Guice, W. A., and Boatner, C. H. (1948). *J. Am. Oil Chem. Soc.* **25,** 457–461.

Halloran, H. R., and Cavanagh, G. C. (1960). *Poult. Sci.* **39,** 18–25.

Harden, M., and Yang, S. P. (1973). *Cotton Gin Oil Mill Press* **74**(7), 22–25.

Harper, G. A. (1963). *Oil Mill Gazet.* **68,** 25–26 and 28–29.

Harper, G. A. (1964). *Oil Mill Gazet.* **69,** 10–14.

Harper, G. A. (1966). *Proc. Cottonseed Process. Clin. 15th, 1966,* ARS 72–56, pp. 56–60.

Heinstein, P. F., Smith, F. H., and Tove, S. B. (1962). *J. Biol. Chem.* **237,** 2643–2646.

Heinstein, P. F., Herman, D. L., Tove, S. B., and Smith, F. H. (1970). *J. Biol. Chem.* **245,** 4658–4665.

Hess, D. C. (1977). *Cereal Foods World* **22,** 98–102 and 105.

Heywang, B. W., and Bird, H. R. (1954). *Poult. Sci.* **33,** 851–854.

Heywang, B. W., and Bird, H. R. (1955). *Poult. Sci.* **34,** 1239–1247.
Heywang, B. W., and Vavich, M. G. (1965). *Poult. Sci.* **44,** 84–89.
Heywang, B. W., Bird, H. R., and Altschul, A. M. (1955). *Poult. Sci.* **34,** 81–90.
Heywang, B. W., Heidebrecht, A. A., and Kemmerer, A. R. (1965). *Poult. Sci.* **44,** 573–577.
Holley, K. T., Harms, W. S., Storherr, R. W., and Gray, S. W. (1955). *Ga., Agric. Exp. Stn., Mimeogr. Ser.* [N.S.] **12,** pp. 1–27.
Hollon, B. F., Waugh, R. K., Wise, G. H., and Smith, F. H. (1958). *J. Dairy Sci.* **41,** 286–294.
Hove, E. L., and Hove, Z. (1944). *J. Biol. Chem.* **156,** 623–632.
Inoyatov, A., Chaprasova, L. V., Talipov, Sh., T., Dzhiyanbaeva, R. Kh., and Ismoilov, A. I. 1974a). *Deposited Doc.* VINITI 971–74, 1, 5 pp. (in Russian); *Chem. Abstr.* **86,** 182549s 1977).
Inoyatov, A., Chaprasova, L. V., Talipov, Sh. T., Dzhiyanbaeva, R. Kh., and Biktimorov, L. (1974b). *Deposited Doc.* VINITI 970–74, 7 pp. (in Russian); *Chem. Abstr.* **86,** 182550k (1977).
Jolad, S. D., Wiedhoff, R. M., and Cole, J. R. (1975). *J. Pharm. Sci.* **64,** 1889–1890.
Jonassen, H. B., and Demint, R. J. (1955). *J. Am. Oil Chem. Soc.* **32,** 424–426.
Jones, R. T. (1976). *Cotton Gin Oil Mill Press* **77**(13), 22–23.
Jonston, C., and Watts, A. B. (1964). *Poult. Sci.* **43,** 957–963.
Kadan, R. S., Freeman, D. W., Ziegler, C. M., Jr., and Spadaro, J. J. (1979). *J. Food Sci.* **44,** 1522–1524.
Katsui, G., and Kato, K. (1954). *Vitamins* **7,** 746–748.
Kemmerer, A. R., Heywang, B. W., and Vavich, M. G. (1961). *Poult. Sci.* **40,** 1045–1048.
Kemmerer, A. R., Heywang, B. W., Vavich, M. G., and Phelps, R. A. (1963). *Poult. Sci.* **42,** 893–895.
King, T. J., and De Silva, L. B. (1968). *Tetrahedron Lett.* pp. 261–263.
Koltun, S. P., Decossas, K. M., Pominski, J., Pons, W. A., Jr., and Patton, E. L. (1959). *J. Am. Oil Chem. Soc.* **36,** 349–352.
Kuhlmann, M. F. (1861). *C.R. Hebd. Seances Acad. Sci.* **53,** 444–452.
Lambou, M. G., Shaw, R. L., Decossas, K. M., and Vix, H. L. E. (1966). *Econ. Bot.* **20,** 256–267.
Larrick, G. P. (1964). *Fed. Regist.* **29,** 1801–1802.
Lewis, C. F. (1965). *Proc. Conf. Cottonseed Protein Concentrates, 7th, 1964* ARS 72–38, pp. 198–203.
Lipstein, B., and Bornstein, S. (1964a). *Poult. Sci.* 43, 686–693.
Lipstein, B., and Bornstein, S. (1964b). *Poult. Sci.* **43,** 694–701.
Lipstein, B., and Bornstein, S. (1966). *Poult. Sci.* **45,** 651–661.
Lipstein, B., and Bornstein, S. (1972a). *Feedstuffs* **44**(2), 31–32.
Lipstein, B., and Bornstein, S. (1972b). *Feedstuffs* **44**(3), 38–39.
Longmore, J. (1886). *J. Soc. Chem. Ind., London* **5,** 200–205.
Lukefahr, M. J., and Houghtaling, J. E. (1975). *Proc. Beltwide Cotton Prod. Res. Conf., 1975* pp. 93–94.
Lyman, C. M. (1966). *Proc. Conf. Inactiv. Gossypol Miner. Salts, 1966* pp. 104–116.
Lyman, C. M., and Widmer, C. (1966). *Proc. Conf. Inactiv. Gossypol Miner. Salts, 1966* pp. 43–53.
Lyman, C. M., El-Nockrashy, A. S., and Dollahite, J. W. (1963). *J. Am. Oil Chem. Soc.* **40,** 571–575.
Lyman, C. M., Cronin, J. T., Trant, M. M., and Odell, G. V. (1969). *J. Am. Oil Chem. Soc.* **46,** 100–104.
McClure, M. A. (1971). *J. Chromatogr.* **54,** 25–31.
McMichael. S. C. (1954). *Agron. J.* **46,** 527–528.
McMichael, S. C. (1959). *Agron. J.* **51,** 630.

Mace, M. E., Bell, A. A., and Stipanovic, R. D. (1974). *Phytopathology* **64,** 1297–1302.

Mace, M. E., Bell, A. A., and Beckman, C. H. (1976). *Can. J. Bot.* **54,** 2095–2099.

Manevich, E. F., Sadykov, A. S., and Ismailov, A. I. (1964a). *Nauchn. Tr., Tashk. Gos. Univ.* **263,** 112–116.

Manevich, E. F., Ismailov, A. I., and Sadykov, A. S. (1964b). *Nauchn. Tr., Tashk. Gos. Univ.* **263,** 117–121.

Marchlewski, L. P. (1899). *J. Prakt. Chem.* **60,** 84–90.

Marchlewski, L. P., Wilson, E. S., and Steward, E. (1896). British Patent 9477/1896.

Marchlewski, L. P., Wilson, E. S., and Steward, E. (1897). German Patent 98,587.6/5.97.

Margalith, P. (1967). *Appl. Microbiol.* **15,** 952–953.

Markman, A. L., and Kolesov, S. N. (1956). *Zh. Prikl. Khim.* **29,** 242–252.

Markman, A. L., and Rzhekhin, V. P. (1965). Gossypol and its derivatives. English translation TT 68-50335, vii. (Available from Clearinghouse for Federal Scientific and Technical Information, Springfield, Virginia.)

Markman, A. L., and Zalesov, Y. P. (1961). *Maslob.-Zhir. Prom'st.* **27,** 19–21.

Martin, J. H. (1959). *Proc. Conf. Chem. Struct. React. Gossypol and Nongossypol Pigments Cottonseed, 1959* pp. 71–90.

Martinez, W. H. (1959). *Proc. Conf. Chem. Struct. React. Gossypol and Nongossypol Pigments Cottonseed, 1959* pp. 52–59.

Martinez, W. H. (1969). *Proc. Conf. Protein-Rich Food Prod. Oilseeds, 1968* ARS 72–71, pp. 33–39.

Martinez, W. H., and Frampton, V. L. (1958). *J. Agric. Food Chem.* **6,** 312.

Martinez, W. H., and Hopkins, D. T. (1975). *In* "Protein Nutritional Quality of Foods and Feeds" (M. Friedman, ed.), Part 2, pp. 355–374. Dekker, New York.

Martinez, W. H., Frampton, V. L., and Cabell, C. A. (1961). *J. Agric. Food Chem.* **9,** 64–66.

Martinez, W. H., Berardi, L. C., Frampton, V. L., Wilcke, H. L., Greene, D. E., and Teichman, R. (1967). *J. Agric. Food Chem.* **15,** 427–432.

Martinez, W. H., Berardi, L. C., and Goldblatt, L. A. (1970). *J. Agric. Food Chem.* **18,** 961–968.

Mattson, F. H., Martin, J. B., and Volpenhein, R. A. (1960). *J. Am. Oil Chem. Soc.* **37,** 154.

Meinke, W. W., and Reiser, R. R. (1962). *Proc. Conf. Cottonseed Protein Anim. Man, 6th, 1960* ARS 72-24, pp. 60–65.

Menaul, P. (1923). *J. Agric. Res.* **26,** 233–237.

Miller, R. F., and Adams, R. (1937). *J. Am. Chem. Soc.* **59,** 1736–1738.

Milner, M. (1965). *Proc. Conf. Cottonseed Protein Concentrates, 7th, 1964* ARS 72–38, pp. 7–13.

Miravalle, R. J. (1969). *Proc. Conf. Protein-Rich Food Prod. Oilseeds, 1968* ARS 72–71, pp. 76–80.

M. M. (1965). *WHO/FAO/UNICEF-Protein Advisory Group News Bull.* pp. 68–73.

Moiseev, V. V., Tselykh, I. M., Piotrovskii, K. B., Kharitonov, A. G., and Gromova, G. N. (1973). *Otkrytiya. Izobret., Prom. Obraztsv. Tovarnye Znaki* **50**(30), 85; U.S.S.R. Patent 390,133; *Chem. Abstr.* **83,** 134816t (1975).

Moore, A. T., and Rollins, M. L. (1961). *J. Am. Oil Chem. Soc.* **38,** 156–160.

Muller, L. L., Jacks, T. J., and Hensarling, T. P. (1976). *J. Am. Oil Chem. Soc.* **53,** 598–602.

Murti, K. S., and Achaya, K. T. (1975). "Cottonseed Chemistry and Technology," pp. 1–348. Publications & Information Directorate. CSIR, New Delhi, India.

Muzaffaruddin, M., and Saxena, E. R. (1966). *J. Am. Oil Chem. Soc.* **43,** 429–430.

Narain, R., Lyman, C. M., Deyoe, C. W., and Couch, J. R. (1961). *Poult. Sci.* **40,** 21–25.

Nassar, M. A. (1969). M.S. Thesis, Azhar University, Cairo, Egypt.

National Coordinating Group on Male Antifertility Agents (1978). *Chin. Med. J. (Peking Engl. Ed.)* **4,** 417–428.

Nazarova, I. P., Glushenkova, A. I., and Markman, A. L. (1976). *Khim. Prir. Soedin.* **5,** 607-609; *Chem. Abstr.* **86,** 122923z (1977).

Norris, F. A. (1959). U.S. Patent 2,915,538.

O'Connor, R. T., Von der Haar, P., DuPre, E. F., Brown, L. E., and Pominski, C. H. (1954). *J. Am. Chem. Soc.* **76,** 2368-2373.

Olson, R. L. (1973). *Oil Mill Gazet.* **77**(9), 7-8.

Osovetskii, M. A., Baranovskaya, G. M., and Eruslanova, L. P. (1966). *Lakokras. Mater. Ikh Primen.* pp. 74-76.

Pack, F. C., and Goldblatt, L. A. (1955). *J. Am. Oil Chem. Soc.* **32,** 551-553.

Pack, F. C., and Goldblatt, L. A. (1956). U.S. Patent 2,746,864.

Pack, F. C., Moore, R. N., and Bickford, W. G. (1954). U.S. Patent 2,685,597.

Patwari, J., and Rao, S. D. T. (1963). *Indian J. Technol.* **1,** 435-436.

Phelps, R. A. (1962). *Proc. Semi-Annu. Meet. AFMA Nutr. Counc., 1962* pp. 10-14.

Phelps, R. A. (1966a). *World's Poult. Sci. J.* **22,** 86-112.

Phelps, R. A. (1966b). *Proc. Conf. Inactiv. Gossypol Miner. Salts, 1966* pp. 5-10.

Piotrovskii, K. B., Gromova, G. N., Isanova, L. M., and Gol'dberg, A. O. (1975). *Kauch. Rezina* **3,** 33-35.

Podolskaya, M. Z. (1936). *Tr., Vses. Nauchno-Issled. Inst. Zhirov* pp. 77-87.

Podolskaya, M. Z. (1939). *Tr., Vses. Nauchno-Issled. Inst. Zhirov* pp. 61-72 (English summary, pp. 71-72); *Chem. Abstr.* **36,** 7064 (1942).

Pominski, C. H., and Von der Haar, P. (1951). *J. Am. Oil Chem. Soc.* **28,** 444-446.

Pominski, C. H., Castillon, L. E., Von der Haar, P., Brown, L. E., and Damare, H. (1951a). *J. Am. Oil Chem. Soc.* **28,** 352-353.

Pominski, C. H., Miller, C. B., Von der Haar, P., O'Connor, R. T., Castillon, L. E., and Brown, L. E. (1951b). *J. Am. Oil Chem. Soc.* **28,** 472-475.

Pons, W. A., Jr. (1977). *J. Assoc. Off. Anal. Chem.* **60,** 252-259.

Pons, W. A., Jr., and Guthrie, J. D. (1949). *J. Am. Oil Chem. Soc.* **26,** 671-676.

Pons, W. A., Jr., and Hoffpauir, C. L. (1955). *J. Am. Oil Chem. Soc.* **32,** 295-300.

Pons, W. A., Jr., and Hoffpauir, C. L. (1957). *J. Assoc. Off. Agric. Chem.* **40,** 1068-1080.

Pons, W. A., Jr., Hoffpauir, C. L., and Hopper, T. H. (1953a). *J. Agric. Food Chem.* **1,** 1115-1118.

Pons, W. A., Jr., Murray, M. D., LeBlanc, M. F. H., Jr., and Castillon, L. E. (1953b). *J. Am. Oil Chem. Soc.* **30,** 128-132.

Pons, W. A., Jr., Pominski, J., King, W. H., Harris, J. A., and Hopper, T. H. (1959a). *J. Am. Oil Chem. Soc.* **36,** 328-332.

Pons, W. A., Jr., Berardi, L. C., and Frampton, V. L. (1959b). *J. Am. Oil Chem. Soc.* **36,** 337-339.

Pons, W. A., Jr., Kuck, J. C., and Frampton, V. L. (1961). *J. Am. Oil Chem. Soc.* **38,** 104-107.

Popova, V. J., Jusupova, I. U., and Rzhekhin, V. P. (1966). *Maslob.-Zhir. Promost.* **32,** 17-19.

Raju, P. J., and Cater, C. M. (1967). *J. Am. Oil Chem. Soc.* **44,** 465-466.

Rakhmanov, R. R., and Yakubov, A. M. (1957). *Dokl. Akad. Nauk Uzb. SSR* pp. 51-55.

Ramaswamy, H. N., and O'Connor, R. T. (1969). *J. Agric. Food Chem.* **17,** 1406-1408.

Rao, K. V., and Sarma, R. V. (1945). *Curr. Sci.* **14,** 270-271.

Reeves, R. G., and Beasley, J. O. (1935). *J. Agric. Res.* **51,** 935-944.

Rice, J. V. (1952). U.S. Patent 2,607,687.

Ridlehuber, J. M., and Gardner, H. K., Jr., (1974). *J. Am. Oil Chem. Soc.* **51,** 153-157.

Rigdon, R. H., Crass, G., Ferguson, T. M., and Couch, J. R. (1958). *AMA Arch. Pathol.* **65,** 228-235.

Rincon, R., Smith, F. H., and Clawson, A. J. (1978). *J. Anim. Sci.* **47,** 865-873.

Robison, W. L. (1934). *Ohio, Agric. Exp. Stn., Bull.* **534,** 1-44.

Rong-xi, D., and Rong-hua, D. (1978). *Acta Biol. Exp. Sinica* **11,** 22.
Royce, H. D. (1933). *Oil Soap* **10,** 123–125.
Royce, H. D., Harrison, J. R., and Hahn, E. R. (1941). *Oil Soap* **18,** 27–29.
Rzhekhin, V. P., Belova, A. B., Tros'ko, U. I., Koneva, Y. A., Borshchev, S. T., Vlasov, V. I., Rozenshtein, G. V., and Tadzhibaev, G. T. (1961). *Maslob.-Zhir. Prom'st.* **27,** 26–29.
Rzhekhin, V. P., Belova, A. B., and Chudnovskaya, A. M. (1963). *Tr., Vses. Nauchno-Issled. Inst. Zhirov* pp. 70–77.
Rzhekhin, V. P., Danilova, T. A., Mironova, A. N., and Novikova, L. V. (1968). *Maslob.-Zhir. Prom'st.* **34**(9), 12–14; *Chem. Abstr.* **70,** 9358a (1969).
Rzhekhin, V. P., Mironova, A. N., Danilova, T. A., and Kyiz, E. P. (1969). *Maslob.-Zhir. Prom'st.* **35**(1), 9–11; *Chem. Abstr.* **70,** 97142d (1969).
Rzhekhin, V. P., Belova, A. B., Chudnovskaya, A. M., and Buta, L. F. (1971). *Tr., Vses. Nauchno-Issled. Inst. Zhirov* **28,** 184–192; *Chem. Abstr.* **78,** 133364d (1973).
Sadykov, A. S. (1965). *J. Sci. Ind. Res.* **24,** 77–81.
Sadykov, A. S., Ismailov, A., and Uzebekova, D. (1959). *Dokl. Akad. Nauk Uzb. SSR* pp. 40–43.
Scheiffele, E. W., and Shirley, D. A. (1964). *J. Org. Chem.* **29,** 3617–3620.
Schramm, G., and Benedict, J. H. (1958). *J. Am. Oil Chem. Soc.* **35,** 371–373.
Schwartze, E. W., and Alsberg, C. L. (1924). *J. Agric. Res.* **28,** 191–198.
Scrimshaw, N. S. (1965). *Proc. Conf. Cottonseed Protein Concentrates, 7th, 1964* ARS 72-38, pp. 129–131.
Seshadri, T. R., and Chander, K. (1962). U.S. Patent 3,043,856.
Shaw, R. L. (1967). *Science* **156,** 168.
Shieh, T. R., Matthews, E., Wodzinski, R. J., and Ware, J. H. (1968). *J. Agric. Food Chem.* **16,** 208–211.
Shirley, D. A. (1966). *Proc. Conf. Inactiv. Gossypol Miner. Salts, 1966* pp. 11–16.
Shirley, D. A., and Dean, W. L. (1955). *J. Am. Chem. Soc.* **77,** 6077–6079.
Shirley, D. A., and Dean, W. L. (1957). *J. Am. Chem. Soc.* **79,** 1205–1207.
Shukurov, Z. Sh., Ismailov, A. I., and Dzhamalova, G. V. (1978). *Maslob.-Zhir. Prom'st.* No. 2, pp. 17–18; *Chem. Abstr.* **88,** 122573b (1978).
Skutches, C. L., Herman, D. L., and Smith, F. H. (1974). *J. Nutr.* **104,** 415–423.
Smirnova, M. I. (1936). *Tr. Prikl. Bot. Genet. Sel., Ser.* **111,** No. 15, 227–240 (in English, p. 240); *Chem. Abstr.* **31,** 5842 (1937).
Smith, F. H. (1961). *Nature (London)* **192,** 888–889.
Smith, F. H. (1963). *J. Am. Oil Chem. Soc.* **40,** 60–61.
Smith, F. H. (1965). *J. Am. Oil Chem. Soc.* **42,** 145–147.
Smith, F. H. (1972). *J. Agric. Food Chem.* **20,** 803–804.
Smith, F. H., and Clawson, A. J. (1966). *Proc. Conf. Inactiv. Gossypol Miner. Salts, 1966* pp. 60–75.
Smith, H. A. (1957). *Am. J. Pathol.* **33,** 353–365.
Smith, K. J. (1970). *J. Am. Oil Chem. Soc.* **47,** 448–450.
Smith. L. A. (1962). "Blueprint for Cotton Research Area XX. Cottonseed." Natl. Cotton Counc. Am., Memphis, Tennessee.
Stansbury, M. F., Pons, W. A., Jr., and Den Hartog, G. T. (1956). *J. Am. Oil Chem. Soc.* **33,** 282–286.
Stansbury, M. F., Cirino, V. O., and Pastor, H. P. (1957). *J. Am. Oil Chem. Soc.* **34,** 539–544.
Stephens, S. G. (1958). *N.C., Agric. Exp. Stn., Tech. Bull.* **131.** 1–32.
Stipanovic, R. D., Bell, A. A., and Howell, C. R. (1973). *J. Am. Oil Chem. Soc.* **50,** 462–463.
Stipanovic, R. D., Bell, A. A., Mace, M. E., and Howell, C. R. (1975). *Phytochemistry* **14,** 1077–1081.

Storherr, R. W., and Holley, K. T. (1954). *J. Agric. Food Chem.* **2,** 745–747.

Summers, J. C., Mead, B., and Thurber, F. H. (1953). "Experiments Conducted in Baking Department (June 1951 to June 1953)." Okla. Agric. Mech. Coll., Sch. Tech. Training, Okmulgee, Oklahoma.

Swain, R. B., and O'Connor, D. E. (1975). U.S. Patent 3,895,003.

Swain, R. B., and O'Connor, D. E. (1976). U.S. Patent 3,965,086.

Talipov, S. T., and Khadeeva, L. A. (1964). *Zh. Anal. Khim.* **19,** 1471–1477.

Thaung, U. K., Gros, A., and Feuge, R. O. (1961). *J. Am. Oil Chem. Soc.* **38,** 220–224.

Thurber, F. H., Vix, H. L. E., Pons, W. A., Jr., Crovetto, A. J., and Knoepfler, N. B. (1954). *J. Am. Oil Chem. Soc.* **31,** 384–388.

Tone, J. N., and Jensen, D. R. (1973). *Trans. Ill. Acad. Sci.* **66,** 42–46.

Vaccarino, C. (1961). *J. Am. Oil Chem. Soc.* **38,** 143–147.

Vaccarino, C. (1965). *Proc. Conf. Cottonseed Protein Concentrates, 7th, 1964* ARS 72-38, pp. 175–184.

Verberg, G., McCall, E. R., O'Connor, R. T., and Dollear, F. G. (1961). *J. Am. Oil Chem. Soc.* **38,** 33–39.

Vermel, E. M., and Kruglyak, S. A. (1963). *Vopr. Onkol.* **9,** 39–43.

Vilenskaya, I. A., Notkin, E. M., Mareev, D. I., Tyurin, V. I., Malysheva, A. A., Borodin, B. V., and Vasil'ev, M. P. (1963). U.S.S.R. Patent 158,996.

Vil'kova, S. N., and Markman, A. L. (1958). *Zh. Prikl. Khim.* **31,** 1548–1553.

Vioque-Pizarro, A., and Malissa, H. (1953). *Mikrochem. Ver. Mikrochim. Acta* **40,** 396–399.

Vix, H. L. E., Spadaro, J. J., Murphey, C. H., Jr., Persell, R. M., Pollard, E. F., and Gastrock, E. A. (1949). *J. Am. Oil Chem. Soc.* **36,** 526–530.

Voelcker, A. (1858). *J. Agric. Soc.* **19,** 420–429.

Von der Haar, P., and Pominski, C. H. (1952). *J. Org. Chem.* **17,** 177–180.

Wachs, W. (1957). *Fette, Seifen, Anstrich.* **59,** 318–321.

Waiss, A. C., Jr., Chan, R. G., Benson, M., and Lukefahr, M. J. (1978). *J. Assoc. Off. Anal. Chem.* **61,** 146–149.

Waldroup, P. W., and Goodner, T. O. (1973). *Poult. Sci.* **52,** 20–27.

Wender, S. H. (1959). *Proc. Conf. Chem. Struct. React. Gossypol and Nongossypol Pigments Cottonseed, 1959* pp. 91–101.

West, J. L. (1940). *J. Am. Vet. Med. Assoc.* **96,** 74–76.

Wilson, F. D., and Shaver, T. N. (1973). *Crop Sci.* **13,** 107–110.

Withers, W. A., and Brewster, J. F. (1913). *J. Biol. Chem.* **15,** 161–166.

Withers, W. A., and Carruth, F. E. (1915). *J. Agric. Res.* **5,** 261–288.

Woronick, C. L., and Grau, C. R. (1955). *J. Agric. Food Chem.* **3,** 706–707.

Yatsu, L. Y., Jacks, T. J., and Hensarling, T. (1970). *J. Am. Oil Chem. Soc.* **47,** 73–74.

Yatsu, L. Y., Hensarling, T. P., and Jacks, T. J. (1974). *J. Am. Oil Chem. Soc.* **51,** 548–550.

Yusupova, I. U., Rzhekhin, V. P., and Rakhomonov, R. R. (1963). *Uzb. Biol. Zh.* **7,** 8–13.

Zamyshlyaeva, A. M., Slozina, G. Z., and Belopol'skii, A. M. (1963). *Tr., Vses. Nauchno-Issled. Inst. Zhirov* **24,** 282–295.

Zharskii, A. M. (1953). *Maslob.-Zhir. Prom'st.* **18,** 23.

Zucker, T. F., and Zucker, L. M. (1947a). *Abstr. Pap., 111th Meet., Am. Chem. Soc.,* 1947 p. 47B.

Zucker, T. F., and Zucker, L. M. (1947b). *J. Am. Oil Chem. Soc.* **24,** (6), p. 28 (abstr.).

CHAPTER 8

Lathyrogens

G. PADMANABAN

I. INTRODUCTION

Lathyrism in human beings is an ancient disease caused by the consumption of the seeds of certain *Lathyrus* species, mainly those of *L. sativus* (chickling vetch or chick-pea). *L. cicera* (flat-podded vetch), and *L. clymenum* (spanish vetchling). These seeds are habitually eaten by large populations in India and Algeria and less commonly in France, Italy, Spain, and other countries. The disease has been reported in a number of countries in Europe, Africa, and Asia. It is prevalent among horses and cattle as well. Human lathyrism continues to be a public health problem in India, where this crippling disease afflicts people living in poorer sections of the country, particularly where there is extreme scarcity of

TOXIC CONSTITUENTS OF PLANT FOODSTUFFS, SECOND EDITION

ISBN 0-12-449960-0

food and *L. sativus* forms the main part of the diet. In man, the symptoms usually manifest suddenly and seem to be precipitated by exposure to a wet environment and overwork. The underlying physiological significance of this environment is not clear. However, studies during the last decade have indicated that β-*N*-oxalyl-L-α,β- diaminopropionic acid, the neurotoxin isolated from *L. sativus* seeds, is probably the chief factor responsible for the manifestation of the disease. Attempts to produce the disease in experimental animals by feeding the *L. sativus* pulse have not yet been successful. However, early workers successfully induced skeletal deformities and changes in mesenchymal tissues in experimental animals after feeding *L. odoratus*. The naturally occurring lathyrogen, β (γ-L-glutamyl) aminopropionitrile, isolated from *L. odoratus* is conspicuously absent in *L. sativus*. Extensive studies have led to the recognition that lathyrism in human beings, which is characterized by spastic paralysis of the legs as a result of neurological lesions of the spinal cord degeneration, is quite different from the skeletal abnormalities induced in experimental animals by feeding *L. odovatus* or β-aminopropionitrile (BAPN). Selye (1957) clearly distinguished between the two forms of lathyrism; he referred to the latter manifestation as osteolathyrism and to the human disease as neurolathyrism.

Studies of experimental neurolathyrism have gained momentum during the last decade, but studies on experimental osteolathyrism have been far more extensive. In addition to BAPN, several synthetic compounds producing osteolathyrism in experimental animals have been discovered. The studies of osteolathyrism have been quite useful for understanding the basic structural features of collagen and elastin in terms of cross-link formation in connective tissues. It has been possible to analyze changes in mesenchymal tissues which are relatively inert but are the site of degenerative changes in human beings. The fact that a chemical agent such as BAPN induces changes in mesenchymal tissue in experimental animals affords an experimental approach toward understanding the etiology of such diseases in human beings, irrespective of whether an osteolathyrogen is the responsible agent for such diseases in human beings. For example, as stated by Levene (1963), "The mechanism of action of BAPN can lead to an understanding of the pathogenesis of Marfan's syndrome, a genetically transmitted condition of which osteolathyrism is a fair phenocopy."

While osteolathyrism, as such, is not a public health problem, neurolathyrism is of particular import, especially in central parts of India. A 1958 epidemiological survey (Dwivedy and Prasad, 1964) indicated that in a single district (Rewa, Madhya Pradesh) there were as many as 25,000 cases of neurolathyrism in a total population of 634,000. During the summer of 1974, an epidemic of lathyrism in Raipur Division of Madhya Pradesh was reported. There has also been a recent incidence of the disease in Bangladesh. The sale of *L. sativus* has been banned in many states in India, but there has not been an effective ban on its cultivation. *Lathyrus sativus* is a hardy crop and survives adverse agricultural conditions.

For this reason it has become a mainstay of some Indian diets, especially under famine conditions. Indian farmers continue to grow it, despite their awareness of its poisonous nature. It is difficult to provide an alternate crop that would grow under the semiarid conditions of these areas, and present efforts are being directed toward distributing *L. sativus* seeds after suitable treatment and developing hybrid varieties of *L. sativus* devoid of the neurotoxin.

Since the publication of the original article on lathyrogens (Sarma and Padmanaban, 1969), a significant amount of work has been carried out on the mode of action of the *L. sativus* neurotoxin. This chapter attempts to update the original material, although the treatment of osteolathyrism is less extensive. A comprehensive review on osteolathyrism appeared in 1974 (Barrow *et al.*, 1974).

II. HISTORY

In an ancient Hindu treatise, "Bhavaprakasa," it was noted that "the triputa pulse causes a man to become lame and it cripples and irritates nerves." Hippocrates was aware that certain peas were toxic to human beings. In the seventeenth century, a ban was imposed on the consumption of chickling vetch in Wurttenberg (Dastur and Iyer, 1959). In 1873, Cantani in Italy coined the term "lathyrism." Surveys conducted as early as 1833 reported patients suffering from neurolathyrism in Indian villages, and several workers have since recorded such cases from time to time.

Various claims have been made concerning the factors responsible for the disease. A toxic amine in the germinated seeds of *L. sativus* was considered the responsible agent by Acton (1922). The presence of phytates, divicine, and alkaloids, the lack of vitamins A, B, and C, as well as virus infection, coupled with generally poor nutrition, have all been implicated at one time or another as causative factors (Dilling, 1920; Strong, 1956). Sadasiram *et al.* (1960) indicated the possible role of manganese. Basu *et al.* (1936) considered neurolathyrism as a deficiency disease due to a low level of tryptophan in the seeds. The seeds of the *Vicia* species were found to be contaminants of a market sample of *L. sativus,* and it was claimed that the toxicity was due to the contaminant rather than to *L. sativus* pulse (Anderson *et al.*, 1924). However, all these reports have been contradicted (Selye, 1957; Gardner, 1959), and no adequate experimental proof has been advanced to substantiate any one claim. Studies by Rao *et al.* (1964) and Bell and Tirimanna (1965) led to the identification of a variety of unusual amino acids in *Lathyrus,* of which β-*N*-oxalyl-L-α,β-diaminopropionic acid, the *L. sativus* neurotoxin, appears to be the main incriminating factor.

Studies on the etiology of osteolathyrism induced in experimental animals by feeding *L. odoratus* have been relatively straightforward and yielded clear-cut results. Vivanco and Jimenez Diaz (1951) proposed that this syndrome should be referred to as "odoratism." Dupuy and Lee (1954) isolated a toxic crystalline substance from *L. pusillus*. This was followed by the characterization of the toxic principle present in *L. odoratus* as β-(γ-L-glutamyl)aminopropionitrile (Schilling and Strong, 1954, 1955). Subsequently, BAPN was recognized to be the active portion of the molecule (Dasler, 1954; Wawzonek *et al.*, 1955). These findings have stimulated the synthesis of several organic nitriles and other compounds that have been screened for their osteolathyrogenic activity. Further studies have been concerned with the action of osteolathyrogens, osteo- and angiolathyritic as well teratogenic effects, in different experimental animals (Barrow *et al.*, 1974).

III. COMPOUNDS WITH OSTEOLATHYROGENIC ACTIVITY

The only case of an established osteolathyrogen occurring in nature is that of β-(γ-L-glutamyl)aminopropionitrile in *L. odoratus,* and the glutamyl residue is not necessary for the biological activity of the compound. Several *N*-acyl derivatives of BAPN were studied for their ability to produce osteolathyrism in rats. It was found that γ-glutamyl (D, L, or DL), glycyl, L-leucyl, and DL-phenylalanyl derivatives were active but that acetyl, succinyl, glutaryl, β-alanyl, DL-pantoyl, and 4-hydroxybutyryl derivatives were not. A free amino group was found to be essential for biological activity *in vivo*. Only those compounds that gave rise to BAPN *in vivo,* as evidenced by the detection of BAPN and cyanoacetic acid in the urine, were found to be active (Ehrhart *et al.*, 1963).

A wide variety of compounds have been scanned for osteolathyrogenic activity. Organic nitriles such as BAPN, ureides such as semicarbazides, hydrazides such as isonicotinic acid hydrazide, and also certain hydrazones possess osteolathyrogenic activity (Levene, 1963). In general, the osteolathyritic activity is greatest with nitriles, followed by the ureides, hydrazides, and hydrazones. A few miscellaneous compounds, such as D-penicillamine, also possess osteolathyrogenic activity. A list of osteolathyrogenic compounds is given in Table I.

A. Toxicity

Rats placed on a 50% diet of sweet pea developed skeletal deformity and aortic rupture (Geiger *et al.*, 1933; Ponseti and Baird, 1952). These lesions were also produced by BAPN at a level of 0.1–0.2% of the diet. The major lesions in osteolathyrism are the induction of a general weakness in mesenchymal tissues and disorders in the growth of cartilage(s) and bone. The effects observed in different tissues of experimental animals were summarized by Rosmus *et al.* (1966) as follows. Although enchondral ossification does occur in lathyritic animals, irregular hyperplastic cartilage is formed in epiphyseal regions with greatly enlarged epiphyseal discs. Malformations of long bones are caused by changes in the epiphyseal discs as well as by the formation of exostoses developing at the sites of attachments of big muscles exposed to continuous tensions. The

TABLE I

COMPOUNDS WITH OSTEOLATHYROGENIC ACTIVITY[a]

Aminoacetonitrile	Acetone semicarbazone
N-Methyleneaminoacetonitrile	Glycine hydrazide
β-Aminopropionitrile	Glutamic acid hydrazide
2-Cyanopropylamine	Cyanoacetic acid hydrazide
β-(γ-L-Glutamyl)aminopropionitrile	*p*-Nitrobenzoic acid hydrazide
Hydrazine	Isonicotinic acid hydrazide
N,N'-Dimethylhydrazine	Nicotinic acid hydrazide
N,N-Dimethylhydrazine	Cyanamide[b]
Thiosemicarbazide	Carbohydrazide[b]
β-Mercaptomethylamine[b]	Thiocarbohydrazide[b]
β,β'-Iminodipropionitrile[b]	Benzoic acid hydrazide

[a] Rosmus *et al.* (1966).
[b] Does not produce all the typical symptoms of osteolathyrism.

chest of lathyritic animals is deformed, the spine shows kyphoscoliotic alterations, and the intervertebral discs are loosened with consequent prolapse of the nucleus pulposis into the spinal canal. Manifestations of the effects of lathyrogens in the vascular system are usually called angiolathyrism. Histologically, the underlying cause is an inhibition of elastic fiber formation in the vascular wall, together with increased fibroblast proliferation and the formation of irregularly arranged collagenous fibers. The result of these changes is a lowered resistance of the vascular wall to stretching and formation of aneurysms.

Species as well as strain differences were noticed in the susceptibility to the action of osteolathyrogens (Coulson *et al.*, 1969; Paik and Lalich, 1970). Osteolathyrogens were also shown to have teratogenic effects (Barrow *et al.*, 1974). There is similarity between copper-deficiency effects in swine and chick and BAPN toxicity in turkeys. Both the effects are characterized by (1) a reduction in aortic tensile strength and a decrease in and modification of aortic elastin, (2) bone deformations, and (3) increased solubility of collagen. Some differences between the effects of copper deficiency and BAPN toxicity were also observed. Ascorbic acid was found to potentiate the effects of copper deficiency but not of BAPN toxicity, and only in presence of ascorbic acid did copper deficiency lead to aortic rupture in turkey poults (Barrow *et al.*, 1974). BAPN toxicity is not counteracted by the addition of dietary copper (Savage *et al.*, 1968).

B. Mode of Action

Early theories on the mode of action of osteolathyrogens were based on monoamine oxidase inhibition, chelation, and an antinicotinamide effect (Levene, 1963). The well-recognized effect of these compounds is that the collagen becomes soft, and the contents of the soluble portion increase in lathyritic collagen. It was also proposed that the collagen effect could be due to alterations in the ground substance. However, it was found that concentrations greater than that required to produce the collagen effect were necessary to induce changes in mucopolysacchride contents of connective tissues (Rosmus *et al.*, 1966). Studies on mucopolysaccharide levels and $^{35}SO_4$ incorporation experiments have yielded variable results depending on the tissue examined. Studies with chick embryo using BAPN have indicated that the cartilage mucopolysaccharide remains unaffected in osteolathyrism (Levene *et al.*, 1966).

There is now conclusive evidence that osteolathyrogens exhibit their effects in connective tissues by inhibiting the initial reaction in cross-link formation. The ability of collagen and elastin fibers to function as major structural components of tissues is primarily dependent on a system of covalent cross-links between the polypeptide chains of the respective proteins. In addition, cross-linking in elastin involves two unusual amino acids, desmosine and isodesmosine, the biosynthesis of which involves aldehyde intermediates (Partridge *et al.*, 1963; O'Dell *et al.*, 1966). In collagen, such a cross-link occurs between dehydrohydroxylysine and hydroxynorleucine (Mechanic and Tanzer, 1970). Oxidative deamination of peptide-bound lysine and hydroxylysine leads to the formation of an aldol or aldimine-type cross-link through the intermediate formation of the reactive aldehyde group. These condensation reactions lead to the formation of desmosine and isodesmosine in elastin. A similar structure is also expected to be present in collagen.

Levene (1962) showed that whereas 2,4-dinitrophenylhydrazine interacts with normal collagen, it does not do so with lathyritic collagen. The oxidative deamination of peptide-bound lysine gives rise to allysine (ϵ-semialdehyde of α-aminoadipic acid) and is catalyzed by the enzyme lysyl oxidase, which is easily detected in extracts of embryonic cartilage. The enzyme acts on both elastin and collagen substrates (Pinnel and Martin, 1968). This enzyme activity is irreversibly inhibited by BAPN. Lysyl oxidase appears to act on specific lysyl residues involved in cross-link formation and also requires cupric ion for activity (Siegel *et al.*, 1970a). The enzyme activity is undetectable in tissue extracts of copper-deficient animals and thus provides an explanation for the similarity between the effects of copper deficiency and osteolathyrism mentioned earlier. Page and Banditt (1967) have shown previously that BAPN inhibits plasma amine oxidase activity. However, plasma amine oxidase does not act on the lysyl residues of collagen and elastin and is inhibited only at high concentrations of BAPN (Siegel *et al.*, 1970b).

IV. COMPOUNDS WITH NEUROLATHYROGENIC ACTIVITY

The species of *Lathyrus* abound in unusual amino acids. Unusual nitrogenous constituents have also been reported to be present in *Vicia* species, the latter sometimes occurring as a contaminant of the former. Some of these amino acids exhibit neurotoxic properties in experimental animals. However, detailed biochemical and histopathological studies have not been carried out with most of these amino acids, and the only compound for which there is a high probability of involvement in human neurolathyrism is β-*N*-oxalyl-L-α,β-diaminopropionic acid, the neurotoxin isolated from *L. sativus*.

A. Chemistry

Table II gives a list of neurotoxic amino acids isolated from *Lathyrus* and *Vicia* species. β-*N*-Oxalyl-L-α,β-diaminopropionic acid has been isolated from *L. sativus* and other *Lathyrus* species implicated in human neurolathyrism (Rao *et al.*, 1964; Murti *et al.*, 1964). Bell and O'Donovan (1966) isolated the higher homologue, namely, γ-*N*-oxalyl-L-α-diaminobutyric acid, from certain other species of *Lathyrus*. Ressler (1962) isolated β-cyanoalanine from *V. sativa* and *V. angustifolia*, and Bell and Tirimanna (1965) detected its γ-glutamyl derivatives as well in the seeds of 16 species of *Vicia*. When the *Lathyrus* species were scanned for the presence of β-cyanoalanine, Ressler *et al.* (1961) discovered instead another neurotoxic amino acid, L-α,γ-diaminobutyric acid, in the seeds of *L. latifolius* and *L. sylvestris wagneri*. In addition, *Lathyrus* species contain such unusual amino acids as lathyrine (Bell, 1962) and homoarginine (Bell, 1962; Rao *et al.*, 1963); these, however, have not been shown to possess neurotoxic properties. Nagarajan *et al.* (1966) claimed that a phenolic type of compound is present in the husk of *L. sativus* seeds, which can

TABLE II

NEUROTOXIC AMINO ACIDS PRESENT IN *Lathyrus* AND *Vicia* SPECIES

β-Cyanoalanine[a] and γ-glutamyl-β-cyanoalanine[b]	*Vicia sativa* and 15 other species of *Vicia*
α,γ-Diaminobutyric acid[c]	*V. aurantica, L. sylvestris, L. latifolius,* and 11 other species of *Lathyrus*
γ-*N*-Oxalyl-L-,γ-diaminobutyric acid[d]	*Lathyrus sylvestris, L. latifolius,* and 18 other species of *Lathyrus*
β-*N*-Oxalyl-L-α,β-diaminopropionic acid[e]	*Lathyrus sativus, L. clymenum, L. latifolius,* and 18 other species of *Lathyrus*

[a]Ressler (1962).
[b]Bell and Tirimanna (1965).
[c]Ressler *et al.* (1961).
[d]Bell and O'Donovan (1966).
[e]Rao *et al.* (1964).

induce a temporary stupor in monkeys under certain conditions. Rukmini (1969a,b) reported the presence of another neurotoxin in *L. sativus* seeds and gave its tentative structure as *N*-β-D-glycopyranosyl-*N*-L-arabinosyl-α,β-diaminopropionitrile. There have been no follow-ups of some of these reports and studies. The presence of α,β-*N*-dioxalyl-L-α,β-diaminopropionic acid in *L. sativus* has been reported (Rajmohan and Ramachandran, 1972).

In a study of the structural requirements for the neurotoxic action of oxalylamino acids in the 1-day-old chick, Rao and Sarma (1967) found that γ-*N*-oxalyl-L-α,γ-diaminobutyric acid, *N*-oxalylglycine, and *N*-oxalyl-β-alanine are potent neurotoxins. In general, N-substituted oxamic acids, in which the substituent carries a free carboxyl group, were found to be neurotoxic to the 1-day-old chick.

Bell and O'Donovan (1966) demonstrated the reversible exchange of the oxalyl group between the ω- and α-amino groups of oxalylamino acids, probably involving the formation of unstable cyclic intermediates. They detected the presence of the α-oxalylamino acids in trace quantities in *Lathyrus* species in which β-*N*-oxalyl-L-α,β-diaminopropionic acid and its higher homologue have been shown to be the predominant forms that occur naturally.

B. Toxicity

1. Effect of *L. sativus* in Human Beings

The disease generally appears whenever a diet consisting of one-third or one-half of *L. sativus* seeds is consumed for 3–6 months. The symptoms are muscular rigidity, weakness, paralysis of the leg muscles, and, in extreme cases, death. In most cases on record, the onset of the disease is sudden (Shourie, 1945); the patients develop a sudden stiffness of the leg muscles and partial paralysis or total loss of control over the lower limbs. The earliest symptom is a spasmodic contraction in the calf muscles locally known as lodakas, which gradually disappears. In the affected areas, people at different stages of onset of the disease can be seen. In milder cases, there is bending of the knees and difficulty in running. People with more advanced cases walk on toes and require a stick for support. This leads to the two-stick stage, and finally the patients are reduced to crawling. Some apparently normal-looking individuals have an exaggerated knee jerk and ankle clonus. This has been referred to as latent lathyrism. The disease affects mostly young men between 20 and 29 years of age, and the incidence is less frequent among women. The external symptoms of the disease indicate neurological lesions, although detailed histopathological information is not available. German (1960) summarized the pathology of the disease as "microglosis in the anterior horns and lateral cords and partial degeneration of the motor tracts of the spinal cord and anterilateral sclerosis in the dorsolumbar spinal cord."

2. *Effect of the Unusual Amino Acid Components of Lathyrus and Vicia Species in Experimental Animals*

Anderson *et al.* (1924) reported that samples of the *L. sativus* seeds as well as samples of wheat collected from the epidemic areas were contaminated with *Vicia sativa* (common vetch). This observation gained importance from the viewpoint of the etiology of human lathyrism when Ressler (1962) isolated β-cyanoalanine from *V. sativa* and *V. angustifolia*. β-Cyanoalanine fed at a level of 0.075% to the chick resulted in 100% mortality within about 10.5 days; γ-glutamyl-β-cyanoalanine had similar toxicity. However, both were less toxic to the rat (Ressler, 1964). β-Cyanoalanine, when administered to a weanling rat at a dosage of 15 mg/100 gm body weight, caused hyperactivity followed by tremors, convulsions, and rigidity, from which the rats recovered after 4 hr. Subcutaneous injection of this compound at 20 mg/100 gm body weight caused convulsions, rigidity, prostration, and death. However, the role of β-cyanoalanine in human lathyrism is open to question, since in the last detailed survey of epidemic areas in India, the *L. sativus* seed samples consumed were found to be free from *Vicia* contamination (Dwivedi and Prasad, 1964).

α,γ-Diaminobutyric acid has been detected in 13 species of *Lathyrus* but not in *L. sativus* and other species implicated in neurolathyrism. Within 2 days of administration to rats at a level of 68 mg/100 gm body weight, it caused weakness in the hind legs, tremors, convulsive behavior, and death (Ressler *et al.*, 1961).

3. *Detailed Studies with β-N-Oxalyl-L-α,β-diaminopropionic Acid, the L. sativus Neurotoxin*

Since *L. sativus* is the main species implicated in neurolathyrism and since β-*N*-oxalyl-L-α,β-diaminopropionic acid (ODAP) is the only neurotoxic amino acid so far known to be present in these seeds, detailed studies in the author's laboratory have been carried out in experimental animals. The preparation used in these studies could have carried traces of the α-isomer.

Adiga *et al.* (1962) found that ODAP was toxic to microorganisms (*Neurospora crassa, Staphylococcus aureus, Escherichia coli,* and *Candida albicans*). When this compound was injected into 1-day-old chicks weighing 40–45 gm at 20 mg per chick, it caused wryneck, head retraction, and convulsions (Adiga *et al* 1963). It was shown earlier that the crude 30% ethanolic extract of *L. sativus* seed meal caused neurological symptoms in the 1-day-old chick (Roy *et al.*, 1963). At the 10-mg level, the *L. sativus* neurotoxin induced the same symptoms with a lesser degree of severity. The symptoms were always temporary, lasting for 8–12 hr. At the 30-mg level, the symptoms persisted for 24 hr, and the birds eventually died. Nagarajan *et al.* (1965) subsequently showed that ODAP causes neural symptoms and opisthotonus in ducklings and baby pigeons when administered at a dosage of 1.4 mg/gm body weight. However, adult birds do not respond to the toxin, even when it is given intraperitoneally in excess of a toxic dose in terms of body weight. Curtis and Watkins (1965) found the *L. sativus* neurotoxin to be a very potent excitant of the spinal interneurons and Betz cells in the cat spinal cord when administered into the immediate extracellular environment of single neurons. They suggested that the inefficacy of this compound in adult animals may be due to a blood–brain barrier to this toxin. Rao and Sarma (1967) showed that young rats (3–14 days old), guinea pigs (8 days old), and pups (12 days old) also respond to this neurotoxin. Thus, the hypothesis of a blood-brain barrier to this compound in adult animals is in accord with the observations on its potency in young animals and with the concept that the blood–brain barrier is not developed to a significant degree in young animals (Lajtha, 1962); Sperry, 1962). Supporting this contention, Rao *et al.* (1967) found that when ODAP is given to adult monkeys through the lumbar route, thus bypassing the blood–brain barrier, in increasing dosages of 5–25 mg at an interval of 2–3 days, typical paralysis of the hind legs develops gradually. Initially, the animals exhibit a transient weakness of the lower limbs and clonic twitchings of the tail. These become more acute with successive dosages, and finally a subtotal or total paralysis of the hind legs and tail is achieved. Preliminary histological studies of the brain and spinal cord of the affected monkeys exclude the possibility of a local trauma to the lumbosacral cord during the spinal tap. There appears to be a destruction of the nerve cells of the gray matter of the spinal cord accompanied by proliferation of the microglial cells. Rao *et al.* (1967) also showed that intracranial and intracerebral injections of the compound to adult rats and mice cause typical neurological symptoms. Further, adult rats, mice, poultry, and monkeys respond to intraperitoneal injections of the compound when these animals are previously fed with calcium chloride, ammonium chloride, and such drugs as Diamox, salicylic acid, and sulfonamide. These chemicals and drugs are known to induce an "acidotic" condition in the animals (Weisberg, 1962). The photograph in Fig. 1 illustrate the effect of neurotoxin in some of the experimental animals under the different conditions studied.

Mehta *et al.* (1976) have questioned the existence of a blood–brain barrier for the neurotoxin in the young squirrel monkey with a mature blood–brain barrier. These workers showed that a tracer dose (about 250–500 μg/kg body weight) of the *L. sativus* neurotoxin can enter the central nervous system of the subhuman primate and that the radioactivity detected may not be due to entrapped blood.

Thus, the concept of a blood–brain barrier to the neurotoxin (Rao and Sarma, 1967) needs reexamination. The requirement of an acidotic condition for the manifestation of the neurotoxic effects of ODAP, administered intraperitoneally to adult animals, may simply not be related to the breakdown of the blood–brain barrier under these conditions. It was indeed reported that such acidotic adult rats manifested a higher concentration of ODAP in the brain than did

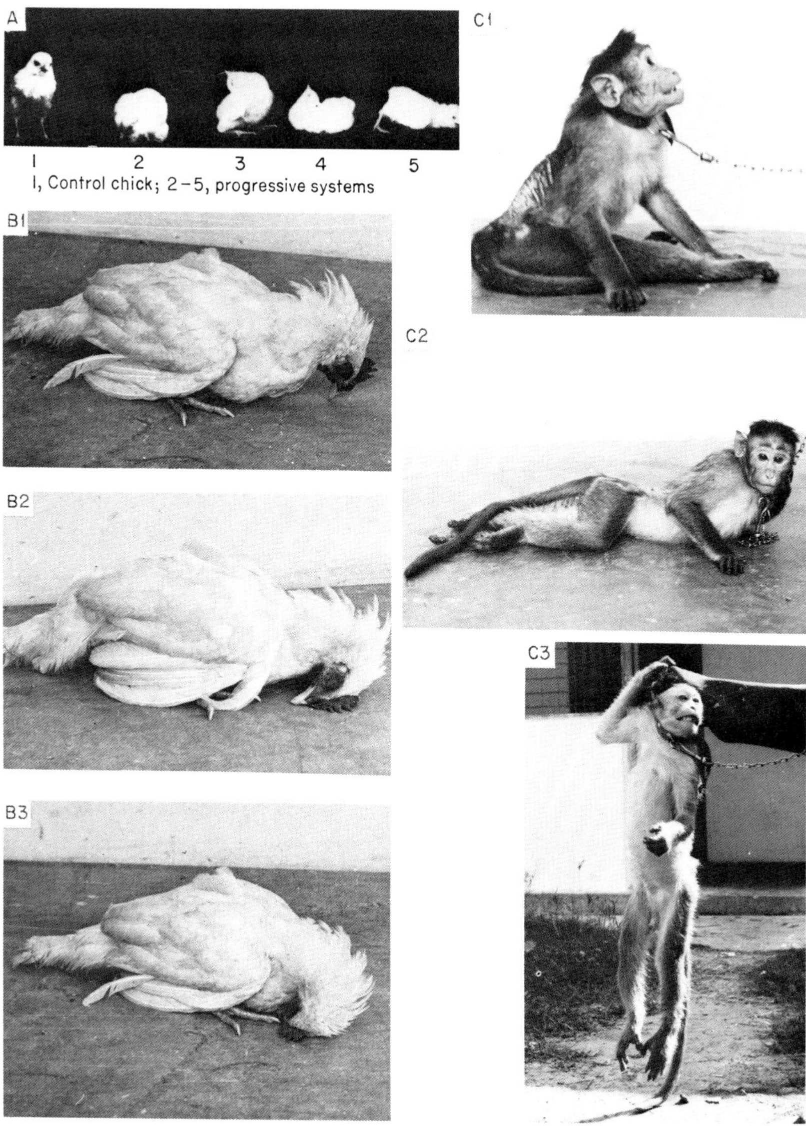

FIG. 1. Effect of β-*N*-oxalyl-L-α,β-diaminopropionic acid in the 1-day-old chick and monkey. The neurotoxin was administered intraperitoneally in (A) and (B). (A) Chick No. 1 is normal, and chicks 2 to 5 show progressive symptoms. (B) Effect of neurotoxin administered to adult birds rendered acidotic with calcium chloride. (C) Effect of toxin administered to monkey through the lumbar route.

control adult animals (Cheema *et al.*, 1969a). It now appears however, that the lack of neurotoxic effects of orally or intraperitoneally administered ODAP to adult animals may also be related to the excretory and detoxication potential of these animals. Adult rats injected intraperitoneally with radioactive ODAP (1 mmole/kg body weight) were found to excrete 50–70% of radioactivity in the urine over a period of 24 hr (Cheema *et al.*, 1971). Mehta *et al.* (1976) found that in the squirrel monkey injected with radioactive ODAP (2–3 μmoles/kg body weight) intravenously, about 20% of the radioactivity was excreted in the urine within an hour. In addition, Cheema *et al.* (1971) detected the keto acid derivative of ODAP in the liver and kidney of ODAP-injected rats. Mehta *et al.* (1976) have not been able to detect the keto acid derivative in the squirrel monkey but reported the presence of an unidentified radioactive 2,4-dinitrophenylhydrazone in the stomach. These workers also reported the detection of the α-isomer of ODAP, which is nontoxic but could arise as a result of nonenzymatic rearrangement during extraction and isolation of ODAP (Bell and O'Donovan, 1966; Wu *et al.*, 1976).

It is clear that further detailed studies on the influence of different physiopathological factors on ODAP excretion and metabolism in experimental animals would help to explain the failure to produce neurolathyrism in adult experimental animals by oral feeding of the pulse or the neurotoxin.

C. Mode of Action

There does not appear to be a common mechanism for the mode of action of neurolathyrogens. Ressler *et al.* (1964) found that pyridoxal hydrochloride can delay the onset and alleviate the symptoms of β-cyanoalanine toxicity and decrease greatly the mortality of rats injected with a lethal dose of β-cyanoalanine. The neurotoxin induces cystathionuria in rats and probably can act as a metabolic inhibitor of the pyridoxal-dependent transsulfurase reaction in the conversion of methionine to cysteine. Pyridoxal does not protect when it is given with β-cyanoalanine, indicating that once the toxicity has set in the vitamin has no effect. Ressler *et al.* (1964) indicated the possibility that the other neurolathyrogens, α,γ-diaminobutyric acid and the oxalylamino acids, may operate through a similar mechanism.

O'Neal *et al.* (1968) found that the neurolathyrogen, L-α,γ-diaminobutyric acid, at a dose of 4.4 mmoles/kg body weight in adult rats caused chronic ammonia toxicity. The animals showed neurotoxic effects in 12–20 hr followed by death in 3–8 days. These workers implicated primary liver damage followed by a secondary brain lesion. L-α,γ-Diaminobutyric acid inhibits ornithine transcarbamylase activity in liver and thus leads to lowered urea synthesis in liver slices. In support of their contention, O'Neal *et al.* (1968) found that the chick, being uricotelic, is less susceptible to L-α,γ-diaminobutyric acid administration than the rat.

Detailed studies on the effect of ODAP in young rats (Cheema *et al.*, 1969b) showed that the neurotoxin causes a small but significant increase in free ammonia concentration of the blood and brain. There is a striking accumulation of glutamine in the brain. The toxin does not interfere with urea synthesis in liver. The susceptibilities of young rats and chickens to ODAP are about the same. A build-up of neurotoxin concentration in the brain is essential for the manifestation of the effects. In addition, ODAP brings about biochemical changes in the brain typical of an excitant amino acid (Cheema *et al.*, 1970). In the convulsing young rats, a decrease in glycogen and high-energy phosphate levels and an increase in inorganic phosphate and lactic acid levels were detected. All these results indicate that the mechanisms of action of L-α,γ-diaminobutyric acid and ODAP are different. Further studies with young rat brain indicated that enhanced protein degradation was possibly responsible for ammonia production (Cheema *et al.*, 1971b), and lysosomal damage was suspected. Evidence for lysosome labilization was obtained in young chicks administered the neurotoxin (Lakshmanan *et al.*, 1971). However, studies with adult monkeys revealed that the primary effect of the neurotoxin may not be lysosome labilization (J. Lakshmanan and G. Padmanaban, unpublished data). When paralysis was induced in adult monkeys within 12 hr by a single intrathecal administration of ODAP, no evidence was obtained for lysosome labilization in the brain of such monkeys killed immediately after the onset of paralysis. However, when the monkeys were allowed to live for a week, labilization of brain lysosomes could be demonstrated. Thus, the early manifestation of brain lysosome labilization in the young animals could be related to the incomplete development of organelle structure and lysosome labilization, and the attendent degradative events probably constitute the secondary phase of the neurotoxin action.

Another possible mechanism of ODAP action has emerged as a result of some recent investigations. It was noticed in the author's laboratory as early as 1965 that the growth inhibitory effects of ODAP on *Neurospora crassa* could be counteracted by glutamic and aspartic acids as well as glutamine and asparagine. An investigation of this aspect was reported by Mehta *et al.* (1972), who found that ODAP is a potent antagonist of L-glutamic acid transport in resting yeast cells, and these workers also suggested the possibility of such an antagonism at the synaptic level. Duque Magalhoes and Packer (1972) found ODAP to inhibit glutamate transport in isolated mitochondria. Lakshmanan and Padmanaban (1974a,b) reported that ODAP at a fairly high concentration of 10^{-3} *M in vitro* inhibits the high-affinity uptake of glutamate by isolated synaptosomes. In further studies (Lakshmanan and Padmanaban, 1977, and unpublished data), a significant concentration of ODAP could be detected in the nerve terminals isolated from young rat brain and adult monkey spinal cord, when these animals manifested the neurotoxic effect as a result of ODAP administration. Synaptosomes isolated from these animals showed a significant alteration in glutamate transport. There is a considerable decrease in the high-affinity uptake of glutamate by these synaptosomes. Conversely, the release of glutamate from such synaptosomes is significantly enhanced. Thus, ODAP localized in synaptosomes as a result of *in vivo* administration has been found to enhance the release and inhibit the reuptake of glutamate in isolated synaptosomes. ODAP added *in vitro* to the isolated system is much less effective than that administered *in vivo*. Isolated synaptosomes do not manifest an uptake system for ODAP. It is considered possible that ODAP administered *in vivo* could reach the terminal by axonal transport, although uptake by glial cells cannot be ruled out at this stage. It would be interesting to examine whether ODAP inhibits glutamate binding to the postsynaptic receptor protein. The existence of receptor proteins for glutamate and aspartate has been demonstrated (Plazas and De Robertis, 1974,1976). ODAP administered *in vivo* was found to be localized in a population of synaptosomes showing high-affinity uptake for glutamate, although the synaptosomes failed to take up ODAP *in vitro*. The net result of ODAP action could be an accumulation of glutamate/aspartate in the synaptic environment, and it is well established that glutamate is both neuroexcitory and neurotoxic (Johnson, 1972). The other possibility is that ODAP behaves like glutamate and mimics the effects. An earlier study, reporting the accumulation of glutamine in the brain of young rats administered ODAP (Cheema *et al.*, 1971b), can also be viewed as reflecting the increased glutamate concentration in the synaptic environment.

Olney *et al.* (1976) reported that ODAP administered intraperitoneally to immature mice induces lesions in the retina, hypothalamus, and lower medulla and that this pattern of damage is similar to that demonstrated in animals after oral or subcutaneous administration of glutamate. Certain differences in the pattern of lesions were also noticed, and these workers suggested that the conformational constraints of the ODAP molecules may cause it to interact more selectively than glutamate at central excitory receptor sites. Thus, it is likely that ODAP may not compete for the glutamate high-affinity uptake system but may compete for or inhibit the binding of glutamate to its receptor protein in the nerve terminal.

V. BIOSYNTHESIS OF LATHYROGENS

It is interesting that the naturally occurring osteolathyrogen, β-(γ-glutamyl)aminopropionitrile, and the neurotoxic amino acids, such as α,γ-diaminobutyric acid, β-cyanoalanine, and the oxalylamino acids, are biosynthetically interrelated. The possible interconversions among the various lathyrogens are shown in Fig. 2.

Nigam and Ressler (1964) obtained evidence indicating that the carbon skeleton of serine is incorporated into β-(γ-glutamyl)aminopropionitrile in *V. sativa* and *L. sylvestris*. The incorporation of labeled serine into the dipeptide is appreciable only in the presence of cyanide. The incorporation of cyanide into serine is

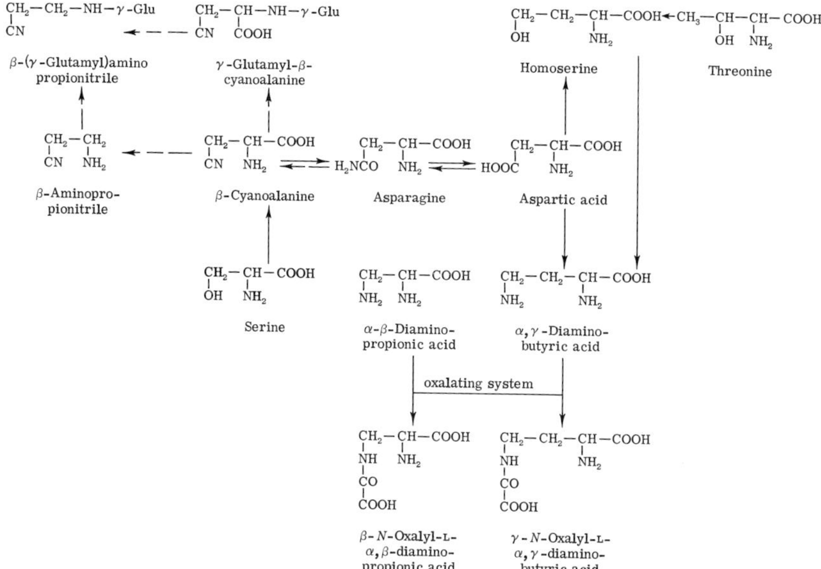

FIG. 2. A biosynthetic scheme for lathyrogens. Dashed arrow, experimental proof not available. Solid arrows, experimentally confirmed.

catalyzed by an enzyme prepared not only from *Vicia sativa* but also from *Lotus tenuis* (Floss *et al.*, 1965) and *E. coli* (Dunnill and Fowden, 1965). It is suggested that β-cyanoalanine may be formed through nonspecific enzyme reactions catalyzed by cysteine sulfhydrase. The ability of diverse systems, including those in which β-cyanoalanine does not occur naturally to incorporate cyanide into serine has been taken to indicate a mechanism that detoxifies cyanide and at the same time leads to a unique pathway of asparagine biosynthesis since the labeled cyanide finally appears in the amide carbon of asparagine. Ressler *et al.* (1963) showed that L-[4$-^{14}$C]cyanoalanine is incorporated into asparagine to the extent of 33.5% in *L. sylvestris*. However, in *Vicia sativa* there appears to be a block in the conversion of β-cyanoalanine to asparagine since the label accumulated in the former when $K^{14}CN$ was used as the precursor. Kessler *et al.* (1961) proposed that β-cyanoalanine itself may be derived from asparagine by a hypothetical enzyme, "amide dehydrasse," although the evidence so far obtained has confirmed only the reverse pathway.

Ressler *et al.* (1961) also suggested that α,β-diaminobutyric acid may be derived from β-cyanoalanine by another hypothetical enzyme, "nitrile reductase." Tschiersch (1964) concluded, on the basis of autoradiographic evidence, that $H^{14}CN$ administration to seedlings of *Vicia sativa* results in the formation of labeled α,γ-diaminobutyric acid, probably mediated through β-cyanoalanine. Nigam and Ressler (1966), using L-[4-^{14}C]-β-cyanoalanine, observed an incorporation of only 0.27% into α,γ-diaminobutyric acid in *L. sylvestris*. However,

L-[^{3}H]homoserine as well as DL-[1-^{14}C]aspartic acid were efficiently incorporated into the diamino acid. The authors concluded that α,γ-diaminobutyric acid is most likely synthesized from both homoserine and aspartic acid, although there is a close relationship between the pathways for the formation of homoserine and α,β-diaminobutyric acid from aspartic acid.

With regard to the biosynthesis of oxalylamino acids, evidence has been obtained in this laboratory (Malathi *et al.*, 1968) indicating that [U-^{14}C]oxalic acid is incorporated as a unit into ODAP in the germinating seedlings of *L. sativus*. An oxalyl-activating enzyme requiring CoA, ATP, and Mg^{2+} has been detected in these seedlings, and, when L-α,β-diaminopropionic acid is included in this system, net information of ODAP can be demonstrated. There is a preferential oxalation of the β-amino group of L-α,β-diaminopropionic acid, and the higher homologues of the diamino acid are ineffective in this system. However, free L-α,β-diaminopropionic acid cannot be detected in the seeds or seedlings, and the biosynthesis of the amino acid fragment of the oxalylamino acid is not yet clear. Bell and O'Donovan (1966) also described preliminary experiments in which the same enzyme system is responsible for the oxalylation of both L-α,β-diaminopropionic acid and L-α,γ-diaminobutyric acid in *Lathyrus* species containing both of the oxalyl derivatives. The formation of the α-oxalyl isomers may proceed by a nonenzymatic isomeric change *in vivo*. In the context of proposing a mechanism for the formation of oxalylamino acids by oxalylation of the corresponding diamino acids, the observations of Nigam and Ressler (1966) must be explained. It has been found that although free α,γ-diaminobutyric acid in *L. sylvestris* is effectively labeled from radioactive homoserine or aspartic acid, the diamino acid moiety of γ-N-oxalyl-L-α,γ-diaminobutyric acid carries a very low amount of radioactivity.

VI. ANALYTICAL PROCEDURES FOR DETECTION AND ESTIMATION

A. Chemical Methods

Since the lathyrogens occurring in nature happen to be amino acids or their derivatives, ninhydrin reagent can be used for their detection and estimation. β-Aminopropionitrile gives a characteristic green color with ninhydrin, a reaction that Garbutt and Strong (1957) developed into a quantitative assay for the osteolathyrogen. In the case of the neurotoxic amino acids, β-cyanoalanine also gives a green color with ninhydrin. The extremely good resolution of acidic and basic constitutents achieved by high-voltage electrophoresis makes this a very useful method for scanning plant or seed extracts for the presence of unusual nitrogeneous constituents. Table III presents the characteristic ionic mobilities

TABLE III

IONIC MOBILITIES AND NINHYDRIN COLOR REACTIONS OF THE NEUROTOXIC AMINO ACIDS[a]

Amino acid	Ionic mobility[b]	Ninhydrin color
α,γ-Diaminobutyric acid	Positively charged at pH 1.9, 3.6, and 6.5; moves faster than arginine; negatively charged at pH 11.5	Brown purple
β-*N*-Oxalyl-L-α,β-diaminopropionic acid	Strong acid; negatively charged at pH 1.9	Purple; does not develop on pretreatment with cupric nitrate
α-*N*-Oxalyl-L-α,β-diaminopropionic acid	Strong acid; moves slightly slower than α-isomer	Gray purple; not affected by pretreatment with cupric nitrate
γ-*N*-Oxalyl-L-α,γ-diaminobutyric acid	Strong acid; uncharged at pH 1.9; negatively charged at pH 3.6; moves fractionally slower than the lower	Purple; does not develop on pretreatment with cupric nitrate
α-*N*-Oxalyl-L-α,γ-diaminobutyric acid	Strong acid; carries a small positive charge at pH 3.6; moves slightly slower than the α-isomer	Purple; not affected by pretreatment with cupric nitrate
β-Cyanoalanine	Uncharged at pH 3.6 and 6.5; positively charged at pH 1.9; moves faster than phosphothreonine and slower than threonine	Green
γ-Glutamyl-β-cyanoalanine	Negatively charged at pH 3.6; moves faster than aspartic acid; positively charged at pH 1.9; moves slower than aspartic acid	—[c]

[a]Based on information contained in papers by Bell (1962, 1964) and Bell and Tirimanna (1965).
[b]The electrophoretic run is usually carried out for 30–40 min at 60–75 V/cm.
[c]Color reaction with ninhydrin has not been reported, but since this compound has an α-amino and α-carboxyl group, it would be expected to give the usual purple color.

and ninhydrin color reactions of the neurotoxic amino acids isolated from *Lathyrus* and *Vicia* species.

ODAP, the principal incriminating factor in human neurolathyrism, can be easily isolated and quantified. When an aqueous extract of the seed meal is applied to a Dowex-50-(H^+) column, ODAP is held very weakly and can be eluted with water or mild acid. The unabsorbed substances, such as sugars and organic acids, are lost in the initial column wash, and further elution with water results in quantitative elution of ODAP from the column. A ninhydrin analysis can be performed directly on the water eluate or after paper electrophoresis. Rajagopal Rao *et al.* (1974) described an enzymatic procedure for the estimation of ODAP as well as other derivatives of diaminopropionic acid. This method involves the liberation of diaminopropionate by acid hydrolysis followed by quantification of diaminopropionate by the specific enzyme diaminopropionate–ammonia lyase.

Methods are now available for synthesizing ^{14}C-labeled ODAP with reasonably high specific activity (Rao, 1976; Haskell and Bowlus, 1976). The compound can also be tritiated and then purified to constant specific activity (Lakshmanan and Padmanaban, 1977). The availability of radioactive ODAP with high specific activity would considerably help in its detection and estimation and in the study of its metabolism in experimental animals.

B. Biological Methods

Bioassay methods for osteolathyrogens based on weight loss, skeletal deformity, neurological manifestations, and angiorrhexis of experimental animals are often misleading. These manifestations do not take place in some animals in spite of the fact that the osteolathyrogens increase the amount of extractable collagen in all cases (Levene, 1961). Making use of this common denominator, Gross *et al.* (1960) developed an assay system using the chick embryo in which osteolathyrogenic activity is defined as that amount which results in increased fragility of the connective tissues and in the extractability of collagen from such tissues. Briefly, groups of 20–30 fertilized eggs of the white leghorn variety were injected at 14 days of incubation with the compound in 0.2 ml of sterile saline through a pinhole into the chorioallantoic membrane. After 48 hr of further incubation, the embryo fragility was measured by determining the weight required to detach the head from the body within a period 10–100 sec. A useful parameter is the fragility index, which is taken as the ratio of the breaking weight for controls to that for experimental embryos. A more reliable parameter involves the measurement of viscosity and hydroxyproline content of the cold saline extracts of minced bones from 10–20 treated embryos. Any compound can be assayed by comparing its effectiveness with that of two different levels of BAPN.

For the biological assay of the neurotoxic amino acids, such as β-cyanoalanine (Ressler, 1962) and α,γ-diaminobutyric acid (Ressler *et al.*, 1961), male weanling rats have been used. When these compounds were administered by stomach tube at a level of 15 mg/100 gm body weight the rats showed typical neurological symptoms, such as convulsions, tremors, weakness of the hind legs, and other symptoms. In this laboratory the 1-day-old chick was found to be a useful experimental animal for screening compounds for neurotoxic effects (Adiga *et al.*, 1963; Rao and Sarma, 1966). Characteristic head retraction, wryneck, and convulsions are induced by intraperitoneal injection of the neurotoxic amino acids into the 1-day-old chick. These methods have limitations in view of the species specificity exhibited by these neurotoxic amino acids, and a more reliable biochemical parameter to assay these compounds has not yet been worked out.

VII. *LATHYRUS SATIVUS* AS FOOD

A. Social and Economic Aspects

The *Lathyrus* species are generally cultivated for purposes of food, fodder, green manure, and ornament. For example, the flowers of the sweet pea (*L. odoratus*) are noted for their range of color, beauty of form, and fragrance. Despite the cultivation of *Lathyrus* in several parts of the world, human lathyrism is a public health problem only in India. In India, *L. sativus* occupies 4% of the total area under pulse crops and constitutes 3% of the total pulse production. About 4 million acres, producing nearly 0.5 million tons of seeds, were under this crop in 1956–1957. About 56 varieties of *L. sativus* are known in India. In addition, contamination with *Vicia* seeds often occurs. The different varieties of *L. sativus* consist of small-, medium-, or large-sized seeds in combinations of gray, black, or mottled color. The pulse is used principally to prepare unleavened bread (chapatis) and is sometimes eaten as paste balls or as a cooked preparation. In some places, the seeds are dehusked and parched before use. Because of its cheapness and ease of cultivation, it is used as an adulterant of other pulses. Mixed with oilcake and salts, the seeds are used as feed for cattle (Sastri, 1962).

Originally, it was common practice for landlords to distribute wages to tenants in the form of *L. sativus* seeds. With the ban imposed on the consumption of *L. sativus* in several states in India, this practice is now believed to be obsolete. However, an effective ban on the cultivation of *L. sativus* is not possible in view of the lack of a suitable alternative crop acceptable to the farmers. Thus, it is of interest whether *L. sativus* itself can be suitably processed so as to free it from the toxin(s) and render it fit for human consumption (see Section VII,C). In this context it is also important to have some knowledge of the nutritive quality of this pulse.

B. Amino Acid Composition

Analysis of the whole seeds (Sastri, 1962) revealed the proximate composition to be as follows: moisture, 10%; protein, 28.2%; fat, 0.6%; carbohydrates, 58.2%; and mineral matter, 3%. This fairly satisfactory proximate composition has led many workers to investigate the amino acid composition of the seed meal.

The values reported are variable, presumably owing to the different varities used and techniques employed. An amino acid analysis of the seed meal after acid hydrolysis was carried out in this laboratory using the automatic amino acid analyzer (Spackman *et al.*, 1958). The results are presented in Table IV. It can be seen that the pulse is rich in lysine and contains fairly good concentrations of the other essential amino acids except methionine. In addition, this pulse was also reported to be deficient in tryptophan (Basu *et al.*, 1936; Ramachandran and Phansalkar, 1956). In this laboratory an attempt has also been made to prepare a protein concentrate from *L. sativus*. The method involves extraction of the seed meal with 0.1 *M* carbonate–bicarbonate (pH 10.0) buffer followed by acidification of the extracted protein to pH 3.0. The precipitated protein is collected and

TABLE IV

AMINO ACID COMPOSITION OF *L. sativus* SEED MEAL AND PROTEIN CONCENTRATE

	Content (% of sample)	
Amino acid	Seed meal[a]	Protein concentrate[b]
Lysine	1.85	5.99
Histidine	1.10	2.18
Arginine	1.41	5.69
Aspartic acid	1.80	7.41
Threonine	0.85	2.78
Serine	1.20	4.02
Glutamic acid	2.25	8.79
Proline	1.46	4.22
Glycine	0.72	2.30
Alanine	0.80	2.92
Half-cystine	Trace	Trace
Valine	0.81	2.76
Methionine	0.35	0.99
Isoleucine	1.01	3.32
Leucine	1.45	5.30
Tyrosine	0.62	2.54
Phenylalanine	1.03	3.29

[a]The protein content of the seed meal based on nitrogen estimation was 24.5%. This value, however, includes the contribution of unusual nonprotein nitrogenous constituents which are present to the extent of 3–4% in the seeds.

[b]The protein content of the protein concentrate based on nitrogen estimation was 70%.

lyophilized. On a laboratory scale, the protein concentrate can be obtained with a yield of 65% and contains 70% protein. Amino acid analysis of the protein concentrate is given in Table IV. It is free of the neurotoxin and can be used as a lysine supplement. It was reported (Sastri, 1962) that at a 10% level of protein intake, *L. sativus* protein has high digestibility (90%) but low biological value (50%). Autoclaving of the seeds and supplementation with methionine are reported to greatly enhance the nutritive value of the protein.

C. Methods for the Removal of Toxin

Attempts have been made to process the *L. sativus* seeds so as to free them of the toxin (Mohan *et al.*, 1966). The methods involve (1) cooking the pulse in an excess of water and draining off the excess water, (2) soaking the pulse overnight in cold water, and (3) steeping the dehusked seeds in hot water. The treatment of the whole seed results in a slightly smaller proportion of the free amino acids being removed as compared to that in the case of dehusked seeds. In the latter case there is almost a complete removal of the neurotoxin. It is claimed that the steeping and boiling processes do not result in a marked change in the protein content of the dehusked seeds. The steeped, dehusked seeds can be easily dried in the sun, and the discrete grains obtained can be ground to a flour. The only drawback in this procedure appears to be the loss of water-soluble essential metabolities, especially the B vitamins. It was recommended that the treated *Lathyrus* be given with certain other foods which will ensure an adequate intake of B vitamins (Anonymous, 1967). Yet another process involves soaking the seeds overnight in water followed by steaming for 30 min and sun-drying the grains. The toxin is removed to an extent of 80%. The loss of thiamine is not appreciable in this treatment as compared to that of the steeping procedures. However, there is a substantial loss of riboflavin and nicotinic acid in all the treatment methods. It has also been found that the binding capacity of the flour, which is essential for the preparation of unleavened bread (chapatis), is lost in this method of processing. However, this can be overcome by mixing the *Lathyrus* flour with that of barley or wheat in the proportion 4:1 (Anonymous, 1968). It may be added that in the normal method of preparing chapatis from untreated seeds there is no appreciable decrease in the toxin content.

In this laboratory, roasting the seeds at 150°C for 20 min resulted in about 85% destruction of the neurotoxin. The applicability of the roasted seed flour to certain food preparations and the effect of roasting on the palatability and nutritive quality of the seeds remain to be investigated. The preparation of protein concentrate free of the neurotoxin from the seed meal raises the possibility of using it as a cheap lysine supplement to cereals since the latter is generally deficient in this essential amino acid.

VIII. CONCLUDING REMARKS

As indicated earlier, osteolathyrism is not known to occur in human beings. However, extensive studies on experimental osteolathyrism have led to a significant understanding of the normal cross-link structure in collagen and elastin. Neurolathyrism is a public health problem, and despite the fact that it has been known for over a century, only in the last decade or so has some picture emerged as to the biochemical basis of this disease. It is highly likely that *L. sativus* neurotoxin ODAP is the chief incriminating factor in the etiology of the disease in human beings. Several questions are still not answered. The toxin is effective in young animals when it is administered orally or intraperitoneally. It is not effective in adult animals unless it is introduced directly into the cerebrospinal fluid. The disease in human beings essentially affects adults. Does a blood-brain barrier exist for the neurotoxin in adult human beings? If it exists, how it is overcome in afflicted patients? Studies with experimental animals indicate that the excretory potential of the animal and mode of metabolism of the neurotoxin in relation to the animal species may prove to be important factors determining the susceptibility of the animal to the neurotoxin effects. Women appear to be less susceptible to the disease. Is there any hormonal influence on the susceptibility of the individual to the neurotoxic manifestation?

ODAP–glutamate interaction at the synaptic level is a promising explanation for the neurotoxic effects of the *L. sativus* toxin. Further studies of this interaction are necessary.

As yet, no cure is available for the disease. The chief preventive measures now advocated are to treat the seeds to render them free of the toxin and thus fit for human consumption. The search for a suitable genetic hybrid variety of *L. sativus* free of the neurotoxin continues. The propagation of such a variety, if suitable from the viewpoint of other agricultural considerations, would perhaps be the ultimate method of eliminating the disease. The preparation of a protein concentrate free of the neurotoxin from *L. sativus* may be economically viable and could prove to be a good lysine supplement to cereal proteins.

References

Acton, H. W. (1922). *Indian Med. Gaz.* **57**, 241–247.

Adiga, P. R., Padmanaban, G., Rao, S. L. N., and Sarma, P. S. (1962). *J. Sci. Ind. Res., Sect. C* **21**, 284–286.

Adiga, P. R., Rao, S. L. N., and Sarma, P. S. (1963). *Curr. Sci.* **32**, 153–155.

Anderson, L. A. P., Howard, A., and Simonsen, J. L. (1924). *Indian J. Med. Res.* **12**, 613–643.

Anonymous (1967). *Nutr. Rev.* **25**, 231–233.

Anonymous (1968). Annual Report, pp. 14–16. Nutr. Res. Lab., Indian Counc. Med. Res., New Delhi.

Barrow, M. W., Simpson, C. F., and Miller, E. J. (1974). *Q. Rev. Biol.* **49**, 101–128.
Basu, K. P., Nath, M. C., Ghani, M. O., and Mukherjee, R. (1936). *Indian J. Med. Res.* **24**, 1027–1042.
Bell, E. A. (1962). *Biochem. J.* **83**, 225–229.
Bell, E. A. (1964). *Nature (London)* **203**, 378–380.
Bell, E. A., and O'Donovan, J. P. (1966). *Phytochemistry* **5**, 1211–1219.
Bell, E. A., and Tirimanna, A. S. L. (1965). *Biochem. J.* **97**, 104–111.
Cheema, P. S., Padmanaban, G., and Sarma, P. S. (1969a). *Indian J. Biochem.* **6**, 146–147.
Cheema, P. S., Malathi, K., Padmanaban, G., and Sarma, P. S. (1969b). *Biochem. J.* **112**, 29–33.
Cheema, P. S., Padmanaban, G., and Sarma, P. S. (1970). *J. Neurochem.* **17**, 1295–1298.
Cheema, P. S., Padmanaban, G., and Sarma, P. S. (1971a). *Indian J. Biochem. Biophys.* **8**, 16–19.
Cheema, P. S., Padmanaban. G., and Sarma, P. S. (1971b). *J. Neurochem.* **18**, 2137–2140.
Coulson, W. F., Linker, A., and Bottcher, E. (1969). *Arch. Pathol.* **87**, 111–116.
Curtis, D. R., and Watkins, J. C. (1965). *Pharmacol. Rev.* **17**, 347–391.
Dasler, W. (1954). *Science* **120**, 307–308.
Dastur, D. K., and Iyer, C. G. S. (1959). *Nutr. Rev.* **17**, 33–36.
Dilling, W. J. (1920). *J. Pharmacol.* **14**, 359–366.
Dunnill, P. M., and Fowden, L. (1965). *Nature (London)* **208,** 1206–1207.
Dupuy, H. P., and Lee, J. G. (1954). *J. Am. Pharm. Assoc.* **43**, 61–62.
Duque-Magalhoes, M. C., and Packer, L. (1972). *FEBS Lett.* **23**, 188–190.
Dwivedi, M. P., and Prasad, B. G. (1964). *Indian J. Med. Res.***52**, 81–116.
Ehrhart, L. A., Lipton, S. H., and Strong, F. M. (1963). *Biochemistry* **2**, 300–304.
Flass, H. G., Hadwiger, L., and Conn, E. E. (1965). *Nature (London)* **208**, 1207–1208.
Garbutt, J. T., and Strong, F. M. (1957). *J. Agric. Food Chem.* **5**, 367–370.
Gardner, A. F. (1959). *Am. J. Clin. Nutr.* **7**, 213–223.
Geiger, B. J., Steenbock, H., and Parsons, H. T. (1933). *J. Nutr.* **6**, 427–442.
German, W. J. (1960). *J. Neurosurg.* **17**, 657–663.
Gross, J., Levene, C. I., and Orloff, S. D. (1960). *Proc. Soc. Exp. Biol. Med.* **105**, 148–151.
Haskell, B. E., and Bowlus, S. B. (1976). *J. Org. Chem.* **41**, 159–160.
Johnson, J. L. (1972). *Brain Res.* **37**, 1–19.
Lajtha, A. (1962). *In* "Neurochemistry" (K. A. C. Elliot, I. H. Page, and J. H. Quastel, eds.), pp. 399–430. Thomas, Springfield, Illinois.
Lakshmanan J., Cheema, P. S., and Padmanaban, G. (1971). *Nature (London), New Biol.* **234**, 156–157.
Lakshmanan, J., and Padmanaban, G. (1974a). *Nature (London)* **249**, 469–471.
Lakshmanan, J., and Padmanaban, G. (1974b). *Biochem. Biophys. Res. Commun.* **58**, 690–698.
Lakshmanan, J., and Padmanaban, G. (1977). *J. Neurochem.* **29**, 1121–1125.
Levene, C. I. (1961). *J. Exp. Med.* **114**, 295–310.
Levene, C. I. (1962). *J. Exp. Med.* **116**, 119–130.
Levene, C. I. (1963). *Fed. Proc., Fed. Am. Soc. Exp. Biol.* **22**, 1386–1388.
Levene, C. I., Kranzler, J., and Franco-Browder, S. (1966). *Biochem. J.* **101**, 435–440.
Malathi, K., Padmanaban, G., Rao, S. L. N., and Sarma, P. S. (1968). *Biochem. Biophys. Acta* **141**, 71–78.
Mechanic, G., and Tanzer, M. L. (1970). *Biochem. Biophys. Res. Commun.* **41**, 1597–1604.
Mehta, T., Hsu, A. F., and Haskell, B. E. (1972). *Biochemistry* **11**, 4053–4062.
Mehta, T., Zarghami, N. S., Cusick, P. K., Parker, A. J., and Haskell, B. E. (1976). *J. Neurochem.* **27**, 1327–1331.
Mohan, V. S., Nagarajan, V., and Gopalan, C. (1966). *Indian J. Med. Res.* **54**, 410–414.
Murti, V. V. S., Seshadri, T. R., and Venkitasubramanian, T. A. (1964). *Phytochemistry* **3**, 73–78.
Nagarajan, V., Mohan, V. S., and Gopalan, C. (1965). *Indian J. Med. Res.* **53**, 269–272.

Nagarajan, V., Mohan, V. S., and Gopalan, C. (1966). *Indian J. Biochem.* **3**, 130–131.
Nigam, S. N., and Ressler, C. (1964). *Biochem. Biophys. Acta* **93**, 339–345.
Nigam, S. N., and Ressler, C. (1966). *Biochemistry* **5**, 3426–3431.
O'Dell, B. L., Elsden, D. F., Thomas, J., Partridge, S. M., Smith, R. H., and Palmer, R. (1966). *Nature (London)* **209**, 401–402.
Olney, J. W., Mishra, C. H., and Rhee, V. (1976). *Nature (London)* **264**, 659–661.
O'Neal, R. M., Chen, C. H., Reynolds, C. S., Meghal, S. K., and Koeppe, R. E. (1968). *Biochem. J.* **106**, 699–706.
Page, R. C., and Benditt, E. P. (1967). *Proc. Soc. Exp. Biol. Med.* **124**, 454–459.
Paik, W. C. W., and Lalich, J. J. (1970). *Arch. Pathol.* **90**, 316–320.
Partridge, S. M., Elsden, D. F., and Thomas, J. (1963). *Nature (London)* **197**, 1297–1298.
Pinnel, S. R., and Martin, G. R. (1968). *Proc. Natl. Acad. Sci. U.S.A.* **61**, 708–176.
Plazas, S. F., and De Robertis, E. (1974). *J. Neurochem.* **23**, 1115–1120.
Plazas, S. F., and De Robertis, E. (1976). *J. Neurochem.* **27**, 889–894.
Ponseti, I. V., and Baird, W. A. (1952). *Am. J. Pathol.* **28**, 1059–1077.
Rajagopal Rao, D., Hariharan, K., and Vijayalakshmi, K. R. (1974). *J. Agric. Food Chem.* **22**, 1146–1148.
Rajmohan, K., and Ramachandran, L. K. (1972). *Proc. Soc. Biol. Chem., Indian* **31**, 9–10.
Ramachandran, M., and Phansalkar, S. V. (1956). *Indian J. Med. Res.* **44**, 501–509.
Rao, S. L. N. (1976). *Biochemistry* **14**, 5218–5221.
Rao, S. L. N., and Sarma, P. S. (1966). *Indian J. Biochem.* **3**, 57–58.
Rao, S. L. N., and Sarma, P. S. (1967). *Biochem. Pharmacol.* **16**, 218–219.
Rao, S. L. N., Ramachandran, L. K., and Adiga, P. R. (1963). *Biochemistry* **2**, 298–300.
Rao, S. L. N., Adiga, P. R., and Sarma, P. S. (1964). *Biochemistry* **3**, 432–436.
Rao, S. L. N., Sarma, P. S., Mani, K. S., Raghunatha Rao, T. R., and Sriramachari, S. (1967). *Nature (London)* **214**, 610–611.
Ressler, C. (1962) *J. Biol. Chem.* **237**, 733–735.
Ressler, C. (1964). *Fed. Proc., Fed. Am. Soc. Exp. Biol.* **23**, 1350–1353.
Ressler, C., Redstone, P. A., and Erenberg, R. H. (1961). *Science* **134**, 188–190.
Ressler, C., Giza, Y. H., and Nigam, S. N. (1963). *J. Am. Chem. Soc.* **85**, 2874–2875.
Ressler, C., Nelson, J., and Pfeffer, M. (1964). *Nature (London)* **203**, 1286–1287.
Rosmus, J., Transvsky, K., and Doyl, Z. (1966). *Biochem. Pharmacol.* **15**, 1405–1410.
Roy, D. N., Nagarajan, V., and Gopalan, C. (1963). *Curr. Sci.* **32**, 116–118.
Rukmini, C. (1969a). *Indian J. Biochem.* **5**, 182–184.
Rukmini, C. (1969b). *Indian J. Chem.* **7**, 1062–1063.
Sadasivam, T. S., Sulochana, C. B., John, V. T., Subbaram, M. R., and Gopalan, C. (1960). *Curr. Sci.* **29**, 86–87.
Sarma, P. S., and Padmanaban, G. (1969). *In* "Toxic Constituents of Plant Foodstuffs" (I. E. Liener, ed.), pp. 267–291. Academic Press, New York.
Sastri, B. N. (1962). "Wealth of India," Vol. VI, pp. 36–47. Counc. Sci. Ind. Res. Publ., New Delhi.
Savage, J. E., Bird, D. W., Reynolds, G., and O'Dell, B. L. (1968). *J. Nutr.* **88**, 15–19.
Schilling, E. D., and Strong, F. M. (1954). *J. Am. Chem. Soc.* **76,** 2848.
Schilling, E. D., and Strong, F. M. (1955). *J. Am. Chem. Soc.* **77**, 2843–2845.
Selye, H. (1957). *Rev. Can. Biol.***16**, 1–82.
Shourie, K. L. (1945). *Indian J. Med. Res.* **33**, 239–247.
Siegel, R. C., Page, R. C., and Martin, G. R. (1970a). *Biochim. Biophys. Acta* **222**, 552–555.
Siegel, R. C., Pinnell, S. R., and Martin, G. R. (1970b). *Biochemistry* **9**, 4486–4492.
Spackman, D. H., Stein, W. H., and Moore, S. (1958). *Anal. Chem.* **30**, 1190–1206.

Sperry, W. M. (1962). *In* "Neurochemistry" (K. A. C. Elliot, I. H. Page, and J. H. Quastel, eds.), pp. 55–84. Thomas, Springfield, Illinois.

Strong, F. M. (1956). *Nutr. Rev.* **14**, 65–67.

Tschiersch, B. (1964). *Phytochemistry* **3**, 365–376.

Vivanco, F., and Jimenez Diaz, C. (1951). *Rev. Clin. Esp.* **40**, 157–163.

Wawzonek, S., Punseti. I. V., Shepard, R. S., and Wiedenmann, L. G. (1955). *Science* **121**, 63–65.

Weisberg, H. F. (1962). "Water, Electrolyte and Acid-Base Balance," pp. 208–259. Williams & Wilkins, Baltimore, Maryland.

Wu, G., Bowlus, S. B., Kim, K. S., and Haskell, B. E. (1976). *Phytochemistry* **15**, 1257–1259.

CHAPTER 9

Favism

J. MAGER, M. CHEVION, AND G. GLASER

TOXIC CONSTITUENTS OF PLANT FOODSTUFFS, SECOND EDITION

ISBN 0-12-449960-0

I. FAVISM AS AN INBORN ERROR OF METABOLISM: EPIDEMIOLOGICAL, GENETIC, AND ENZYMOLOGICAL ASPECTS OF THE DISEASE

A. Brief Description and Epidemiology of the Disease

The term ''favism'' was coined in 1894 by the Italian physician L. Montano to designate an acute hemolytic anemia following ingestion of broad beans (fava beans) or inhalation of pollen of the *Vicia faba* plant (Sansone *et al.*, 1958). It is possible that the ancient Greeks had already recognized the presence of toxic constituents in broad beans. The mathematician Pythagoras founded a religion based on the tenets of the transmigration of souls and the sinfulness of eating the broad bean (Russell, 1965; Waldron, 1973). The first authentic descriptions, however, of this disease in the medical literature date back to the mid-1850s (see Aurichio, 1935). Since then, a rapidly growing number of case reports and clinical studies of favism have contributed to establish its status as a separate nosological* entity and to corroborate the etiological role of fava beans.

In a series of 1211 cases of favism in Italy reviewed by Fermi and Martinetti (1905), 725 were due to ingestion of broad beans, 459 were attributed to inhalation of pollen, whereas the remaining 27 cases were of undetermined origin. In a recent epidemiological survey of 579 cases of favism in Iran, only 4 (0.7%) were thought to be attributable to exposure to fava plant pollen, whereas consumption of broad beans was held to be responsible for the remainder of the cases (Hedayat *et al.*, 1971). Other reports failed to confirm the occurrence of pollen-induced outbreaks of favism (Chung, 1965; Kattamis *et al.*, 1969; Belsey, 1973). Thus, definitive and unequivocal evidence for the causative role of pollen is still lacking. Hemolytic crises due to ingestion of dry or cooked beans were found to be usually of a rather moderate intensity, whereas the most severe attacks were observed after eating of fresh raw fava seeds (Luisada, 1941). In rare instances, plants other than *Vicia faba,* e.g., seeds of *Pisum sativum* (garden peas) and pollen of *Anagyris foetida* and *Verbena hybrida,* have been incriminated as a cause of a hemolytic syndrome closely similar to favism (see Larizza *et al.*, 1960). Favism has been observed in individuals who had previously consumed fava beans with no untoward effects; others developed the disease on first exposure (Angelova and Andrev, 1959; Kattamis *et al.*, 1969; Belsey, 1973). Recurrent attacks of favism are not uncommon (Belsey, 1973).

Epidemiological studies in Italy revealed a seasonal incidence of the disease, characterized by two peaks: in April–May, when the plant blossoms, and in July–August, when the fresh ripe beans appear on the market (Luisada, 1941). In a more recent investigation, Donoso *et al.* (1969) found a pronounced seasonal

*Nosology is defined as the science of the classification of diseases.

peak in the Caspian littoral sea region between the middle of May and the middle of June, whereas in Abadan the highest incidence of favism occurred in April; in both instances, the seasonal peaks coincided with the harvesting of fava beans in the respective regions.

Data accumulated over the past 10 years reveal that favism is much more common in children under the age of 10 than in adults, with the highest frequency observed in the age group of 2–4 years, probably resulting from the first exposure of these infants to the beans. There are quite a few reports, however, on favism occurring during the first year of life, and even in breast-fed infants, apparently because of transmission of the noxious agent through the mother's milk (Angelov and Andrev, 1959; Chung, 1965; Kattamis, 1969, 1971).

The disease shows a marked predilection for the male sex, the male to female ratio varying in different series studied between 21:1 and 2.7:1 (Hedayat *et al.*, 1971; Belsey, 1973).

The clinical picture of favism is governed by the symptomatology inherent in the hemolytic event, the major manifestations being pallor, fatigue, dyspnea, nausea, abdominal or back pain, fever, and chills. The hemolysis is of variable intensity and gravity, the more severe cases being attended by hemoglobinuria and jaundice (see Beutler, 1972) and occasionally by acute renal failure (Schmitz and Fritz, 1968; Symvoulidis *et al.*, 1972). The onset of hemolysis may be extremely rapid and abrupt, especially in cases attributed to pollen inhalation, with symptoms starting within a few minutes following the noxious exposure. In the vast majority of cases, an interval of about 5–24 hr intervenes between the ingestion of the broad beans and the first manifestations of the hemolytic attack, whereas in rare instances the onset of the symptoms may be delayed until the third day or later following the consumption of broad beans (Luisada, 1941; Hedayat *et al.*, 1971).

The course of the disease is usually self-limited, the acute stage lasting 24–48 hr, and is then followed by prompt, spontaneous recovery. In contrast to the usually benign outcome of the disease in adults, a 6–8% case fatality rate was noted in children under 6 years of age (Fermi and Martinetti, 1905). More recently, however, the mortality figures have been greatly reduced on account of the advent of blood transfusion therapy (Crosby, 1956; Hedayat *et al.*, 1971).

B. Geographic Distribution of Favism

Favism exhibits a striking prevalence in the insular and littoral regions in the Mediterranean area and in the Middle East (Sardinia, Sicily, southern provinces of the Italian peninsula, Greece, Rhodes, Cyprus, Turkey, Lebanon, Israel, Iraq, Spain, the Balearic Islands, Algeria, Egypt, and Sudan) (Luisada, 1941; Hedayat *et al.*, 1971; Belsey, 1973; Amin-Zaki *et al.*, 1972; Hassan, 1971). The disease is also frequently encountered in China (Chung, 1965) and Bulgaria (Angelov and Andrev, 1959). Sporadic cases have been reported from Germany (Gehrmann *et al.*, 1963, Johannsen *et al.*, 1968), France (Auquier *et al.*, 1968), Poland (Rockicka-Milewska *et al.*, 1968; Rożynkowa *et al.*, 1970, 1971), Rumania (Schneer, 1968), Yugoslavia (Vince-Ribarić, 1962), and Singapore (Wong, 1972) (see also Table I).

The highest incidence was observed on the island of Rhodes (Kattamis *et al.*, 1969). The disease was found to be unevenly distributed on the island, with regions in which up to 40 cases per 1100 male inhabitants were reported. The overall prevalence of favism in the total population of the island during the years 1952–1965 was found to be 1.7–5.4 per thousand. A similarly high prevalence was recorded in Sardinia, with about five cases of favism per 1000 inhabitants (Crosby, 1956). This highly selective geographic distribution of favism is particularly puzzling since fava beans are grown and consumed almost all over the world as a cheap and popular staple food, distinguished by its relatively high content of carbohydrates and proteins (58 and 25%, respectively) per edible portion of mature dry seeds (see U.S. Department of Agriculture, 1963).

C. Role of Glucose-6-Phosphate Dehydrogenase Deficiency in the Etiology of Favism

The various concepts advanced in the earlier literature on the infectious (see Luisada, 1941), toxic (Gasbarrini, 1915), or immunological (Manai, 1929; Dacie, 1954) etiology of favism appear today to be of historic interest only. These views, apart from being devoid of solid experimental ground, failed to account for some of the most salient features of the disease, namely, its restricted geographic and ethnic distribution and its pronounced familial tendency (Luisada, 1941).

The door to an understanding of the true pathogenetic nature of favism was opened by the elucidation of the nature of the inborn error of metabolism underlying the so-called drug sensitivity, i.e., an abnormal propensity of certain individuals to develop acute hemolysis in response to treatment with primaquine and a variety of other drugs (see Beutler, 1972). Shortly after the discovery by Dern *et al.* (1954) that "primaquine sensitivity" is determined by an intrinsic abnormality of the erythrocytes, it was revealed by Beutler *et al.* (1955, 1957) that the susceptible red blood cells exhibit a relatively low content of reduced glutathione (GSH) and an enhanced rate of GSH destruction on incubation with 1-acetyl-2-phenylhydrazine in the presence of glucose. This so-called glutathione instability of drug-sensitive red blood cells was then shown by Carson *et al.* (1956) to be due to a deficiency of the nicotinamide adenine dinucleotide phosphate (NADP)-linked glucose-6-phosphate dehydrogenase (G6PD) and the resultant incapacity of these cells to maintain an adequate supply of NADPH to cope with the increased demand for GSSG (oxidized glutathione) reduction imposed by the challenging drugs (see Section III). In normal erythrocytes exposed to an oxidant stress, the enhanced rate of GSSG formation is compensated by a parallel increase in the rate of its reduction to GSH as a result of the concomitant stimulation of the G6PD activity, governed by an intrinsic regulatory mechanism (see Yoshida, 1973). The selective vulnerability of the red blood cells to this enzyme deficiency is accounted for by their critical dependence on the pentose phosphate shunt as the sole mechanism for NADPH generation, due to the lack of the alternate pathways for NADPH supply that are present in other cells (Beutler, 1971).

Crosby (1956), in a brief and brillant report concerned with the clinical and epidemiological aspects of favism in Sardinia, was the first to point out the resemblance of this disease to the primaquine-induced hemolytic anemia insofar as in both instances the red blood cells are capable of normal survival unless they are challenged by the noxious agent. He suggested, by analogy to primaquine sensitivity, that a hereditary enzymatic deficiency may be the underlying cause of the susceptibility to favism.

The essential correctness of Crosby's idea was soon substantiated by direct experimental evidence obtained independently by Sansone and Segni (1956, 1957a,b, 1958) in Italy and by Szeinberg *et al.* (1957, 1958a,b) in Israel. These workers found that the GSH content of erythrocytes from persons known to have been affected by favism tends to be significantly lower (mean values below 50 mg %) than in normal individuals (mean range, 60–88 mg %) (Sansone and Segni, 1956; Szeinberg *et al.*, 1957). More significantly, the red cell GSH level was found to decline sharply during the acute phase of favism. (Szeinberg and Chari-Bitron, 1957; Larizza *et al.*, 1958), concomitant with the frequent appearance of methemoglobin and intracellular inclusions called Heinz bodies, similar to those observed in drug-induced hemolysis (Panizon and Pujatti, 1957; Larizza *et al.*, 1960). Furthermore, in all persons with a past history of favism the erythrocyte GSH proved to be unstable in Beutler's acetylphenylhydrazine test *in vitro* (Sansone and Segni, 1957a; Szeinberg *et al.*, 1958a). Finally, and most important, the GSH instability was invariably associated with a pronounced G6PD deficiency of the red blood cells (Sansone and Segni, 1958; Szeinberg *et al.*, 1958b; Larizza *et al.*, 1958; Zinkham *et al.*, 1958).

D. Ethnic Distribution of G6PD Deficiency and Favism

G6PD deficiency is probably the most common genetically determined enzymatic defect in human beings, affecting, according to a rough estimate by Carson (1960), about 100 million people of all races throughout the world. Its geographic distribution closely parallels that of malaria, presumably because of the selective advantage offered by the enzymatic deficiency in increasing the resistance of the red blood cells to infestation by *Plasmodium falciparum* (Allison and

TABLE I

INCIDENCE OF G6PD DEFICIENCY AND OCCURRENCE OF FAVISM IN DIFFERENT ETHNIC GROUPS

	G6PD Deficiency		Favism	
Ethnic groups	Incidence (%)	Reference[a]	Occurrence[b]	Reference[a]
Negroes				
United States	13	1		
Canada (Nova Scotia)	3–27	2		
Congo (Kinshasa)	18–23	3		
Pygmies	4	3		
Bantu	2–4	4		
Nigeria	10	5		
Ghana	24	6		
Gambia	15	6		
Sudan	7.1–8.3	7	+	7
Mediterraneans and Asians				
Sardinia (males)	4–48	8	+	9
Sicily, southern Italy			+	10
Greece (mainland and Crete)	0.7–3	11	+	11
Rhodes (Greeks)	22	12	+	12
Cyprus (Greeks)	7–11	13	+	13
Ashkenazic Jews	0.2			
Oriental Jewish communities (males)				
Kurdish	53	14		15
Iraqi	24	16		17
Iranian	15			
Yemenite	5			
North African	2			
Israeli Arabs	4			
Lebanon	3	18		
Egypt (males)	26	19	+	19,26
Turkey	11	20	+	
Iraq (males)	8.9	21	+	21
Kuwait (adults)	20	22		
Saudi Arabia (adults)	15–24	23		
Iran	10	24	+	25,26
India	6	3		
Philippines	12	3		
China	5.5	27	+	28,29
Singapore	3–4	30	+	30
Papua	6	31		
Melanesia	0–29	32		
Micronesia	0–9	33		

[a]Key to references: 1. Marks and Gross (1959); 2. Langley *et al.* (1969); 3. Motulsky (1960); 4. Charlton and Bothwell (1959); 5. Gilles *et al.* (1960); 6. Allison *et al.* (1961); 7. Hassan (1971); 8. Siniscalco *et al.* (1961); 9. Crosby (1956); 10. Luisada (1941);

Clyde, 1961; Luzatto *et al.*, 1969). The frequency of occurrence of this inborn error of metabolism is widely dissimilar in the different ethnic groups. As shown in Table I, the highest figures of incidence of G6PD deficiency have been recorded, in decreasing order, in some of the oriental Jewish communities of Israel, Sardinians, Cypriot Greeks, American Negroes, and certain African populations. On the other hand, the abnormal trait is extremely rare or virtually absent in northern European nations and among Ashkenazic Jews (of European descent), North American Indians, and Eskimos (see Motulsky, 1960). It should be stated, however, that the available statistical data are on the whole rather incomplete and practically nonexistent in many underdeveloped areas lacking the elementary facilities for the detection of G6PD deficiency.

No detailed statistics are available concerning the ethnic distribution of favism. As may be seen, however, from Table I, the occurrence of favism does not parallel the frequency of the G6PD-deficient trait in the different populations. Most conspicuous in this respect is the complete absence of favism in North American Negroes (see Beutler, 1971).

E. Mode of Inheritance of G6PD Deficiency

G6PD deficiency is transmitted by a gene located in the X chromosome. This mode of inheritance was deduced from extensive family studies (Childs *et al.*, 1958; Szeinberg *et al.*, 1958c; Larizza *et al.*, 1960) and from the parallel segregation pattern observed when the G6PD-deficient trait and some other sex-linked anomaly, such as color blindness (Porter *et al.*, 1962) or hemophilia A (Boyer and Graham, 1965), happened to coexist in the same individual (see also Aebi, 1967). Accordingly, the enzyme deficiency is fully expressed in the hemizygous ($\bar{X}Y$) male, because the mutant gene ($\bar{X}$) is not counteracted by the normal allele (X). On the other hand, full expression is rather uncommon in females, since it depends on the statistically rare occurrence of a homogyzous mutant genotype ($\bar{X}\bar{X}$). In the majority of affected females the G6PD deficiency is found to be of a partial or intermediate nature, in accordance with the expected preponderance of the heterozygous ($\bar{X}X$) constellation (Larizza *et al.*, 1960).

TABLE I (*Continued*)

11. Zannos-Mariolea and Kattamis (1961); 12. Kattamis *et al.* (1969); 13. Plato *et al.* (1964); 14. Szeinberg *et al.* (1958b); 15. Bogair (1951); 16. Szeinberg and Sheba (1960); 17. Efrati (1952); 18. Taleb *et al.* (1964); 19. Ragab *et al.* (1966); 20. Say *et al.* (1965); 21. Amin-Zaki *et al.* (1972); 22. Shaker *et al.* (1966); 23. Gelpi (1965); 24. Walker and Bowman (1959); 25. Hedayat *et al.* (1971); 26. Belsey (1973); 27. Chan *et al.* (1964); 28. Du (1952); 29. Vella (1959); 30. Wong (1972); 31. Ryan and Parsons (1961); 32. Kidson and Gorman (1962); 33. Kidson and Gajdusek (1962).

[b]Plus (+) sign indicates that outbreaks of favism have been reported in the respective populations; blank space indicates that, to our knowledge, no such occurrences have been reported.

By the use of ingenious cytochemical techniques capable of detecting G6PD deficiency in single cells, it could be shown that erythrocytes of females with a heterozygous trait of intermediate enzyme deficiency constitute a mixture of normal and G6PD-deficient cells (Beutler *et al.*, 1962; Sansone *et al.*, 1963). This so-called cellular mosaicism bears out the prediction of Lyon's "X chromosome inactivation" theory (Lyon, 1961). According to this concept, a random loss of functional activity of one of the X chromosome pair (of either paternal or maternal origin) takes place in each individual somatic cell early in the course of morphogenesis, with a perpetuation of the inactivation pattern in the progeny. It follows, therefore, that the degree of susceptibility to drug-induced or favic hemolysis in heterozygous females will be critically dependent on the relative proportions of the G6PD-deficient and normal erythrocytes in their blood.

F. Molecular Characteristics of Normal and Mutant G6PD

During the last decade, G6PD has been the subject of intensive enzymological investigations. The normal enzyme was found to exist in several oligomeric forms (composed of two, four, or six identical subunits), the degree of aggregation depending on a variety of physicochemical factors, e.g. ionic strength, pH, and protein concentration of the enzyme solution (Yoshida, 1966, 1967b; Yoshida and Hoagland, 1970). The catalytically active molecular species that appears to be predominantly in the dimeric form is stabilized by its tight association with NADP or NADPH, whereas in the absence of the coenzymes the apoenzyme tends to dissociate into the inactive monomers (Yoshida, 1973).

Starch gel electrophoresis of normal red cell hemolyzates revealed the existence of two major molecular variants of G6PD, characterized as a fast migrating band A and a slow band B (Boyer *et al.*, 1962). The more common form B is found in Caucasian subjects and in American Negroes, whereas type A occurs among Negroes only (Boyer *et al.*, 1962; Kirkman and Hendrickson, 1963). In subsequent studies, about 80 additional genetic variants of G6PD were differentiated by a combination of various enzymological criteria, such as catalytic rate, electrophoretic mobility, substrate specificity, and particularly the ability to utilize 2-deoxy-D-glucose, K_m values for glucose 6-phosphate and NADP, pH optimum, and heat stability (Kirkman *et al.*, 1964; Yoshida *et al.*, 1971; Beutler, 1971).

In nearly 40 variants that are not associated with any clinical symptoms, the catalytic activity is within the normal range and in one instance (G6PD Hektoen) is even severalfold higher than the normal average level. About 20 variants exhibit severe red cell enzyme deficiency, which manifests as hemolytic anemia in response to oxidative stress by drugs or fava beans. Another group, comprising about 20 G6PD-deficient mutants, is associated with "chronic nonspherocytic congenital hemolytic anemia,'' occurring spontaneously in the absence of any extraneous challenging agent (Yoshida, 1970).

The most common G6PD-deficient mutants are as follows: the Mediterranean type, which is prevalent among Sephardic Jews, Italians, and Greeks; the A type common among Negroes; the Canton type found primarily among southern Chinese and other Oriental populations; and the Debrousse type most frequent among Arabs.

Three variants of human G6PD that were isolated in molecularly homogeneous form and subjected to ''fingerprinting'' were found to differ from one another in a single amino acid residue. Thus, an asparagine residue present in variant B is replaced by aspartic acid in variant A (Yoshida, 1967a). Similarly, G6PD Hektoen was found to differ from the B variant by a substitution of histidine for tyrosine (Yoshida, 1970).

By applying a density gradient centrifugation procedure to separating red blood cells into different age groups, it was found that the G6PD activity in a 5% fraction of youngest cells from a blood sample of the deficient A^- variant was practically identical to that of normal erythrocytes (Yoshida *et al.*, 1967). It was concluded that the G6PD A^- enzyme has normal initial specific activity, but the rate of its inactivation and eventual loss from the cell is considerably enhanced. This conclusion was further corroborated by immunological methods (Marks and Gross, 1959; Piomelli *et al.*, 1968; Yoshida *et al.*, 1968). On the other hand, by the use of similar procedures, the Mediterranean mutation was found to involve both a decrease in the number of molecules and a diminished specific activity of the enzyme (Yoshida *et al.*, 1968).

G. Correlation between the Degree of G6PD Deficiency and the Severity of Clinical Symptomatology

The severity of clinical manifestations among the various G6PD mutants does not correlate well with the differences in the extent of G6PD deficiency, as measured *in vitro* under the conventional assay conditions. Yoshida and Lin (1973) suggested that the apparent inconsistencies could be resolved by taking into account the regulatory mechanisms governing the G6PD activity under the physiological conditions prevailing within the red blood cell. It was found that the activity of the oxidative pentose phosphate shunt in the red cell is strongly suppressed, representing only about 0.1–0.2% of the maximal potential catalytic rate of G6PD, as determined in cell-free hemolyzates. The low intracellular activity of the pentose shunt is attributable to the strong inhibitory effect of NADPH and ATP on the activity of G6PD, which constitutes the rate-limiting step in the overall oxidative pentose phosphate pathway. The various G6PD-deficient mutants differ in their affinities for NADP (K_m), as well as in the degree of their susceptibility to inhibition by ATP and NADPH (K_i). Thus, the spontaneously hemolytic variants (e.g., Manchester, Tripler, or Alhambra) are strongly inhibited by ATP and NADPH, so that under the conditions existing in the human erythrocyte (NADP/NADPH ratio close to 1:50 and 1:5 m*M* ATP) these variants are virtually nonfunctional, although their potential G6PD activity is about 20% of the normal. In contrast, the nonhemolytic mutants (e.g., Mediterranean or A^-) are resistant to inhibition by NADPH and ATP, and, therefore, under simulated physiological conditions, their catalytic activity varies between 30 and 50% of the average activity of the normal major variants (A and B).

H. Differences between Caucasian and Negro Types of G6PD Deficiency

The considerable genetic polymorphism and biochemical heterogeneity of the enzymatic defect brought to light by the above studies may at least partly account for the marked differences and variations in the biological and clinical manifestations of the inborn error observed in different ethnic groups. The differences are particularly striking between the affected American Negroes and Caucasians (Italians, Greeks, Sephardic Jews, etc.). Thus, in G6PD-deficient Negro males, the range of the erythrocytic G6PD activity is about 10–21% of the normal mean, whereas in Caucasian males the enzyme activity is only 0–6% of the normal level (Marks and Gross, 1959; Marks and Banks, 1965). This disparity in the extent of the enzyme deficiency seems to be correlated with the relatively milder clinical course of the drug-induced hemolysis usually observed in Negroes (see Beutler, 1972; Pannaciulli *et al.*, 1965). Furthermore, the G6PD activity of the young red cell population (following a hemolytic crisis) is considerably elevated in Negroes but not in Caucasians (see Marks and Banks, 1965). The enzyme deficiency appears to be more widely distributed in the different tissues of affected Caucasians than in Negroes (Brunetti *et al.*, 1960; Chan *et al.*, 1965; Marks *et al.*, 1959). The vestigial G6PD activity in deficient Caucasians is usually represented by the variant B, whereas in deficient Negroes the enzyme shows invariably A-type characteristics (Boyer *et al.*, 1962).

I. Possible Role of Genetic Determinants Other Than G6PD Deficiency in the Pathogenesis of Favism

The existence of an extracorpuscular factor was postulated by Stamatoyannopoulos *et al.* (1966) to explain the relatively low incidence of favism in G6PD-deficient subjects, its total absence in American Negroes, as well as its bizarre and apparently unpredictable mode of occurrence, which shows no correlation with the frequency of exposure to *Vicia faba* or the degree of enzyme deficiency. A study conducted by these authors on the incidence of favism in three different areas of Greece showed that overt episodes of favism do not occur at random in G6PD-deficient subjects, but there is a tendency for regional and familial aggregation of cases. Thus, in the area of Karditsa with a frequency of G6PD deficiency of about 27%, favism is rare, whereas on Corfu Island, with an approximately 5% incidence of this enzyme deficiency, favism is of relatively frequent occurrence. In both areas, the consumption of fava beans is very common. The authors concluded that the family data they collected are consistent with "the hypothesis of Mendelian segregation of an autosomal gene which in the heterozygous state enhances the susceptibility to favism of G6PD-deficient individuals." The functional role of the hypothetic gene remains to be defined. Its influence on the susceptibility of the red blood cells to hemolysis or alternatively its modifying effect on the absorption, detoxication, or excretion of the causative agent of favism are among the various possibilities to be considered (Stamatoyannopoulos *et al.*, 1966; see also Tarlov *et al.*, 1962). A preferential association between favism and certain acid phosphatase phenotypes (types A and C) has been described (Bottini *et al.*, 1971). However, the pathogenetic significance of this correlation is not clear.

J. Methods of Detection of G6PD Deficiency

1. *Glutathione Stability Test*

The test designed by Beutler (1957) is based on his observation that the GSH content of G6PD-deficient but not of normal erythrocytes declines markedly in the course of aerobic incubation with acetylphenylhydrazine and glucose under standardized conditions. The recommended method for glutathione determination is the procedure of Ellman (1959) as adapted by Beutler *et al.* (1963).

2. *Spectrophotometric Determination of G6PD*

The assay is based on the measurement of the rate of increase of absorbance at 340 nm due to formation of NADPH in a reaction system consisting of glucose 6-phosphate, NADP, buffer solution, and suitably diluted hemolyzate (see Carson, 1960).

3. *Methemoglobin Reduction Test*

The assay introduced by Brewer *et al.* (1962) is based on the observation of Dawson *et al.* (1958) that the rate of methemoglobin reduction by G6PD-deficient erythrocytes in the presence of methylene blue is considerably slower than normal. A modification using Nile blue sulfate (instead of methylene blue), a dye that does not permit interaction between cells, was designed for the detection of intermediary states of G6PD deficiency in heterozygous females (Beutler and Baluda, 1963).

4. *Indicator-Linked Screening Methods*

In these tests, the G6PD-linked reduction of NADP is measured indirectly by following the rate of reduction of a suitable dye serving as an artificial terminal hydrogen acceptor. The original procedure of Motulsky and Campbell-Kraut (1961), in which the decolorization time of brilliant cresyl blue is determined, was modified by the use of 3-(4,5-dimethylthiazolyl-1,2)-2,5-diphenyltetrazolium bromide (MTT), which is reduced by the G6PD-linked system to a purple insoluble formazan derivative. The latter reagent can be used in a spot-test technique (Fairbanks and Beutler, 1962).

5. *Fluorescent Screening Methods*

This screening procedure, introduced by Beutler and Mitchell (1968), takes advantage of the fact that NADPH (generated by G6PD) fluoresces when illuminated with ultraviolet light, whereas NADP does not. A small volume of blood is hemolyzed with saponin and mixed with a buffered solution of suitable amounts of glucose 6-phosphate, NADP, and GSSG. Following incubation for 5–10 min, the mixture is spotted on filter paper and examined under an ultraviolet lamp for fluorescence. An automated, quantitative version of this method adapted to use with the Technicom AutoAnalyzer instrument has been described (Dickson *et al.*, 1973).

6. *Ascorbate–Cyanide Test*

This test, devised by Jacob and Jandl (1966), depends on the ability of ascorbate to undergo oxidation in the presence of oxyhemoglobin with formation of H_2O_2. When catalase is inhibited by cyanide, the generated H_2O_2 can be detoxified only by GSH peroxidase. Any enzymatic deficiency interfering with the proper functioning of the GSH peroxidase system (G6PD, GSH, GSH reductase, and GSH peroxidase) causes H_2O_2 to accumulate. The hydrogen peroxide reacts with hemoglobin, producing hemochromes with resultant brown discoloration detectable by visual inspection.

II. THE SELECTIVE TOXICITY OF FAVA BEANS: SEARCH FOR THE CAUSATIVE AGENT OF FAVISM

A. Effect of Ingestion of Fava Beans on the Life Span of G6PD-Deficient Erythrocytes

Efforts to obtain direct experimental evidence for the selective toxicity of broad beans for G6PD-deficient cells are greatly handicapped by the lack of a susceptible laboratory animal and the potential health risk inherent in the induction of a hemolytic disorder in sensitive volunteers. To circumvent this difficulty, some authors have adopted the experimental stratagem of Dern *et al.* (1954), consisting of the transfusion of ^{51}Cr-labeled G6PD-deficient erythrocytes into normal compatible recipients and the determination of the effect of orally administered broad beans on the survival of the tagged cells. These studies have yielded conflicting results in the hands of different investigators. Thus, both Greenberg and Wong (1961) and Davies (1962) failed to detect any significant decrease in the life span of the transfused erythrocytes following ingestion of broad beans by healthy recipients. The negative outcome of these trials, however, is open to criticism and may be ascribed perhaps to the inadequacy of the experimental conditions, such as insufficient size of the challenging dose of the fava beans or loss of their noxious activity due to cooking or prolonged storage. On the other hand, Panizon and Vullo (1962) concluded from their well-documented experiments that fresh fava beans or juice prepared from them induced in the majority of cases a definite shortening of the survival of the transfused G6PD-deficient erythrocytes. According to the authors' estimate, however, the effective dose of the fava beans was surprisingly high, i.e., about 50,000 times larger, on a weight basis, than the minimal hemolytic dose of primaquine (Panizon and Vullo, 1962). Furthermore, the hemolytic effect of the fava beans was much more pronounced when fava-sensitive individuals (in the early stage of recovery from a favic crisis) rather than normal volunteers served as recipients. These results, therefore, seem to indicate that G6PD-deficiency and ingestion of fava beans are not sufficient by themselves to bring about a full-blown hemolytic crisis and point to the adjuvant role of an extracorpuscular factor in the pathogenesis of this disease (Panizon and Vullo, 1961, 1962; Panizon, 1967).

B. Studies with Crude Fava Bean Extracts

Attempts to ascertain unequivocally the presence of a noxious factor in fava beans with a specific effect on drug-sensitive erythrocytes have thus far met with a moderate degree of success.

Mela and Perona (1959) noted a decrease of GSH in G6PD-deficient red blood cells incubated in the presence of broad bean juice prepared by squeezing with a hydraulic press. Walker and Bowman (1960) reported that crude aqueous extracts of fava beans, as well as saline extracts of fava pollen and pistils (Bowman and Walker, 1961), caused a rapid and significant drop in the GSH level of sensitive erythrocytes but had little or no effect on normal erythrocytes. The activity of the broad bean extracts was destroyed by boiling for 90 min. No effect was detected with extracts prepared from peas or other leguminous seeds tested.

Results similar to those described above, though less clear-cut, were also obtained by Contu *et al.* (1961). Furthermore, Panizon and Zacchello (1965) found that fava bean juice or whole bean homogenates induced a fall of GSH and a definite impairment of the survival of ^{51}Cr-labeled G6PD-deficient red blood cells transfused into normal recipients, whereas normal erythrocytes remained unaffected. Panizon and his collaborators (Panizon, 1967; F. Panizon, personal communication, 1968) isolated from fava beans a material soluble in lipid solvents that induced a very marked decline of the intraerythrocytic GSH accompanied by formation of methemoglobin and Heinz bodies; in contrast to primaquine, this material also acted on GSH in the absence of hemoglobin. The authors concluded that their data are incompatible with an allergic pathogenesis of favism, a view still held by certain investigators (Carcassi, 1958; Kantor *et al.*, 1962). They also suggested that fava beans and primaquine give rise to hemolysis through an essentially similar mechanism. More recently, Bottini and his associates (Bottini *et al.*, 1970; Bottini, 1973) described the separation of crude fava bean extracts into two distinct fractions, both capable of oxidizing GSH in pure solution and in G6PD-deficient erythrocytes.

Various serum abnormalities were observed in patients during an acute favic episode, such as atypical antibodies against their own erythrocytes (Marcolongo *et al.*, 1950) or antibodies against fava bean extracts (Kantor and Arbesman, 1959; Kantor *et al.*, 1962). Some investigators described the occurrence in saline extracts of broad beans of a heat-labile substance causing agglutination of both normal and G6PD-deficient erythrocytes (Roth and Frumin, 1960; Greenberg and Wong, 1961) and counteracted by a normal serum factor residing in the IgA fraction of the γ-globulins (Creger and Gifford, 1952). Frumin and his coworkers reported that the factor neutralizing the fava bean hemagglutinin was absent in the serum of a number of favic patients tested both in the acute hemolytic stage and during the remission (Perera and Frumin, 1965; Nathan *et al.*, 1974). These findings are at variance with the observation of Greenberg and Wong (1961) that both favic and normal sera exhibited the ability to inhibit hemagglutination by fava bean extracts. The paucity of the data does not permit a critical assessment of their significance and their possible relevance to the pathogenesis of favism.

C. Fractionation of Fava Bean Extracts

In a study carried out in our laboratory, the quest for the causative agent of favism in broad bean extracts was guided by its presumed discriminatory capacity for oxidizing GSH in G6PD-deficient but not in normal erythrocytes, when incubated *in vitro* in the presence of glucose. In fact, several fractions conforming to this criterion could be isolated from aqueous broad bean extracts by use of ion-exchange chromatography. Some of the purified fractions were sparingly soluble in water and exhibited in neutral solution a rapid loss of GSH-oxidizing activity, concomitant with a change in their spectral characteristics. The structural instability of these substances appeared to be inherent in their tendency to undergo spontaneous oxidation in air, since the deterioration could be largely prevented by storage under nitrogen (S. Bien, M. Noom, G. Glaser, A. Roigin, and J. Mager, unpublished results). The properties of the active fractions were reminiscent of those described for some pyrimidine derivatives known to occur in fava beans in the form of aglycones of the β-glycosides termed "vicine" and "convicine." Our subsequent work, therefore, was concerned primarily with exploring the possible role of these aglycones in the causation of favism.

D. Structure and Properties of Vicine, Convicine, and Their Aglycones

Vicine was first isolated by Ritthausen and Kreusler (1870) from *Vicia sativa* seeds by a method involving extraction with dilute H_2SO_4 and precipitation with $HgSO_4$; the final yield of the crystalline material was about 0.35% (Ritthausen, 1876). Vicine was subsequently found in other species of *Vicia* including *Vicia faba* (Winterstein and Somló, 1933), beet juice, and peas (Schulze, 1891; Bendich and Clements, 1953). The glycosidic nature of this compound was recognized by Ritthausen (1896), who also succeeded in isolating the aglycone divicine (Ritthausen, 1899a,b). The pyrimidine nucleoside structure was assigned to vicine by Johnson (1914) and confirmed by Levene (1914). The correct formulation, however, of vicine as 2,6-diamino-4,5-dihydroxypyrimidine 5-(β-D-glucopyranoside) (**I**) was arrived at only several decades later by Bendich and Clements (1953).

Convicine was also discovered by Ritthausen (1881) in *Vicia sativa*. It was identified by Johnson (1914) as a β-glycoside of isouramil, and the correct position of the glycosidic bond was established by Bendich and Clements (1953). The formulation of convicine as 2,4,5-trihydroxy-6-aminopyrimidine 5-(β-D-glucopyranoside) (**II**) was confirmed by unambiguous evidence (Bien *et al.*, 1968).

The aglycones divicine (**III**) and isouramil (**IV**) can be obtained from the respective glycosides (vicine and convicine) by mild acid hydrolysis or by enzymatic splitting with β-glucosidase (emulsin). In recent years, several synthetic

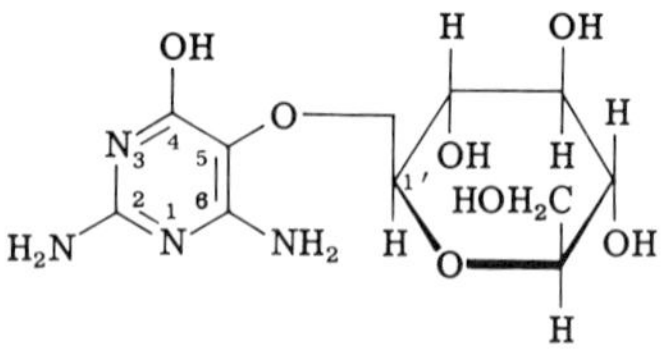

Vicine
(I)

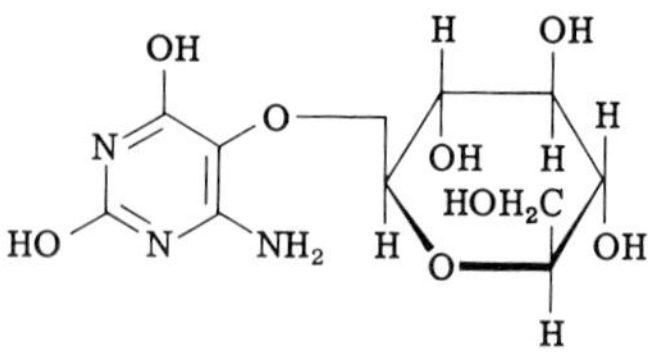

Convicine
(II)

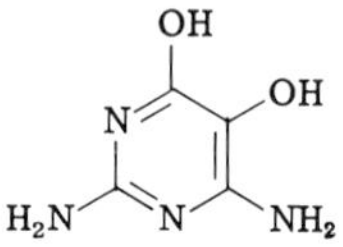

Divicine
(III)

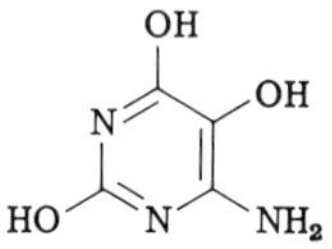

Isouramil
(IV)

procedures for preparing these compounds have also been developed (Davoll and Laney, 1956; McOmie and Chesterfield, 1956; Chesterfield *et al.*, 1964; Bien *et al.*, 1968; Ikeda *et al.*, 1973).

Vicine and convincine, as well as the corresponding aglycones, react with the Folin–Ciocalteu phenol reagent, yielding a blue color similar to that produced by tyrosine or tryptophan. This property was employed by Higazi and Read (1974) as a basis for developing a quantitative assay of vicine in plant material and blood. In our hands, however, the method failed to fulfill the authors' claim for its specificity and, therefore, did not lend itself to direct determination of the pyrimidine glycosides in natural materials. Both divicine and isouramil reduce vigorously alkaline solutions of 2,6-dichlorophenolindophenol, phosphomolybdate, or phosphotungstate and elicit an intense blue color reaction with ammoniacal ferric chloride solution, indicative of the presence of an enolic hydroxyl group in the molecule. The aglycones are highly unstable in the presence of oxygen; the rate of their oxidative breakdown is most rapid at alkaline pH and falls off with decreasing pH values. At room temperature, the half-lives of divicine and isouramil in neutral solutions are of the order of 30–40 min. The breakdown of the pyrimidine aglycones is accelerated by traces of copper (Cu^{2+}) and other heavy metals; both compounds are almost instantaneously destroyed by boiling (Bendich and Clements, 1953; M. Chevion *et al.*, results to be published).

The closely similar ultraviolet spectra of divicine and isouramil are characterized by a peak at 280 nm (reduced form), shifting upon exposure to oxygen to a less prominent absorption maximum at 255 nm (oxidized form) (Razin *et al.*, 1968). The disappearance of the characteristic peak and the concomitant increase

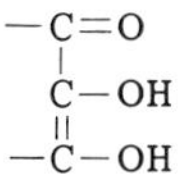

$$\begin{array}{l} -\mathrm{C}{=}\mathrm{O} \\ \quad | \\ \ \ \mathrm{C}-\mathrm{OH} \\ \quad \| \\ -\mathrm{C}-\mathrm{OH} \end{array} \qquad\qquad \begin{array}{l} -\mathrm{C}{=}\mathrm{O} \\ \quad | \\ \ \ \mathrm{C}-\mathrm{OH} \\ \quad \| \\ -\mathrm{C}-\mathrm{NH_2} \end{array}$$

A 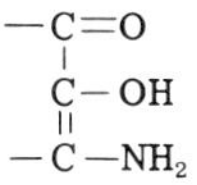 B

in nonspecific ''end absorption'' occurring in the course of the above-mentioned oxidative decomposition of these compounds are suggestive of the rupture of the pyrimidine ring structure.

Bendich and Clements (1953) pointed out in their monumental study that some of the distinctive features of divicine and its cogeners, namely, their powerful reducing properties, spectral characteristics, and molecular instability, are strikingly similar to those of ascorbic acid. Furthermore, these authors inferred from a study of different substitutions that the common structural denominator underlying these properties is a carbonyl-conjugated enediol (**A**) or aminoenol (**B**) system. Consequently, all the characteristic properties of the aglycones are abolished by substitution of the hydroxyl group at C-5, such as that represented by the glycosidic linkage present in vicine and convicine. In fact, these glycosides show none of the reducing properties of their aglycones, are remarkably heat stable in solution, and their ultraviolet spectra differ significantly from those of their constituent pyrimidines (Bendich and Clements, 1953; Bien *et al.*, 1968; S. Bien, M. Noam, G. Glaser, A. Razin, and J. Mager, unpublished results).

E. Effects of Divicine and Isouramil on Red Cell Metabolism *in Vitro*: Synergistic Interaction of Isouramil and Ascorbic Acid

Incubation of human red cell suspensions in phosphate-buffered isotonic saline (pH 7.4), supplemented with isouramil or divicine (to be referred to as aglycones), resulted in a rapid fall of their GSH level, followed by a slower decline of their ATP content (Mager *et al.*, 1965). Addition of glucose prevented the injurious action of the aglycones on normal erythrocytes but had no protective influence on G6PD-deficient cells. This behavior of the aglycones contrasted with the virtual inertness displayed in the same system by convicine and vicine, in which the readily oxidizable enolic hydroxyl group at C-5 of the pyrimidine moiety is blocked by the β-glycosidic bond. The activity of the aglycones, as gauged by their effects on the GSH and ATP contents of the erythrocytes, was roughly 20–30 times higher than that of acetylphenylhydrazine (APH). In contrast, the structurally related pyrimidine, dialuric acid, was relatively ineffective in the test system used, presumably due to its pronounced tendency to interact with GSH by producing an addition compound with a characteristic absorption peak at 305 nm (Patterson *et al.*, 1949).

In contradistinction to APH and primaquine, which require the presence of hemoglobin for catalyzing GSH oxidation (Beutler *et al.*, 1957), the fava pyrimidine aglycones were able to oxidize GSH in pure solution. The oxidation of GSH by these compounds proceeded to completion even at a tenfold or higher molar ratio of GSH to pyrimidine; it was considerably enhanced by shaking in air and completely suppressed in an atmosphere of nitrogen. The nonstoichiometric nature, as well as the oxygen dependence of the oxidation of GSH by the aglycones, suggested that this reaction is mediated by a catalytic oxidoreduction mechanism, similar to that of the well-known GSH–ascorbic acid redox system (Borsook *et al.*, 1937). This analogy is further emphasized by the observation that sodium ascorbate at relatively high concentrations (5–10 n*M*) was capable of producing a marked decrease (50–80%) in the GSH content of G6PD-deficient erythrocytes incubated in phosphate-buffered saline solution (pH 7.4) for 3 hr at 37°C (see also Waller and Benöhr, 1973; Prins and Loos, 1969. Moreover, a mixture of 0.05 m*M* isouramil and 0.5 m*M* ascorbate caused an approximately 70% decline of the intracellular GSH level under similar incubation conditions, whereas the same amounts of each compound alone were totally ineffective in this system. Essentially similar results were obtained with normal human or rabbit erythrocytes, provided that the incubation was carried out in the absence of glucose. The supraadditive nature of the combined affect of isouramil and ascorbate on erythrocyte GSH was shown to represent a net outcome of a series of reactions, namely, oxidation of GSH by its direct interaction with the rapidly autoxidizable isouramil and the concomitantly generated H_2O_2, as well as indirectly by a cyclic system involving oxidation of ascorbate to dehydroascorbate by isouramil and the renewed reduction of dehydroascorbate to ascorbate by GSH (Razin *et al.*, 1968). It is not unlikely that such a synergistic interrelationship may play a crucial role in the pathogenesis of favism.

In a subsequent series of experiments (G. Glaser *et al.*, results to be published) designed to test the effect of isouramil on red cell survival, glucose-deprived rabbit erythrocytes were labeled with ^{51}Cr and incubated in phosphate-buffered saline at 37°C for different periods of time in the presence of various amounts of isouramil. The cells were then reinfused into the donor rabbits, and their ^{51}Cr half-life was determined. It was found that incubation of the erythrocytes with 2 m*M* isouramil for 75 min resulted in a reduction of their ^{51}Cr half-life to less than 5% of the normal value (9 hr and 9 days, respectively). Lower levels of isouramil or shorter incubation periods gave rise to correspondingly less pronounced impairment of the red cell survival. Furthermore, here again the effect of isouramil was found to be synergistically enhanced by addition of ascorbate.

Shortly after the publication of our study (Mager *et al.*, 1965), the earlier investigations of Lin and Ling (1962a,b,c) on the possible relation of vicine to favism, came to our attention through summaries appearing in *Chemical Abstracts* in 1966. The Formosan workers (Lin and Ling, 1962b) observed a transient hemoglobinuria occurring in puppies 3 hr after the oral administration of vicine (0.2 gm/kg body weight), which was isolated from *Vicia faba* by a modified procedure of Levene (1914; Lin and Ling 1962a). These authors found also that vicine exerts some minor inhibitory effects on the activities of glucose-6-phosphate and 6-phosphogluconate dehydrogenases in human red cell hemolysates (Lin and Ling, 1962c). However, the significance of these effects in the pathogenesis of favism is rather doubtful.

In another paper, Lin (1963) described some chemical interactions between divicine and sulfhydryl compounds. With an excess of GSH relative to divicine, a maximal absorption appeared at 305 nm, whereas with a mixture of divicine and cysteine two peaks, at 285 and 245 nm, were observed. In both instances, the amounts of sulfhydryl compounds that disappeared were severalfold higher, on a molar basis, than the quantity of divicine added. In our opinion, the complex absorbing at 305 nm plays no essential part in mediating the catalytic oxidation of GSH by divicine, but rather represents a side reaction analogous to that observed with alloxan (Patterson *et al.*, 1949). Furthermore, according to our data, the spectrum of the pyrimidine–cysteine mixture appears to be determined by the molecular structure of divicine per se stabilized by the reducing action of cysteine (Razin *et al.*, 1968).

F. Possible Etiological Role of Divicine and Isouramil in Favism

The overall pattern of metabolic disturbances resulting from incubation of G6PD-deficient red cells with the aglycones of vicine and convicine is essentially identical to that elicited by treatment with acetylphenylhydrazine (Beutler *et al.*, 1957; Mager *et al.*, 1965). The powerful capacity for oxidizing GSH exhibited by the pyrimidine aglycones *in vitro* as well as the observed deleterious effect of isouramil on red cell survival are consistent with a possible causative role of these substances in precipitating the favic crises. The free aglycones may arise from the parent fava glycosides either in the beans or in the digestive tract through the hydrolytic action of β-glucosidase. The conceivable vicissitudes in the availability of the requisite conditions for enzymatic release of the aglycones from the glycosides, as well as the particular lability of these compounds, might account for the puzzling irregularity that characterizes the occurrence of favism in susceptible individuals, irrespective of the degree and frequency of their exposure to the noxious agent (see Luisada, 1941).

It may be pertinent to mention in this connection that divicine has been in the past named as the causative agent of neurolathyrism (Anderson *et al.*, 1925) because of its parenteral toxicity to experimental animals (Kleiner, 1912). This view, however, was discredited by the finding that oral administration of divicine at a level as high as 1% of the diet to rats (Lee, 1950) and chicks (Arscott and Harper, 1963) produced no adverse effects other than growth retardation. As pointed out by Liener (1966), it was probably because of this lack of specific toxicity of divicine by oral route that its possible significance as the causative principle of favism has escaped the attention of the earlier investigators in this field.

G. Effect of 3,4-Dihydroxyphenylalanine (Dopa) on Erythrocyte GSH and Critical Evaluation of Its Postulated Role in the Etiology of Favism

Kosower and Kosower (1967) put forward the hypothesis that 3,4-L-dihydroxyphenylalanine (dopa) may be one of the active principles responsible for the ability of fava beans to induce hemolysis in G6PD-deficient individuals. This substance, known to be a moderately strong reducing agent, is present in broad beans in substantial amounts [about 0.25% of the fava pods (Guggenheim, 1913)], mainly in the free state and partly in the form of its β-glycosidic derivative (Pridham and Saltmarsh, 1963; Andrews and Pridham, 1965). Kosower and Kosower (1967) found that significant losses of GSH occurred in G6PD-deficient erythrocytes, when incubated at 37°C for 3 hr in a glucose-containing medium supplemented with dopa in amounts ranging from 0.75 to 3 μmoles/ml. In contrast, oxidation of GSH by dopa in normal red blood cells was demonstrable only in the absence of glucose.

Careful scrutiny of the data of Kosower and Kosower reveals that the amounts of GSH that disappeared (were oxidized) were related to the amounts of dopa added by a roughly 1:10 molar ratio. Thus, contrary to the authors' claim, these results do not seem to support the notion of a nonstoichiometric (catalytic) oxidation of GSH by dopa. The latter conclusion is also in line with our observation that no appreciable oxidation of GSH took place when a mixture of GSH (2 m*M*) and dopa (4 m*M*) in 0.01 *M* phosphate buffer (pH 7.4) was incubated for 30 min at 37°C with continuous shaking in air. Moreover, comparative experiments showed that dopa at a concentration as high as 10 m*M* failed to affect the GSH level in normal washed human erythrocytes incubated for 3 hr at 37°C in the absence of added glucose, whereas 1 m*M* isouramil caused almost complete dissappearance of the intracellular GSH under the same conditions. On the other hand, combined addition of 1 m*M* dopa and 0.2 m*M* isouramil resulted in nearly 80% destruction of the erythrocytic GSH, whereas each compound alone was without perceptible effect (Razin *et al.*, 1968).

Similar results were obtained in experiments performed with glucose-starved rabbit erythrocytes. In addition, however, it was found that, although dopa potentiated the oxidant action of isouramil on the intracellular GSH, it failed to enhance the effect of isouramil in shortening the survival of ^{51}Cr-tagged erythrocytes treated *in vitro* and reinfused into the donor rabbit (G. Glaser *et al.*, results to be published).

More recently, the hypothesis of Kosower and Kosower (1967 was endorsed by Beutler (1970) and modified to suggest that dopaquinone, the oxidation product of dopa, rather than dopa itself is the active factor responsible for the fava bean-induced hemolysis. This claim is based on the observation that a mixture of dopa (0.15 m*M*) and tyrosinase (presumed to generate dopaquinone) caused a rather inconspicuous decrease of the GSH content in G6PD-deficient but not in normal erythrocytes. However, the postulated role of dopa in the pathogenesis of favism was not borne out by the outcome of an experiment *in vivo* indicating that the survival of ^{51}Cr-labeled G6PD-deficient erythrocytes transfused into a normal individual was not impaired by repeated intravenous administration of dopa to the recipient (Gaetani *et al.*, 1970). Furthermore, as pointed out by Beutler himself, since L-dopa is being used extensively in rather large doses for treating Parkinson's disease, some of the Parkinsonian patients (with coexistent G6PD-deficiency) in favism-prone areas would be at risk of developing hemolytic crises in the course of therapy. To our knowledge, however, so far no single case has been reported in the literature to substantiate this expectation.

III. THE MECHANISM OF THE BIOCHEMICAL LESION UNDERLYING RED CELL DESTRUCTION IN DRUG-INDUCED HEMOLYSIS AND FAVISM

The hemolytic effect of the noxious drugs on G6PD-deficient erythrocytes appears to be attributable to their ability to function as reversible redox systems mediating the oxidation of the intracellular GSH (Emerson *et al.*, 1949; Beutler *et al.*, 1957). This property is shared also by divicine and isouramil, the pyrimidine aglycones of the fava bean glycosides vicine and convicine. In the normal red blood cell the oxidant effect of the drug is readily overcome by the coordinate action of the NADPH-generating pentose phosphate pathway and the NADPH-linked GSSG-reductase according to the following reaction scheme: [Eqs. (1)–(3)]:

$$\text{Glucose 6-phosphate} + \text{NADP}^+ \xrightarrow{\text{dehydrogenase}} \text{6-phosphogluconate} + \text{NADPH} + \text{H}^+ \tag{1}$$

$$\text{6-Phosphogluconate} + \text{NADP}^+ \xrightarrow{\text{dehydrogenase}} \text{ribulose 5-phosphate} + \text{NADPH} + \text{H}^+ + \text{CO}_2 \tag{2}$$

$$\text{GSSG} + \text{NADPH} + \text{H}^+ \xrightarrow{\text{reductase}} \text{2GSH} + \text{NADP}^+ \tag{3}$$

$$\text{2GSH} + \text{H}_2\text{O}_2 \xrightarrow{\text{GSH peroxidase}} \text{GSSG} + \text{2H}_2\text{O} \tag{4}$$

Under physiological conditions, the vestigial G6PD activity and perhaps also the limited capacity of GSSG reductase to use NAD as an alternate hydrogen donor (Francoeur and Denstedt, 1954) enable the enzyme-deficient erythrocyte to maintain an adequate level of GSH compatible with a nearly normal or moderately reduced survival (Brewer *et al.*, 1961). This precarious metabolic equilibrium, however, breaks down under the stress conditions imposed by the oxidant compound. The resultant irreversible oxidation of GSH and the attendant catabolism of GSSG (Beutler, 1957) seem to constitute the major metabolic lesion leading to the eventual destruction of the enzyme-deficient erythrocyte. The validity of this concept, implying a vital role of GSH in preserving the structural integrity of the red blood cells, is strongly supported by the finding that the virtual absence of GSH in the blood cells of individuals affected with an inborn defect of its biosynthesis predisposes them to drug-induced hemolysis and favism (Oort *et al.*, 1961; Waller and Gerok, 1964; Boivin and Galand, 1965; Prins *et al.*, 1966; Minnich *et al.*, 1971). Similarly, congenital GSSG-reductase deficiency likewise manifests itself by drug sensitivity (Loehr and Waller, 1962; Waller *et al.*, 1965, 1969).

A major manifestation of the oxidant action of the drugs both *in vivo* and *in vitro* is the formation of methemoglobin and the concomitant appearance of Heinz bodies, which, according to Allen and Jandl (1961), represent a product of hemoglobin denaturation resulting from oxidation of its SH groups with concurrent formation of a mixed glutathione disulfide and loss of the heme group (see also Srivastava and Beutler, 1970; Bunn and Jandl, 1966; Jacob and Winterhalter, 1970; Jacob, 1970; Rachmilewitz *et al.*, 1969; Nagel and Ranney, 1973).

Cohen and Hochstein (1961, 1963, 1964) indicated that the oxidant drug or its active metabolite interacts with oxyhemoglobin, producing hydrogen peroxide. The relatively low but potentially harmful levels of peroxide cannot be efficiently destroyed by catalase and are normally eliminated through the action of GSH peroxidase (Mills, 1957, 1959, 1960; Mills and Randall, 1958), which catalyzes the following reaction [Eq. (4)]:

$$2\text{GSH} + \text{H}_2\text{O}_2 \xrightarrow{\text{GSH-peroxidase}} \text{GSSG} + 2\text{H}_2\text{O} \tag{4}$$

The sustained operation of this system is ensured by the concomitant regeneration of GSH, mediated by the NADPH-linked GSSG-reductase. Thus, the integrated pathway consisting of the oxidative pentose phosphate shunt (as a source of NADPH supply), GSSG-reductase, and GSH-peroxidase serves to detoxify the hydrogen peroxide, so as to obviate its deleterious effects on the red cell membrane and hemoglobin (see also Cohen, 1966; Flohé and Brand, 1969).

This concept, assigning a vital function to GSH-peroxidase in protecting the cell from the oxidative insult by the peroxide-forming drugs, has gained additional support from the recognition of a hereditary deficiency of this enzyme and its causative role in certain cases of spontaneous or drug-induced hemolytic anemia (T. Necheles *et al.*, 1968; Steinberg *et al.*, 1970; T. F. Necheles *et al.*,

1970; Boivin *et al.,* 1969, 1970). Furthermore, the recent discovery that GSH-peroxidase contains selenium as an integral and catalytically essential component of its molecule (Rotruck *et al.,* 1973; Flohé *et al.,* 1973) has led to the understanding of the biochemical mechanism underlying the protective effect of dietary selenium against hydrogen peroxide-induced hemolysis (Rotruck *et al.,* 1972).

Jacob and Jandl (1962a,b), in studying the effects of SH-binding compounds (*p*-hydroxymercuribenzoate, *N*-ethylmaleimide) on erythrocytes, emphasized the essentiality of the surface sulfhydryl groups for the structural intactness and normal survival of these cells. It should be pointed out, however, that contrary to the typical thiol reagents used in the above studies, primaquine and related drugs (Panizon and Zacchello, 1966; Beutler, 1966), as well as the fava bean pyrimidine aglycones (Mager *et al.,* 1965), do not induce an overt lysis *in vitro* but appear to exert their deleterious effect *in vivo* by rendering the red cells vulnerable to destruction by the reticuloendothelial system in the liver and spleen (Rifkind, 1965, 1966; Beutler, 1971). Consequently, it is not clear to what extent the conclusions drawn from the model experiments of Jandl and his associates are applicable to drug-induced hemolysis and favism.

Some investigators (Kosower *et al.,* 1969; Flohé *et al.,* 1971) suggested that in certain instances free radicals, rather than H_2O_2, generated in the course of the metabolism of the noxious agent, may interact with GSH and protein thiols with the resultant formation of GSSG and protein S—S linkages.

Other studies have been concerned with the possible derangement of the energy-yielding metabolism as part of the mechanism underlying the drug-induced red cell hemolysis. It was observed by several authors that aerobic incubation of erythrocyte suspensions in the presence of primaquine or acetylphenylhydrazine resulted in a pronounced inhibition of glycolysis (Loehr and Waller, 1961; Kosower *et al.,* 1964) and a progressive decrease in the ATP level of the cells (Mohler and Williams, 1961; Loehr and Waller, 1961; Mager *et al.,* 1964). Essentially similar effects were obtained on incubating the red cells in the presence of divicine or isouramil (Mager *et al.,* 1965). Addition of glucose obviated the deleterious effects of the drugs in normal but not in G6PD-deficient erythrocytes. The primary site of the antimetabolic action of APH was traced to hexokinase (Kosower *et al.,* 1964; Mager *et al.,* 1964). Furthermore, the APH-induced inhibition of hexokinase was shown to be mediated by GSSG formation, thus representing a particular case of so-called disulfide poisoning previously described by Eldjarn and Bremer (1962).

While the potential significance of hexokinase inhibition in shortening the life span of the red cells need scarecely be elaborated, it remains to be seen whether this metabolic derangement plays an essential part in the actual mechanism of red cell destruction occurring in drug-induced hemolysis and favism (see Brewer *et al.,* 1964).

A number of enzymatic alterations in G6PD-deficient erythrocytes were reported to occur independently of their exposure to noxious drugs (Schrier *et al.*, 1958, 1959; Larizza *et al.*, 1958). Observations on the decrease in the activities of NADPH-diaphorase (Jaffé, 1963), phosphomonoesterase (Oski *et al.*, 1963; Bottini and Modiano, 1965), and pyrophosphatase (Scheuch *et al.*, 1961; Brunetti *et al.*, 1962a,b) appear to be of particular interest. The reduced activity of these enzymes, which are known to be SH dependent, may be due to the inclement environment created by the diminished GSH level. Furthermore, it has been shown that GSSG inhibits the activity of a variety of enzymes, such as glucose-6-phosphate dehydrogenase, inorganic pyrophosphatase, triosephosphate dehydrogenase (Scheuch and Rapoport, 1962), hexokinase (Eldjarn and Bremer, 1962; Mager *et al.*, 1964), and ATPase (Kutscher, 1961).

IV. CONCLUDING REMARKS

The data reviewed in this chapter clearly indicate that, despite the considerable progress achieved in the research on favism, there are still serious gaps in our understanding of the pathogenesis of this disease. Particularly perplexing is the inadequacy of our current knowledge to account for the sporadic and rather capricious incidence of favism, as well as the absence of a clear-cut correlation between the degree of exposure of the susceptible individuals to the noxious principle of the fava plant and the occurrence of the hemolytic syndrome. It appears reasonable to surmise that the epidemiology of favism is governed not only by genetic factors, but also by a number of environmental determinants, such as variations in the amount of the toxic principle present in different varieties of the *Vicia faba* plant, as well as differences in the food and cooking habits of the G6PD-deficient subjects in the various populations.

The major and most urgent issue, however, is the definitive establishment of the chemical identity of the causative agent of favism. It is obvious that the achievement of this goal would greatly contribute to the elucidation of other aspects of this disease and might also provide a rational basis for a prophylactic and therapeutic approach.

ACKNOWLEDGMENTS

Part of the original work of the authors referred to in the text and the preparation of this chapter were supported by Grant P15/181/16 from the World Health Organization and by an award from the Chief Scientists' Office of the Ministry of Health, Israel.

REFERENCES

Aebi, H. E. (1967). *Annu. Rev. Biochem.* **36**, 271.

Allen, D. W., and Jandl, J. H. (1961). *J. Clin. Invest.* **40**, 454.

Allison, A. C., and Clyde, D. F. (1961). *Br. Med. J.* **1**, 1346.
Allison, A. C., Charles, L. J., and McGregor, I. A. (1961). *Nature (London)* **190**, 1198.
Amin-Zaki, L., El-Din, S., and Kubba, K. (1972). *Bull. W. H. O.* **47**, 1.
Anderson, L. A. P., Howard, A., and Simonsen, J. L. (1925). *Indian J. Med. Res.* **12**, 613.
Andrews, R. S., and Pridham, J. B. (1965). *Nature (London)* **205**, 1213.
Angelov, A., and Andrev, I. (1959). *Vop. Pediat. Akus. Ginekol.* No. 2, p. 7 (cited after Belsey, 1973).
Arscott, G. H., and Harper, J. A. (1963). *J. Nutr.* **80**, 251.
Auquier, L., Paolaggi, J. B., and Dastugue, B. (1968). *Sem. Hop.* **44**, 2037.
Aurichio, L. (1935). *Rass. Clin.-Sci.* **13**, 20
Belsey, M. A. (1973). *Bull. W. H. O.* **48**, 1–13.
Bendich, A., and Clements, G. C. (1953). *Biochim. Biophys. Acta* **12**, 462.
Beutler, E. (1957). *J. Lab. Clin. Med.* **49**, 84.
Beutler, E. (1970). *Blood* **36**, 523.
Beutler, E. (1971). *Semin. Hematol.* **8**, 311.
Beutler, E. (1972). *In* "The Metabolic Basis of Inherited Disease" (J. B. Stanbury, J. B. Wyngaarden, and D. B. Fredrickson, eds.), 3rd ed., p. 1358. McGraw-Hill, New York.
Beutler, E., and Baluda, M. C. (1963). *Blood* **22**, 323.
Beutler, E., and Mitchell, M. (1968). *Blood* **32**, 816.
Beutler, E., Dern, R. J., Flanagan, C. L., and Alvin, A. S. (1955). *J. Lab. Clin. Med.* **45**, 286.
Beutler, E., Robson, M., and Buttenwieser, E. (1957). *J. Clin. Invest.* **36**, 617.
Beutler, E., Yeh, M., and Fairbanks, V. F. (1962) *Proc. Natl. Acad. Sci. U.S.A.* **48**, 9.
Beutler, E., Duron, O., and Kelly, B. M. (1963) *J. Lab. Clin. Med.* **61**, 882.
Bien, S., Salemnik, G., and Zamir, L. (1968). *J. Chem. Soc.* **C** p. 496.
Bogair, N. (1951). *Harofeh Haivri* **24**, 72 (in Hebrew).
Boivin, P., and Galand, C. (1965). *Nouv. Rev. Fr. Hematol.* **5**, 707.
Boivin, P., Rogé, G., and Guéroult, N. (1969). *Enzymol. Biol. Clin.* **10**, 68.
Boivin, P., Hakim, J., and Blery, M. (1970). *Presse Med.* **78**, 171.
Borsook, H., Davenport, H. W., Jeffreys, C. E. P., and Warner, R. C. (1937). *J. Biol. Chem.* **117**, 237.
Bottini, E. (1973). *J. Med. Genet.* **10,** 154.
Bottini, E., and Modiano, G. (1965). *Acta Paediatr. Lat.* **18,** 242.
Bottini, E., Lucarelli, P., Spennati, G. F., Businov, L., and Palmarino, R. (1970). *Clin. Chim. Acta* **30**, 831.
Bottini, E., Lucarelli, P., Agostino, R., Palmarino, R., Businco, L., and Antognosi, G. (1971). *Science* **171**, 409.
Bowman, J. E., and Walker, D. G. (1961). *Nature (London)* **189,** 555.
Boyer, S. H., and Graham, J. B. (1965). *Am. J. Hum. Genet.* **17,** 320.
Boyer, S. H., Porter, I. H., and Weilbacher, R. G. (1962). *Proc. Natl. Acad. Sci. U.S.A.* **48**, 1868.
Brewer, G. J., Tarlov, A. R., and Kellermeyer, R. W. (1961). *J. Lab. Clin. Med.* **58**, 217.
Brewer, G. J., Tarlov, A. R., and Alving, A. S. (1962). *J. Am. Med. Assoc.* **180**, 386.
Brewer, G. J., Powell, R. D., Swanson, S. H., and Alving, A. S. (1964). *J. Lab. Clin. Med.* **64**, 601.
Brunetti, P., Rosetti, R., and Broccia, G. (1960). *Rass. Fisiopatol. Clin. Ter.* **32**, 338.
Brunetti, P., Grignani, F., and Ernisli, G. (1962a). *Acta Haematol.* **27**, 146.
Brunetti, P., Grignani, F., and Ernisli, G. (1962b). *Acta Haematol.* **27**, 246.
Bunn, H. F., and Jandl, J. H. (1966) *Proc. Natl. Acad. Sci. U.S.A.* **56**, 974.
Carcassi, U. (1958). "Eritroenzimopatie e anemie emolitiche." Omnia Med., Pisa.
Carson, P. E. (1960). *Fed. Proc., Fed. Am. Soc. Exp. Biol.* **19**, 995.
Carson, P. E., Flanagan, C. L., Ickes, C. E., and Alving, A. S. (1956). *Science* **124**, 484.

Chan, T. K., Todd, D., and Wong, C. C. (1964). *Br. Med. J.* **2**, 102.

Chan, T. K., Todd, D., and Wong, C. C. (1965). *J. Lab. Clin. Med.* **66**, 937.

Charlton, R. W., and Bothwell, T. H. (1959). *S. Afr. J. Med. Sci.* **24,** 88.

Chesterfield, J. H., Hurst, D. T., McOmie, J. F. W., and Tute, M. S. (1964). *J. Chem. Soc.* p. 1001.

Chevion, M., Navok, T., Glaser, G., and Mager, J. (1979). To be published.

Childs, B., Zinkham. W., Browne, E. A., Kimbro, E. L., and Torbert, J. V. (1958). *Bull. Johns Hopkins Hosp.* **102**, 21.

Chung, S. F. (1965). *Paediatr. Indonesia* **5,** Suppl., 880.

Cohen, G. (1966). *Biochem. Pharmacol.* **15**, 1775.

Cohen, G., and Hochstein, P. (1961). *Science* **134**, 1756.

Cohen, G., and Hochstein, P. (1963). *Biochemistry* **2**, 1420.

Cohen, G., and Hochstein, P. (1964). *Biochemistry* **3**, 895.

Contu, J., Pitzus, F., Lenzerini, L., and Marcolongo, R. (1961). *Rass. Fisiopatol. Clin. Ter.* **33**, 42.

Creger, W. P., and Gifford, H. (1952). *Blood* **7**, 721.

Crosby, W. H. (1956). *Blood* **11**, 91.

Dacie, J. V. (1954). "The Hemolytic Anemias," p. 354. Churchill, London.

Davies, P. (1962). *Q. J. Med.* **31**, 157.

Davoll, J., and Laney, D. H. (1956). *J. Chem. Soc.* p. 2124.

Dawson, J. P., Thayer, W. W., and Desforges, J. F. (1958). *Blood* **13**, 1113.

Dern, R. J., Weinstein, I. M., Leroy, G. V., Talmage, D. W., and Alving, A. S. (1954). *J. Lab. Clin. Med.* **43**, 303.

Dickson, L. G., Johnson, C. B., and Johnson, D. R. (1973). *Clin. Chem.* **19**, 301.

Donoso, G., Hedayat, H., and Khayatyan, H. (1969). *Bull. W. H. O.* **40**, 513.

Du, S. D. (1952). *Chin. Med. J.* **70**, 17.

Efrati, P. (1952). *Dapim Refu.* **11**, 134 (in Hebrew).

Eldjarn, L., and Bremer, J. (1962). *Biochem. J.* **84**, 286.

Ellman, G. L. (1959). *Arch. Biochem. Biophys.* **82**, 70.

Emerson, C. P., Ham, T. H., and Castle, W. B. (1949). *In* "Conference on the Preservation of Formed Elements and of the Proteins of the Blood" (C. J. Potthoff, ed.), p. 115. Am. Natl. Red Cross, Washington, D.C.

Fairbanks, V. F., and Beutler, E. (1962) *Blood* **20**, 591.

Fermi, C., and Martinetti, P. (1905). *Ann. Ig.* **15**, 75.

Flohé, L., and Brand, J. (1969). *Biochim. Biphys. Acta* **191**, 541.

Flohé, L., Niebch, G., and Reiber, H. (1971). *Z. Klin. Chem. Klin. Biochem.* **9**, 431.

Flohé, L., Günzler, W. A., and Schocke, H. H. (1973). *FEBS Lett.* **32**, 132.

Francoeur, M., and Denstedt, O. F. (1954). *Can. J. Biochem. Physiol.* **32**, 663.

Gaetani, G., Salvidio, E., Panaciulli, I., Ajmar, F., and Paravidino, G. (1970). *Experientia* **26**, 785.

Gasbarrini, A. (1915). *Policlinico, Sez. Prat.* **22**, 1505.

Gehrmann, von G., Sturm, A., and Amelung, D. (1963). *Dtsch. Med. Wochenschr.* **88**, 1865.

Gelpi, A. P. (1965). *Blood* **25,** 486.

Gilles, H. M., Watson-Williams, J., and Taylor, B. G. (1960). *Nature (London)* **185**, 257.

Glaser, G., Weissberg, J., Chevion, M., and Mager, J. (1979). To be published.

Greenberg, M. S., and Wong, H. (1961). *J. Lab. Clin. Med.* **57**, 733.

Guggenheim, M. (1913). *Z. Phys. Chem.* **88**, 276.

Hassan, M. M. (1971). *J. Trop. Med. Hyg.* **74**, 187.

Hedayat, Sh., Rahbar, S., Mahbooli, E., Ghaffarpour, M., and Sobhi, N. (1971). *Trop. Geogr. Med.* **23**, 149–157.

Higazi, M. L., and Read, W. W. C. (1974). *J. Agric. Food. Chem.* **22,** 570.

Ikeda, K., Sumi, T., Yokoi, K., and Mizuno, Y. (1973). *Chem. Pharm. Bull.* **21**, 1327.

Jacob, H. S. (1970). *Semin. Hematol.* **7**, 341.
Jacob, H. S., and Jandl, J. H. (1962a). *J. Clin. Invest.* **41**, 779.
Jacob, H. S., and Jandl, J. H. (1962b). *J. Clin. Invest.* **41**, 1514.
Jacob, H. S., and Jandl, J. H. (1966). *N. Engl. J. Med.* **274**, 1162.
Jacob, H. S., and Winterhalter, K. (1970). *Proc. Natl. Acad. Sci. U.S.A.* **65**, 697.
Jaffé, E. R. (1963). *Blood* **21**, 561.
Johannsen, L. P., Witt, I., and Künzer, W. (1968). *Dsch. Med. Wochenschr.* **93**, 2463.
Johnson, T. B. (1914). *J. Am. Chem. Soc.* **36**, 337.
Kantor, S. Z., and Arbesman, L. (1959). *J. Allergy* **30**, 114.
Kantor, S. Z., Pinkhas, J., and Djaletti, M. (1962). *J. Allergy* **33**, 390.
Kattamis, C. A. (1969). *Ann. Soc. Belg. Med. Trop.* **49**, 289.
Kattamis, C. A. (1971). *Arch. Dis. Child.* **46**, 741.
Kattamis, C. A., Kyriazakon, M., and Chaidas, S. (1969). *J. Med. Genet.* **6**, 34.
Kidson, C., and Gajdusek, D. C. (1962). *Aust. J. Sci.* **25**, 61.
Kidson, C., and Gorman, J. G. (1962). *Am. J. Phys. Anthropol.* **20**, 357.
Kirkman, H. N., and Hendrickson, E. M. (1963). *Am. J. Hum. Genet.* **15**, 241.
Kirkman, H. N., McGurdy, P. R., and Naiman, I. L. (1964). *Cold Spring Harbor Symp. Quant. Biol.* **29**, 361.
Kleiner, I. S. (1912). *J. Biol. Chem.* **11**, 443.
Kosower, N. S., and Kosower, E. M. (1967). *Nature (London)* **215**, 286.
Kosower, N. S., Vanderhoff, G. A., and London, I. M. (1964). *Nature (London)* **201**, 684.
Kosower, N. S., Song, K. R., and Kosower, E. M. (1969). *Biochim. Biophys. Acta* **192**, 15.
Kutscher, H. (1961). *Folia Haematol. (Leipzig)* **78**, 360.
Langley, G. B., Todd, F. R., and Bishop, A. J. (1969). *Can. Med. Assoc. J.* **100**, 973.
Larizza, P., Brunetti, P., Grignani, F., and Ventura, S. (1958). *Haematologica* **43**, 205.
Larizza, P., Brunetti, P., and Grignani, F. (1960). *Haematologica* **45**, 1, 129.
Lee, J. G. (1950). *J. Nutr.* **40**, 587.
Levene, P. A. (1914). *J. Biol. Chem.* **18**, 305.
Liener, I. E. (1966). *Publ.* **1354**, 47–50.
Lin, J. Y. (1963). *J. Formosan Med. Assoc.* **62**, 777.
Lin, J. Y., and Ling, K. H. (1962a). *J. Formosan Med. Assoc.* **61**, 484; *Chem. Abstr.* **65**, 4143 (1966).
Lin, J. Y., and Ling, K. H. (1962b). *J. Formosan Med. Assoc.* **61**, 490; *Chem. Abstr.* **65**, 4144 (1966).
Lin, J. Y., and Ling, K. H. (1962c). *J. Formosan Med. Assoc.* **61**, 579.
Loehr, G. W., and Waller, H. D. (1961). *Dsch. Med. Wochenschr.* **86**, 27.
Loehr, G. W., and Waller, H. D. (1962). *Med. Klin. (Munich)* **36**, 1521.
Luisada, A. (1941). *Medicine (Baltimore)* **20**, 229.
Luzatto, L., Usanga, E. A., and Reddy, S. (1969). *Science* **164**, 839.
Lyon, M. F. (1961). *Nature (London)* **190**, 372.
McOmie, J. F. W., and Chesterfield, J. H. (1956). *Chem. Ind. (London)* **2**, 1453.
Mager, J., Razin, A., Hershko, A., and Izak, G. (1964). *Biochem. Biophys. Res. Commun.* **17**, 703.
Mager, J., Glaser, G., Razin, A., Izak, G., Bien, S., and Noam, M. (1965). *Biochem. Biophys. Res. Commun.* **20**, 235.
Manai, A. (1929). "Il favismo." Edit. Stamp. Libreria Ital. e Stran., Sassari.
Marcolongo, F., Carcassi, U., and Cusudda, U. (1950). *Sci. Med. Ital.* **1**, 625.
Marks, P. A., and Banks, J. (1965). *Ann. N. Y. Acad. Sci.* **123**, 198.
Marks, P. A., and Gross, R. T. (1959). *J. Clin. Invest.* **38**, 2253.
Marks, P. A., Gross, R. T., and Hurwitz, R. E. (1959). *Nature (London)* **183**, 1266.
Mela, C., and Perona, G. P. (1959). *Boll. Soc. Ital. Biol. Sper.* **35**, 146.

Mills, G. C. (1957). *J. Biol. Chem.* **229,** 189.
Mills, G. C. (1959). *J. Biol. Chem.* **234,** 502.
Mills, G. C. (1960). *Arch. Biochem. Biophys.* **86,** 1.
Mills, G. C., and Randall, H. P. (1958). *J. Biol. Chem.* **232**, 589.
Minnich, V., Smith, M. B., Brauner, M. J., and Majerus, P. W. (1971). *J. Clin. Invest.* **50**, 507.
Mohler, D. N., and Williams, W. J. (1961). *J. Clin. Invest.* **40**, 1753.
Motulsky, A. G. (1960). *Hum. Biol.* **32**, 28.
Motulsky, A. G., and Campbell-Kraut, J. M. (1961). *Proc. Conf. Genet. Polymorphism Geogr. Variations Dis., 1960* p. 159.
Nagel, R. E., and Ranney, H. M. (1973). *Semin. Hematol.* **10**, 269.
Nathan, R. D., Pachtman, E. A., Fiorelli, G., and Frumin, A. M. (1974). *Am. J. Clin. Pathol.* **61**, 462.
Necheles, T. F., Boles, T., and Allen, D. (1968). *J. Pediat.* **72**, 319.
Necheles, T. F., Steinberg, M. H., and Cameron, D. (1970). *Br. J. Haematol.* **19**, 605.
Oort, M., Loos, J. A., and Prins, H. K. (1961). *Vox Sang.* **6**, 370.
Oski, F. A., Shahidi, N. T., and Diamond, L. K. (1963). *Science* **139**, 409.
Panizon, F. (1967). *Minerva Pediatr.* **19**, 1391.
Panizon, F., and Pujatti, G. (1957). *Stud. Sassaresi* **35**, 105.
Panizon, F., and Vullo, C. (1961). *Acta Haematol.* **26**, 337.
Panizon, F., and Vullo, C. (1962). *Haematologica* **47**, 205.
Panizon, F., and Zacchello, F. (1965). *Acta Haematol.* **33**, 129.
Panizon, F., and Zacchello, F. (1966). *Atti Soc. Med.-Chir. Padova Fac. Med. Chir. Univ. Padova* **41**, 5.
Pannaciulli, I., Tizianello, A., Ajmar, F., and Salvidio, E. (1965). *Blood* **25**, 92.
Patterson, J. W., Lazarow, A., and Levey, S. (1949). *J. Biol. Chem.* **177**, 197.
Perera, C. B., and Frumin, A. M. (1965). *Bibl. Haematol.* **23,** 589.
Piomelli, S., Corash, L. M., Davenport, D. D., Miraglia, G., and Amorosi, E. L. (1968). *J. Clin. Invest.* **47**, 940.
Plato, C. C., Rucknagel, D. L., and Gershowitz, H. (1964). *Am. J. Hum. Genet.* **16**, 267.
Porter, I. H., Schulze, J., and McKusick, V. A. (1962). *Ann. Hum. Genet.* **26**, 107.
Pridham, J. B., and Saltmarsh, M. J. (1963). *Biochem. J.* **87,** 218.
Prins, H. K., and Loss, J. A. (1969). *In* "Biochemical Methods in Red Cell Genetics" (Y. Y. Yunis, ed.), p. 123. Academic Press, New York.
Prins, H. K., Oort, M., Loos, J. A., Zurcher, C., and Beckers, T. (1966). *Blood* **27**, 145.
Rachmilewitz, E. A., Peisach, J., Bradley, T. B., Jr., and Blumberg, W. E. (1969). *Nature (London)* **222**, 248.
Ragab, A. H., El-Alfi, O. S., and Abboud, M. A. (1966). *Am. J. Hum. Genet.* **18,** 21.
Razin, A., Hershko, A., Glaser, G., and Mager, J. (1968). *Isr. J. Med. Sci.* **4**, 862.
Rifkind, R. A. (1965). *Blood* **26**, 433.
Rifkind, R. A. (1966). *Am. J. Med.* **41**, 711.
Ritthausen, H. (1876). *Ber. Dtsch. Chem. Ges.* **9,** 301.
Ritthausen, H. (1881). *J. Prakt. Chem.* [N.S.] **24**(2), 202.
Ritthausen, H. (1896). *Ber. Dtsch. Chem. Ges.* **29**, 2108.
Ritthausen, H. (1899a). *J. Prakt. Chem.* **59,** 480.
Ritthausen, H. (1899b). *J. Prakt. Chem.* **59,** 482.
Ritthausen, H., and Kreusler, U. (1870). *J. Prakt. Chem.* [N.S.] **2**(2), 333
Rokicka-Milewska, R., Maj, S., and Jablonska-Skwiecinska, E. (1968). *Pediat. Pol.* **43**, 621.
Roth, K. L., and Frumin, A. M. (1960). *J. Lab. Clin. Med.* **56**, 695.
Rotruck, J. T., Pope, A. L., Ganther, H. E., and Hoekstra, W. C. (1972). *J. Nutr.* **102**, 689.

Rotruck, J. T., Pope, A. L., Ganther, H. E., Swanson, A. B., Hafeman, D. G., and Hoekstra, W. G. (1973). *Science* **179**, 588.

Rożynkowa, D., Gebala, A., and Zagórski, Z. (1970). *Pol. Med. J.* **9**, 70.

Rożynkowa, D., Matwijow, J., and Zagórski, Z. (1971). *Genet. Pol.* **12**, 588.

Russell, B. (1965). "History of Western Philosophy," 2nd ed., p. 50. Allen & Unwi, London.

Ryan, B. P. K., and Parsons, I. C. (1961). *Nature (London)* **192**, 477.

Sansone, G., and Segni, G. (1956). *Boll. Soc. Ital. Biol. Sper.* **32**, 456.

Sansone, G., and Segni, G. (1957a). *Lancet* **2**, 295.

Sansone, G., and Segni, G. (1957b). *Boll. Soc. Ital. Biol. Sper.* **33**, 1057.

Sansone, G., and Segni, G. (1958). *Boll. Soc. Ital. Biol. Sper.* **34**, 327.

Sansone, G., Piga, A. M., and Segni, G. (1958). Il favismo. *Minerva Med.* ,

Sansone, G., Rasore-Quartino, A., and Veneziano, G. (1963). *Pathologica* **4**, 371.

Say, B., Ozand, P., and Berkel, I. (1965). *Acta Paediatr. Scand.* **54**, 319.

Scheuch, D., and Rapoport, S. M. (1962). *Acta Biol. Med. Ger.* **8**, 31.

Scheuch, D., Kahrig, C., Ockel, E., Wagenknecht, L., and Rapoport, S. M. (1961). *Nature (London)* **190**, 631.

Schmitz, K., and Fritz, K. W. (1968). *Med. Welt* **23**, 1419.

Schneer, J. H. (1968). *Acta Haematol.* **40**, 44.

Schrier, S. L., Kellermeyer, R. W., and Alving, A. S. (1958). *Proc. Soc. Exp. Biol. Med.* **99**, 354.

Schrier, S. L., Kellermeyer, R. W., Carson, P. E., Ickes, C. E., and Alving, A. S. (1959). *J. Lab. Clin. Med.* **54**, 232.

Schulze, E. (1891). *Hoppe-Seyler's Z. Physiol. Chem.* **15**, 140.

Shaker, Y., Onsi, A., and Aziz, R. (1966). *Am. J. Hum. Genet.* **18**, 609.

Siniscalco, M., Bernini, L., Latte, B., and Motulsky, A. G. (1961). *Nature (London)* **190**, 1179.

Srivastava, S. K., and Beutler, E. (1970) *Biochem. J.* **119**, 353.

Stamatoyannopoulos, G., Fraser, G. R., Motulsky, A. G., Fessas, P., Akrivakis, A., and Papayonnopoulou, T. (1966). *Am. J. Hum. Genet.* **18**, 253.

Steinberg, M., Brewer, M., and Necheles, T. (1970). *Arch. Intern. Med.* **125**, 302.

Symvoulidis, A., Voudiclaris, S., Mountokalakis, T., and Pougounias, H. (1972). *Lancet* **2**, 819.

Szeinberg, A., and Chari-Bitron, A. (1957). *Acta Haematol.* **18**, 229.

Szeinberg, A., and Sheba, C. (1960). Personal communication to Motulsky (1960).

Szeinberg, A., Sheba, C., Hirshorn, N., and Bodonyi, E. (1957). *Blood* **12**, 603.

Szeinberg, A., Asher, Y., and Sheba, C. (1958a). *Blood* **13**, 348.

Szeinberg, A., Sheba, C., and Adam, A. (1958b). *Nature (London)* **181**, 1256.

Szeinberg, A., Sheba, C., and Adam, A. (1958c). *Blood* **13**, 1043.

Taleb, N., Loiselet, J., and Guorra, F. (1964). *C.R. Hebd. Seances. Acad. Sci.* **258,** 5749.

Tarlov, A. R., Brewer, G. J., Carson, P. E., and Alving, A. S. (1962). *Arch. Intern. Med.* **109**, 209.

U.S. Department of Agriculture (1963). *U.S., Dep. Agric., Agric. Handb.* **8**.

Vella, F. (1959). *Med. J. Austr.* **2**, 196.

Vella, F., and Ibrahim, S. A. (1962). *Sudan M. J.* **1,** 136 (cited after Hassan, 1971).

Vince-Ribarić, V. (1962). *Lijech. Vjesn.* **84**, 151.

Waldron, H. A. (1973). *Br. Med. J.* **2**, 667.

Walker, D. G., and Bowman, J. E. (1959). *Nature (London)* **184**, 1325.

Walker, D. G., and Bowman, J. E. (1960). *Proc. Soc. Exp. Biol. Med.* **103**, 476.

Waller, H. D., and Benöhr, H. C. (1973). *In* "Erythrocytes, Thrombocytes, Leukocytes" (E. Gerlach, ed.), p. 200. Thieme, Stuttgart.

Waller, H. D., and Gerok, W. (1964). *Klin. Wochenschr.* **42**, 948.

Waller, H. D., Loehr, G. W., Zysno, E., Gerok, W., Voss, D., and Strauss, G. (1965). *Klin. Wochenschr.* **43**, 413.

Waller, H. D., Benoehr, H. C., and Waumans, P. (1969). *Klin. Wochenschr.* **47**, 25.
Winterstein, A., and Somló, F. (1933). *In* "Handbuch der Pflanzenanalyse" (G. Klein, ed.), p. 362. Springer-Verlag, Berlin and New York.
Wong, H. B. (1972). *J. Singapore Paediatr. Soc.* **14**, 17.
Yoshida, A. (1966). *J. Biol. Chem.* **241**, 4966.
Yoshida, A. (1967a). *Proc. Natl. Acad. Sci. U.S.A.* **57**, 838.
Yoshida, A. (1967b). *Biochem. Genet.* **1**, 81.
Yoshida, A. (1970). *J. Mol. Biol.* **52**, 483.
Yoshida, A. (1973). *Science* **179**, 532.
Yoshida, A., and Hoagland, V. D., Jr. (1970). *Biochem. Biophys. Res. Commun.* **40**, 1167.
Yoshida, A., and Lin, M. (1973). *Blood* **41**, 877.
Yoshida, A., Stamatoyannopoulos, G., and Motulsky, A. G. (1967). *Science* **155**, 97.
Yoshida, A., Stamatoyannopoulos, G., and Motulsky, A. G. (1968). *Ann. N.Y. Acad. Sci.* **155**, 868.
Yoshida, A., Beutler, E., and Motulsky, A. G. (1971). *Bull. W. H. O.* **45**, 243.
Zannos-Mariolea, L., and Kattamis, C. (1961). *Blood* **18**, 34.
Zinkham, W. H., Lenhard, R. E., Jr., and Childs, B. (1958). *Bull. Johns Hopkins Hosp.* **102**, 169.

CHAPTER 10

Allergens

FRANK PERLMAN

I. INTRODUCTION

Allergens hold a somewhat anomalous position in a discussion of naturally occurring food toxins. The true toxins are undesired constituents of some foods and exhibit their effects on anyone who consumes them. The severity of such toxic effects is roughly proportional to the quantity consumed. On the other hand, allergens are usually normal food constituents, and the abnormality rests in

TOXIC CONSTITUENTS OF PLANT FOODSTUFFS, SECOND EDITION

ISBN 0-12-449960-0

the individual who has an altered reactivity (allergy) to such otherwise innocuous substances. The intensity of the reaction depends on the degree of hypersensitivity of the person consuming the food rather than on the quantity consumed. Thus, an almost infinitesimal amount of food in a highly allergic individual may cause violent symptoms, whereas in less sensitive persons ingestion of a much larger quantity may result in only mild manifestations of allergy. A normal (nonallergic) individual experiences no ill effects after eating these foods.

A discussion of food allergens must include some comments on the clinical disorders commonly produced. Such disorders are the results of an immunological reaction. Therefore, a few brief, pertinent comments are offered on the antibodies involved and on the results of antigen–antibody interaction.

II. CLINICAL DISORDERS

A. General Consideration of Symptoms

The symptoms are numerous and varied, and almost any tissue of the body may act as the "shock organ." Such symptoms therefore are peculiar to the organ involved (Table I). The organs most commonly affected are the skin and the respiratory tract, and these may account for 90% of the disorders of allergy. The gastrointestinal tract is less frequently involved, and the symptoms may simulate acute or chronic medical or surgical disorders. Food allergy affecting the central nervous system may produce such manifestations as "migraine" and even convulsive seizures. On rare occasions allergy to foods may be manifested in the genitourinary tract, in the cardiovascular system, and in special sense organs. Aberrant behavior in patients may be related to such allergic disorders. This is often the result of continued, vague discomfort that is difficult to relate to organic disease by the patient or his physician. In infants and small children, the refusal to eat certain foods or the complaints of vague abdominal distress are sometimes mistaken for food dislikes or willfulness. These misinterpretations may result in domestic conflict and in the development of behavior patterns, which complicate symptoms and even overshadow the underlying food allergy.

B. Factors Influencing Symptoms

Patients may be exposed to allergens in many ways during their normal day-to-day activities. Exposure to allergens may be by simple contact, by inhalation, particularly among food handlers, by injection, particularly of biological products (viral vaccines made from chick embryo), but most frequently by ingestion. There was some controversy as to the capacity of sensitizing the fetus transuterinely (Ratner, 1928). However, in recent years it has been clearly demonstrated that the fetus is capable of producing a specific immunoglobulin of allergy

TABLE I

COMMON DISORDERS ATTRIBUTED TO FOOD ALLERGENS

Shock organ	Disease or symptoms
Skin	
Superficial layer	Eczema
Deeper layer	Urticaria (hives)
Mucous membranes	Angioedema (eye, lip, tongue, throat)
Respiratory tract	
Nose	Allergic rhinitis (hay fever, "sinus trouble")
Bronchi	Asthma
Alveoli	Allergic pneumonitis
Gastrointestinal tract	Numerous symptoms mimicking acute and chronic medical and surgical disorders (abdominal distress, vomiting, and diarrhea are most common)
Central nervous system	Migraine-type headaches, convulsions
Special sense organs	Conjunctivitis, internal eye disorders, ear disturbances with ringing and dizziness
Skeletal structures	Acute and chronic joint swelling (Zussman, 1966)
Genitourinary system (rarely involved in allergy)	Bladder inflammation—proteinuria
Psychic	Irritability, lassitude, altered behavior pattern in children and even in adults, when food as the cause is overlooked or ignored

(IgE) during early gestation. Likewise, the newborn may be sensitized through the breast milk of the mother after she has eaten food proteins of small molecular size, such as egg albumin, which passes through the intestinal tract without alteration of the allergenic determinants (Shannon, 1922; Donnally, 1930; Smith and Bain, 1931). Thus, infants have been found to react to foods that they had not ingested previously and that are unrelated to their present diet. The manner of exposure to the food does not necessarily determine the symptoms elicited or the organ affected. For example, ingestion of wheat frequently produces skin eruption or chronic respiratory symptoms, especially nasal congestion. In general, simple contact with the offending allergen results more frequently in skin eruption, whereas inhalation causes nasal and bronchial symptoms. Ingestion or injection of markedly allergenic foods in highly susceptible individuals sometimes causes minimal local organ tissue reaction and yet may result in profound systemic effects with all the symptoms of shock, as well as severe hives and mucous membrane swelling.

The symptoms vary in speed of onset, severity, and duration. They may occur within several minutes, and, like the thunder that follows on the heels of lightning, will be violent. If the cause is not recognized and treated promptly, the results may be fatal. Symptoms occurring hours after exposure to the food are milder and more readily controlled. Some foods, which produce mild allergic symptoms and which are ingested daily (milk and cereals), may not be immediately identified by the patient and may be overlooked by the most diligent physician.

In those cases of mild allergy to foods, diverse factors are of importance. Ingestion of large quantities of offending foods permits sufficient amounts of unaltered allergen to be absorbed from the intestinal tract to produce obvious symptoms, whereas ingestion of small quantities may produce minimal effects, thus obscuring or confusing the history. The additive effect of consumption of several allergenic foods, especially those that are botanically related, results in symptoms that are not manifested by any one alone. The presence of allergy during certain seasons due to other causes sometimes permits the seasonal food to produce symptoms that would not occur if the allergic state were "in balance." Thus, for example, during the summer pollen hay fever season, overindulgence in some mildly allergenic seasonal fruit will produce symptoms, whereas at another season of the year this fruit would be well tolerated.

The state of the intestinal tract itself has a strong influence on the capacity of certain foods to cause allergic reactions (Gruskay and Cooke, 1955). The capacity of certain food allergens to pass through the intestinal wall with little alteration has been attributed to a low level of digestive enzymes. This may be true in the case of early feeding of grains and egg in infants but has been disproved in the case of adults. Concurrent ingestion of irritating foods such as onion, radishes, and condiments can potentiate or precipitate the reaction or prolong their symptoms.

III. IMMUNOLOGICAL ASPECTS

Allergic reactions to foods are spontaneously occurring disorders peculiar to certain human beings and are less frequently demonstrable in lower animals. These phenomena display a strong hereditary tendency and are immunological disorders, the result of antigen (allergen)–antibody (reagin) interaction. Human allergy to food differs from the phenomenon of experimental anaphylaxis or induced immunity, and some of these differences are listed in Table II (Perlman, 1964).

The most common immunological events producing symptoms due to foods may be divided into immediate hypersensitivity, or humoral immunity (type I), and delayed hypersensitivity, or cellular immunity (type IV). Infrequently, "immune complex disease," a type III hypersensitivity, results from massive injections of food or from the inhalation of organic dusts containing food allergens and fungi. Rarely is type II or cytotoxic-type hypersensitivity manifested in food

TABLE II

SOME DIFFERENCES BETWEEN HUMAN ALLERGY AND EXPERIMENTAL ANAPHYLAXIS OR INDUCED IMMUNITY

Human allergy	Experimental anaphylaxis or induced immunity
1. Evidence of a hereditary factor	1. No evidence of a hereditary factor
2. Difficult to induce	2. Readily produced in laboratory animal
3. Shock organ varies with individuals and at varying times in the same individual	3. Shock organ is specific for each species of experimental animal
4. Complete desensitization cannot be produced	4. Complete nonresponsiveness can be produced
5. Allergic antibody (reagin) characteristic to it	5. Anaphylactic antibody characteristic to it
a. Identified among the immunoglobulins as IgE	a. Usually identified among the immunoglobulins as IgG
b. Detectable by *in vitro* methods radioallergosorbent test (RAST)	b. Detectable by various *in vitro* methods
c. Detectable *in vivo* by sensitizing skin of human and lower primates (PK reaction)	c. Detectable *in vivo* by sensitizing tissue of lower animals (passive cutaneous anaphylaxis—PCA)
d. Thermolabile	d. Thermostable
e. Does not cross the placental barrier	e. Readily crosses the placental barrier
f. Detectable in the patient for years or a lifetime	f. Lasts only a short time in the animal (weeks to months)

allergy. Complement, which is involved in type II and III reactions, does not appear to play any role in the more common types of food allergy, producing type I and IV reactions.

The role of the lymphocytes in allergic reactions has emerged recently and is gaining increasing importance in attempts to understand the mechanisms of these complex and varied immunological phenomena. The B lymphocyte (bone marrow derived or bursa equivalent) is identified with antibody production and with type I (humoral) reaction. The T lymphocyte (thymus dependent) is identified with type IV (cellular) reaction. There is further evidence of interaction between the B and T lymphocytes. Thus, the understanding of the complexity of the processes requires continual awareness of the new developments, which are beyond this discussion.

A. Genetic Control of the Immune Response

The genetics of the allergic reaction is becoming clearer. In human allergy, it has long been evident through population sampling that this allergic disorder is a Mendelian dominant characteristic. Recent immunohistological studies confirm genetic controls and influence. The term "human lymphocyte antigen" (HLA), although now represented by similar systems in other animals, is still used to identify all the genes controlled by this HLA chromosomal region on chromosome 6 (the major histocompatibility complex). Because of the many emerging and changing concepts and terminology, no attempt at a more detailed discussion of this facet of genetics is merited in a discussion of food allergens. It must be pointed out, however, that, although the capacity to react to food allergens is genetically controlled, the specific food offenders are the result of many influences both in the allergen itself and in the allergic person as well as in the manner and degree of the person's exposure.

B. Antibodies

Humoral antibodies include IgG, IgM, IgA, and IgE. The B lymphocyte is identified with the production of these antibodies but is influenced by the T lymphocyte both in stimulation and suppression. It is the IgE that is most frequently identified with what is termed "reaginic serum" and is most closely associated with immediate allergic reaction to food (type I, or immediate hypersensitivity). The characteristics of IgE are listed in Table III and compared with those of IgG, the most familiar immunoglobulin. Two major differences are worthy of particular mention, namely, the thermolability of the reaginic antibodies and their capacity to sensitize passively the skin of nonallergic human beings and subhuman primates (Perlman and Layton, 1967). It is this skin-sensitizing capacity that is used in detecting the offending food allergen. Since reaginic antibodies are found in the serum, and thus in the tissues, their presence

TABLE III

IMMUNOGLOBULINS IN ALLERGY: THEIR DIFFERENT PHYSICAL, BIOCHEMICAL, AND BIOLOGICAL PROPERTIES

Characteristics	IgG	IgE	Reagin
Molecular weight	140,000	200,000	
Sedimentation value	7 S	8.2 S	
Carbohydrate contents	2.9 %	8.2 %	
Concentration in serum	700–1500 mg%	10–70 mg%	
Complement fixation	Positive	Negative	
Thermostability	Stable	Labile	Labile
Tissue of highest concentration	Serum	Skin and mucosa	Skin and mucosa
Passively fixed to human skin	Negative	Positive	Positive
Crosses placenta	Positive	Negative	Negative
Associated with symptoms of food allergy	Negative	Positive	Positive

can be demonstrated by the introduction of the specific allergen, as will be described in the section on skin testing (Section V,B,1). The presence of reagin can be further demonstrated by transferring serum from the allergic person to the skin of nonallergic persons or subhuman primates (Layton *et al.*, 1962a–d). After 24–48 hr introduction of the specific allergen at this transferred serum site will elicit an immediate wheal and flare reaction similar to the response on direct skin testing of the allergic donor (Fig. 1).

The reagin titer (IgE) is not always correlated with the degree of the patient's distress, nor can the presence of reagin in the serum of a person with large, direct skin reactions always be demonstrated by passive transfer. It is often observed that abstaining from foods for a period of months to years results in the loss of skin reactivity and, by the same token, a reduction in the reaginic titer as well as in the symptoms themselves. The patient finds that he can eat such foods without distress. The child in this case is often said to have "outgrown" his food allergy. This is especially common in children who as infants were highly allergic to milk and, by the time they have reached school age, can tolerate this food. On the other hand, resumption of frequent exposure to some foods rekindles symptoms and again makes skin tests strongly positive. This has been seen particularly with egg sensitivity.

FIG. 1. Transfer of reaginic serum of a corn-sensitive patient to the right cheek of a macacus monkey. The challenge was made by injecting intravenously with corn allergen extract after first preparing the animal with an intravenous injection of Evans blue dye. The resulting reaction at the passively sensitized site was evident in 10 min from extravasation of serum containing the dye. [Taken from Layton and Yamanaka (1962) and reproduced with the permission of C. V. Mosby Co.]

The sequence of events of this type of immediate hypersensitivity (type I) begins with the allergen reaching and reacting with the reaginic antibodies fixed to target cells (primarily mast cells), particularly in the shock organ, such as the lungs and the skin. It has been postulated that as the allergen, which is polyvalent, bridges two IgE antibodies attached to the cell wall, the enzyme system is activated in these mast cells, which expel their granules. It is these granules which then release "mediator." Such mediators are usually vasoactive amines and polypeptides, among which histamine is the most important to be identified with the typical clinical symptoms. Other mediators include slow-reactive substance (SRS-A), eosinophil chemotactic factor (ECRF), the kinins, heparin, and others having less clear and less significant clinical importance for this discussion.

Other humoral antibodies listed previously are of minimal importance. In type II and type III hypersensitivity to foods and contaminants in foods, IgG is the primary immunoglobulin involved. This together with the antigen and in the presence of complement results in certain diseases. These include disorders resulting from inhalation of certain vegetable and fungal products (Pepys, 1969). Immunoglobulin G is found in human serum in large quantities induced by a variety of antigenic stimuli. For example, the specific IgG antibody directed to allergens in bovine serum albumin, to milk, and to other food allergens can usually be detected by direct *in vitro* precipitin methods (Heiner and Sears, 1962). This may well represent a response to the ingestion or injection of antigenic material (Holland *et al.*, 1962), but it is not identified with typical allergic symptoms. After repeated injection of antigen, the IgG has been called "blocking antibodies" and is detected in patients who have been given a series of such injections, such as pollen extracts and possibly foods and biologicals. Thus, the level of such specific IgG has little if any correlation to symptoms of food allergy.

Other immunoglobulins (IgA, IgM, and IgD) play no known role in the allergic diseases under consideration. They are involved in other common and obscure immunological processes unrelated to our present subject and are mentioned only in passing for the sake of completeness.

Cellular antibodies are associated with delayed-type hypersensitivity (type IV). They are not readily detected in the patient's serum but rather are found in the committed T cells (thymus dependent). Characteristically, they are involved in the clinical response, which is slow to develop (hours to days). They are seen in certain dermatological disorders, especially among food handlers, and as the dermatitis produced by certain vegetation, such as poison oak, and in hypersensitivity lung disease due to inhalation of organic dusts. The disorders are the result of cellular response to the contact or inhalation of such organic dusts of foods, chemical additives, as well as contaminants in the form of fungi, bacteria, and arthropods. Because of the slow cellular antibody response and the lack of circulating (humoral) antibodies, the method of detecting the offending allergen is by patch testing (Fisher, 1973). More detailed discussion of the immunological methods of detection for this type of hypersensitivity is presented in section V.

IV. ANTIGENS (FOOD ALLERGENS)

A. General Considerations

Almost any food has the capacity to produce an allergic reaction. Some are frequently the cause of symptoms, whereas a few, although commonly consumed, are rarely implicated. Lists of offenders vary with the experience of each clinician, but there is almost universal agreement on a few articles of diet. Among the plant foods these include grains, legumes, and nuts. As the list lengthens it begins to vary on the basis of experience of the allergist himself. Regional customs, abundance of certain foods, as well as ethnic or religious dietary customs, are among the reasons for variation in the lists. The quality and reliability of the allergenic extracts for testing may determine the enthusiasm with which food testing is employed in confirming the clinical history.

B. Biological and Botanical Relationships

Common allergenicity among botanically or biologically related foods has been studied superficially and for the most part clinically. Often, the physician eliminates long lists of such related foods; however, more careful studies have raised doubts as to the validity of such arbitrary elimination of essential articles of diet (Piness *et al.*, 1940). Foods of a single family may possess some common offending allergens, but there will be quantitative differences so that there is a great abundance of an allergen in one and only minimal amount of the same allergen in another. However, the clinical experience rather than the biochemical analysis is the final basis for determination of the food as the cause of symptoms. This is best illustrated among the legumes and among grains, the members of which are botanically closely related but allergenically show variation (Figs. 2 and 3).

C. Allergenic Specificity within Individual Plants

Within the plant, the various anatomical structures vary widely in the contents and character of their allergens. One tissue may show no resemblance to the other, immunologically or clinically. Thus, although a leaf and a root may be nonallergenic, the pollen and the seed, even in minute quantities, may produce marked reactions (Layton *et al.*, 1962a–d).

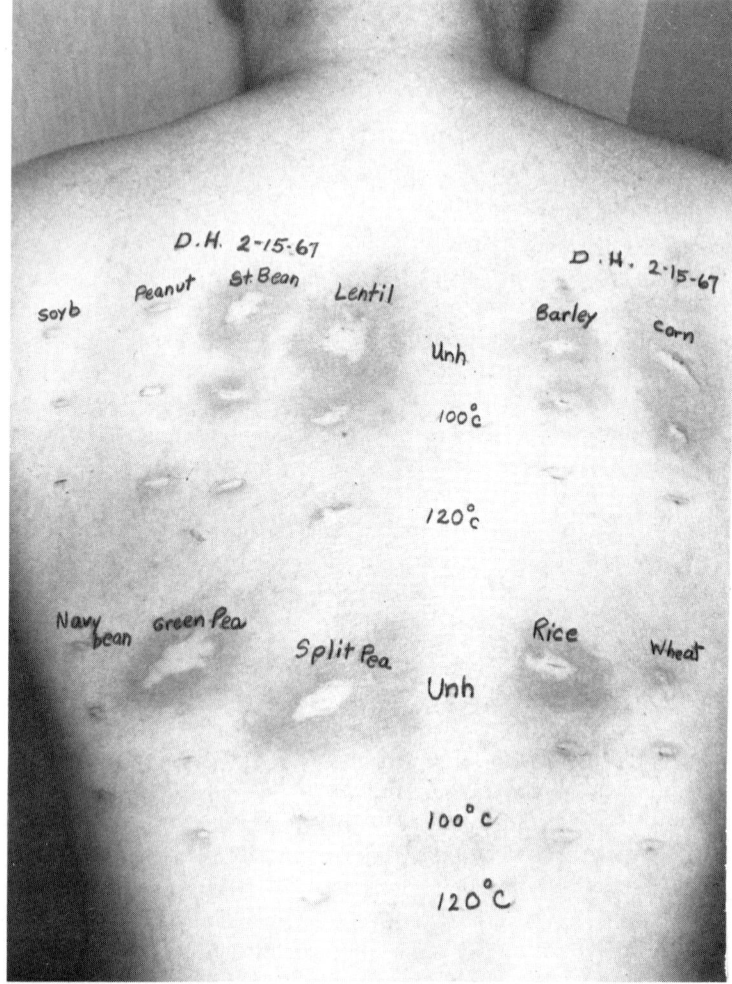

FIG. 2. Reactions to various legumes (left back) and grains (right back) showing effect of heat on those extracts that in an unheated state gave strong skin test reactions. This patient, who in childhood was clinically allergic to several of the foods, especially peas and beans, is now able to eat canned peas without ill effect.

D. Nature of Food Allergens

The allergenic components in foods are multiple and complex and are unique to each vegetable under consideration. The specific allergens in each food have at best been grossly characterized. Some biochemical studies have

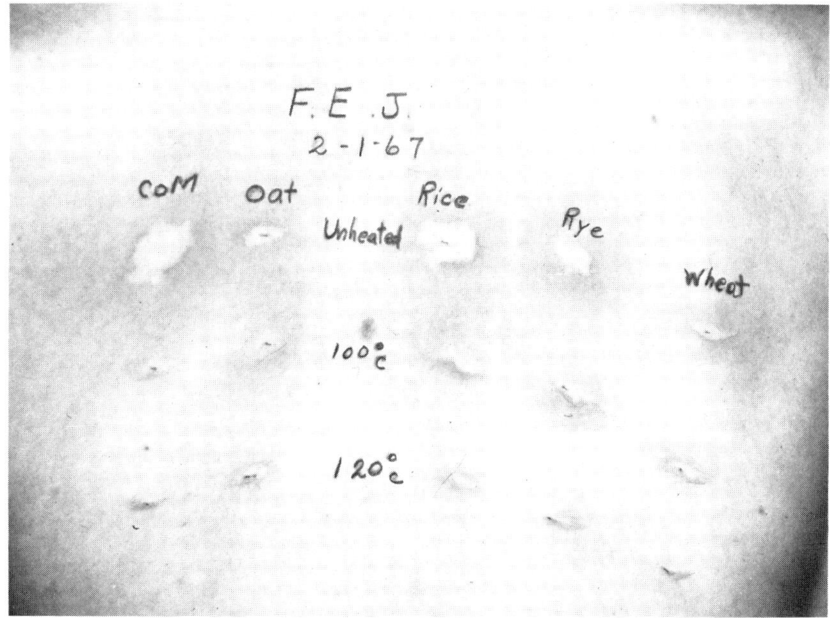

FIG. 3. Various degrees of skin test reaction to grains with reduction of such reaction after heating to 100°C for 30 min and complete abolition of reaction after heating to 120°C for 20 min.

been done on a number of foods (Schulz and Anglemier, 1964). From these studies it is evident that the allergens are large molecular weight substances that are not dialyzable and are most often identified with the protein fraction (Engelfried, 1940). An albumin fraction is commonly present. These proteins can be concentrated by salting-out with ammonium sulfate or precipitated out at 0°–4°C with 75% acetone without denaturation. Pure lipid fractions of foods do not cause reactions either clinically or immunologically, and, if any skin test reaction does occur in highly sensitive individuals, one must suspect some contamination of this lipid fraction with protein. This may be true of pure carbohydrate fractions, which are reported by some to cause symptoms, especially in the case of pure starch obtained from grain or sugar from beet. The size of the proteinaceous allergen may have an influence on its stability and its antigenic determinants, as well as on its capacity to pass through the intact intestinal tract without complete loss of its effect on the shock organ.

The study of the green pea illustrates the complexity of isolating the allergenic fraction. Briefly, the aqueous extract of this legume is divided into three fractions [albumin, and two globulins named legumin and vicillin (Danielsson, 1950)]. Only the albumin fraction gave strong skin test reactions and increased in content as the pea ripened. With further purification of this albumin fraction by Sephedex G-75 gel filtration, six peaks were identified, but only peak five of this albumin fraction demonstrated allergenicity, as shown by positive wheal and flare on known pea-allergic persons. Furthermore, by the use of polyacrylamide gel electrophoresis at pH 6.8, even peak 5 was shown to contain at least five distinct bands of widely differing electrophoretic mobility, further complicating the search for the specific antigenic determinant.

E. Stability of Food Allergens

Influences on food allergens begin in the course of ripening and continue through harvesting, storing, processing, and preparation in the kitchen. In order for the allergic reaction to be set off, some allergenic identity must be retained through ingestion and digestion, and a sufficient quantity of such material must be intestinally absorbed to reach the shock organ containing a specific reaginic antibody (Walzer, 1927). This is indeed a tortuous route, and some foods are so fragile that allergenicity is lost even in the course of harvesting and processing. With these foods, only the growers and processers themselves are affected. Other foods are so stable or are handled with such care that they become the most common ingestant allergen to account for symptoms. In the succeeding comments, a detailed discussion of the various allergenic foods will point out these differences. However, a few comments on the influence of heat alone illustrate in general the complexity of the problem.

Some years ago, Malkin and Markow (1938) found no difference in 91% of tests made with raw versus cooked foods. Our recent studies done on the effect of heat with grains, legumes, and other foods (Figs. 2, 3, 4, and 5) show no consistency in the stability of allergens, as measured by the skin test (Table IV). On testing some foods, the skin reactions were abolished at 120°C, whereas with others there was little or no alteration in activity. Within the family of legumes, the allergenicity of split pea and peanut was unaffected by temperatures above boiling, whereas in the green pea the allergen was readily destroyed at this temperature. This would suggest that ripening produces a higher quantity of stable allergens. However, when both the green and the ripe pea and peanut all withstood high temperatures in another subject, the results suggested a quantitative effect with perhaps a reduction in the amount of allergenic substances that, in a highly sensitive individual, still elicited a strong skin test reaction. In general, the most highly reactive person retained skin reactivity even with heated material. With our present knowledge, any attempt at drawing conclusions as to the stability of allergens would be entirely arbitrary. It is not possible to account for the varying results obtained with experimentally heated food allergens without some knowledge of the number and character of the allergenic determinants on these food proteins and their relationship to various specific reaginic antibodies in the aller-

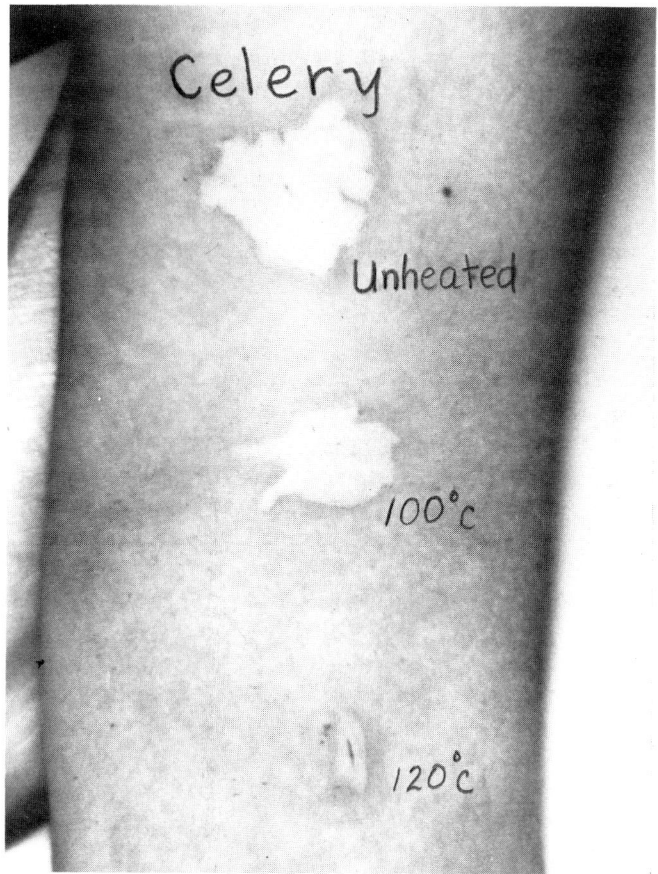

FIG. 4. Effect of heat on celery extract, showing slight reduction of allergenicity after heating to 100°C for 30 min and abolition of skin test reaction after heating to 120°C for 20 min.

gic subject. The same questions are left unanswered with respect to other physical and chemical influences on allergic stability and their effects on the capacity of such foods to produce symptoms in allergic individuals. Thus, the food en route from the natural source to the shock organ of the patient is an unpredictable allergen, and the allergic person himself, who is also allergenically capricious, multiplies the complexity of the problem. A few foods show invariable thermolability. This is best exemplified in coffee, which at one time was thought to be allergenic because of the thermolabile chlorogenic acid. This has been disproved (Layton, *et al.*, 1966). With a few foods the heated product produced skin test reactions, whereas the unheated material did not (Malkin and Markow, 1938). Such a surprising phenomenon would suggest that heat

TABLE IV

EFFECT OF HEAT ON ALLERGENICITY OF SOME FOODS AS MEASURED BY SKIN TESTING[a]

	Corn			Rice			Rye			Wheat		
Patient	Unh	100°C	120°C	Unh	100°C	120°C	Unh	100°C	120°C	Unh	100°C	120°C
M.M.	+++	+++	+++	+++	+++	++	+++	+	+	±	±	±
P.H.	+++	+++	+++	+++	++	+	+	±	±	++++	+	−
P.P.	++++	++	±	++++	++	±	Not tried			++++	+	−
J.A.	++	+	−	++++	+++	−	Not tried			Not tried		
F.J.	++++	+++	−	++++	+++	−	++++	+++	−	+	−	−
E.E.H.	+++	++	+	++++	++++	−	+++	++	−	+++	+	±
D.H.[b]	+++	++	+	++++	++	−	Not tried			++	−	−

	String bean			Green pea			Split pea			Peanut		
Patient	Unh	100°C	120°C	Unh	100°C	120°C	Unh	100°C	120°C	Unh	100°C	120°C
D.H.[b]	+++	+	−	++++	−	−	++++	−	−	−	−	−
J.B.[c]	+	−	−	++++	+++	+++	++++	+++	+++	++++	++++	+++
E.H.	++++	++++	+++	+++	+	−	+++	++	++	+++	+++	++

Patient	Egg			Banana			Celery			Potato		
	Unh	100°C	120°C	Unh	100°C	120°C	Unh	100°C	120°C	Unh	100°C	120°C
J.B.	++++	++++	+++									
S.K.	+++	++	+−									
W.R.	+++	+++	+									
P.H.	+++	++	+									
R.K.				+++	±	−						
E.E.H.				++	−	−						
M.M.	+++	++	+−				+++	+	−	++++	++	−
H.R.	+++	++	−				+++	+	−			

[a]Comparison of portions of same sample after 100°C for 30 min, 120°C for 20 min, and unheated (unh).
[b]D.H. tests shown in Fig. 2. After 15 years of eliminating legumes, he can now eat canned peas and beans without ill effects.
[c]Peanut extract heated to 180°C for 15 min still gave ++ reaction.

Egg
Unh
100°c
120°c
P.H. 2-4-67

Mustard
Unh.
100°c
120°c
W.H. 2-8-67

Potato
Unh.
100°c
120°c
M.M. 2-7-67

Banana
Unh.
100°c
120°c
M.W. 1-27-67

may alter the molecular configuration of the food, interrupt enzymatic and bacterial action, or merely "uncover" the allergen. This would result in the production of an active allergenic substance (Cook, 1942).

These varied and contradicting comments must make evident the paucity of our knowledge of allergens in foods which, when finally absorbed, result in symptoms. To attempt to correlate what occurs in human beings after ingestion of a palatable food with an immunologic phenomenon as indicated in the skin test employing carefully prepared allergenic extracts must indeed leave many doubts that would challenge the validity of unqualified translation of the skin test to symptoms. The subsequent detailed discussion of various foods of vegetable origin will shed a little more light on the subject but will also demonstrate further the complexity of the problem.

V. DETECTION OF FOOD ALLERGENS

A. Subjective Methods

1. History

This is the simple but painstaking method in which the patient recounts his symptoms in relation to food intake. Although it is time-consuming, it is a most rewarding method for the experienced interviewing physician. It may, however, be quite misleading to one who has not had sufficient experience in interviewing on this subject.

2. Diet Diary

In this method, a careful listing of the ingested foods is made, together with a record of symptoms experienced by the patient. This is done for days or even weeks. Later the physician attempts to correlate food intake and symptoms.

3. Elimination Diets

Diets are reduced to a few basic foods, and the relation to the relief of symptoms is noted (Row, 1954; Van Metre *et al.*, 1968).

FIG. 5. Effect of heat (100°C for 30 min and 120°C for 20 min) on skin test reactivity of food allergens as compared with a portion of the same sample unheated. Reactions were abolished with all four foods (egg, mustard, potato, and banana) at 120°C, but only potato and, to a lesser degree, banana were reduced in activity after heating to 100°C.

4. Provocative Tests

In this method, the suspected foods are well disguised and are ingested in an attempt to reproduce symptoms (Goldman *et al.*, 1963). This can be a hazardous procedure when highly allergenic foods are consumed by extremely allergic persons (cottonseed, shellfish, nuts, buckwheat).

Formalized provocative food testing by ingestion and by injection is practiced by a group of physicians (Lee *et al.*, 1969). This is a painstaking procedure requiring the preparation of the patient by abstinence from the foods to be tested. When the food is given or injected the patient is observed closely for the development of symptoms, which may be subtle or acute. Although clinicians practicing this form of food testing find it to be informative, there is still some controversy as a result of several investigative committee studies (Committee Report, 1974).

A browning effect is common among many foods containing proteins and reducing sugars. These changes are of two types: (1) enzymatic and (2) nonenzymatic (Millard effect) (Reed, 1966; Reynolds, 1963, 1965). It has been shown with grains (wheat and rye) that such discoloration is accompanied by reduction in allergenicity, as demonstrated by a reduction of skin test reactions. This change occurs rapidly when allergenic extracts are prepared with buffered saline to which 0.4% phenol has been added for bateriostasis. Extraction with a stabilizer using 50% glycerol or with the use of reducing agents, such as sodium formaldehyde sulfoxalate (Strauss and Spain, 1940), prevents such enzymatic or oxidative changes. Foods prepared commercially for display and for consumption are carefully handled and treated in this manner for the sake of appearance.

Since all these subjective methods rely on the patient himself, who must recount some correlation between ingestion and symptoms, it is evident that all are fraught with misinterpretations and oversights. This is illustrated by the case of a young man who continued to have unaccountable, severe hives and swelling of the throat. When an episode occurred after an oriental meal many foods were suspected, but when another such episode followed, the eating of a roll covered with sesame seeds the probable offending food became more obvious. A subsequent skin test with extract of sesame seed elicited large wheal and flare, confirming the subjectively unrevealed cause.

Other problems may be encountered in the use of subjective methods for detecting foods that produce less explosive reactions and that are ingested daily (milk, cereals, fruits, and vegetables). Such foods are often overlooked in their relation to chronic, persistent symptoms. Infants and small children, unable to express the relation of food to discomfort, may be misjudged as willful. Even an adult may have his symptoms worsened by being placed on a bland diet containing large quantities of milk or grain for what is presumed to be an irritable intestinal tract but proves to be a disturbance due to food that he has been directed to consume in large quantities. Too often the subjective methods for the detection of food allergens result in frustration and confusion. With reports appearing in popular magazines on the subject of food allergy, this condition has become a catchall for multiple, vague complaints and has resulted in faddist diets.

B. Objective Methods

1. Skin Test

The most common objective procedure for the diagnosis of an offending food allergy is still the skin test, although there are a number of factors that detract from its value or reliability. Such testing depends on the presence in the skin of reaginic antibodies (IgE) which, upon introduction of an offending food allergen, result in a wheal and flare reaction (see Section III). Although the mucous membrane of the conjunctivae, nose, mouth, and rectum is sometimes used experimentally, the skin remains the most convenient site for such diagnostic procedures. Skin testing is usually performed either by simple superficial scarification or by injection into the epidermis. A simple scratch test is best done first since it is safe from any systemic reaction. Such a test is significant if a wheal and flare develop. However, in less sensitive individuals or with the use of weaker extracts, it may often be negative. The intradermal test is much more sensitive but may produce false-positive reactions with allergens containing nonspecific irritants. In addition, this method can be hazardous in inexperienced hands and in sensitive subjects, especially when highly allergenic foods are used, such as seeds, seafoods, and buckwheat (Lamson, 1924, 1929; Benson, 1940). The systemic reaction may be severe, and death has been reported from the intradermal injection of 0.02 ml of extract of cottonseed allergen (Harris and Shure, 1950). In such cases, a superficial scratch test would have elicited a strong positive local reaction and precluded the need for performing the more hazardous intradermal test.

Some controversy persists over the value of skin testing with food allergens (Freedman, 1959). A positive wheal and flare are after all immunological phenomena and need not necessarily identify the cause of the patient's symptoms. Like many other procedures in the course of clinical diagnosis, it is valuable and acceptable as a bit of added evidence in the process of arriving at an etiological diagnosis. In this perspective a positive skin test reaction has value. Negative skin test reactions often occur in spite of a positive history. This results in the condemnation of the diagnostic procedure when in reality the food allergen used in testing may have been biologically (allergenically) inert (Perlman, 1959). It is regrettable that there is as yet no generally accepted method for preparing skin test material or for objectively assaying its quality and reliability.

Acceptable methods for the preparation of food allergens suitable for accurate skin testing must fulfill the following requirements: (1) The food containing the allergen must be subjected to as little denaturation as possible so that the final product will give a strong, positive wheal and flare in allergic persons; (2) the final allergen must be free of nonspecific irritants; (3) the final allergen must be free of contaminating allergens, which may give a false-positive reaction; and (4) the final product must be a concentrate in stable form, preferably in a dry state, which can be readily reconstituted for testing.

Various methods have been proposed, such as simple extraction and filtering of the juice followed by preservation under toluene or in glycerol. Another method is the preserving of foods in the frozen state to be used only when needed and then by preparing a simple extract (Ancona and Schumacher, 1954; Josephson and Glaser, 1963). Some food allergens sold for testing are merely the powdered, dried material. These methods do not provide stable or reliable allergens and few are free from nonspecific irritants. A relatively simple method for satisfying the above requirements is the precipitating out of important allergenic proteinaceous fractions with ammonium sulfate or with organic, water-miscible solvents such as acetone. When this is done at 0°–4°C a highly concentrated protein fraction is obtained, which is usually free of any significant nonspecific irritants and can be kept indefinitely in the dried, powdered state and reconstituted for testing as needed. In general, greater accuracy of testing and correlation with the patient's history occurs with foods that are known to produce immediate and violent symptoms, particularly seeds, grains, and a few isolated fruits and vegetables. Similarly, testing with these foods to confirm the history must be done with caution because of the marked local reaction and sometimes even severe systemic effect in such highly sensitive persons.

2. *In Vitro and Indirect Testing*

In recent years, *in vitro* testing has been developed for confirming the patient's history, although it is less accurate and less reliable than direct skin testing of the allergic patient. It involves the use of serum from allergic subjects containing antibodies specific to the suspected food allergen.

The transfer of the serum from an allergic person to the skin of a nonallergic person by the method commonly called "P.K." (Prausnitz and Küstner, 1921) and to lower primates by the method of Layton *et al.* (1962b) is less sensitive and therefore less reliable than the direct testing of the patient. One method of *in vitro* testing employs the measurement of histamine release from sensitive leukocytes challenged with the suspected food. This has been modified and has limited value (Lichtenstein and Osler, 1964). Others have used the measurement of histamine release from sensitized monkey and human lung tissue (Malley and Perlman, 1970). Still another method is based on the degranulation of a patient's mast cells (Shelley, 1963) or the degranulation of rat mast cells (Haddad and Korotzer, 1972). All of these methods listed are, for the most part, experimental and as yet of no practical usefulness.

The radioallergosorbent test (RAST) introduced in recent years for detecting reaginic antibodies directed to foods offers a more reliable method of *in vitro* testing (Aas and Johanssen, 1971; Hoffman and Haddad, 1974; Yunginger and Gleich, 1975). This method correlates most closely with direct skin testing and offers promise of becoming the moethod of choice in any indirect patient testing. This method may also become useful for standardizing the allergens in foods.

3. *Obsolete and Infrequently Used Methods*

A number of proposed methods for detecting offending foods have been discarded as being of questionable diagnostic value. These include (a) the pulse acceleration test (Corwin *et al.*, 1961), (b) the leukopenic index measured by a drop in leukocyte count following the ingestion of a suspected food, (c) increased proteose in the urine, (d) X-ray changes in the upper intestinal tract due to the spasm following the ingestion of a suspected food incorporated in barium. They are enumerated here because these methods emerge periodically in the literature.

The cytotoxic test for the diagnosis of food allergy is discussed in current literature (Bryan and Bryan, 1971) and is enthusiastically employed by some clinicians. Such testing relies on the direct microscopic observation of the action of the allergic patient's serum in the presence of the specific offending food antigen and detecting the activity and viability of the patient's neutrophilic leukocytes. Although there is some rationale for this test, few investigators have been able to confirm the reproducibility or value of this cytotoxic testing for food allergy. The criticism of such a procedure is based on a number of factors, including the difficulty of performance with any degree of accuracy, the quality and purity of the food allergens used, the low level of antibodies in the serum specific for many foods, and the uncritical correlation of these tests with the clinical problems.

In summary, it may be said that well-established evidence of food allergy often fails to correlate with acceptable subjective and objective diagnostic procedures. The diagnosis of food allergy remains a clinical decision, and the various diagnostic methods are only confirmatory. It is this frequent discrepancy between tests and symptoms that has created controversy over the merits and the results of diagnostic procedures and even over the entire question of foods producing many of the commonly suspected symptoms.

VI. SPECIFIC FOOD ALLERGENS

A. Cereal Grains

The family Gramineae comprises a group of common allergenic foods, including barley, corn, oat, rice, rye, and wheat. Grains can cause symptoms by inhalation and by contact, as well as by ingestion. These affect the farmer, the baker, the housewife, as well as the consumer. The entities known as ''baker's eczema'' and ''baker's asthma'' are well-accepted occupational diseases. These common food allergens are the basic articles of diet among diverse nationalities. The frequency with which one grain may become allergenic depends on its importance in the diet, and yet some are more common offenders regardless of diet changes in our drifting world population. The allergens in each grain may differ as to physiocochemical characteristics as well as clinical importance, and yet it is common to find a subject reacting on skin testing to many of the grains (Daussant and Grabar, 1966). In spite of this, it is common clinical practice to substitute one grain for another in order to gain relief from symptoms (Rowe, 1954).

1. Wheat

This grain is the most common of the offending allergens in the group, and its elimination often is axiomatic in outlining empirical diets for allergic infants and children. It is so ubiquitous that there is great difficulty in eliminating it from the daily diet. The usual breads cannot be made without the gluten of wheat, and the so-called wheat-free bread is often a misnomer. The allergenic fraction is probably albumin, and other fractions including gliadin (Goldstein *et al.*, 1969) and gluten have, in our experience, often failed to give positive skin test reactions in wheat-sensitive patients. Although gluten itself causes well-characterized gastrointestinal disorders, these are related to malabsorption problems and not to allergy. The allergen in wheat is reduced in activity by temperatures above 120°C, although such temperatures within the moist loaf of bread are never attained. Wheat extract for testing rapidly turns black on standing and quickly loses skin test reactivity. This oxidative change can be lessened by preserving the extract in 50% glycerol or by extracting wheat with solutions containing reducing agents such as sodium formaldehyde sulfoxalate (Strauss and Spain, 1940).

2. Corn

This grain also is a common food allergen (Rinkle and Randolph, 1951) and has produced severe clinical symptoms. In areas where Mexican and American corn meal dishes are becoming popular, acute episodes are more frequently reported. The allergen in corn, unlike that in wheat, does not readily oxidize on standing in solution. It withstands the higher baking temperatures with less loss of skin test reactivity.

3. *Rice*

This cereal is often prescribed empirically to allergic persons as a substitute for wheat and other grains. Although such a diet change has proved successful, allergy to rice frequently occurs. It is quite probable that in the polishing process to which this grain is subjected prior to consumption in the United States, a great deal of the allergenic portion is removed. Unpolished rice and wild rice are more likely to produce symptoms. Although the allergenic extract of rice is not visibly oxidized by storage, heat treatment reduces the skin test reactivity to a significant degree.

4. *Rye*

Again, this cereal has been substituted for bread stuffs made from wheat, but in some individuals it, too, has produced allergic symptoms. Rye extract, like wheat, oxidizes and turns black on standing in saline solution and loses allergenicity, as evidenced by a reduced skin test reaction.

5. *Oat*

This grain cereal provides a good substitute for wheat and other grains in allergic individuals. It is acceptable as a cereal and as a basic ingredient for baked goods. Oat is less frequently an allergen, either clinically or on skin testing.

6. *Barley*

This grain frequently causes allergic reactions both in its natural form and as a malt in flavorings and beverages. Symptoms are often overlooked when this grain appears as malt, but the allergenic extract is resistant to oxidation and shows much less effect from heating to temperatures above 100°C.

7. *Buckwheat*

This food does not, in the true sense, belong to the group of grains, although it is consumed in a like manner. It belongs to the family Polygonaceae and is botanically related to rhubarb. Buckwheat on occasion produces a violent reaction after ingestion and, in areas such as the Orient where the straw is used for head rest, it has resulted in asthma and hay fever symptoms (Matsumura *et al.*,

1964). Attempts to detect its importance by skin test or by provocative trial must be carried out with extreme caution, since the allergen has on occasion resulted in life-threatening and even fatal reactions (Lamson, 1924).

Two additional interesting observations are offered on grain allergen. It has often been observed that Asians coming to the Willamette Valley of Oregon, where grass pollen reaches high levels of concentration in the air, after several years develop for the first time grass pollen hay fever. How much a predominantly rice diet predisposes a potentially allergic person to the development of clinical symptoms due to grass pollen hay fever is conjectural. Another observation is the frequency with which persons giving strong reactions to grass pollens also react strongly to rice and other grain extracts. These two observations suggest the possibility that in this plant food the pollen, which is represented by the male germ plasm, may share some common allergenic determinants with the mature seed (Layton *et al.*, 1962).

B. Vegetables

Vegetables comprise a large group of unrelated allergenic foods that vary in their recorded importance, depending on the clinical experiences of the allergists. The large size and diversity of the list are compounded by the fact that within a single vegetable the edible portions may differ in allergenicity. Thus, for some the allergen may be the leaf, for others the root, and for many others the bloom or seed. In addition, each plant portion may vary in degree of allergenicity as it is altered in stages of ripening. Further, a vegetable eaten raw may be highly allergenic, whereas after thorough cooking some thermolabile allergens are evidently sufficiently denatured to make that vegetable innocuous. This thermolability is demonstrated with such highly allergenic vegetables as carrots, potatoes, squash, and to some degree celery (Fig. 4).

Again, it must be pointed out that the significance of the skin test reaction is limited because testing is usually done with fresh, raw, carefully handled material. Yet on routine skin testing it is observed that positive reactions to vegetables may be multiple and may agree with the clinical experiences. Often reactions occur to various leafy vegetables that are only remotely related botanically. Frequently, such patients also react to extract of the tobacco leaf.

Legumes as a group illustrate many of the previous general remarks about vegetable allergens and deserve some detailed comments. As a botanical group, the legumes show some common allergenicity, yet there is a wide variation both qualitatively and quantitatively (Perlman, 1966). Severe symptoms after ingestion may occur with one, and yet no reactions occur with another. The peanut represents the most highly allergenic member. Again, various stages of ripening reveal changes in allergenicity. Thermostability is more evident among legumes eaten in the ripened stages, as seen with the peanut, lentil, and split pea. However, in highly allergic individuals the green pea retains its allergenicity even after being subjected to a temperature of 120°C (patient J.B. in Table IV).

Soybean, which has been successfully used as a milk substitute (Ratner *et al.*, 1955), occasionally proves to be highly allergenic in older children and adults (Fries, 1966a). Some employees in the plywood industry, where soybean is used in the preparation of glue, have severe symptoms due to inhalation or contact with this substance. Such reactions occur in spite of the high temperature used in processing the soybean for this use, thus demonstrating the stability of the allergen.

C. Fruits

Some of what has been said of vegetables may be applied to fruits. However, these foods differ in that a single anatomical structure of the plant is consumed and usually in the ripe stage, when the sugars predominate. Fruits are eaten raw and often include both the skin and the seed, where some believe the allergen is concentrated (Hale, 1960). Thermolability of some fruit allergens is demonstrated by the tolerance for canned products, such as berries, whereas the raw fruit causes severe symptoms (Tuft and Blumstein, 1942). We have found that such fresh fruits consumed in the pollen season potentiate otherwise controlled symptoms, yet these fruits in themselves do not produce such symptoms. Freedman *et al.* (1962) implicated chlorogenic acid, a constituent of all fruits. This observation, however, has been disproved by Layton *et al.* (1966).

With strawberries, notoriously reputed to cause hives, a thorough, quick washing with scalding water followed by chilling makes them innocuous without destroying taste or texture. It has not been determined whether this urticarigenic (hive-producing) substance is removed or destroyed by heat, or whether it is merely a surface contaminant.

Exotic fruits, such as banana, pineapple, and mango, periodically are strong and isolated fruit allergens. In the case of banana and pineapple, heat processing greatly reduces such antigenicity (Fries and Glazer, 1950) (Fig. 5). It would indeed be intriguing to identify the specific allergen in such fruits, which in infants causes highly specific, high-titer reaginic antibodies after only a single dietary exposure.

D. Nuts, Seeds, and Beans

This group is arbitrarily divided into (1) nuts as popularly identified, (2) seeds of mustard, cotton, and flax, and (3) beans of castor and coffee plants. [Coffee and chocolate are included under beverages (see Section II,E,1). Not included are the cereals, legumes, condiments, and other miscellaneous allergens of seed origin.] These are among the most highly allergenic food groups. They account for some serious and even life-threatening reactions. Cottonseed, particularly

FIG. 6. Piles of burlap bags ready for use in local hauling of castor bean pomace (in the piles). These bags are patched repeatedly and find their way into the used bag market for hauling potatoes, green coffee, etc. Taken from Layton *et al.* (1965) and reproduced with the permission of S. Karger AG, Basel, Switzerland.

when tested using a minute quantity, injected into the skin has resulted in death within minutes (Harris and Shure, 1950; Benson, 1940). A few members, such as castor bean, have been studied extensively and thoroughly, yet our understanding of the allergenic protein is at best sketchy (Layton *et al.*, 1961; Spies, 1967).

Castor bean acts as an allergenic inhalant in the course of its processing for fertilizer. It may contaminate processing plants, laboratories, and even the sacks that are later used for other foods, in which case it is the true culprit, although other foods have been implicated (Fig. 6). Castor bean has caused severe allergic reactions among employees and even family members when contaminated clothing was brought home (Layton *et al.*, 1962a). Highly purified castor oil is not known to cause allergic reactions, again indicating that the protein fraction and not the lipid fraction contains the allergen.

The seeds of mustard, cotton, and flax are common and potent allergens. Mustard is readily identified because of its characteristic taste and because of the immediate and severe onset of symptoms after ingestion. Cottonseed and flaxseed often appear subtly in breadstuffs, in some cereals, and even in cow's milk through the animal feed. They may produce violent symptoms that are not immediately accounted for. However, here, too, the lipid or hydrogenated fat from these seeds does not cause allergic reactions, so that arbitrary elimination of such fats from cooking or baking is generally unwarranted. The allergenic protein fraction is somewhat reduced by heat but not completely destroyed (Fig. 5).

Seeds of sesame, poppy, caraway, anise, and others often are added to bakery products. They also produce acute allergic reactions and may often elude detection because of the small quantity that is ingested and that can yet have marked clinical effect. The same severe reaction has been noted with coconut and with some less frequently consumed exotic nuts such as macadamia, cashew, and others.

E. Miscellaneous Food Allergens

1. Beverages

Beverages include a variety of unrelated foods, such as chocolate, coffee, tea, roasted grains, beer, wines, and alcoholic distillates. In spite of the fact that one or more of these beverages is consumed by everyone, there are remarkably few authenticated instances of allergic reactions to them.

a. Chocolate. Chocolate is popularly associated with such allergic reactions as skin disorders, headaches, and gastrointestinal symptoms. It is used as a beverage, a solid food, a flavor, and even a cosmetic and medicinal (cocoa butter). In its preparation for human consumption, the cocoa bean is subjected to high temperature, which reduces its allergenicity (Fries, 1966b). Therefore, when it is eaten and during skin testing, it rarely acts as an allergen (Maslansky and Wein, 1971).

b. Coffee. This beverage is consumed throughout the world and yet in its thoroughly roasted state is rarely responsible for the production of true allergic symptoms. In the green state, coffee is a potent allergen. Handlers of the green coffee at the roasting plant, have severe symptoms, yet they can drink the brew made from the roasted bean without any ill effect. It is evident that the allergen in green coffee is thermolabile at roasting temperatures (Layton *et al.*, 1965). Inadequate roasting temperatures can leave active allergens in the center of the coffee bean which even later brewing will not alter and which will result in symptoms. For a time, chlorogenic acid was thought to be the allergen in coffee and was characterized as a small molecular weight dialyzable and thermolabile fraction (Freedman *et al.*, 1962). However, there is now strong evidence that the

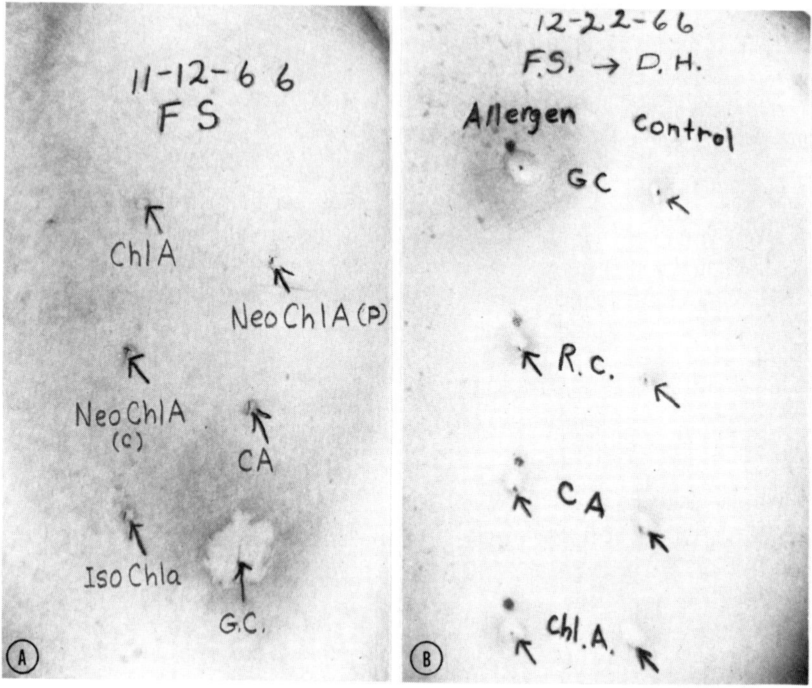

FIG. 7. Skin test reactions with various fractions of coffee, demonstrating that the allergen is in the nondialyzable retentate and is thermolabile. Pure chlorogenic acid failed to give skin reaction. (A) Direct testing with chlorogenic acid from coffee (ChlA), neochlorogenic acid from prunes [Neo ChlA (p)], neochlorogenic acid from coffee [NeoChlA (c)], caffeic acid (CA), isochlorogenic acid from coffee (IsoChla), and green coffee retentate (GC). Only the nondialyzable green coffee (retentate) reacted. (B) Indirect (passive transfer) testing using serum of patient in (A) on a nonallergic recipient: green coffee (GC), roasted coffee (R.C.), caffeic acid (CA), chlorogenic acid (pure), (CHl.A.).

positive skin-reacting allergen was a contaminant of this "pure" chlorogenic acid and not the chlorogenic acid itself. The allergen of green coffee, although thermolabile, is a large molecular weight, nondialyzable substance (Layton *et al.*, 1966) (Fig. 7).

c. Substitutes and Adulterants of Coffee. These include chickory root, grains, legumes, and chestnuts. They are usually thoroughly roasted, which may well account for the infrequency with which they are encountered as causes of allergy, even among individuals allergic to the grains or legumes.

d. Tea. This beverage is rarely encountered as an allergen. In our experience, it is most often consumed as the "black," fermented form. Quite probably among those who handle, process, and package tea there occurs allergic reactions. This author has not encountered such persons (Uragoda, 1970).

e. Carbonated "Soft" Beverages. These contain herbs and other natural flavors as well as artificial coloring and flavoring. Periodically, they are implicated in an allergic reaction, and provocative tests may be positive. The allergen is not easily detected. It is undoubtedly a complex, and may involve chemical additives rather than the naturally occurring food allergen.

f. Beer. This commonly consumed alcoholic beverage may possess several allergens, including malt (barley), yeast, and hops. The last may well have definite allergenic quality, since among hop growers and handlers allergic reactions are not infrequently observed in the form of skin eruption and asthma. Beer at most is subjected to a pasturizing temperature, which is not sufficient to denature the allergens present.

g. Wine. This alcoholic beverage may contain not only the fermented sugars of the fruit but also some of the proteinaceous material of the skin, flesh, and seed, as well as the mold responsible for the fermentative process. In spite of the amount of wine consumed, allergic symptoms rarely occur. Those symptoms attributed to wine consumption are usually the result of the pharmacological actions of the various contaminants as well as of the higher alcohols and total volume ingested.

h. Alcoholic Distillates. These are made from grains and fruits. The allergenic material is minimal, if present at all, in the final distillate. Reactions, even in the form of hives or respiratory distress, are probably the result of the pharmacological action of the alcohol, since this may trigger a reaction in a susceptible allergic individual.

2. *Condiments and Flavors*

Condiments and flavors comprise a vast and unrelated group of foods. Some have been discussed among other food allergens, including nuts and seeds. Such condiments, including caraway, anise, sesame, etc., show allergenicity similar to that discussed above in connection with true nuts and seeds. Many flavors are not derived from natural foods but rather are chemicals and hold no place in this discussion. The vast number of spices from a wide group of unrelated plants may produce reactions in isolated instances and are usually detected because of their suddenness and severity and because they are readily identified in the course of their ingestion. The distinct and sometimes unpleasant flavor of some result in symptoms of "intolerance" and not of true allergy. Onion and garlic flavorings are most often singled out.

3. Yeast and Molds

These foods are usually classified as inhalent allergens and yet are often constituents of foods. Yeast ingested in breadstuff and particularly in high concentration in health foods or diet supplements has caused severe reactions with gastrointestinal symptoms and severe skin eruption. In some highly sensitive persons a small amount of yeast in bread causes persistent symptoms. Temperatures of baking do not reach high enough levels in the center of the load to inactivate the allergen in yeast.

Molds encountered primarily in the preparation of cheese are usually represented by the various types of *Penicillium*. The mold contents differ with the variety of cheese, and sensitivity to mold justifies elimination of this food. It might be well to point out that allergy to this *Penicillium* mold does not necessarily suggest allergy to the penicillin antibiotic agent, or vice versa, any more than milk and beef allergy are the same.

4. Pollens

The male germ plasm of vegetable origin, like molds, is classified as an inhalant allergen. However, it is encountered in foods not only as a minor contaminant (honey) but also as a primary ingredient in certain "health foods." Although the nutrient value of pollens has not been substantiated, their allergenic effect on some susceptible, unsuspecting persons has resulted in violent symptoms with manifestations of transient hives and even profound shock. Surprisingly, such pollens may be chiefly from plants not commonly associated with hay fever. In view of the unproved worth of pollen as a nutrient and its frequent, unpredictable, adverse effects, there is some question about its acceptability as a health food.

5. Vegetable Gums

Vegetable gums are included chiefly for the sake of completeness. They are widely used as food additives and are a dominant ingredient in some confectioneries. In cosmetics, particularly hair spray, gums have produced inhalant symptoms of asthma and, as contactants, have produced local dermatitis. As ingestants, vegetable gums from various sources have rarely been implicated as food allergens (Gelfand, 1949).

VII. SUMMARY

This review of allergens in foods reveals quite clearly the paucity of our knowledge about the chemical and physical characteristics of the basic material producing these unique immunological reactions. The findings offered in some detail are based on observations of the patient and on biological tests that employ the action of allergen and reagin on the skin of human beings and subhuman primates as well as *in vitro* procedures. The allergens used were carefully prepared from the raw material and did not always resemble the allergen in food, which had been processed, ingested, digested, and absorbed in order to act on the organs manifesting the symptoms. Yet in most instances a good correlation existed between the skin tests with carefully prepared materials and the symptoms caused by the allergen, which reached the shock organ after ingestion. This would suggest that the allergenic determinants were at least qualitatively the same in spite of any quantitative change or alteration of the whole protein molecule.

Although it is essential that more be known about the food allergens, some superficial observations of their various physiochemical characteristics and their allergenic qualities have been presented here. It will take the combined skills and efforts of food scientists, biochemists, immunologists, and clinical allergists to provide a better understanding of allergens in foods which in certain individuals, and in general, act as violent "toxins." (Quod ali cibus est aliis fiat acre venenum.*—Lucretius, 95–55 B.C.)

REFERENCES

Aas, K., and Johanssen, S. G. (1971). *J. Allergy Clin. Immunol.* **48,** 134–142.

Ancona, G. R., and Schumacher, I. C. (1954). *Calif. Med.* **80**, 181–184.

Benson, R. L. (1940). *J. Allergy* **11**, 145–146.

Bryan, W. T., and Bryan, M. P. (1971). *Otolaryngol. Clin. North Am.* **4**, 523–534.

Caplin, I., Bronsky E. A., Crozier, W., Dickstein, B., Exline, L., Hale, R., Haynes, J. T., Hurst, A., Josephson, B., Mount, B., Perlman, F., and Sostheim, R. (1973). *Ann. Allergy* **31,** 375–381.

Committee Report (1974). *Ann. Allergy* **33,** 164–166.

Cooke, R. A. (1942). *Ann. Intern. Med.* **16**, 71.

Corwin, A. H., Hamburger, H., and Dukes-Dobos, F. N. (1961). *Ann. Allergy* **19**, 1300–1311.

Danielsson, C. E. (1950). *Acta Chem. Scand.* **4**, 762–771.

Daussant, J., and Grabar, P. (1966). *Ann. Inst. Pasteur Paris* **110**, Suppl., 79–83 (abstr.).

Donnally, H. H. (1930). *J. Immunol.* **19**, 15–40.

Engelfried, J. J. (1940). *J. Allergy* **11**, 569.

Fisher, A. A. (1973). "Contact Dermatitis." Lea & Febiger, Philadelphia, Pennsylvania.

Freedman, S. O., Siddiqi, A. I., Knupey, J. H., and Sehon, A. H. (1962). *Am. J. Med. Sci.* **244,** 548–555.

Freedman, S. S. (1959). *Pediatr. Clin. North Am.* **6**, 853–865.

Fries, J. H. (1966a). *J. Asthma Res.* **3**, 209–211.

Fries, J. H. (1966b). *Ann. Allergy* **24**, 484–491.

Fries, J. H., and Glazer, I. (1950). *J. Allergy* **21**, 169–175.

Gelfand, H. H. (1949). *J. Allergy* **20**, 311–321.

Gleich, G. J., Larso, J. B., Jones, R. T., and Baer, H. (1974). *J. Allergy Clin. Immunol.* **53,** 158–169.

*"What is food to one may be fierce poison to another."

Goldman, A. S., Anderson, D. W., Jr., Sellers, W. A., and Saperstein, S. (1963). *Pediatrics* **32**, 425–443.

Goldstein, G. B. *et al.* (1969). *J. Allergy* **44**, 37–50.

Gruskay, F. L., and Cooke, R. E. (1955). *Pediatrics* **16**, 763–769.

Haddad, Z. H., and Korotzer, J. L. (1972). *J. Allergy Clin. Immunol.* **49**, 210–218.

Hale, R. (1960). *Ann. Allergy* **18**, 270–275.

Harris, M. C., and Shure, N. (1950). *J. Allergy* **21**, 208–216.

Heiner, D. C., and Sears, J. W. (1962). *Am. J. Dis. Child.* **103**, 634–654.

Hoffman, D. R., and Haddad, Z. H. (1974). *J. Allergy Clin. Immunol.* **54**, 165–173.

Holland, N. H., Hong, R., Davis, N. C., and Clark, D. W. (1962). *J. Pediatr.* **61**, 181–195.

Josephson, B. M., and Glaser, J. (1963). *Ann. Allergy* **21**, 33–40.

Lamson, R. W. (1924). *J. Am. Med. Assoc.* **82**, 1091–1098.

Lamson, R. W. (1929). *J. Am. Med. Assoc.* **93**, 1775–1778.

Layton, L. L., and Yamanaka, E. (1962). *J. Allergy* **33**, 217–275.

Layton, L. L., Dante, B. T., Moss, L. K., Dye, N. H., and DeEds, F. (1961). *J. Am. Oil Chem. Soc.* **38**, 405–410.

Layton, L. L., Lee, S., Yamanaka, E., Greene, F. C., and Green, T. W. (1962a). *Int. Arch. Allergy Appl. Immunol.* **20**, 257–261.

Layton, L. L., Lee, S., and DeEds, F. (1962b). *Proc. Soc. Exp. Biol. Med.* **108**, 623–626.

Layton, L. L., Yamanaka, E., and Green, F. C. (1962c). *J. Allergy* **11**, 276–280.

Layton, L. L., Yamanaka, E., and Green, T. W. (1962d). *J. Allergy* **33**, 232–235.

Layton, L. L., Panzani, R., Greene, F. C., and Corse, J. W. (1965). *Int. Arch. Allergy Appl. Immunol.* **28**, 116–127.

Layton, L. L., Panzani, R., and Corse, J. W. (1966). *J. Allergy* **38**, 268–275.

Lee, C. H., Williams, R. I., and Binkley, E. L. (1969). *Arch. Otolaryngol.* **90**, 87–94.

Lichtenstein, L. M., and Osler, A. G. (1964). *J. Exp. Med.* **120**, 507–530.

Malkin, J. I., and Markow, M. (1938). *J. Allergy* **10**, 337–341.

Malley, A., and Perlman, F. (1970). *J. Allergy* **45**, 14–29.

Maslanky, L., and Wein, G. (1971). *Conn. Med.* **35**, 5.

Matsumura, T., Tateno, K., Yugami, S., and Kuroumi, T. (1964). *J. Asthma Res.* **1**, 219–227.

Pepys, J. (1969). "Hypersensitivity Diseases of the Lungs Due to Fungi and Other Organic Dusts." Karger, Basel.

Perlman, F. (1959). *J. Allergy* **30**, 24–34.

Perlman, F. (1964). *In* "Symposium on Foods—Proteins and Their Reactions." (H. W. Schulz and H. F. Anglemier, eds.), Chapter 19, pp. 423–439. Avi Publ., Westport, Connecticut.

Perlman, F. (1966). *Food Technol.* **20**, 1438–1442.

Perlman, F., and Layton, L. L. (1967). *J. Allergy* **39**, 205–213.

Piness, G. I., Miller, H., Carnahan, H. D., Altose, A. R., and Hawes, R. C. (1940). *J. Allergy* **11**, 251–265.

Prausnitz, C., and Kustner, H. (1921). *Zentralbl. Bakteriolog. Parasitenkad. Infektionskr. Hyg., Abt. I: Orig.* **86**, 160–169.

Ratner, B. (1928). *Am. J. Dis. Child.* **36**, 277–288.

Ratner, B., Utrecht, S., Crawford, L. V., Malone, H. J., and Retsina, M. (1955). *Am J. Dis. Child.* **89**, 187–193.

Reed, G. (1966). "Enzymes in Food Processing," 1st ed., pp. 186–192. Academic Press, New York.

Reynolds, T. M. (1963). *Adv. Food Res.* **12**, 1–52.

Reynolds, T. M. (1965). *Adv. Food Res.* **14**, 167–283.

Rinkle, H. J., and Randolph, T. C. (1951). "Food Allergy." Thomas, Springfield, Illinois.

Rowe, A. H. (1954). *Q. Rev. Allergy Appl. Immunol.* **8**, 391–403.

Schulz, H. W., and Anglemier, A. F., eds. (1964). "Symposium on Foods—Proteins and Their Reactions." Avi Publ., Westport, Connecticut.

Shannon, W. R. (1922). *Minn. Med.* **5,** 137–143.

Shelley, W. B. (1963). *Ann. N. Y. Acad. Sci.* **103**, 427–435.

Smith, F. S., and Bain, K. (1931). *J. Allergy* **2**, 282–284.

Spies, J. R. (1967). *Ann. Allergy* **25**, 29–34.

Strauss, M. B., and Spain, W. C. (1940). *J. Allergy* **12**, 61–62.

Tuft, L., and Blumstein, G. I. (1942). *J. Allergy* **13**, 574–582.

Uragoda, C. G. (1970). *Br. J. Ind. Med.* **27**, 181–182.

Van Metre, T. E., Anderson, A. S., Barnard, J. H., Bernstein, I. L., Shafee, F. H. Crawford, L. V., and Wittig, H. J. (1968). *J. Allergy* **41**, 195–208.

Walzer, M. (1927). *J. Immunol.* **14**, 143–174.

Yunginger, J. W., and Gleich, G. J. (1975). *Pediatr. Clin. North Am.* **22**, 3–15.

Zussman, B. M. (1966). *South. Med. J.* **59,** 935–939.

CHAPTER 11

Naturally Occurring Carcinogens

GERALD N. WOGAN AND WILLIAM F. BUSBY, JR.

I. INTRODUCTION

Several hundred chemicals have been shown to be carcinogenic (i.e., induce malignant tumors) when administered experimentally to animals. Representatives of many chemical classes with diverse structures are included among the known carcinogens. These experimental findings, together with epidemiological evidence on patterns of cancer incidence in human populations that has accumulated over the past two or three decades, have led to the widely accepted postulate that many, perhaps most, human cancers result from exposure to environmental chemicals. This mounting evidence has produced an era of acute public awareness of the possible importance of environmental chemicals to cancer and other human health problems. Much attention has been paid to those chemicals such as food additives and contaminants, industrial chemicals, and pollutants that are introduced into the environment intentionally or accidentally as a result of human activity.

TOXIC CONSTITUENTS OF PLANT FOODSTUFFS, SECOND EDITION

ISBN 0-12-449960-0

Increased public attention has stimulated greatly expanded efforts to conduct carcinogenesis bioassays on a large number of previously untested industrial and other chemicals that are encountered by people in various ways. It is highly probable that through these efforts we will identify additional carcinogens in the environment. In assessing the total health impact of environmental chemicals, including their possible role in the causation of human cancers, and in devising strategies for minimizing their impact, it is important to recognize that many chemicals of importance in this regard are of natural, not human, origin.

Such naturally occurring chemicals can occur in a variety of sources, but most are encountered by man as components of either foods or beverages. In some circumstances, carcinogens are present as normal constituents of plants or plant products used for foods, herbal medicines, or teas. Other naturally occurring carcinogens are not ordinarily present, but appear as accidental contaminants of foods or beverages, as in the case of compounds produced by spoilage fungi growing on stored food crops.

The purpose of this chapter is to provide a summary of current information on known carcinogens of natural origin and of the available evidence bearing on assessment of their significance to man. In organizing a review of a broad field such as this, a problem inevitably arises in delineating the scope of the discussion, since an exhaustive review is neither feasible nor desirable for the present purpose. All of those normal constituents of plants used for foods (or otherwise ingested) for which there is well-documented evidence of carcinogenicity are included. Aflatoxins and other microbial products are also discussed in some detail, because they have been more extensively studied than any other natural carcinogen and because aflatoxin B_1 represents one of the most potent carcinogens known. Furthermore, the aflatoxins are the only group of widely distributed carcinogens for which quantitative data relating human exposure to cancer incidence are available. On the other hand, important carcinogens in the classes of N-nitrosamines and N-nitrosamides are not included even though they can also occur as food contaminants. Although compounds of these chemical types can result from natural processes (e.g., nitrosation of amines or amides present in foods), they have been excluded on the grounds that current evidence indicates that they are present or formed mainly in processed meats, fish, or other animal products and are therefore beyond the scope of this book.

II. MYCOTOXINS AND OTHER MICROBIAL PRODUCTS

A. Aflatoxins

The aflatoxins are a group of secondary fungal metabolites that are potent animal toxins and carcinogens and have been epidemiologically implicated as environmental carcinogens in man.

Epizootic outbreaks of hepatic necrosis resulting in the deaths of thousands of turkey poults ("turkey X disease"), ducklings, and chicks in England in 1960 and 1961 were eventually traced to feed contamination, specifically a shipment of Brazilian peanut meal used as a protein supplement in poultry feed. This meal, termed Rossetti meal because of the name of the ship in which it was imported, proved to be both toxic and carcinogenic and was found to be contaminated with the common fungus *Aspergillus flavus* (Sargeant *et al.*, 1961, 1963). The active principles were extracted and isolated from *A. flavus* cultures by groups in England and The Netherlands (van der Zijden *et al.*, 1962; Nesbitt *et al.*, 1962) and chemically identified by a group in the United States (Asao *et al.*, 1963). Coincidentally, an outbreak of hepatoma in hatchery-reared rainbow trout in Washington State in 1960 was also traced to aflatoxin contamination of the fish feed.

Because of the biological potency of the aflatoxins and the fact that aflatoxin-producing mold strains are widely dispersed in air and soil and are capable of growth on a variety of natural substrates, extensive studies relating to the occurrence, biological activity, metabolism, and biochemical mode of action have been carried out.

A monograph (Goldblatt, 1969) and several reviews (Detroy *et al.*, 1971; Wogan and Pong, 1970; Wogan, 1973; Butler, 1974) provide an overview of the aflatoxin literature.

1. Chemistry and Occurrence

The aflatoxins are highly substituted coumarins containing a fused dihydrofurofuran moiety and are named because of their production by the fungus *Aspergillus flavus* (*A. flavus* toxins). Four major aflatoxins are produced. Aflatoxins B_1 and B_2 (AFB_1 and AFB_2) are so designated because of their strong blue fluorescence under ultraviolet light; aflatoxins G_1 and G_2 (AFG_1 and AFG_2) fluoresce greenish-yellow. Their chemical structures are shown in Fig. 1.

The aflatoxins are soluble in methanol, chloroform, and other polar solvents but are only sparingly soluble in water (10–30 μg/ml). The toxins strongly absorb ultraviolet light (362–363 nm) with extinction coefficients varying from 14,700 for AFB_2 to 21,800 for AFB_1. Fluorescence emission is at 425 nm for AFB_1 and AFB_2 and at 450 nm for AFG_1 and AFG_2.

The aflatoxins are produced by only a few strains of *A. flavus* and *A. parasiticus*. Those strains capable of toxin production generally synthesize only two or three aflatoxins under a given set of conditions, one of which is always AFB_1, the most potent toxin and carcinogen of the group. Aflatoxins in the G series almost always occur in lesser amounts than AFB_1 (Wogan, 1977).

Requirements for toxin production are relatively nonspecific, since the mold can produce aflatoxins on virtually any food or synthetic media that will support

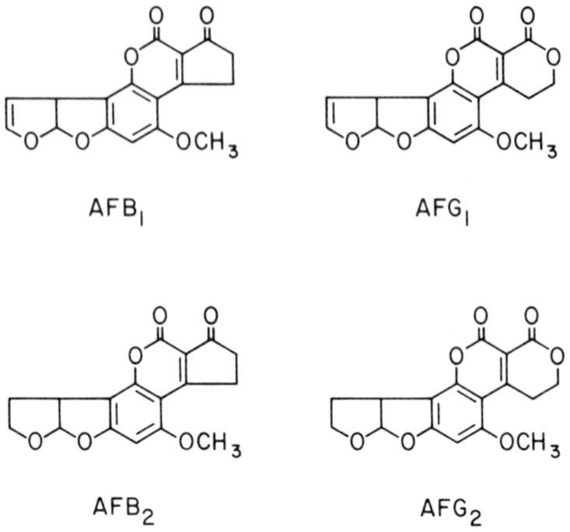

FIG. 1. Structures of aflatoxins.

growth. This is evidenced by the diverse types of foodstuffs contaminated with aflatoxin. Most contamination is associated with grains and nuts that have not been adequately dried at harvest and that have been stored at relatively high temperatures. In the United States, AFB_1 has been most often identified in commercial samples of peanuts, various nuts, cottonseed, corn, and figs (Stoloff, 1976). In Uganda, over 70% of the bean samples were contaminated (Alpert *et al.*, 1971). Maize, peanuts, cereal grains, peas, and cassava were contaminated to a lesser degree, although 50% of the cassava samples and 20–30% of the sorghum and peanut samples contained over 1 mg aflatoxin per kilogram sample. Peanuts, corn, chili peppers, and millet samples from Thailand markets were numerically the most contaminated commodities, with an incidence greater than 10% (Shank *et al.*, 1972a).

A second source of aflatoxin exposure is through the consumption of milk from animals being fed AFB_1-contaminated feed. In this instance, AFM_1, a product of AFB_1 metabolism, is excreted in the milk. This topic has been thoroughly reviewed (Purchase, 1972).

Several minor aflatoxins may also be produced by the growth of certain *A. flavus* strains. AFB_{2a} and AFG_{2a}, the 2-hydroxy derivatives of AFB_1 and AFG_1, respectively, were isolated from fungal cultures (Dutton and Heathcote, 1968). Small quantities of AFM_1 were also detected.

2. *Carcinogenicity*

Aflatoxins have been shown to have carcinogenic activity in many species of animals, including rodents, nonhuman primates, birds, and fish. The liver was the organ principally affected, with the toxins inducing hepatocellular carcinomas and other tumor types. However, under some circumstances, a significant incidence of tumors at sites other than the liver has been recorded. These sites were dependent on the specific aflatoxin used, species, strain, and toxin dose level.

The majority of published information on aflatoxin carcinogenicity involves experiments in rats with AFB_1, in which this toxin has a high order of potency. In addition to experiments dealing with dose–response characteristics, influences of such factors as route, dosing regimen, strain, sex, and age have been investigated. Also, effects of various other modifying factors on carcinogenic responses have been evaluated, including diet, hormonal status, liver injury, microsomal enzyme activity, and concurrent exposure to other carcinogenic agents. The aflatoxin carcinogenesis literature has been reviewed in detail (Wogan, 1973).

Considerable variation exists among rat strains with respect to sensitivity to carcinogenic effects of low levels of AFB_1. However, a dietary level of 1.0 ppm consistently induced liver carcinoma at a high incidence in several rat strains even when feeding was not continued throughout the entire period of observation, as shown by representative data in Table I. In contrast, inbred mice showed no evidence of carcinogenic response at this level. In fact, a dietary level of 150 ppm was ineffective in inducing liver tumors in a random-bred mouse strain fed that level for their entire life span. On the other hand, infant mice of an F_1 hybrid strain (C57B1 × C3H) developed a high incidence of hepatocellular carcinomas when given repeated injections of AFB_1 during the perinatal period (Vesselinovitch *et al.*, 1972).

The primary target organ for AFB_1 in most rat strains is the liver. Hepatocellular carcinomas and cholangiocarcinomas were induced, among other lesions. Tumors of other tissues have been observed in AFB_1-treated rats. These included carcinomas of the glandular stomach and mucinous adenocarcinomas of the colon. Both are infrequently observed in rats treated with carcinogens, and although only a relatively small number have been observed in aflatoxin-treated animals, a causative association has been suggested (Newberne and Wogan, 1968). More recent experiments suggested that the incidence of colon tumor may be enhanced by vitamin A deficiency (Newberne and Rogers, 1973).

Elevated incidences (23–57%) of renal epithelial neoplasms were reported in male Wistar rats given 0.25–1 ppm AFB_1 in the diet for 147 days. Liver tumors appeared in 62–86% of the animals (Epstein *et al.*, 1969).

Published evidence of the carcinogenicity of aflatoxin B_1 in nonrodent species is summarized in Table II. Liver carcinomas were induced in subhuman primates in several experiments. Although the total number of positive responses was small, it seemed clear that AFB_1 was an effective carcinogen for the liver of these species. Among the remaining animal species that have been studied, the rainbow trout and duck appeared to be of comparable sensitivity, responding to levels of 4–30 ppb in the diet. Effective levels in the guppy and ferret were in the

TABLE I

HEPATOCARCINOGENICITY OF AFLATOXIN B_1 IN RODENTS

Species	Dosing regimen	Duration of treatment	Period of observation	Liver tumor incidence	Reference
Rat, Fischer	1.0 ppm in diet	33 wk	52 wk	3/6	Svoboda *et al.* (1966)
Rat, Fischer	1.0 ppm in diet	41–64 wk	Same	18/21	Wogan and Newberne (1967)
Rat, Fischer	55 ppb in diet	71–97 wk	Same	20/25	Wogan *et al.* (1974)
Rat, Porton	1.0 ppm in diet	20 wk	90 wk	19/30	Butler (1969)
Rat, Wistar	1.0 ppm in diet	21 wk	87 wk	12/14	Epstein *et al.* (1969)
Mouse, Swiss	150 ppm in diet[a]	20 mo	Same	0/60	Wogan (1973)
Mouse, C57B1/6NB	1.0 ppm in diet	20 mo	Same	0/30	Wogan (1973)
Mouse, C3HfB/HEN	1.0 ppm in diet	20 mo	Same	0/30	Wogan (1973)
Mouse, hybrid F_1, 4 days old	6.0 μg/gm body weight	3 doses i.p.	80 wk	16/16	Vesselinovitch *et al.* (1972)

[a] A mixture of aflatoxins B_1 and G_1 was used in this experiment.

TABLE II

HEPATOCARCINOGENICITY OF AFLATOXIN B_1 IN NONRODENT SPECIES

Species	Dosing regimen	Duration of treatment	Preiod of observation	Liver tumor incidence	Reference
Monkey, Rhesus (male)	1.655 gm total[a]	5.5 yr	8.0 yr	1/1	Gopalan *et al.* (1972)
Monkey, Rhesus (female)	1.655 gm total	5.5 yr	10.75 yr	1/1	Tilak (1975)
Monkey, Rhesus (female)	0.504 gm total	6.0 yr	8.0 yr	1/1	Adamson *et al.* (1973)
Monkey, Rhesus (male and female)	99–842 mg total	3.8–6.0 yr	4.2–6.0 yr	3/3	Adamson *et al.* (1976)
Marmoset	3.0 mg total	50–55 wk	Same	1/3	Lin *et al.* (1974)
	5.04–5.84 mg total[b]	87–94 wk	Same	2/3	
Tree shrew (male and female)	24–66 mg total	74–172 wk	Same	9/12	Reddy *et al.* (1976)
Ferret	0.3–2.0 ppm in diet	28–37 mo	Same	7/9	Butler (1969)
Duck	30 ppb in diet	14 mo	Same	8/11	Carnaghan (1965)
Rainbow trout	4 ppb in diet	12 mo	Same	15%	Sinnhuber *et al.* (1968)
	8 ppb in diet	12 mo	Same	40%	
	20 ppb in diet	12 mo	Same	65%	
Rainbow trout embryos	0.5 ppm in water	1 hr	296–321 days	38%	Sinnhuber and Wales (1974)
Salmon	12 ppb in diet[c]	20 mo	Same	50%	Wales and Sinnhuber (1972)
Guppy	6 ppm in diet	11 mo	Same	7/113	Sato *et al.* (1973)

[a] A mixture of aflatoxins B_1 and G_1 was used in this experiment.
[b] These animals were infected simultaneously with hepatitis virus.
[c] This diet also contained 50 ppm cyclopropenoid fatty acids.

parts per million range, and the salmon responded only when exposed concurrently to AFB_1 and cyclopropene-containing fatty acids that have a cocarcinogenic property in trout.

Numerous reports have dealt with factors of various kinds that modify the carcinogenic response of rats to AFB_1, including experimental alteration of nutritional and endocrine status, various types of liver insults, and simultaneous exposure to aflatoxins and other pharmacologically active compounds.

Because of the possibility that exposure to aflatoxins could occur in some human populations also suffering from malnutrition, the influence of nutritional status on aflatoxin carcinogenesis in rats has received attention by a number of investigators. The effect of dietary protein has been evaluated with somewhat contradictory results. Madhavan and Gopalan (1968) found that rats fed a low-protein (5%) diet developed liver tumors at a lower incidence than controls fed a 20% protein ration. On the other hand, Newberne *et al.* (1966) found that diets containing 9% protein resulted in a higher incidence of liver tumors in a shorter period of time than a diet containing 22% protein when both groups of rats were intubated with 375 μg AFB_1 per animal. The reasons for these disparate results are unclear, and this important problem requires further investigation.

Rogers and Newberne (1969, 1971) focused attention on the possible importance of simultaneous occurrence of marginal insufficiency of dietary lipotropic agents and aflatoxin exposure, and studied these conditions in rats. In contrast to the protective effect observed with a toxic dose of AFB_1, animals with a lipotrope deficiency developed liver tumors much earlier and at a higher frequency than did control animals treated with a carcinogenic regimen. The mechanisms responsible for these interactions have not been identified.

Simultaneous administration of AFB_1 with other pharmacologically active compounds, including other carcinogens, has been investigated. Reddy and Svoboda (1972) studied the effects of lasiocarpine, a pyrrolizidine alkaloid that is carcinogenic and strongly hepatotoxic (see Section V) on carcinogenesis in rats by feeding AFB_1 at 2 ppm. They found that the alkaloid did not prevent initiation of liver tumors by aflatoxin but did alter the pathogenic pattern in which the tumors developed. McLean and Marshall (1971) demonstrated a protective action of phenobarbitone against the carcinogenicity of AFB_1. Continuous administration of phenobarbitone in drinking water during a 9-week period in which AFB_1 was fed at 5 ppm in the diet resulted in a lower incidence and the delayed appearance of liver tumors in rats when compared with rats fed aflatoxin alone. These investigators suggested that phenobarbitone induced liver microsomal enzymes that metabolized the aflatoxin to noncarcinogenic products. Similar results were obtained by Swenson *et al.* (1977).

The hormonal status of the animal exerts a profound effect on AFB_1 carcinogenesis. Goodall and Butler (1969) studied the effects of hypophysectomy on rats fed a diet containing 4 ppm AFB_1. Whereas 14/14 intact animals developed liver tumors in 49 weeks, no tumors developed in 14 hypophysectomized animals in the same period, despite the fact that they received AFB_1 at a higher rate than the controls. Newberne and Williams (1969) reported that rats fed diets containing 0.2 ppm AFB_1 and 4 ppm diethylstilbestrol developed fewer liver tumors (20% incidence) than animals fed AFB_1 alone (71% incidence). Castration prevented the development of liver tumors in carbon tetrachloride-treated rats of both sexes (Cardeilhac and Nair, 1973). The animals were treated with ten doses of AFB_1 and carbon tetrachloride simultaneously over a 166-day period. Intact rats of both sexes treated with AFB_1 alone developed tumors. The effects of insulin, growth hormone, and adrenocorticotropin (ACTH) on AFB_1-induced liver tumors were investigated by Chédid *et al.* (1977). Growth hormone administration did not alter the 100% tumor incidence obtained with AFB_1 alone. The group treated with insulin plus AFB_1 had only a 60% rate of tumor development and a 20% incidence of myelocytic leukemia, whereas ACTH treatment (4 units/week for 20 weeks) totally blocked tumor formation. High rates of malignant lymphoma and lymphoid hyperplasia were observed in the ACTH-treated group.

A carcinogenic potency series for aflatoxins B_1, B_2, and G_1 was established in the rat (Butler *et al.*, 1969; Wogan *et al.*, 1971). AFB_1 was more carcinogenic than AFG_1, which in turn was more potent than AFB_2. This carcinogenic relationship was also verified in the rainbow trout (Ayres *et al.*, 1971). Aflatoxin M_1 has been demonstrated to be carcinogenic in the rat and rainbow trout. Synthetic, racemic AFM_1 induced a single liver tumor after 96 weeks in a group of 29 rats treated with a total dose of 1 mg AFM_1 (Wogan and Paglialunga, 1974). Studies with rainbow trout (Sinnhuber *et al.*, 1974) resulted in a 60% incidence of liver tumors after feeding 16 ppb AFM_1 in the diet for 12 months.

3. *Metabolism*

Aflatoxins are primarily metabolized by the microsomal mixed-function oxidase system, a complex organization of cytochrome-coupled, O_2- and NADPH-dependent enzymes located mainly on the endoplasmic reticulum of liver cells but also present in kidney, lungs, skin, and other organs. These enzymes oxidatively metabolize a wide variety of foreign or xenobiotic compounds, with the net result being the detoxification of the parent compound by the formation of various hydroxylated derivatives, which in turn are conjugated with sulfate or glucuronic acid to form water-soluble glucuronide or sulfate esters. These conjugates can then be readily excreted in the urine or bile. During the course of metabolism, certain highly reactive metabolites may be generated which have the capacity to react covalently with various nucleophilic centers in cellular macromolecules such as DNA, RNA, and protein. This "activation" reaction poses a biological hazard to the cell and constitutes a plausible theory of the manner in which certain compounds exert toxic and carcinogenic effects (Miller, 1970; Jollow *et al.*, 1977).

The established pathways of AFB_1 metabolism along with certain likely, but unproved, reactions are depicted in Fig. 2. Many of these reactions share a commonality with mechanisms by which polynuclear aromatic hydrocarbons, a large class of lung and skin carcinogens, are metabolized (Jerina and Daly, 1974; Sims, 1976). Recent comprehensive reviews of aflatoxin metabolism have been published (Campbell and Hayes, 1976; Shank, 1977a).

a. Aflatoxin Metabolites and Pathways. i. AFM_1. Ring hydroxylation of AFB_1 at the 4 position (the 10 position according to recent IUPAC recommendations) forms AFM_1. This metabolite was first detected in the milk of cows ingesting AFB_1 (Allcroft and Carnaghan, 1963; Masri *et al.*, 1967). It was produced *in vitro* from AFB_1 by liver microsomes from a variety of species, including human beings (Masri *et al.*, 1974a; Roebuck and Wogan, 1977; Merrill and Campbell, 1974) and excreted in the urine of AFB_1-treated sheep (Allcroft *et al.*, 1966) and monkeys (Dalezios and Wogan, 1972) as well as in the bile of rats (Bassir and Osiyemi, 1967). AFM_1 was also detected in the urine of human beings consuming AFB_1-contaminated peanut butter (Campbell *et al.*, 1970). The glucuronide of AFM_1 was reported from the tissue and excreta of chickens (Mabee and Chipley, 1973). AFM_1 is both toxic and carcinogenic.

ii. AFB_{2a}. This hemiacetal metabolite is produced from AFB_1 by hydration of the 2,3-vinyl ether double bond resulting in hydroxylation at the 2 position. It was readily formed from AFB_1 under mildly acidic conditions (Pohland *et al.*, 1968) and *in vitro* with chick, duckling, guinea pig, and mouse microsomes (Patterson and Roberts, 1970, 1971a). AFB_{2a} was apparently unstable under physiological conditions (pH 7.4) and bound to protein and hepatic microsomes (Patterson and Roberts, 1970; Gurtoo and Dahms, 1974). The AFB_{2a} was thought to cleave to form dialdehyde derivatives, which then covalently bound to the amino groups of proteins by Schiff base formation (Gurtoo and Campbell, 1974).

iii. AFP_1. This metabolite was produced by the O-demethylation of AFB_1 and was the major excretory product in the urine of AFB_1-treated rhesus monkeys (Dalezios *et al.*, 1971), where it was present as glucuronide and sulfate conjugates. It was formed *in vitro* by mouse, monkey, and human microsomes (Roebuck and Wogan, 1977).

iv. AFQ_1. Aflatoxin Q_1 is formed from AFB_1 by ring hydroxylation of the carbon atom β to the carbonyl function of the cyclopentenone ring. It was identified as a major *in vitro* AFB_1 metabolite using monkey liver (Masri *et al.*, 1974b) and human liver microsomes (Büchi *et al.*, 1974). It was also produced *in vitro* by chicken, rat, and mouse microsomes (Masri *et al.*, 1974a; Roebuck and Wogan, 1977).

v. Aflatoxicol (AFL). Reduction of the cyclopentenone carbonyl function of

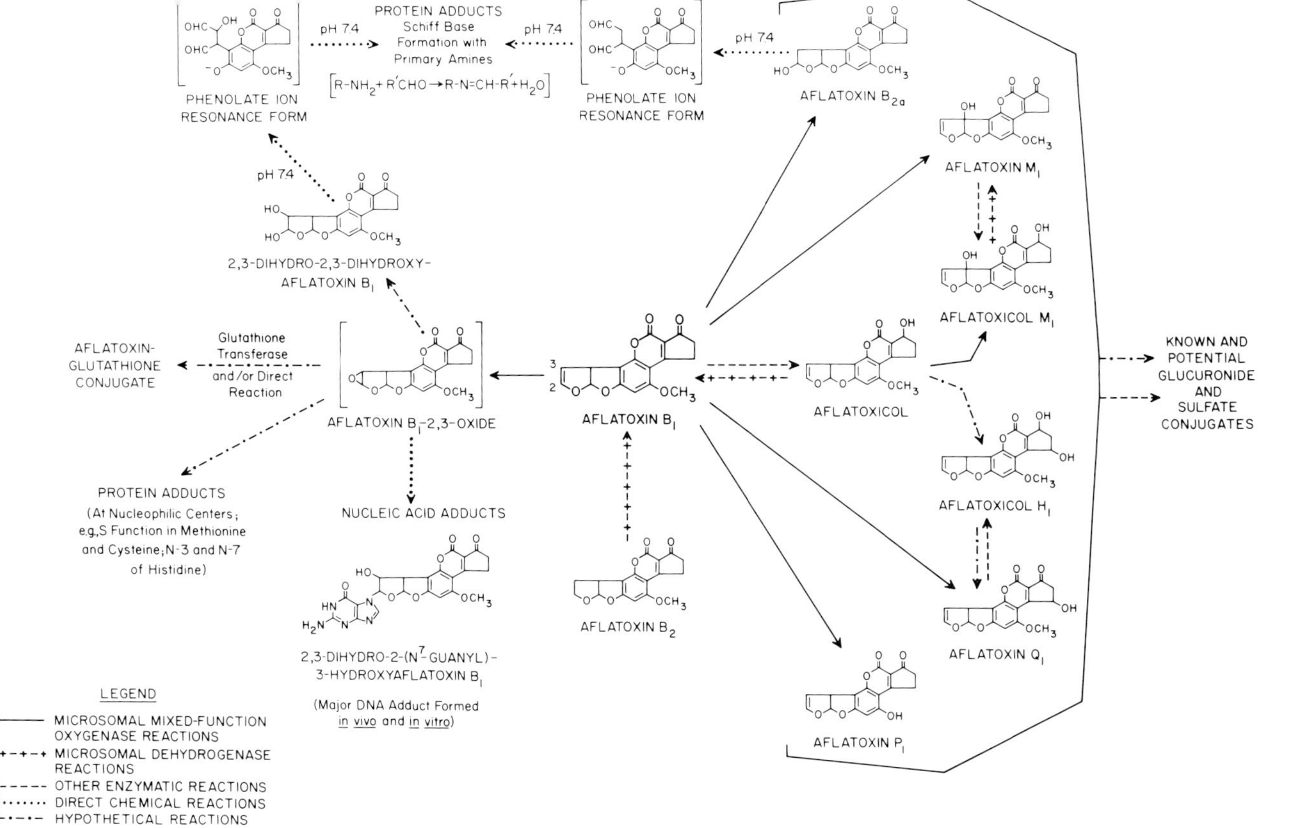

FIG. 2. Metabolic transformation of aflatoxin B_1.

AFB_1 by an NADPH-dependent cytoplasmic enzyme produced aflatoxicol (Patterson and Roberts, 1971a). It would, therefore, not be formed in *in vitro* microsomal incubations that lack the soluble enzyme fraction. Rabbit and bird liver homogenates were active aflatoxicol producers, whereas rodent and sheep preparations were inactive (Patterson and Roberts, 1972). Fish liver fractions were also efficient metabolizers of AFB_1 to aflatoxicol (Schoenhard *et al.*, 1976), especially when oxicative metabolism was blocked by carbon monoxide (Salhab and Edwards, 1977a). A reverse reaction, which generated AFB_1 from aflatoxicol, was attributed to a microsomal aflatoxicol dehydrogenase activity, which utilized NADP as a hydrogen acceptor (Salhab and Edwards, 1977b). This dehydrogenase activity was not inhibited by carbon monoxide, as was the microsomal mixed-function oxidase system, and was detected in the human, monkey, dog, rat, and mouse. The relatively high mutagenic potential of aflatoxicol (Wong and Hsieh, 1976) may be due to its metabolism to AFB_1.

vi. Aflatoxicol H_1 ($AFLH_1$). Salhab and Hsieh (1975) reported a new aflatoxin metabolite, which was formed from AFB_1 using both the microsomes and a soluble enzyme preparation from human or monkey liver. The compound is a dihydroxyl derivative of AFB_1 with substitutions at the cyclopentenone carbonyl function and the β-carbon (Fig. 2). $AFLH_1$ was formed from AFQ_1 with a soluble enzyme preparation. It was nontoxic in the chicken embryo test. Salhab and Edwards (1977b) demonstrated the formation of $AFLH_1$ from aflatoxicol with human, monkey, rat, and mouse liver homogenates. $AFLH_1$ may arise from the oxidation of aflatoxicol as well as the reduction of AFQ_1, but direct confirmation of this reaction has not been reported.

vii. Aflatoxicol M_1 ($AFLM_1$). Oxidation of aflatoxicol with a dog liver microsomal preparation produced $AFLM_1$ (Salhab *et al.*, 1977). This compound proved to be the M_1 derivative of aflatoxicol and was generated by a microsomal mixed-function oxidation analogous to the conversion of AFB_1 to AFM_1. $AFLM_1$ was also formed by the reduction of AFM_1 by rabbit liver cytosol enzymes, which were shown capable of reducing AFB_1 to aflatoxicol (Salhab and Edwards, 1977b). The back-reaction ($AFLM_1$ to AFM_1) was catalyzed by a human liver NADP-dependent, carbon monoxide-insensitive microsomal dehydrogenase previously shown to convert aflatoxicol to AFB_1.

viii. AFB_2. Roebuck *et al.* (1978) reported the formation of AFB_1 when AFB_2 was incubated with duckling liver postmitochondrial supernatant. Swenson *et al.* (1977) noted that the same aflatoxin derivative (2,3-dihydro-2,3-dihydroxyaflatoxin B_1) was released by acid hydrolysis from the nucleic acids of both AFB_1- and AFB_2-treated rats. On this basis, they postulated an AFB_2 to AFB_1 conversion by desaturation of carbons 2 and 3 of AFB_2. This pathway could account for the weakly toxic and carcinogenic properties of AFB_2.

ix. Aflatoxin–glutathione conjugate. A polar, fluorescent, ninhydrin-positive compound was detected in *in vitro* incubations of AFB_1, rat liver microsomes, and reduced glutathione. No compound was detected in the absence of microsomes or glutathione (Raj *et al.*, 1975).

b. Macromolecular Binding of Aflatoxin. The covalent incorporation of aflatoxin into nucleic acids and proteins is now considered to be an important mechanism by which AFB_1 initiates its toxic and carcinogenic effects. The known and postulated pathways for the macromolecular binding of AFB_1 are summarized in Fig. 2.

It was recognized early that AFB_{2a}, formed by metabolic hydroxylation of AFB_1, could spontaneously bind covalently to proteins at physiological pH (Patterson and Roberts, 1970; Gurtoo and Dahms, 1974). AFB_{2a} was essentially nontoxic; therefore, this reaction seemed to lack relevance to the disruption of cellular function and has not been investigated further.

In vitro incubation of AFB_1, rat liver microsomes, and appropriate cofactors with RNA, DNA, and polynucleotides resulted in AFB_1 binding to these nucleic acids at levels as high as one aflatoxin residue per 30 nucleotides (Garner *et al.*, 1972; Garner, 1973). Swenson *et al.* (1973) isolated 2,3-dihydro-2,3-dihydroxyaflatoxin B_1 by mild acid hydrolysis of RNA adducted *in vitro* with liver microsomes and AFB_1. No binding was observed with AFB_2, demonstrating that 2,3 unsaturation of the terminal furan ring was a requisite for activation. This information suggested the hypothesis that, on chemical grounds, aflatoxin B_1 2,3-oxide was the reactive precursor and probably the ultimate carcinogenic metabolite of AFB_1. The 2,3-dihydro-2,3-dihydroxyaflatoxin B_1 was also isolated from liver DNA and rRNA of rats treated with AFB_1 (Swenson *et al.*, 1974). Although binding of AFB_1 to liver proteins was observed, the level was only 4–7% of that observed with nucleic acids. Attempts to isolate the highly reactive 2,3-oxide have been unsuccessful. Swenson *et al.* (1975) synthesized a more stable model compound, aflatoxin B_1 2,3-dichloride. This electrophilic analogue of the epoxide was considerably more potent than AFB_1 as a bacterial mutagen and in the induction of various rat and mouse tumors. The AFB_1 dichloride also reacted spontaneously with DNA, RNA, protein, and amino acids with nucleophilic centers, such as cysteine, histidine, and lysine. The AFB_1 dichloride was hydrolyzed in aqueous solutions to yield the 2,3-dihydro-2,3-dihydroxyaflatoxin B_1, which proved especially reactive toward protein, presumably via dialdehyde derivatives forming Schiff bases with the primary amino groups of protein (Swenson *et al.*, 1975). The major DNA adduct formed by AFB_1 *in vitro* (approximately 90% of the AFB_1 bound to DNA) has been isolated and identified as 2,3-dihydro-2-(N^7-guanyl)-3-hydroxyaflatoxin B_1 (Essigmann *et al.*, 1977). This adduct is also the major product formed *in vivo* in rat liver (Lin *et al.*, 1977; Croy *et al.*, 1978).

The relative biological hazard posed by the formation of aflatoxin adducts with different classes of cellular macromolecules is unknown. However, in studies comparing *in vivo* macromolecular binding of AFB_1 (toxic) and AFB_2 (relatively nontoxic) to rat liver DNA, rRNA, and protein, Swenson *et al.* (1977) noted that AFB_2 bound to the nucleic acids at approximately 1% the level of AFB_1, whereas protein binding was 35–70% of that observed with AFB_1. This suggested that aflatoxin–protein adducts were relatively unimportant in the manifestation of toxic and carcinogenic effects.

c. Factors Affecting Metabolism. As discussed previously, many factors, such as phenobarbital treatment, sex, strain, dietary factors, and hormonal status, alter the biological response to aflatoxins in rats. The same factors quantitatively affect the metabolism of aflatoxin as well. There has been very good correlation, with regard to the parameters studied, between the toxicity and carcinogenicity of AFB_1 and the capacity of animals to bind AFB_1 to liver DNA under *in vivo* conditions.

Phenobarbital treatment of rats stimulated *in vitro* AFB_1 metabolism in terms of increased formation of the hydroxylated aflatoxin metabolites, AFM_1 and AFP_1 (Schabort and Steyn, 1969; Patterson and Roberts, 1971b). It also resulted in an approximately three fold increase in binding of AFB_1 to DNA *in vitro* (Gurtoo and Bejba, 1974). The *in vivo* situation was in direct opposition to the *in vitro* results. Garner (1975) noted a 73% decrease in AFB_1 binding to liver DNA and a 40% decrease in binding to liver rRNA after phenobarbitone treatment. No significant difference in AFB_1 binding to protein was observed. Similar, but more pronounced, results were obtained by Swenson *et al.* (1977). The reason for this apparent contradiction may be the lack of cytosol enzymes (e.g., the glutathione transferases) and glutathione in the microsomal preparations used for *in vitro* assays, which would otherwise serve as a detoxification pathway and shunt "activated" AFB_1 (the aflatoxin B_1 2,3-oxide) away from binding to cellular nucleic acids (Neal and Godoy, 1976). Glutathione transferase activity was inducible by phenobarbital (Hales and Neims, 1977).

Garner and Wright (1975) compared the kinetics of *in vivo* AFB_1 binding to DNA, rRNA, and protein in a sensitive and a relatively insensitive species (rat and hamster, respectively). In addition, target (liver) and nontarget (kidney) organ macromolecular binding was quantitated. Generally, binding of AFB_1 to rat liver DNA and rRNA reached higher levels in a shorter time and was removed (repaired) at a more rapid rate than in hamster liver. Equivalently low levels of binding in rat and hamster liver nucleic acids were reached after 48 hr. Similar levels of binding were noted in the rat and hamster kidney at all time points. It should be noted, however, that a single acutely toxic, but not carcinogenic, dose was administered in this instance. It would be important to determine whether or not a carcinogenic regimen (repeated or continuous dosing of AFB_1) could alter the pattern and persistence of AFB_1-macromolecular binding.

Hormonal status and sex were significant determinants of aflatoxin response.

Hypophysectomy reduced AFB_1 binding to DNA to about one-half of the control values (Swenson *et al.*, 1977). Smaller effects were noted with rRNA binding, and no effect on protein binding was observed. Gurtoo and Motycka (1976) investigated the effect of sex, castration, and hormone replacement therapy on the metabolism of AFB_1 to AFM_1 and AFQ_1 and of AFB_1 binding to DNA *in vitro* and *in vivo*. Female rat liver microsomes had approximately one-half the activity of male rat microsomes in the *in vitro* metabolism assays. Ovariectomy did not alter this activity. Castration of male rats, however, resulted in AFB_1 metabolism being reduced to approximately the levels observed with female rat microsomes. Testosterone treatment of castrated males enhanced microsomal activity to nearly normal levels. Treatment of intact females with testosterone also increased the *in vitro* metabolic activity of microsomes, often to levels greater than those of intact males. Binding of AFB_1 to liver DNA *in vivo* was investigated in a single experiment using intact and testosterone-treated female rats. An intact female rat bound 30% less AFB_1 to DNA than an intact male. In confirmation of the *in vitro* experiments, the levels of AFB_1 binding in testosterone-treated female rats were similar to those in male rats.

Only limited data on the effect of dietary factors on AFB_1 metabolism are currently available. Protein deficiency, induced by feeding rats a 5% casein diet for 16 days, reduced AFB_1 binding to DNA by about 70% (Preston *et al.*, 1976).

Profound strain differences in rats were observed with the *in vitro* metabolism of AFB_1 (Gurtoo and Motycka, 1976). Sprague–Dawley rats were approximately 2.5 times more efficient metabolizers of AFB_1 (in terms of AFQ_1 production and AFB_1–DNA binding) than Long–Evans rats. Wistar and Fischer rats were intermediate in this respect.

d. Excretion and Tissue Distribution. The excretion and distribution of AFB_1 in animals are significant in terms of monitoring exposure to environmental aflatoxins, in determining tissue residues of aflatoxin for half-life studies, and as a means of studying the mechanisms of aflatoxin metabolism. Very few quantitative distribution or balance studies have been performed, with most reports being concerned with mammals and birds of economic importance.

Excretion patterns of AFM_1 were studied in the milk, feces, and urine of sheep (Allcroft *et al.*, 1966; Nabney *et al.*, 1967; Stoloff *et al.*, 1971) and excretion and tissue distribution of ^{14}C-labeled AFB_1 were examined in chickens (Mabee and Chipley, 1973). Kinetic studies of aflatoxin excretion in the urine, feces, and milk of lactating cows revealed that a threshold dietary level of 46 ppb AFB_1 was required before AFM_1 was detectable in milk (Polan *et al.*, 1974).

Quantitative excretion and distribution data for ring-labeled AFB_1 were determined in the rat (Wogan *et al.*, 1967). Approximately 90% of a single i.p. AFB_1 dose was excreted or present in intestinal tract contents within 24 hr. Feces represented the principal excretory route, accounting for up to 75% of the dose, with urine containing an additional 20%. Only 0.3–0.5% of the dose was recovered as CO_2. Maximal levels of aflatoxin radioactivity occurred in tissues (mainly the liver) 30 min after dosing. After 24 hr approximately one-half of the total retained AFB_1 dose was still present in the liver.

Urinary excretion patterns were studied in male Rhesus monkeys given a single i.p. injection of ring-labeled [^{14}C]AFB_1 (Dalezios and Wogan, 1972). A maximal rate of excretion was attained within 1 hr and 35% of the total dose was recovered in the urine after 4 days. Approximately 15% of this activity was chloroform soluble and included AFM_1, AFP_1, AFB_1, and several unidentified metabolites. The glucuronide of AFP_1 represented the major water-soluble component, with lesser amounts of the AFP_1-sulfate conjugate being present. Aflatoxin residues were primarily retained by the liver (6% of the original dose after 4 days) with much lower levels detectable in other tissues. Similar experiments using orally administered [^{14}C]AFB_1 adsorbed on casein showed that 80% of the total dose was excreted in the urine and feces after 1 week (Dalezios *et al.*, 1973). This pattern was noted with both high (0.4 mg/kg) and low (0.15 mg/kg) AFB_1 doses. Detectable radioactivity was still present in the urine and blood after 5 weeks. The proportion of chloroform-soluble metabolites in the urine was initially high (74% in the first day) but decreased rapidly to 10% or less of the total urinary metabolites by 3 days. Unlike the study with i.p.-administered AFB_1 (Dalezios and Wogan, 1972), no unconjugated AFP_1 was detected in urine, and considerably reduced levels of AFP_1 conjugates were present. Maximal fecal excretion of aflatoxin residues was noted 2 to 3 days postdosing. AFM_1 and AFB_1 were the major organic-soluble components, accounting for 15% of the total fecal radioactivity. Again the liver retained significant quantities of aflatoxin residues (approximately 1% of total dose after 5 weeks).

4. *Aflatoxins and Human Liver Cancer*

Aflatoxins are among the few chemically identified and widely disseminated environmental carcinogens for which quantitative estimates of human exposure have been systematically sought. Despite the fact that significant differences in responsiveness are known to exist among animal species, it is reasonable to assume that man might respond to either acute or chronic effects of the toxins in the event that exposure takes place through contamination of dietary components. It also seems reasonable to postulate that the character and intensity of the human response might vary depending on such factors as age, sex, nutritional status, concurrent exposure to other agents (e.g., viral hepatitis or parasitic infestation), as well as the level and duration of exposure to aflatoxins.

As information has accumulated on various aspects of the aflatoxin problem, it has become apparent that the risk of exposure to aflatoxins is much less in technologically developed countries than in developing ones. The lower risk is attributable to the combined effects of several factors contributing to the prevention of contamination of foods or food raw materials. The use of such agricultural practices as rapid postharvest drying of crops and controlled storage conditions tends to reduce mold damage in general and thereby also to reduce the likelihood of aflatoxin contamination.

In societies not equipped technologically to apply such practices, the risk of aflatoxin exposure is clearly much greater. Since their discovery, reports have occasionally been made of the identification of aflatoxins in many kinds of human foods collected in various parts of the world. Although these findings indicated the widespread geographic nature of the problem, they provided little quantitative information on human exposure, because the samples were randomly collected and no actual intake studies were performed.

Information concerning human exposure to aflatoxins has become available from epidemiological studies in Uganda (Alpert *et al.*, 1968, 1971), the Philippines (Campbell and Salamat, 1971), Swaziland (Keen and Martin, 1971), Kenya (Peers and Linsell, 1973), Thailand (Shank *et al.*, 1972b,c), and Mozambique (van Rensburg *et al.*, 1974). Also, AFM_1 was detected in the urine of Filipinos consuming peanut butter heavily contaminated with approximately 500 ppb AFB_1 (Campbell *et al.*, 1970). It was estimated that 1–4% of the ingested AFB_1 was excreted in the urine as AFM_1. All of these investigations were designed to obtain estimates of aflatoxin intake by populations in which primary carcinoma of the liver occurs at different incidences. Aflatoxin carcinogenesis in human beings has been reviewed by Wogan (1975) and Shank (1977b).

Summary data from four of these studies are presented in Table III, which contains information on aflatoxin ingestion and liver cancer incidence, arranged in order of increasing values for each parameter. It can be seen that aflatoxin ingestion varied over a range of values from 3 to 222 ng/kg body weight per day. Estimated liver cancer incidence values extended from a minimum of 2.0 to a maximum of 35.0 cases per 100,000 population per year. There was a positive association between the two parameters, in that high intake values were consistently associated with high incidence rates. The association was most apparent in connection with incidence rates for adult men, for which larger numbers of cases were involved, consequently yielding more precise estimates of disease incidence.

Taken together, these data provide strong circumstantial evidence for a causal relationship between aflatoxin ingestion and liver cancer incidence in human beings. Although this evidence cannot be regarded as proof that aflatoxins are the single cause of liver cell carcinoma in man, the data are sufficient to indicate that exposure to the carcinogen elevates the incidence of this form of cancer and therefore warrant continued investigations into effective means for monitoring and control of their occurrence as food contaminants.

TABLE III

Summary of Current Evidence on Aflatoxin Ingestion and Liver Cancer Incidence

Population	Dietary aflatoxin intake (ng/kg body wt./day)	Cases of liver cancer[a] (No./100,000/year)					
		In adults (>15 years)				In total population	
		Men		Women		Both sexes	
		No.	Incidence	No.	Incidence	No.	Incidence
Kenya (high altitude)[b]	3–5	1	3.1	0	0		
Thailand (Songkhla)[c]	5–8	—	—	—	—	2	2.0
Swaziland (Highveld)[d]	5–9	9	7.1	2	1.4		
Kenya (medium altitude)[b]	6–8	13	10.8	6	3.3		
Kenya (low altitude)[b]	10–15	16	12.9	9	5.4		
Swaziland (Middleveld)[d]	8–15	24	14.8	5	2.2		
Swaziland (Lebombo)[d]	15–20	4	18.7	0	0		
Thailand (Ratburi)[c]	45–77	—	—	—	—	6	6.0
Swaziland (Lowveld)[d]	43–54	35	26.7	7	5.6		
Mozambique[e]	222	—	35.0	—	15.7	—	25.4

[a]Periods covered were 1 year in Thailand; 3 years in Mozambique; 4 years each in Kenya and Swaziland.
[b]Peers and Linsell (1973).
[c]Shank *et al.* (1972b,c).
[d]Keen and Martin (1971).
[e]Van Rensburg *et al.* (1974).

B. Sterigmatocystin and Related Compounds

Sterigmatocystin is a toxic and carcinogenic metabolite of the mold *Aspergillus versicolor*. It was first isolated in 1954 from laboratory cultures of the organism and has been fully characterized with respect to its chemical structure (Fig. 3). Various aspects of the chemistry, biosynthesis, and biological activity of sterigmatocystin and structurally related mycotoxins have been reviewed by Hamasaki and Hatsuda (1977).

The possible occurrence of sterigmatocystin as a food contaminant was suggested when it was isolated from cultures of *Aspergillus versicolor*, *A. nidulans*, and a *Bipolaris* species identified in a survey of cereal and legume crops of South Africa for toxin-producing molds (Holzapfel *et al,* 1966). In later studies on other food-spoilage fungi, aflatoxin-producing strains of *Aspergillus flavus* were found also to produce derivatives of sterigmatocystin, including *O*-methylsterigmatocystin (Burkhardt and Forgacs, 1968), and aspertoxin (Rodricks, 1969). The structures of these compounds are shown in Fig. 3.

These compounds are structurally related to aflatoxins in that both contain the dihydrofurofuran configuration fused to a six-membered ring. This structural feature seems to be an important determinant of the toxicity and carcinogenicity of aflatoxins and perhaps also for other members of the series. Recent investigations have revealed that *Aspergillus versicolor* strains are capable of producing a large number of difuroxanthones and difuroanthraquinones, all of which share this structural similarity (Hamasaki and Hatsuda, 1977). These compounds have not been investigated with respect to their possible toxicity, carcinogenicity, or occurrence as food contaminants.

The carcinogenic properties of sterigmatocystin were first demonstrated by Dickens *et al.* (1966), who found that repeated subcutaneous injections of sterigmatocystin solutions to rats induced sarcomas at the injection sites. Subsequently, Purchase and van der Watt (1968, 1970) showed that feeding or gavage of the compound to rats for 52 weeks caused liver cell carcinomas in a

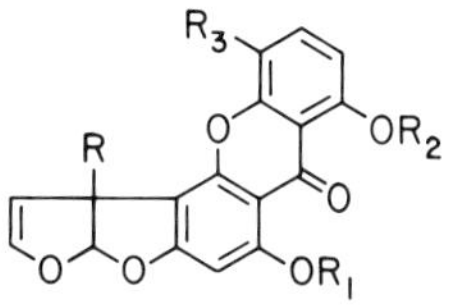

	R	R_1	R_2	R_3
Sterigmatocystin	H	CH_3	H	H
Aspertoxin	OH	CH_3	CH_3	H
O-Methylsterigmatocystin	H	CH_3	CH_3	H

FIG. 3. Structures of sterigmatocystin and congeners.

large proportion of rats surviving doses of 1.5–2.25 mg/rat per day. The potency of sterigmatocystin as a carcinogen was further shown by its ability to induce squamous cell carcinomas and other lesions when applied topically to the skin of rats (Purchase and van der Watt, 1973). In relation to mechanism of action, it is important that aflatoxin B_1 is inactive under these conditions, as noted by van der Watt (1974). Fujii *et al.* (1976) reported that a single subcutaneous injection of sterigmatocystin to newborn mice induced tumors of the liver and lung. The response was dose-related over a range of 0.5–5 μg/gm body weight.

Sterigmatocystin has been found in only a small number of samples of heavily molded wheat and green coffee beans. Its distribution is seemingly limited, since analytical methodology for its detection is available and has been applied fairly extensively in surveillance programs (Stoloff, 1976).

C. Luteoskyrin and Cyclochlorotine

Luteoskyrin (Fig. 4) and cyclochlorotine (Fig. 5) are toxic metabolites of *Penicillium islandicum* Sopp. Although there are no data based on direct chemical analysis concerning the occurrence of these mycotoxins in foods, *P. islandicum* has occasionally been reported as one of the major isolates of rice and other grains in Japan, Ethiopia, and South Africa (Enomoto and Ueno, 1974). Investigations of the mold and its metabolites were stimulated by experimental demonstrations that, when cultures of it were fed to animals, severe liver damage occurred, and by the apparent frequency with which this organism could be found in grains used for human foods.

Cyclochlorotine is a cyclic peptide of high lethal potency to animals. Chronic oral administration to mice at doses of 40–60 μg/animal per day produced a low incidence of liver tumors and reticuloendotheliomas, indicating weak carcinogenic activity. Cirrhosis and hemorrhagic necrosis were also observed (Uraguchi *et al.*, 1972).

Luteoskyrin is a lipid-soluble bisanthraquinone that is also hepatotoxic and weakly carcinogenic for the liver of mice. Feeding of diets containing the pure compound at levels adequate to provide daily intakes of 50–500 μg/animal

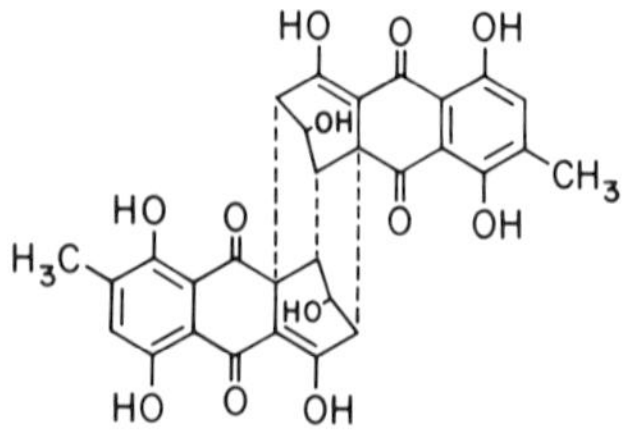

FIG. 4. Luteoskyrin.

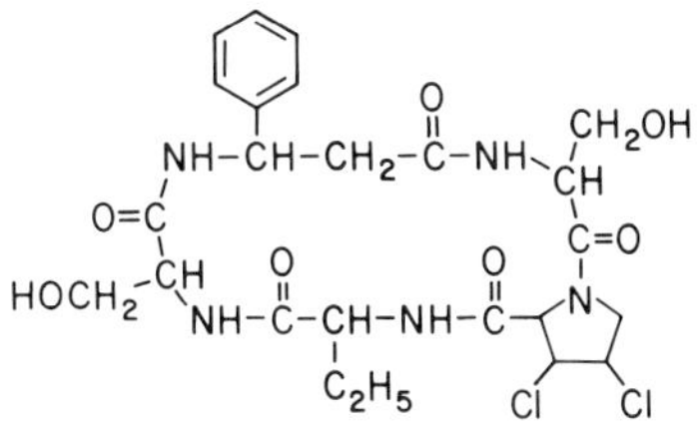

FIG. 5. Cyclochlorotine.

resulted in liver cell adenomas and carcinomas, in addition to liver reticuloendotheliomas (Uraguchi *et al.*, 1972).

D. Other Microbial Products

1. Griseofulvin

Griseofulvin is an antifungal antibiotic produced by *Penicillium griseofulvum*. It is used in the treatment and prophylaxis of human mycotic infections, being administered at doses up to 1.0 gm/kg body weight per day for long periods of time. The extent to which it might occur in moldy food crops is unknown, and its presence in human foods has not been reported. Diets containing griseofulvin at levels of 1% fed 12–16 months induced hepatomas in rats (DeMatteis *et al.*, 1966). Mice injected with four doses of the compound over a period of 1–21 days of age also developed liver tumors after 1 year (Epstein *et al.*, 1967). There are no reports of possible consequences of administration of large doses of this compound to people over long periods of time.

2. Products of Actinomycetes and Bacteria

Several substances produced by *Streptomyces* species were shown experimentally to be carcinogenic. However, none of these compounds was tested by oral administration, and their importance as possible food contaminants remains to be determined. These compounds include actinomycin D, which induces sarcomas on repeated subcutaneous injection of microgram doses and also induces malignant mesenchymal tumors of the peritoneum following repeated intraperitoneal injection (Svoboda *et al.*, 1970). Mitomycin C has similar properties (Weisburger *et al.*, 1975).

Streptozotocin and elaiomycin are also antibiotics produced by *Streptomyces* species. Both are potential alkylating agents, and both have been investigated for carcinogenic properties. Streptozotocin produces permanent diabetes in experimental animals by virtue of a cytotoxic action on the β cells of the pancreas. Because of this property, attention has been focused on the compound as a possible chemotherapeutic agent to treat pancreatic islet cell carcinoma in man. As is often the case with cancer chemotherapeutic agents, the compound also was found to have carcinogenic properties. Single intravenous injections into rats or hamsters induced high incidences of tumors of the renal cortex (Arison and Feudale, 1967) or of the skin (Sibay and Hayes, 1969). Elaiomycin also produces kidney tumors, as well as intestinal and central nervous system tumors, in rats (Schoental, 1969).

Ethionine, the *S*-ethyl analogue of methionine, is a metabolite product of several bacteria, including *Escherichia coli*. Chronic administration of high levels (0.25%) of ethionine in the diet of rats induces high incidences of liver cell carcinomas (Schoental, 1976). It is possible that ethionine could be synthesized by gut microflora and hence create an opportunity for absorption of this carcinogen from the intestinal tract, but no evidence exists that this, in fact, occurs to any significant extent.

III. CYCASIN

Cycasin is the glucoside of methylazoxymethanol, the structure of which is shown in Fig. 6. This compound is one of a related series of azoxyglycosides that occur as normal tissue constituents of cycads, an ancient family of gymnosperms, which were widely distributed during the Mesozoic period. Cycads living today are essentially limited to tropical and subtropical areas, but occasionally extend into temperate zones, and are found in the southern United States, Japan, and Australia. A total of nine genera of the plants, some with many species, have been identified, including *Cycas* (East Africa to Japan), *Zamia* (Caribbean area), *Macrozamia* and *Bowenia* (Australia), *Encephalartos* and *Stangeria* (southern Africa), *Microcycas* (Cuba), and *Dioon* and *Ceratozamia* (Mexico).

Cycad plants are well adapted to survive adverse weather conditions, such as drought and hurricanes. Roots, seeds, and stems contain high levels of starch and

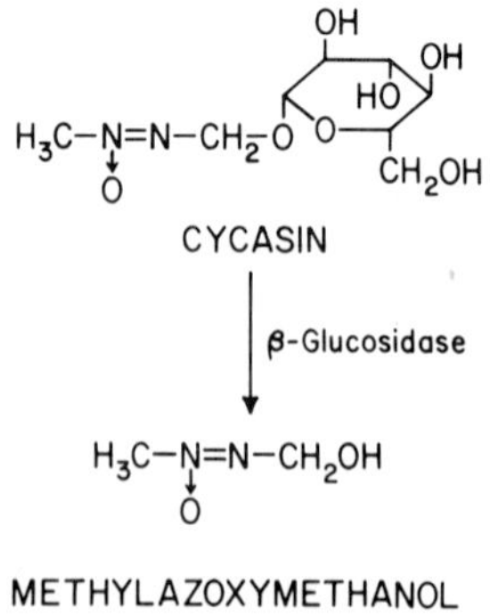

OH
OH
HO
$H_3C-\underset{\downarrow O}{N}=N-CH_2-O$ O
CH_2OH

CYCASIN

β-Glucosidase

$H_3C-\underset{\downarrow O}{N}=N-CH_2OH$

METHYLAZOXYMETHANOL

FIG. 6. Cycasin and methylazoxymethanol (MAM).

can therefore provide emergency sources of food when agricultural crops are destroyed. In this way, they are used as sources of food and also are folk medicines in many parts of the world (Whiting, 1963). Various aspects of the cycasin field were reviewed in detail by Laqueur and Spatz (1968) and Laqueur (1977).

In addition to cycasin, several other glycosides have been identified in related plant species, all of which contain the same aglycone, methylazoxymethanol (MAM), but differ in the sugar moiety. The aglycone can be liberated through enzymatic hydrolysis by β-glycosidase, present in many plants and bacteria. This constitutes an important aspect of the toxicity and carcinogenicity of cycasin, since MAM is the active portion of the molecule (Matsumoto and Strong, 1963; Kobayashi and Matsumoto, 1965). The concentration of cycasin or related compounds in various plants was reported to vary from traces to 4% of the dry weight (Pakekar and Dastur, 1965).

Discovery of the carcinogenicity of cycasin-containing foods resulted from experiments in which rats fed diets containing crude cycad meal developed tumors of liver, kidney, intestine, and lung (Laqueur *et al.*, 1963). A similar preparation was also carcinogenic for guinea pigs, causing liver tumors after intermittent feeding of diets containing 5% cycad meal (Spatz, 1964). The husks of cycad kernels, which are said to be chewed without detoxification by Guamanians, induced liver and kidney tumors when fed to rats at 0.5–2.0% of the diet.

In subsequent experiments, cycasin, or more specifically its aglycone MAM, has been identified as a versatile carcinogen. Experiments in which the pure materials were employed can be summarized in the following way: to indicate the pattern of responses that emerges with respect to potency, tumor sites, and the variables of age and species (see Laqueur, 1977).

Whereas cycasin is a potent carcinogen for rats when administered orally, it is inactive when given by parenteral administration and can be quantitatively recovered unchanged from urine and feces under these conditions. Moreover, cycasin is inactive when given orally to germfree rats. The target organs for cycasin carcinogenesis in rats can be affected by dose and dosage regimen. Adult hamsters given a single dose of cycasin by stomach tube developed intestinal adenomas and carcinomas, and adult mice responded to the same treatment with formation of hepatomas and lung adenomas. These data indicate that hydrolytic liberation of the aglycone is a necessary step in activation of the compound, hydrolysis being accomplished through the action of a β-glucosidase of intestinal bacteria.

That MAM is in fact the active moiety of the cycasin molecule has been amply demonstrated by experiments using MAM or MAM-acetate, a derivative with greater chemical stability which can be deacetylated by serum esterases. This compound is carcinogenic to conventional rats when administered by parenteral or oral routes and is also carcinogenic to germfree rats. Under both conditions, high incidences of liver, intestinal, and renal tumors are induced. MAM is carcinogenic for the guinea pig, in which it induces the same spectrum of tumors but at lower frequency than in rats. Adult hamsters given a single intravenous injection of MAM developed tumors of the colon, as did rats (Zedeck and Sternberg, 1974). Direct infusion of MAM into the large intestine also induced carcinoma of the colon of rats (Narisawa and Nakano, 1973).

Injection of cycasin into newborn rats, mice, and hamsters induces hepatomas in a large proportion of animals; tumors at other sites have been observed in rats but not in other species. These findings seemingly contradict the conclusion that cycasin is inactive unless it passes through the gut, where hydrolysis could occur. This discrepancy has been resolved, however, by the demonstration that newborn rats have a high concentration of β-glucosidase in skin and other tissues and therefore have the capacity for liberating MAM from cycasin. The enzyme decreases by 5–8 days of age, and it is no longer demonstrable by the twenty-fifth day of life.

Cycasin and MAM readily cross the placenta in pregnant rats, and their presence in fetuses can be demonstrated by chromatography. Feeding cycasin to pregnant rats during the gestation period causes the induction of intestinal and brain tumors in offspring surviving beyond 6 months of age.

Methylazoxymethanol bears an obvious chemical similarity to the synthetic carcinogen dimethylnitrosamine, and in many respects the pathological responses elicited by these two compounds are also similar. The suggestion was made (Miller, 1964) that the two compounds are metabolically converted to the same alkylating agent. Alkylation of rat liver nucleic acids and proteins by cycasin and MAM was demonstrated *in vitro* (Matsumoto and Higa, 1966) and *in vivo* (Shank and Magee, 1967). In both circumstances, 7-methylguanine was identified as a prominent alkylation product, and the similarities between its formation in rats treated with MAM and those treated with dimethylnitrosamine were demonstrated. MAM-Acetate also inhibited protein synthesis and microsomal oxidase activity in rat liver in a manner similar to that caused by dimethylnitrosamine (Lundeen *et al.*, 1971). Cycasin and MAM are mutagenic to bacteria and *Drosophila* and induce chromosomal aberrations in onion root tip cells. They therefore share many of the kinds of biological activities possessed by other alkylating agents.

The potential importance of azoxyglycosides as public health hazards is obvious from the evidence of their carcinogenic potency briefly summarized in the foregoing sections. The extent to which they might contribute to the etiology of human disease would depend, among other factors, on the extent to which exposure occurs.

Populations using cycads for food were cognizant of toxic properties of various parts of the plant long before the toxic agent was identified. Traditional methods of preparation of flour from cycad material were developed so that such processes as fermentation, heating, water extraction, or sun-drying detoxified the final product (Whiting, 1963). Material processed in traditional fashion by Guamanians was found experimentally to be nontoxic and noncarcinogenic to rats (Yang *et al.,* 1968). Thus far, no studies have been reported that would indicate that populations eating cycad material have any higher incidence of neoplastic diseases. Hirono *et al,* (1970) found no association between cycad toxicity and patterns of cancer incidence among the inhabitants of the Myako Islands, who subsisted on cycads during a period of recurrent, severe typhoons late in 1959. However, the period of observation may not have been long enough to permit detection of fluctuation of incidence in a relatively small population at risk.

IV. PYRROLIZIDINE ALKALOIDS

Pyrrolizidine alkaloids are constituents of some plant species in the genus *Senecio* and also of many other plant genera. The content of alkaloids in different plants varies from traces to about 5% of the dry weight. As is often the case with plant constituents, concentration differs in various plant parts and is influenced by such factors as season, climate, and soil conditions.

Interest in the pyrrolizidine alkaloids (sometimes referred to as *Senecio* alkaloids) originally developed because of livestock losses from liver and lung lesions in animals grazed on pastures in which plants containing them were growing. The interest increased with the discovery that some members of this group of alkaloids are carcinogenic and the possibility that they may be involved in human liver diseases, including kwashiorkor, venoocclusive disease, and primary liver cancer, which often occur in populations using alkaloid-containing plants as foods and folk medicines (Rose, 1972). Human exposure might also occur indirectly if residues of the alkaloids or their metabolites appear in tissues or products (e.g., milk) of animals used for food (e.g., Dienzer *et al.,* 1977). Three relevant reviews comprehensively summarize the extensive literature on the chemistry and biological properties of these alkaloids (Bull *et al.,* 1968; McLean, 1970; Schoental, 1976).

Despite the fact that plants containing them are unrelated phylogenetically, these alkaloids all belong to a single chemical group containing the pyrrolizidine nucleus. Bull *et al.* (1968) list 102 such compounds of known structure, and the list continues to grow as additional compounds are identified. Not all of these compounds are toxic, but those that are share certain structural features thought to be related to their metabolic activation, as will be discussed later.

With respect to their various biological properties, the pyrrolizidine alkaloids have been investigated in many different test systems, and some generalizations are justified for purposes of the present discussion. In field cases of poisoning of domestic animals, a variety of syndromes have been observed, prominent features of which include acute and chronic liver damage, lung damage, neurological symptoms, and a hemolytic syndrome. The kind of syndrome developing depends on the species of animal, type of plant consumed, total amount of alkaloid consumed, and pattern of consumption.

In experimental studies, the pattern of toxicity depends mainly on the dose of alkaloid employed. Single high doses cause acute liver necrosis, whereas smaller doses give rise to a chronic liver lesion characterized by megalocytosis that progresses long after dosing has been discontinued. Smaller doses also cause progressive lung damage. Young animals are more susceptible than adults to chronic liver toxicity.

Up to the present time, about 30 pyrrolizidine alkaloids have been found to be hepatotoxic, mostly through experimental studies in rodents. There is also circumstantial evidence for the hepatotoxicity of some of these alkaloids in man. Some plant materials known to contain alkaloids of this group, the identity of which is unknown or only partly established, have also been shown to be carcinogenic in experimental animals. Five pure alkaloids have been found to be carcinogenic to rats, including retrorsine and its N-oxide, isatidine, monocrotaline, petasitenine, and lasiocarpine. Lasiocarpine is an ester of heliotridine; the remainder are esters of retronecine. The structures of these compounds are shown in Fig. 7.

Retrorsine administered orally to rats at doses up to 30 mg/kg body weight induced tumors of the liver and other tissues, and similar effects were noted for isatidine (Schoental *et al.*, 1954). Monocrotaline induced liver carcinomas in rats given weekly doses by gastric intubation of 25 mg/kg for 4 weeks and then 8 mg/kg for 38 weeks (Newberne and Rogers, 1973). Petasitenine induced hemangioendotheliomas in the livers of rats when it was administered continuously in the drinking water at a concentration of 0.01% (Hirono *et al.*, 1977). Lasiocarpine was found to produce malignant tumors of the liver, skin, and intestine in rats injected intraperitoneally with the alkaloid at a dose of 7.8 mg/kg twice weekly for 4 weeks and then once a week for 52 weeks (Svoboda and Reddy, 1972).

All of the toxic alkaloids, including those with carcinogenic activity, are ester derivatives of 1-hydroxymethyl-1,2-dehydropyrrolizidine, esterification being possible at positions 1 and 7 (Fig. 8). Some alkaloids are open diesters, such as lasiocarpine, whereas in others the esters are linked to form a cyclic structure, as in the cases of retrorsine, isatidine, monocrotaline, and petasitenine. Structural requirements for toxicity can be summarized as follows. The ring nucleus must be unsaturated in the 1,2 position, the nucleus must be esterified, and one of the ester side chains must contain a branched chain.

Although the toxic pyrrolizidine alkaloids are chemically rather stable compounds, they are known to undergo several kinds of metabolic transformation in

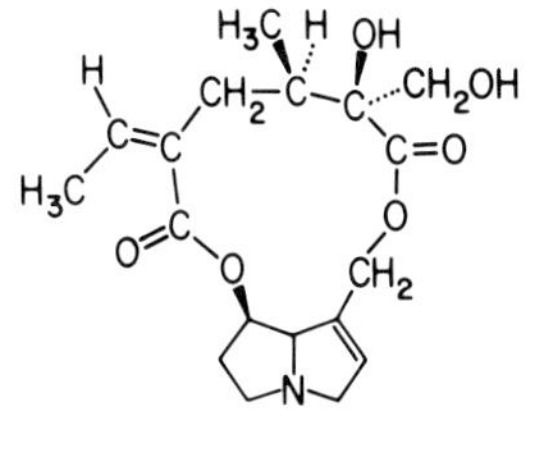

RETRORSINE 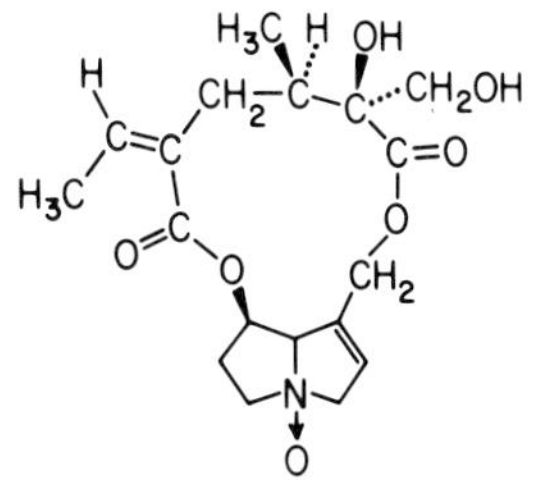 ISATIDINE

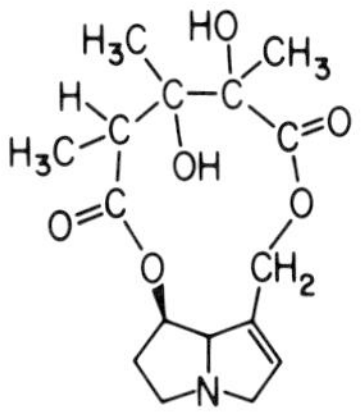

MONOCROTALINE 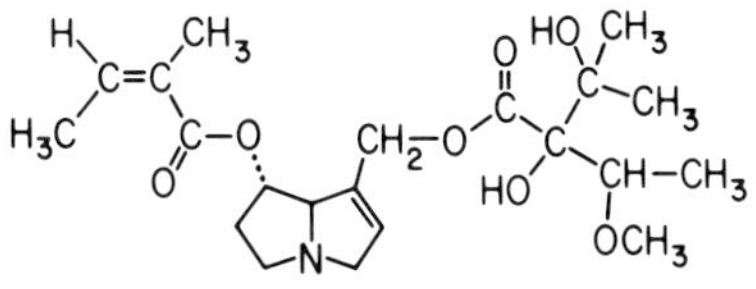 LASIOCARPINE

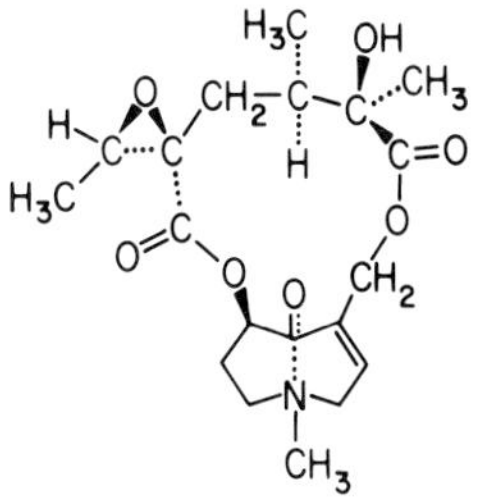

PETASITENINE 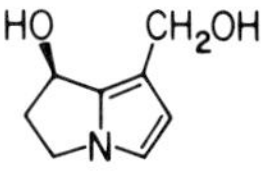 DEHYDRORETRONECINE

FIG. 7. Structures of carcinogenic pyrrolizidine alkaloids.

animal tissues. Important among these are the formation of N-oxides, hydrolysis of the ester linkages, and conjugation reactions.

A further metabolic pathway thought to represent an activation mechanism in the formation of proximate or ultimate carcinogenic or toxic forms of the alkaloids has been characterized and is summarized in Fig. 8. Current evidence suggests that some or all of the cytotoxic effects of the unsaturated compounds are caused by pyrrolic metabolites (dehydroalkaloids) formed by enzymatic dehydrogenation in liver. The dehydroalkaloid would represent the initial product

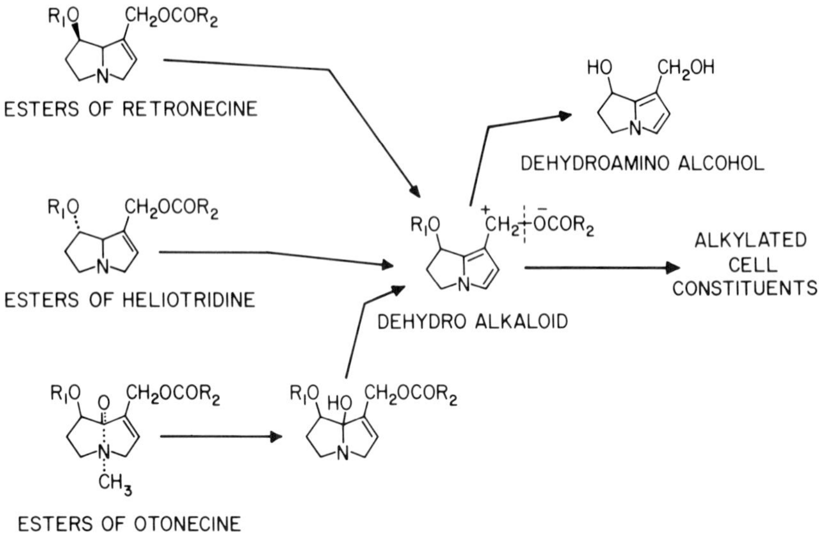

FIG. 8. Proposed pathways of metabolic activation of pyrrolizidine alkaloids.

formed from esters of retronecine (e.g., retrorsine, isatidine, monocrotaline, petasitenine), of heliotridine (e.g., lasiocarpine), or of otonecine. The dehydroalkaloids are highly reactive alkylating agents that react immediately with cell constituents to give covalently bound derivatives or that hydrolyze to yield the dehydroamino alcohol.

One such metabolite, dehydroretronecine (Fig. 7), the main water-soluble derivative of a number of widely distributed alkaloids, has been shown to be carcinogenic. Biweekly subcutaneous injection of this compound into rats at a dose of 20 mg/kg for 4 months and then 10 mg/kg for 8 months induced malignant rhabdomyosarcomas at the injection sites (Allen *et al.*, 1975). Biochemical studies revealed that this compound actively alkylated glutathione and cysteine *in vitro,* suggesting that interactions with sulfhydryl groups may play a role in its cytotoxicity (Robertson *et al.*, 1977).

Although it is clear that pyrrolizidine alkaloids pose a serious health threat to animals by virtue of their acute and chronic toxicity, their significance to human health is more difficult to assess. The best evidence of toxicity in human beings is the well-documented association between the use of folk medicines, some of which include plants known to contain alkaloids of this group, and the occurrence of venoocclusive disease in children in Jamaica. This malady has largely been eradicated by successful educational efforts to discourage traditional practices resulting in exposure. Less severe toxicity may very well occur in other cultures, the customs of which include similar uses of folk medicines. However, there is no evidence of elevated cancer incidence in any such populations, so it remains unclear whether these compounds cause a significant increase in cancer risk in human populations exposed to them.

V. BRACKEN FERN

The bracken fern (*Pteridium aquilinum*) is widely distributed throughout the world. Poisoning of livestock grazing on fresh or dried bracken has been recognized by veterinary practitioners for many years. Two very different sets of symptoms can be observed in poisoned animals. Symptoms of thiamine deficiency are sometimes involved, resulting from the presence in the plant of thiaminase; these are readily reversed by replacement therapy with the vitamin. The second type, usually seen in cattle or sheep grazing on bracken, include many of the signs of poisoning by ionizing radiation or radiomimetic chemicals. The characteristics of this group and early attempts to isolate and identify the active agents in bracken were summarized by Evans (1968). Several lines of evidence indicate that bracken fern also contains potent carcinogenic substances, since benign and malignant tumors have been induced by feeding the fern to a variety of laboratory animal species and cattle.

The initial observations of Evans and Mason (1965) that rats fed dried bracken developed tumors of the intestinal tract have been confirmed and extended. Benign and malignant adenomas and adenocarcinomas of the ileum and also papillomas and carcinomas of the urinary bladder were induced in the rats by feeding dried bracken collected in Canada (Schacham *et al.*, 1970) or in Turkey (Price and Pamukcu, 1968; Pamukcu and Price, 1969).

Carcinogenicity of Japanese bracken was investigated by Hirono and coworkers. Fresh bracken (''warabi'') used for human food was fed to rats for 4 months. All animals surviving for 7 months or longer developed tumors of the intestinal tract. Cooking in hot water (with or without alkali) according to Japanese custom reduced, but did not eliminate, the carcinogenic properties of the fern (Hirono *et al.*, 1970, 1972). Distribution of carcinogenic material in various parts of the plant was studied by Hirono *et al.* (1973), who found that the active principles appeared to be concentrated in the rhizome and curled fronds to a greater extent than in the stalk, as indicated by the relative potency of these fractions in producing intestinal tumors in rats. The commonly eaten portion of the plant (the frond) contained high levels of the carcinogenic agents. It is interesting that no carcinogenic activity was detected by feeding to rats other Japanese plants used in similar fashion as foods (Hirono *et al.*, 1972). Subsequently, Mori *et al.* (1977) reported that bracken carcinogen(s) appeared to be heat stable, since drying of the plant at 70–90°C did not alter its potency in producing tumors of the ileum of rats.

The response of rats to bracken carcinogenesis can be modified by secondary factors, as illustrated by the fact that the incidence of intestinal and bladder tumors was markedly increased by the simultaneous administration of bracken and thiamine (Pamukcu *et al.*, 1971). Pamukcu *et al.* (1977) reported that disulfiram, calcium chloride, and butylated hydroxyanisole significantly inhibited carcinogenesis of the intestinal tract in rats fed these agents simultaneously with bracken. Calcium chloride and polyvinylpyrrolidone had similar protective action in tumors of the urinary tract.

With respect to carcinogenicity in other laboratory animals, feeding of bracken fern has been found to induce intestinal tumors in the guinea pig, hamster, and Japanese quail (Evans *et al.*, 1967) and leukemia and lung tumors in mice (Pamukcu *et al.*, 1972). Tumors of the urinary bladder of mice were induced by surgical implantation of cholesterol pellets impregnated with bracken extracts (Pamukcu *et al.*, 1970).

Available data also indicate an important role of bracken fern in the etiology of urinary bladder tumors in cattle. Patterns of disease incidence have been associated epidemiologically with incidence of chronic bovine hematuria and also with the geographic distribution of the fern (Pamukcu, 1963). Bladder carcinomas and papillomas were induced experimentally in cattle by feeding of the plant for periods of 9 months to 4 years or more (Pamukcu *et al.*, 1967; Price and Pamukcu, 1968). Occurrence of tumors was invariably preceded by hematuria.

Toxic and carcinogenic substances derived from bracken are passed into the milk (Evans *et al.*, 1972) and urine (Pamukcu *et al.*, 1966) of cattle being fed the plant. It is not known whether the substances involved were present initially in the bracken or were metabolic products.

Up to the present time, attempts to isolate and characterize the carcinogen(s) in bracken have been unsuccessful. Two main approaches have been utilized in these attempts. One has consisted of gross compositional analysis aimed at the identification of compounds of known carcinogenicity as major components of the plant. A number of classes of compounds have been identified by this approach, including organic acids, flavonoids, indanones, catecholamines, pteroquilin, and pterolactam. Many of these are common components of plant tissues, and none is known to be carcinogenic, although some are toxic.

The second line of investigation has been to fractionate bracken and conduct bioassays on various fractions as they are isolated and purified. This approach has produced some suggestive results but no definitive information. Leach *et al.* (1971) isolated a substance that was toxic to mice and that was said also to be mutagenic and carcinogenic to mice. Evidence for the latter effects was not published. The same group carried the chemical characterization further and presented preliminary evidence (Evans and Osman, 1974) that the substance isolated was identical to shikimic acid, a common constituent of plants. They also presented evidence which suggested that authentic shikimic acid as well as the compound isolated by them were carcinogenic to mice. These findings were not confirmed by Hirono *et al.* (1977), who found no evidence of carcinogenicity of shikimic acid in rats under conditions in which bracken fern is strongly active. There is no other published information on this point.

In other studies of similar purpose, Wang *et al.* (1976) isolated from bracken fern a condensed tannin that was toxic to mice by injection and induced bladder tumors when cholesterol pellets containing it were surgically implanted into urinary bladders of mice. Tannins are not carcinogenic to rodents when administered orally, and the significance of these findings to the carcinogenic properties of ingested bracken is unclear. Saito *et al.* (1975) isolated a number of 1-indanone derivatives from Japanese bracken but found them to be inactive as carcinogens for mice, whereas the material from which they were isolated had potent activity in producing intestinal and other tumors.

The existence of carcinogenic substances in bracken fern, clearly established by these various experimental approaches, has important public health implications. The fern is consumed in significant quantities in several areas of the world, notably in Japan and in the northeastern United States and Canada (as "fiddleheads"). Epidemiological studies are needed to determine whether this practice is in any way associated with patterns of cancer incidence in populations living in those areas. Indirect exposure might also occur, since the active material is readily passed into the milk of cows fed the fern, as well as through the placenta in mice (Evans *et al.*, 1972). These findings have important implications for human beings, since they indicate that *in utero* exposure of the fetus could result when pregnant women consume the fern. Postulated exposure of infants could occur through mothers' milk or through the milk of cattle grazing on pastures where the plant grows. Exposure through these routes thus extends the opportunity for contact with the carcinogen far beyond direct consumption, and further studies on the problem are clearly warranted.

VI. OTHER PLANT CARCINOGENS

A. Safrole and Related Compounds

Safrole (4-allyl-1,2-methylenedioxybenzene) is a component of many essential oils, such as star anise oil, camphor oil, and sassafras oil. It also occurs in oil of mace, ginger, California bay laurel, and cinnamon leaf oil (Homburger and Boger, 1968). Until recently, it had been used as a flavoring agent in soft drinks and other food products. Its use for these purposes was discontinued after it was discovered that the compound has hepatocarcinogenic activity in rodents.

Toxicity and carcinogenesis evaluations have been conducted on safrole and also on two structurally related compounds, isosafrole and dihydrosafrole. The LD_{50} values in rats for safrole, isosafrole, and dihydrosafrole, are 1950, 1340, and 2260 mg/kg, respectively. Carcinogenic properties of safrole were established by the induction of liver tumors in rats fed high levels of the com-

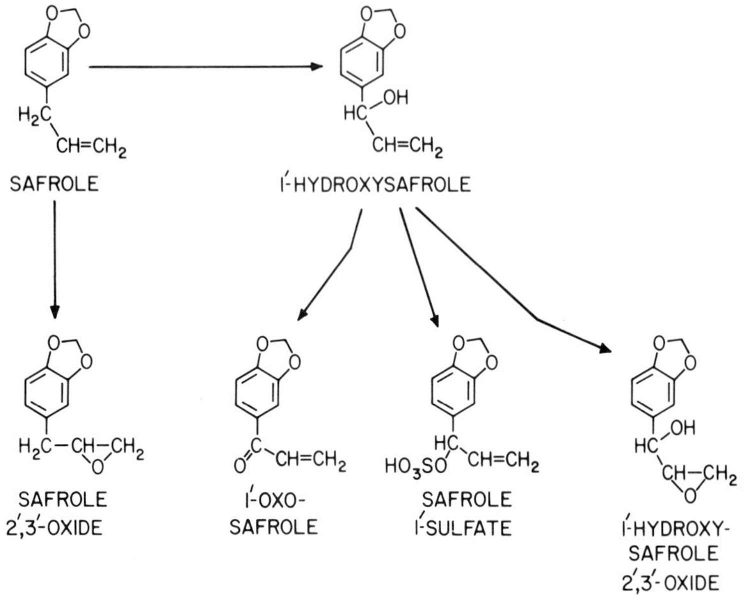

FIG. 9. Structures of the proximate and possibly ultimate electrophilic carcinogenic metabolites of safrole in the rat and mouse.

pound for 2 years. Animals fed 0.5% safrole developed malignant liver tumors at high incidence (14/50 animals), whereas lower levels (0.1%) induced benign tumors at lower incidence (Long *et al.*, 1963; Hagan *et al.*, 1965). An interesting structure–activity pattern emerges from similar studies on dihydrosafrole and isosafrole. Isosafrole fed at 0.5% of the diet induced hepatocellular carcinomas in 3/50 rats and is therefore a weaker hepatocarcinogen for the rat than safrole. On the other hand, dihydrosafrole fed at 1.0% of the diet induced no liver tumors in rats but did produce papillary epidermoid carcinomas of the esophagus at a 75% incidence; 50% of the tumors were judged to be malignant. The response was dose-related, since the compound at 0.5% induced esophageal tumors at 74% incidence (35% malignant), and 0.25% at 20% incidence (5% malignant) (Homburger and Boger, 1968; Long and Jenner, 1963).

Thus, in rats, safrole and isosafrole appear to be hepatocarcinogens, but dihydrosafrole is carcinogenic for the esophagus. The same pattern does not appear to hold for the mouse. In two inbred mouse strains, all three compounds induced hepatomas, and no esophageal tumors were reported (Innes *et al.*, 1969). The three compounds were not of equal potency, however. When administered by stomach tube (464 mg/kg daily) for 4 weeks and then in the diet (1112 ppm) for 78 weeks, safrole and dihydrosafrole induced hepatomas in 27/33 and 10/34 mice, respectively. Isosafrole administered at a lower dose (215 mg/kg and then 517 ppm) induced hepatomas in 6/34 mice.

It has also been shown that subcutaneous injection of safrole into infant mice (on days 1, 7, 14, and 21 days of age) induced hepatomas (58%) and pulmonary adenocarcinomas in survivors killed at 1 year of age (Epstein *et al.*, 1970).

Studies on the metabolism of safrole have revealed probable activation pathways through which the compound is converted to its proximate and possibly ultimate carcinogenic forms. These are summarized in Fig. 9. When safrole is administered to rats, hamsters, guinea pigs, or mice, significant quantities are converted to 1′-hydroxysafrole, which is excreted in urine in conjugated form (Borchert *et al.*, 1973b). This hydroxylated metabolite subsequently proved to be a more potent carcinogen than the parent compound when fed to rats (Borchert *et al.*, 1973a). Further metabolism of 1′-hydroxysafrole to its sulfate and to 1′-hydroxysafrole 2′,3′-oxide by rat and mouse liver preparations was demonstrated (Wislocki *et al.*, 1976). Formation of 1′-oxosafrole *in vivo* is indicated by urinary excretion of adducts formed with secondary amines by rats and guinea pigs given safrole or 1′-hydroxysafrole. Evidence for the formation of safrole 2′,3′-oxide is provided by excretion of 2′,3′-dihydroxysafrole in the urine of rats and guinea pigs given safrole by injection (see Wislocki *et al.*, 1976). All of these derivatives proved to be more potent carcinogens and mutagens than the parent compounds, suggesting that they are probably the ultimate reactive forms of safrole (Wislocki *et al.*, 1977).

B. Antithyroid Substances

Goiterogenic substances are of interest in the present context not because of the weight of evidence that they are in themselves carcinogenic but rather because carcinogenesis in experimental animals can be strongly influenced by various kinds of endocrine imbalance. On this basis, goiterogens deserve attention as possible carcinogenic hazards to man. Endemic goiter results from environmental factors, and within affected populations the incidence of thyroid carcinoma is increased. The presence of goiterogenic materials in foods thus contributes to the background environment leading to the increased frequency with which goiter arises.

A variety of antithyroid substances occurs in plants (Kingsbury, 1964). Thiourea has been identified in seeds of some *Laburnum* species. When fed to rats, thiourea induces thyroid carcinomas and also malignant tumors of the eyelid and acoustic duct (Purves and Griesbach, 1947; Rosin and Ungar, 1957). Another important antithyroid substance is goitrin (L-5-vinyl-2-thiooxazolidone), which is studied in more detail in Chapter 4. This compound, which blocks iodine uptake, has been isolated from several *Brassica* species, including kale, brussels sprouts, turnip, broccoli, rape, and kohlrabi. Rapeseed meal induced thyroid adenomas in rats (Griesbach *et al.*, 1945), and its toxicity was reduced by hot-water extraction and supplementation of the diet with additional iodine. Many other foods are known to contain other antithyroid materials, which may contribute to the general goiterogenic background (the reader is referred to Chapter 4 for further information).

C. Cyclopropene Fatty Acids

Fatty acids containing the cyclopropene moiety occur in some plants of the order Malvales (Phelps *et al.*, 1965). One member of this group, the cotton plant, is pertinent to the present discussion since cottonseed oil and meal are used for animal and human foods. Two cyclopropene acids, sterculic and malvalic acids (Fig. 10), occur in cottonseed oil. In crude oil their content is 1–2% but is reduced to 0.5% or less in refined edible oils. Malvalic acid is present in larger concentration than sterculic acid. The oil of the Java olive (*Sterculia foetida*) contains nearly 70% cyclopropene fatty acids, principally sterculic acid, and has been an important source of the compounds for experimental purposes.

There is no evidence that these compounds are carcinogenic. However, a number of reports indicate that simultaneous feeding to rainbow trout of these compounds and aflatoxin B_1 or 2-acetylaminofluorene increases tumor incidence and decreases the latent period for tumor formation (Lee *et al.*, 1968). Further research is required to establish the extent to which this enhancement occurs in other experimental carcinogenesis systems.

D. Compounds Active by Parenteral Routes

Many natural products have been reported to have carcinogenic properties on the basis of local tumor induction resulting from parenteral, usually subcutane-

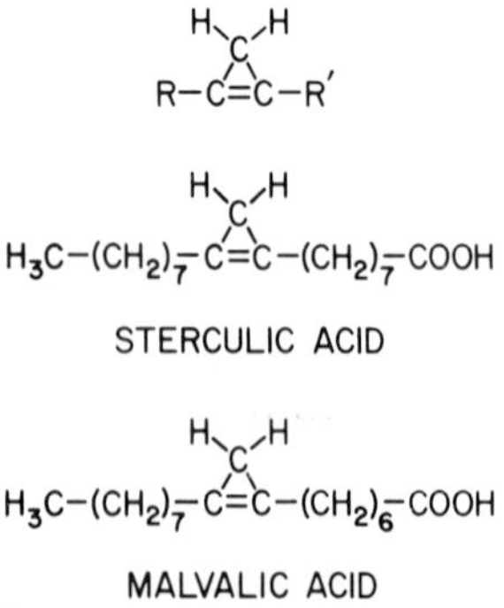

$$R-C(CH_2)=C-R'$$

$$H_3C-(CH_2)_7-C(CH_2)=C-(CH_2)_7-COOH$$

STERCULIC ACID

$$H_3C-(CH_2)_7-C(CH_2)=C-(CH_2)_6-COOH$$

MALVALIC ACID

FIG. 10. Structures of cocarcinogenic cyclopropene fatty acids.

ous, injection. Although most such evidence falls outside the scope of this review, some substances in this class have proved useful as model carcinogens for structure–activity or mechanism of action studies and thus deserve brief consideration.

Tannic acid is a powerful hepatotoxic agent when administered topically or by injection (Korpassy, 1961). In one set of experiments, about 56% of rats surviving tannic acid injection for longer than 100 days developed liver tumors. Moreover, simultaneous dosing with tannic acid and 2-acetylaminofluorene induced more liver tumors and a greater degree of malignancy than did the synthetic carcinogen alone (Mosonyi and Korpassy, 1953). Because commercial preparations of tannic acid, such as were used in these experiments, are poorly characterized mixtures of substances, it is not possible to identify the active components in these studies.

Compounds that cause local sarcoma induction after subcutaneous injection include cholesterol and its oxidation products. The relevance of this evidence to the problem of evaluating carcinogenic risk to man is uncertain.

VII. SUMMARY AND PERSPECTIVES

It is difficult to formulate generalizations about the possible importance of these naturally occurring substances in the etiology of human cancers. A wide spectrum of qualitatively and quantitatively different problems is generated by the presence of these substances in the environment to which man is exposed. It is self-evident that it would be highly desirable to minimize or eliminate human exposure to all such substances simply on the basis of possible carcinogenic risk to man, scarce though the evidence may be for such a risk in some cases. It will be extremely difficult to attain this objective under the best circumstances and will require intensive long-range programs. Some of the main difficulties encountered in the design and implementation of such programs become apparent on even a cursory analysis of several parameters of the general problem.

Direct evidence of carcinogenic risk to man, based on knowledge of human responsiveness, is lacking in all cases. Epidemiological evidence linking aflatoxin exposure with elevated incidence of liver cancer is suggestive of an etiological role for these agents, but such evidence is entirely lacking with regard to the other substances. Judgments concerning carcinogenic risk to man must therefore rest on estimates of human exposure and evidence of carcinogenicity in animals.

Exposure is potentially widespread with regard to aflatoxins and other mold metabolites but is probably more restricted in the cases of cycasin or pyrrolizidine alkaloids because of the traditional use of food or herbal medicines containing the carcinogens.

Evidence for carcinogenicity in animals is equally variable among the substances under consideration. With regard to potency, it is clear that aflatoxins, methylazoxymethanol, and the active agent in bracken fern are potent carcinogens in one or more animal species. The evidence is less clear for carcinogenic activity of the pyrrolizidine alkaloids, and the significance of those substances that are active only by parenteral administration is still more difficult to estimate. With respect to tissue and organ specificity, some substances are highly site specific, such as aflatoxins and pyrrolizidine alkaloids that attack mainly the liver, whereas others (methylazoxymethanol, bracken fern toxin) are extremely versatile animal carcinogens.

Each substance or class of substance will therefore present its own set of problems in attempts to intervene and reduce exposure. Such attempts will be successful in some instances only by such major attainments as improved agricultural technologies or changing traditional uses of foods or herbal medicines. Such programs require and deserve the great expenditures of effort involved in their implementation.

References

Adamson, R. H., Correa, P., and Dalgard, D. W. (1973). *J. Natl. Cancer Inst.* **50**, 549–553.

Adamson, R. H., Correa, P., Sieber, S. M., McIntire, K. R., and Dalgard, D. W. (1976). *J. Natl. Cancer Inst.* **57**, 67–78.

Allcroft, R., and Carnaghan, R. B. A. (1963). *Vet. Rec.* **75,** 259–263.

Allcroft, R., Rogers, H., Lewis, G., Nabney, J., and Best, P. E. (1966). *Nature (London)* **209**, 154–155.

Allen, J. R., Hsu, I. C., and Carstens, L. A. (1975). *Cancer Res.* **35**, 997–1002.

Alpert, M. E., Hutt, M. S. R., and Davidson, C. (1968). *Lancet* **1**, 1265–1267.

Alpert, M. E., Hutt, M. S. R., Wogan, G. N., and Davidson, C. S. (1971). *Cancer* **28**, 253–260.

Arison, R. N., and Feudale, E. L. (1967). *Nature (London)* **214**, 1254–1255.

Asao, T., Büchi, G., Abdel-Kader, M. M., Chang, S. B., Wick, E. L., and Wogan, G. N. (1963). *J. Am. Chem. Soc.* **85**, 1706–1707

Ayres, J. L., Lee, D. J., Wales, H. H., and Sinnhuber, R. O. (1971). *J. Natl. Cancer Inst.* **46,** 561–564.

Bassir, O., and Osiyemi, F. (1967). *Nature (London)* **215**, 882.

Borchert, P., Wislocki, P. G., Miller, J. A., and Miller, E. C. (1973a). *Cancer Res.* **33**, 575–589.

Borchert, P., Miller, J. A., Miller, E. C., and Shires, T. (1973b). *Cancer Res.* **33**, 590–600.

Büchi, G. H., Muller, P. M., Roebuck, B. D., and Wogan, G. N. (1974). *Res. Commun. Chem. Pathol. Pharmacol.* **8**, 585–592.

Bull, L. B., Culvenor, C. C. J., and Dick, T. T. (1968). "The Pyrrolizidine Alkaloids: Their Chemistry, Pathogenicity and Other Biological Properties." North-Holland Publ., Amsterdam.

Burkhardt, H. J., and Forgacs, J. (1968). *Tetrahedron* **24**, 717–720.

Butler, W. H. (1969). *In* "Aflatoxin: Scientific Background, Control and Implications" (L. A. Goldblatt, ed.), pp. 223–236. Academic Press, New York.

Butler, W. H. (1974). *In* "Mycotoxins" (I. F. H. Purchase, ed.), pp. 1–28. Am. Elsevier, New York.

Butler, W. H., Greenblatt, M., and Lijinsky, W. (1969). *Cancer Res.* **29**, 2206–2211.

Campbell, T. C., and Hayes, J. R. (1976). *Toxicol. Appl. Pharmacol.* **35**, 199–222.

Campbell, T. C., and Salamat, L. (1971). *In* "Mycotoxins in Human Health" (I. F. H. Purchase, ed.), pp. 271–280. Macmillan, New York.

Campbell, T. C., Craedo, J. P., Bulatao-Jayme, J., Salamat, L., and Engel, R. W. (1970). *Nature (London)* **227**, 403–404.
Cardeilhac, P. T., and Nair, K. P. C. (1973). *Toxicol. Appl. Pharmacol.* **26**, 393–397.
Carnaghan, R. B. A. (1965). *Nature (London)* **208**, 308.
Chédid, A., Bundeally, A. E., and Mendenhall, C. L. (1977). *J. Natl. Cancer Inst.* **58**, 339–349.
Croy, R. G., Essigmann, J. M., Reinhold, V. N., and Wogan, G. N. (1978). *Proc. Natl. Acad. Sci. U.S.A.* **75**, 1745–1749.
Dalezios, J. I., and Wogan, G. N. (1972). *Cancer Res.* **32**, 2297–2303.
Dalezios, J. I., Wogan, G. N., and Weinreb, S. (1971). *Science* **171**, 584–585.
Dalezios, J. I., Hsieh, D. P. H., and Wogan, G. N. (1973). *Food Cosmet. Toxicol.* **11**, 605–616.
DeMatteis, F., Donnelly, A. J., and Runge, W. J. (1966). *Cancer Res.* **26**, 721–726.
Detroy, R. W., Lillehoj, E. B., and Ciegler, A. (1971). *In* "Microbial Toxins" (S. Kadis, A. Ciegler, and S. J. Ajl, eds.), Vol. VI, pp. 3–178. Academic Press, New York.
Dickens, F., Jones, H. E. H., and Wayneforth, H. B. (1966). *Br. J. Cancer* **20**, 134–144.
Dienzer, M. L., Thomason, P. A., Burgett, D. M., and Isaacson, D. L. (1977). *Science* **95**, 497–499.
Dutton, M. F., and Heathcote, J. G. (1968). *Chem. Ind. (London)* pp. 418–421.
Enomoto, M., and Ueno, I. (1974). *In* "Mycotoxins" (I. F. H. Purchase, ed.), pp. 303–326. Am. Elsevier, New York.
Epstein, S. S., Andrea, J., Joshi, S., and Mantel, N. (1967). *Cancer Res.* **27**, 1900–1906.
Epstein, S. S., Bartus, B., and Farber, E. (1969). *Cancer Res.* **29**, 1045–1050.
Epstein, S. S., Fujii, K., Andrea, J., and Mantel, N. (1970). *Toxicol. Appl. Pharmacol.* **16**, 321–334.
Essigmann, J. M., Croy, R. G., Nadzan, A. M., Busby, W. F., Reinhold, V. N., Büchi, G., and Wogan, G. N. (1977). *Proc. Natl. Acad. Sci. U.S.A.* **74**, 1870–1874.
Evans, I. A. (1968). *Cancer Res.* **28**, 2252–2261.
Evans, I. A., and Mason, J. (1965). *Nature (London)* **208**, 913–914.
Evans, I. A., and Osman, M. A. (1974). *Nature (London)* **250**, 348–349.
Evans, I. A., Widdop, B., and Barber, G. D. (1967). *Br. Emp. Cancer Campaign Res., Annu. Rep.* pp. 411–412.
Evans, I. A., Jones, R. S., and Mainwaring-Burton, R. (1972). *Nature (London)* **237**, 107–108.
Fujii, K., Kurata, H., Odashima, S., and Hatsuda, Y. (1976). *Cancer Res.* **36**, 1615–1617.
Garner, R. C. (1973). *Chem.-Biol. Interact.* **6**, 125–129.
Garner, R. C. (1975). *Biochem. Pharmacol.* **24**, 1553–1556.
Garner, R. C., and Wright, C. M. (1975). *Chem.-Biol. Interact.* **11**, 123–131.
Garner, R. C., Miller, E. C., and Miller, J. A. (1972). *Cancer Res.* **32**, 2058–2066.
Goldblatt, L. A. (1969). "Aflatoxin." Academic Press, New York.
Goodall, C. M., and Butler, W. H. (1969). *Int. J. Cancer* **4**, 422–429.
Gopalan, C., Tulpule, P. G., and Krishnamurthi, D. (1972). *Food Cosmet. Toxicol.* **10**, 519–521.
Griesbach, W. E., Kennedy, T. H., and Purves, H. D. (1945). *Br. J. Exp. Pathol.* **26**, 18–24.
Gurtoo, H. L., and Bejba, N. (1974). *Biochem. Biophys. Res. Commun.* **61**, 735–742.
Gurtoo, H. L., and Campbell, T. C. (1974). *Mol. Pharmacol.* **10**, 766–789.
Gurtoo, H. L., and Dahms, R. (1974). *Res. Commun. Chem. Pathol. Pharmacol.* **9**, 107–118.
Gurtoo, H. L., and Motycka, L. (1976). *Cancer Res.* **36**, 4663–4671.
Hagan, E. C., Jenner, P. M., Jones, W. I., Fitzhugh, O., Long, E. L., Brouwer, J. G., and Webb, W. K. (1965). *Toxicol. Appl. Pharmacol.* **7**, 18–24.
Hales, B. F., and Neims, A. H. (1977). *Biochem. Pharmacol.* **26**, 555–556.
Hamasaki, T., and Hatsuda, Y. (1977). *In* "Mycotoxins in Human and Animal Health" (J. V. Rodricks, C. W. Hesseltine, and M. A. Mehlman, eds.), pp. 597–607. Pathotox Publishers, Inc., Park Forest South, Illinois.
Hirono, I., Shibuya, C., Fushimi, K., and Haga, M. (1970). *J. Natl. Cancer Inst.* **45**, 179–188.

Hirono, I., Shibuya, C., Shimizu, M., and Fushimi, K. (1972). *J. Natl. Cancer Inst.* **48,** 1245–1249.

Hirono, I., Fushimi, K., Mori, H., Miwa, T., and Haga, M. (1973). *J. Natl. Cancer Inst.* **50,** 1367–1371.

Hirono, I., Fushimi, K., and Matsubara, N. (1977). *Toxicol. Lett.* **1**, 9–10.

Holzapfel, C. W., Purchase, I. F. H., Steyn, P. S., and Gouws, L. (1966). *S. Afr. Med. J.* **40,** 1100–1101.

Homburger, F., and Boger, E. (1968). *Cancer Res.* **28**, 2372–2374.

Innes, J. R. M., Ulland, B. M., Valerio, M. G., Petrucelli, L., Fishbein, L., Hart, E. R., Pallotta, A. J., Bates, R. R., Falk, H. L., Gart, J. J., Klein, M., Mitchell, I., and Peters, J. (1969). *J. Natl. Cancer Inst.* **42,** 1101–1114.

Jerina, D. A., and Daly, J. W. (1974). *Science* **185**, 573–582.

Jollow, D. J., Kocsis, J. J., Snyder, R., and Vainio, H. (1977). "Biological Reactive Intermediates; Formation, Toxicity and Inactivation." Plenum, New York.

Keen, P., and Martin, P. (1971). *Trop. Geogr. Med.* **23**, 44–53.

Kingsbury, J. M. (1964). "Poisonous Plants in the United States and Canada." Prentice-Hall, Englewood Cliffs, New Jersey.

Kobayashi, A., and Matsumoto, H. (1965). *Arch. Biochem. Biophys.* **110**, 373–380.

Korpassy, B. (1961). *Prog. Exp. Tumor Res.* **2,** 245–290.

Laqueur, G. L. (1977). *Adv. Mod. Toxicol.* **3**, 231–261.

Laqueur, G. L., and Spatz, M. (1968). *Cancer Res.* **28**, 2262–2267.

Laqueur, G. L., Michelsen, O., Whiting, M. G., and Kurland, L. T. (1963). *J. Natl. Cancer Inst.* **31**, 919–951.

Leach, H., Barber, G. D., Evans, I. A., and Evans, W. C. (1971). *Biochem. J.* **124**, 13P.

Lee, D. J., Wales, J. H., and Sinnhuber, R. O. (1968). *Cancer Res.* **28**, 2312–2318.

Lin, J. J., Liu, C., and Svoboda, D. J. (1974). *Lab. Invest.* **30,** 267–278.

Lin, J. J., Miller, J. A., and Miller, E. C. (1977). *Cancer Res.* **37,** 4430–4438.

Long, E. L., and Jenner, P. M. (1963). *Fed. Proc., Fed. Am. Soc. Exp. Biol.* **22,** 275.

Long, E. L., Nelson, A. A., Fitzhugh, O. G., and Hansen, W. H. (1963). *Arch. Pathol.* **75,** 594–604.

Lundeen, P. B., Banks, G. S., and Ruddon, R. W. (1971). *Biochem. Pharmacol.* **20**, 2522–2527.

Mabee, M. S., and Chipley, J. R. (1973). *Appl. Microbiol.* **25**, 763–769.

McLean, A. E. M., and Marshall, A. (1971). *Br. J. Exp. Pathol.* **52**, 322–329.

McLean, E. K. (1970). *Pharmacol. Rev.* **22**, 429–483.

Madhavan, T. V., and Gopalan, C. (1968). *Arch. Pathol.* **85**, 133–137.

Masri, M. S., Lundin, R. E., Page, J. R., and Garcia, V. C. (1967). *Nature (London)* **215,** 753–755.

Masri, M. S., Booth, A. N., and Hsieh, D. P. H. (1974a). *Life Sci.* **15**, 203–212.

Masri, M. S., Haddon, W. F., Lundin, R. E., and Hsieh, D. P. H. (1974b). *J. Agric. Food Chem.* **22,** 512–515.

Matsumoto, H., and Higa, H. H. (1966). *Biochem. J.* **98,** 20c–22c.

Matsumoto, H., and Strong, F. M. (1963). *Arch. Biochem. Biophys.* **101,** 299–310.

Merrill, A. H., and Campbell, T. C. (1974). *Toxicol. Appl. Pharmacol.* **27**, 210–213.

Miller, J. A. (1964). *Fed. Proc., Fed. Am. Soc. Exp. Biol.* **23**, 1361–1362.

Miller, J. A. (1970). *Cancer Res.* **30**, 559–576.

Mori, H., Kato, K., Ushimaru, Y., Kato, T., and Hirono, I. (1977). *Gann* **68**, 517–520.

Mosonyi, M., and Korpassy, B. (1953). *Nature (London)* **171**, 791.

Nabney, J., Burbage, M. B., Allcroft, R., and Lewis, G. (1967). *Food Cosmet. Toxicol.* **5**, 11–17.

Narisawa, T., and Nakano, H. (1973). *Gann* **64**, 93–97.

Neal, G. E., and Godoy, H. M. (1976). *Chem.-Biol. Interact.* **14**, 279–289.

Nesbitt, B. F., O'Kelly, J., Sargeant, K., and Sheridan, A. (1962). *Nature (London)* **195**, 1062–1063.

Newberne, P. M., and Rogers, A. E. (1973). *J. Natl. Cancer Inst.* **50**, 439–448.
Newberne, P. M., and Williams, G. (1969). *Arch. Environ. Health* **19**, 489–498.
Newberne, P. M., and Wogan, G. N. (1968). *Cancer Res.* **28,** 770–781.
Newberne, P. M., Harrington, D. H., and Wogan, G. N. (1966). *Lab. Invest.* **15**, 962–969.
Pakekar, R. S., and Dastur, D. K. (1965). *Nature (London)* **206**, 1363–1364.
Pamukcu, A. M. (1963). *Ann. N.Y. Acad. Sci.* **108**, 939–947.
Pamukcu, A. M., and Price, J. M. (1969). *J. Natl. Cancer Inst.* **43**, 275–281.
Pamukcu, A. M., Olson, C., and Price, J. M. (1966). *Cancer Res.* **26,** 1745–1753.
Pamukcu, A. M., Goksoy, S. K., and Price, J. M. (1967). *Cancer Res.* **27**, 917–924.
Pamukcu, A. M., Yalciner, S., Price, J. M., and Bryan, G. T. (1970). *Cancer Res.* **30**, 2671–2674.
Pamukcu, A. M., Wattenberg, L. W., Price, J. M., and Bryan, G. T. (1971). *J. Natl. Cancer Inst.* **47**, 155–159.
Pamukcu, A. M., Erturk, E., Price, J. M., and Bryan, G. T. (1972). *Cancer Res.* **32**, 1442–1445.
Pamukcu, A. M., Yalciner, S., and Bryan, G. T. (1977). *Cancer* **40**, 2450–2454.
Patterson, D. S. P., and Roberts, B. A. (1970). *Food Cosmet. Toxicol.* **8**, 527–538.
Patterson, D. S. P., and Roberts, B. A. (1971a). *Food Cosmet. Toxicol.* **9**, 829–837.
Patterson, D. S. P., and Roberts, B. A. (1971b). *Biochem. Pharmacol.* **20**, 3377–3383.
Patterson, D. S. P., and Roberts, B. A. (1972). *Experientia* **28**, 929–930.
Peers, F. G., and Linsell, C. A. (1973). *Br. J. Cancer* **27**, 473–484.
Phelps, R. A., Shenstone, F. S., Kemmerer, A. R., and Evans, R. J. (1965). *Poult. Sci.* **44**, 358–394.
Pohland, A. E., Cushmac, M. E., and Andrellos, P. J. (1968). *J. Assoc. Off. Anal. Chem.* **51**, 907–910.
Polan, C. E., Hayes, J. R., and Campbell, T. C. (1974). *J. Agric. Food Chem.* **22**, 635–638.
Preston, R. S., Hayes, J. R., and Campbell, T. C. (1976). *Life Sci.* **19**, 1191–1198.
Price, J. M., and Pamukcu, A. M. (1968). *Cancer Res.* **28**, 2247–2251.
Purchase, I. F. H. (1972). *Food Cosmet. Toxicol.* **10**, 531–544.
Purchase, I. F. H., and van der Watt, J. J. (1968). *Food Cosmet. Toxicol.* **6**, 555–556.
Purchase, I. F. H., and van der Watt, J. J. (1970). *Food Cosmet. Toxicol.* **8**, 289–295.
Purchase, I. F. H., and van der Watt, J. J. (1973). *Toxicol. Appl. Pharmacol.* **26**, 274–281.
Purves, H. D., and Griesbach, W. E. (1947). *Br. J. Exp. Pathol.* **28**, 46–53.
Raj, H. G., Santhanam, K., Gupta, R. P., and Venkitasubramanian, T. A. (1975). *Chem.-Biol. Interact.* **11**, 301–305.
Reddy, J. K., and Svoboda, D. J. (1972). *Arch. Pathol.* **93,** 55–60.
Reddy, J. K., Svoboda, D. J., and Rao, M. S. (1976). *Cancer Res.* **36,** 151–160.
Robertson, K. A., Seymour, J. L., Hsia, M. T., and Allen, J. R. (1977). *Cancer Res.* **37**, 3141–3144.
Rodricks, J. V. (1969). *J. Agric. Food Chem.* **17**, 457–461.
Roebuck, B. D., and Wogan, G. N. (1977). *Cancer Res.* **37**, 1649–1656.
Roebuck, B. D., Siegel, W. G., and Wogan, G. N. (1978). *Cancer Res.* **38,** 999–1002.
Rogers, A. E., and Newberne, P. M. (1969). *Cancer Res.* **29,** 1965–1972.
Rogers, A. E., and Newberne, P. M. (1971). *Nature (London)* **229**, 62–63.
Rose, E. F. (1972). *S. Afr. Med. J.* **46**, 1039–1043.
Rosin, A., and Ungar, H. (1957). *Cancer Res.* **17**, 302–305.
Saito, M., Umeda, M., Enomoto, M., Hatanaka, Y., Natori, S., Yoshihira, K., Fukuoka, M., and Kuroyanagi, M. (1975). *Experientia* **31**, 829–831.
Salhab, A. S., and Edwards, G. S. (1977a). *J. Toxicol. Environ. Health* **2**, 583–587.
Salhab, A. S., and Edwards, G. S. (1977b). *Cancer Res.* **37**, 1016–1021.
Salhab, A. S., and Hsieh, D. P. H. (1975). *Res. Commun. Chem. Pathol. Pharmacol.* **10,** 419–430.
Salhab, A. S., Abramson, F. P., Geelhoed, G. W., and Edwards, G. S. (1977). *Xenobiotica* **7,** 401–408.

Sargeant, K., Sheridan, A., O'Kelly, J., and Carnaghan, R. B. A. (1961). *Nature (London)* **192,** 1096–1097.

Sargeant, K., Carnaghan, R. B. A., and Allcroft, R. (1963). *Chem. Ind. (London)* pp. 53–55.

Sato, S., Matsushima, T., Tanaka, N., Sagimura, T., and Takashima, F. (1973). *J. Natl. Cancer Inst.* **50**, 767–778.

Schabort, J. C., and Steyn, M. (1969). *Biochem. Pharmacol.* **18**, 2241–2252.

Schacham, P., Philp, R. B., and Göwdey, C. N. (1970). *Am. J. Vet. Res.* **31**, 191–197.

Schoenhard, G. L., Lee, D. J., Howell, S. E., Pawlowski, N. E., Libbey, L. M., and Sinnhuber, R. O. (1976). *Cancer Res.* **36**, 2040–2045.

Schoental, R. (1969). *Nature (London)* **221**, 765–766.

Schoental, R. (1976). *In* "Chemical Carcinogens" (C. E. Searle, ed.), pp. 626–689. Am. Chem. Soc., Washington, D.C.

Schoental, R., Head, M. A., and Peacock, P. R. (1954). *Br. J. Cancer* **8**, 458–465.

Shank, R. C. (1977a). *J. Toxicol. Environ. Health* **2**, 1229–1244.

Shank, R. C. (1977b). *Adv. Mod. Toxicol.* **3,** 291–318.

Shank, R. C., and Magee, P. N. (1967). *Biochem. J.* **105**, 521–527.

Shank, R. C., Wogan, G. N., Gibson, J. B., and Nondasuta, A. (1972a). *Food Cosmet. Toxicol.* **10,** 61–69.

Shank, R. C., Gordon, J. E., Wogan, G. N., Nondasuta, N., and Subhamani, B. (1972b). *Food Cosmet. Toxicol.* **10**, 71–84.

Shank, R. C., Bhamarapravati, N., Gordon, J. E., and Wogan, G. N. (1972c). *Food Cosmet. Toxicol.* **10**, 171–179.

Sibay, T. M., and Hayes J. A. (1969). *Lancet* **2**, 912.

Sims, P. (1976). *In* "Screening Tests in Chemical Carcinogenesis" (R. Montesano, H. Bartsch, and L. Tomatis, eds.), pp. 211–223. Int. Agency Res. Cancer, Lyon, France.

Sinnhuber, R. O., and Wales, J. H. (1974). *Fed. Proc., Fed. Am. Soc. Exp. Biol.* **33**, 247.

Sinnhuber, R. O., Wales, J. H., Ayres, J. L., Engebrecht, R. H., and Amend, D. L. (1968). *J. Natl. Cancer Inst.* **41**, 711–718.

Sinnhuber, R. O., Lee, D. J., Wales, J. H., Landers, M. K., and Keyl, A. C. (1974). *J. Natl. Cancer Inst.* **53**, 1285–1288.

Spatz, M. (1964). *Fed. Proc., Fed. Am. Soc. Exp. Biol.* **23,** 1384–1385.

Stoloff, L. (1976). *In* "Mycotoxins and Other Fungal Related Food Problems" (J. V. Rodricks, ed.), pp. 23–50. Am. Chem. Soc., Washington, D.C.

Stoloff, L., Dantzman, J., and Armbrecht, B. H. (1971). *Food Cosmet. Toxicol.* **9**, 839–846.

Svoboda, D. J., and Reddy, J. K. (1972). *Cancer Res.* **32**, 908–912.

Svoboda, D. J., Grady, H. J., and Higginson, J. (1966). *Am. J. Pathol.* **49,** 1023–1051.

Svoboda, D. J., Reddy, J. K., and Harris, C. (1970). *Cancer Res.* **30**, 2271–2279.

Swenson, D. H., Miller, J. A., and Miller, E. C. (1973). *Biochem. Biophys. Res. Commun.* **53,** 1260–1267.

Swenson, D. H., Miller, E. C., and Miller, J. A. (1974). *Biochem. Biophys. Res. Commun.* **60,** 1036–1043.

Swenson, D. H., Miller, J. A., and Miller, E. C. (1975). *Cancer Res.* **35,** 3811–3823.

Swenson, D. H., Lin, J., Miller, E. C., and Miller, J. A. (1977). *Cancer Res.* **37,** 172–181.

Tilak, T. B. G. (1975). *Food Cosmet. Toxicol.* **13**, 247–249.

Uraguchi, K., Saito, M., Noguchi, Y., Takahasi, K., Enomoto, M., and Tatsuno, T. (1972). *Food Cosmet. Toxicol.* **10**, 193–207.

van der Watt, J. J. (1974). *In* "Mycotoxins" (I. F. H. Purchase, ed.) pp. 369–382. Am. Elsevier, New York.

van der Zijden, A. S. M., Koelensmid, W. A. A. B., Boldingh, J., Barrett, C. B., Ord, W. O., and Philp, J. (1962). *Nature (London)* **195**, 1060–1062.

van Rensburg, S. J., van der Watt, J. J., Purchase, I. F. H., Coutinho, L. P., and Markham, R. (1974). *S. Afr. Med. J.* **48,** 2508a–2508d.

Vesselinovitch, S. D., Mihailovich, N., Wogan, G. N., Lombard, L. S., and Rao, K. V. N. (1972). *Cancer Res.* **32**, 2289–2291.

Wales, J. H., and Sinnhuber, R. O. (1972). *J. Natl. Cancer Inst.* **48**, 1529–1530.

Wang, C. Y., Chiu, C. W., Pamukcu, A. M., and Bryan, G. T. (1976). *J. Natl. Cancer Inst.* **56**, 33–36.

Weisburger, J. H., Griswold, D. P., Jr., Prejean, J. D., Casey, A. E., Wood, H. B., and Weisburger, E. K. (1975). *In* "The Ambivalence of Cytostatic Therapy" (E. Grundmann and R. Gross, eds.), pp. 1–17. Springer-Verlag, Berlin and New York.

Whiting, M. G. (1963). *Econ. Bot.* **17**, 271–302.

Wislocki, P. G., Borchert, P., Miller, J. A., and Miller, E. C. (1976). *Cancer Res.* **36**, 1686–1695.

Wislocki, P. G., Miller, E. C., Miller, J. A., McCoy, E. C., and Rosenkranz, H. S. (1977). *Cancer Res.* **37**, 1883–1891.

Wogan, G. N. (1973). *Methods Cancer Res.* **7,** 309–344.

Wogan, G. N. (1975). *Cancer Res.* **35**, 3499–3502.

Wogan, G. N. (1977). *Adv. Mod. Toxicol.* **3**, 263–290.

Wogan, G. N., and Newberne, P. M. (1967). *Cancer Res.* **27**, 2370–2376.

Wogan, G. N., and Paglialunga, S. (1974). *Food Cosmet. Toxicol.* **12**, 381–384.

Wogan, G. N., and Pong, R. S. (1970). *Ann. N.Y. Acad. Sci.* **174,** 623–635.

Wogan, G. N., Edwards, G. S., and Shank, R. C. (1967). *Cancer Res.* **27,** 1729–1736.

Wogan, G. N., Paglialunga, S., and Newberne, P. M. (1974). *Food Cosmet. Toxicol.* **12,** 681–685.

Wong, J. J., and Hsieh, D. P. H. (1976). *Proc. Natl. Acad. Sci. U.S.A.* **73**, 2241–2244.

Yang, M. G., Sanger, V. L., Mickelsen, O., and Laqueur, G. L. (1968). *Proc. Soc. Exp. Biol. Med.* **127**, 1171–1175.

Zedeck, M. S., and Sternberg, S. S. (1974). *J. Natl. Cancer Inst.* **53**, 1419–1421.

CHAPTER 12

Toxic Factors Induced by Processing*

SHMUEL YANNAI

I. INTRODUCTION

The main goal of many food-processing operations is to improve the quality of foodstuffs and to ensure their safety for the consumer. However, some processes may induce the formation of materials that are potentially harmful. This unfortunate situation has been recognized for many years, and in the last decade much more information on the nature of chemical changes caused by the processing of foods has become available. Because of advanced analytical techniques many more toxic agents, hitherto unrecognized, have been discovered in processed foods. Some of the latter compounds were also identified in simple model sys-

*Dedicated to the memory of Dr. Leo Friedman, who was one of the authors in the first edition of this book and whose research contributed so much to current knowledge in this field.

TOXIC CONSTITUENTS OF PLANT FOODSTUFFS, SECOND EDITION

ISBN 0-12-449960-0

tems simulating foods that had been subjected to processing conditions resembling those employed in the food industry. Some of these materials may form during storage or cooking in the kitchen.

Other toxic compounds may occur in foodstuffs for reasons not related to processing, home cooking, or storage. First, certain plants are toxic because of their genetic constitution. Such toxic materials are discussed in other chapters of this book. Second, toxic compounds may find their way into crops as a result of various agricultural practices (e.g., residues from pesticides, fertilizers, and growth-regulating materials. Third, poisons may be absorbed into crops—especially some that show unusual capacity to accumulate them—from soils that are exceptionally rich in them (e.g., nitrate, selenium, and a variety of toxic metals). Some of these and other toxic factors may come from contaminated water or air. In this chapter we shall not deal with ingredients belonging to any of the latter three groups, which exist in foodstuffs when still in the raw state.

When dealing with potentially hazardous food components, one should bear in mind that no toxic material is detrimental to health at any level of consumption. Even the most toxic compounds can be tolerated, without any detectable untoward effects, provided that their level in the daily diet is low enough. Moreover, most investigators currently believe that this dose-related action is also true for proved carcinogens and that they also probably have a ''no-effect level'' (Coulston, 1977), although the latter is very hard to establish experimentally (Lijinsky, 1977). Hence, the presence of a toxic factor per se does not necessarily mean that the foodstuffs containing it should be considered unsafe.

Another point should be raised in order to avoid a possible misunderstanding. In many reports and review articles on the effects of processing, the terms ''nutritional'' and ''toxicological'' are used interchangeably with references to changes that might take place. When a food product fed to animals causes a decreased growth performance—or other pathological symptoms—compared to animals fed the unprocessed food item, this biological effect may result from (a) partial or complete destruction or removal of one or more of the susceptible nutrients, such as certain vitamins; (b) rendering of nutrients indigestible or unabsorbable, due to interaction of labile components with other ingredients of the food in question or to changes in molecular structure, which make them metabolically unavailable; or (c) appearance of harmful principles that may bring about anatomical and/or physiological abnormalities that do not exist in the unprocessed food. The first two types of changes are obviously of a nutritional nature, as the unavailable nutrients can be replenished and the nutritive value fully restored. However, abnormalities reflecting true toxicity symptoms cannot be relieved by supplementing the diet with any of the known nutrients. In this chapter only the latter, truly toxic materials and processes whereby they emerge will be discussed. In some cases, however, there is as yet no clear-cut distinction between nutritional and toxicological effects because their mode of action is not fully understood (e.g., the influence of some of the products of the so-called browning reactions, as will be explained later).

II. HEAT TREATMENT

Heating operations associated with cooking, frying, toasting, evaporation (of water or extraction solvents), sterilization, and similar processes, even mild ones such as pasteurization, may bring about a variety of chemical changes in the treated foods. Many of these changes are physiologically undesirable, as some of them involve heat-susceptible nutrients, especially vitamins and amino acids (Paulus, 1976; Anonymous, 1975a; Hollingsworth and Martin, 1972; Ford and Shorrok, 1971; FAO/WHO Expert Group, 1965; Carpenter *et al.*, 1962; Gonzalez del Cueto *et al.*, 1960; Calhoun *et al.*, 1960; Liener, 1958). Consequently, the nutritional quality of the heated food is reduced. Still other changes lead to the formation of potentially hazardous compounds that may produce toxic effects if their level in the food under consideration and the quantity ingested are high enough. Such processes will be discussed in more detail later in this section (in sub-sections A, B, and C). Numerous investigators observed that similar chemical changes might also take place during storage periods (Melnick and Oser, 1949; Carpenter *et al.*, 1963; Ben-Gera and Zimmerman, 1964) and that they become more pronounced under unfavorable conditions, i.e., elevated temperature and relative humidity (Lea, 1958), and sometimes on exposure to light, oxygen (Lightbody and Fevold, 1948; Ellis, 1959), and certain catalysts that may promote these reactions (Coulter *et al.*, 1951; Milner and Thompson, 1954). It appears that prolonged, mild heat treatments (i.e., at a low temperature) can bring about changes comparable to those encountered when the susceptible food is subjected to more severe heating of a short duration (Fricker, 1975).

A. Nitrogenous Compounds and Carbohydrates

Many of the changes brought about by food processing involving heat treatments and storage involve nitrogenous compounds, which are in most cases proteins and free amino acids, carbohydrates, and some of their derivatives. Probably the most important of these changes, and certainly the most extensively investigated ones, result from the so-called nonenzymatic browning or Maillard reaction (Henry and Kon, 1958; Liener, 1958; Lea, 1958). Research reports published in recent years suggest that, apart from the decrease in nutritive value resulting from the unavailability of amino acids and destruction of other food components such as ascorbic acid (Ellis, 1959), some of the browning reaction products are actually toxic (Fragne and Adrian, 1967; Bender, 1972; Adrian, 1973; Mauron, 1975). Work done with simple model systems suggests that heating such amino acids as lysine, glutamic acid, and alanine with glucose at

100°C in the presence of air can also induce the formation of N-nitrosamines, which have been shown to be potent carcinogens and mutagens (Devik, 1967). The latter compounds and their appearance in foods containing nitrite will be discussed in Section VI.

The Maillard reaction will not be described here in full. The interested reader will find the following texts on this subject comprehensive and instructive: Danehy and Pigman (1951), Hodge (1953), Ellis (1959), Burton and McWeeny (1964), Reynolds (1963, 1965), McWeeny *et al.* (1969), Spark (1969), Hurst (1972), and Feeney *et al.* (1975).

One of the early reviews on nonenzymatic browning that also emphasizes practical technological and nutritional aspects is a paper by Danehy and Pigman (1951). In this article the authors describe the series of reactions occurring in foods as well as in different model systems, the physical conditions that favor or inhibit these reactions, and their outcome in each case. In the older studies the question whether the Maillard reaction products were toxic received little attention, since nearly all the published work was related to the products of the reaction only in terms of the nutritional availability of the amino acids participating in it.

The harmful effects of the browning reaction were first described by Lang *et al.* (1959, cited by Lee *et al.*, 1975), who suggested that the decrease in nutritional value brought about by the browning reaction was due not only to the degradation of amino acids, but also to the appearance of toxic materials. Supplementation of diets containing heat-treated protein with essential amino acids (Rao *et al.*, 1963) failed to restore their biological quality. This fact seems to support the hypothesis suggested in the earlier reports that the browned amino acids, or their residues in polypeptide chains, are not only nutritionally unavailable but also physiologically harmful. Browned apricots fed to rats brought about a decrease in weight gain, feed efficiency, serum glucose, and urea levels and caused increases in water consumption, serum glutamic–oxaloacetic transaminase and glutamic–pyruvic transaminase activities, and hypertrophy of the liver and kidney (Lee *et al.*, 1975). The browned dehydrated apricots were extracted with ethanol, which was later evaporated to dryness, and the extract was separated into water- and ether-soluble fractions. Each fraction was then evaporated to dryness and incorporated into rats' diets. The water-soluble fraction was found to be responsible for the adverse physiological effects of the browned apricots, while the ether-soluble fraction was probably devoid of any physiological significance. The browning reaction products obtained by heating a model system containing the amino acids and sugars of the same composition as in apricots exerted physiological effects that resembled those observed in the animals fed the browned apricot-containing diet. This seems to indicate that the physiological effects of the latter were indeed due to nonenzymatic browning.

Adrian and his coinvestigators claimed that the toxic products evolving from the Maillard reaction were the early products, called "premelanoidins." These are soluble compounds, whereas the final products, called "melanoidins," are insoluble and have no physiological activity (Adrian, 1973). The "premelanoidins" are toxic to animals and also to microorganisms. The formation of these early products is described in an earlier paper by the same author (Adrian, 1972). When active carbonyl group-containing compounds (mostly aldoses, ketoses, and fatty acid oxidative breakdown products) react with free amino group-containing species (such as proteins and free amino acids) a variety of reaction products evolve, depending on the nature of the participating reactants and the prevailing conditions. The most important series of events leading to the formation of premelanoidins involves reducing sugars and free amino groups in proteins (e.g., the ϵ-amino function of lysine residues). The resulting carbonylamino condensation product, N-substituted amino sugar, undergoes an Amadori rearrangement, and one of the most important derivatives formed is N-substituted 1-amino-1-deoxy-2-ketose. Further reactions, the nature of which depends on the amounts of oxygen, water, and other substances present, the temperature and pH of the system, etc., are mostly degradative. The resulting products, formed by fission, oxidation, and dehydration, include reductones, furfurals and their oxidation derivatives, and a variety of other cyclic compounds. Many of these products are nitrogen-free, and similar products can also form during heat treatment of sugars alone (Adrian, 1972). The latter process, in which simple sugars, sugar acids, and other polyhydroxy compounds such as ascorbic acid undergo heat degradation in the absence of amino acids or proteins, is referred to as "caramelization" and is catalyzed under acidic or alkaline conditions (Eskin *et al.,* 1971).

The toxicological effects in rats of premelanoidins, described by Adrian (1973), include a slight hypertrophy of the liver (by 25%) and a drop in its nitrogen content, a marked hypertrophy of the cecum (by 100%), a decrease in the reproduction and lactation performance, and allergic reactions. Pregnant rats maintained on a balanced diet containing 0.2% of premelanoidins from degraded glucose–glycine complex showed a feed consumption comparable to that of the controls. However, compared with the control animals their weight gain during gestation was 27% lower, and they exhibited a fivefold increase in the frequency of fetal resorptions. Despite the fact that they had fewer pups per litter, the pups' average weight at birth and weaning was lower than that of the controls. The decrease in weight of the weanlings can probably be attributed to the poor lactation of the mothers exposed to premelanoidins. In a more recent paper on the influence of premelanoidins on the reproduction performance of the rat (Adrian and Susbielle, 1975) the authors reported similar findings and, in addition, a twofold increase in the number of stillborn pups and hypertrophy of the kidneys, spleen, and cecum.

Adrian (1973) cited studies showing that the severity of toxic effects exerted by the Maillard reaction products are proportional to the degree of heat treatment applied. The incidence of mortality from liver necrosis in rats fed overheated milk powder increased greatly when the milk dehydration was executed by drum-drying. Other reports cited by Adrian prove that the Maillard reaction products are also toxic to certain microorganisms and that mold and yeast cells show growth inhibition and cellular abnormalities resulting from exposure to premelanoidins. Thus, the poor growth performance of animals given heat-damaged materials is interpreted by Adrian as indicating that these materials not only are of a low nutritive value, but are actually toxic.

Lang (1960, cited by Mauron, 1975) reported that the browning reaction products formed upon heating of ribonucleic acid with glucose caused reduction in rats' growth rate and feed efficiency. RNA browned by heating in the presence of glucose was even more toxic than browned L-lysine and DL-tryptophan, their LD_{50} values being 0.49 mg/kg body weight for the former (Lang and Schaeffner, 1964, cited by Lee *et al.*, 1975) and 4.1 and 6.1 mg/kg, respectively, for the two latter amino acids (Krug *et al.*, 1959, cited by Lee *et al.*, 1975).

Furthermore, Aranjo Neto *et al.* (1974) showed that *Candida utilis,* the cells of which are exceptionally rich in ribonucleic acid (12%), which had been stored at temperatures ranging from 10° to 30°C and relative humidity exceeding 70% for 3 to 4 years, brought about a decrease in utilization of casein, although the casein-containing diet was supplemented with methionine. It appears, therefore, that RNA is likely to produce harmful compounds even under relatively mild storage conditions. In view of the high RNA content of most microorganisms used to make single-cell protein (SCP), special care should be taken in evaluating the wholesomeness of such foodstuffs.

1. Furfural Compounds

As mentioned earlier, an important group of premelanoidins, which have been shown to exert toxic effects in animals, is composed of the furfural compounds. Small amounts of furfurals are present in many foods, including bread, coffee, processed fruits and fruit juices, and certain alcoholic beverages (e.g., whiskey). The most common furfurals occurring in foodstuffs subjected to heat treatments are furfural (**I**) and 5-hydroxymethylfurfural (**II**), which form mainly from pentoses and hexoses, respectively (Rice, 1972). As much as 1% (on a dry weight basis) of 5-hydroxymethylfurfural was produced in concentrated apricot purée

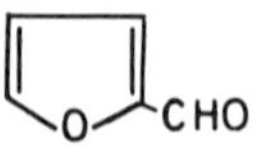

O CHO

Furfural (I)

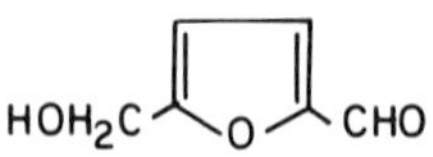

HOH_2C O CHO

5-Hydroxymethylfurfural (II)

after 8 months storage at 45°C (Lee *et al.*, 1975). Furfural fed to rats produced liver cirrhosis (Nakahara *et al.*, 1941, cited by Rice, 1972).

In a chronic feeding study in rats it was found that a level of 250 mg/kg body weight of 5-hydroxymethylfurfural did not produce any adverse effect, and its estimated LD_{50} value was greater than 1 gm/kg (Lang and Baessler, 1971, cited by Lee *et al.*, 1975); however, an intravenous injection of 100 mg of 5-hydroxymethylfurfural per kilogram body weight stimulated intestinal motility because of its choleretic and diuretic effect.

In a more recent study (Feron, 1972) it was found that injections of furfural alone (0.2 ml of 1.5% solution administered weekly for 36 weeks) caused slight hyperplasia of the tracheobronchial epithelium of hamsters. This shows that furfural has an irritating effect, but it was not tumorigenic under the experimental conditions in the author's laboratory. When benzo[*a*]pyrene alone was administered, 41 out of the 62 hamsters developed respiratory tumors and 2 animals (3%) showed peritracheal sarcomas. However, when the animals were challenged with both benzo[*a*]pyrene and furfural, the metaplastic changes occurred earlier, a few more bronchial and peripheral squamous cell carcinomas were observed, and peritracheal sarcomas appeared in 33% of the animals. These findings suggest that furfural augments the carcinogenic action of benzo[*a*]pyrene, although there is still doubt whether oral administration of both substances will also cause potentiation of the carcinogenic effect of benzo[*a*]pyrene by furfural. Therefore, investigation of the possible effect of ingested furfural as a cocarcinogen seems to be very important.

2. *Reductones*

Another group of premelanoidins, called reductones, which are formed by heating reducing sugars in the presence of amino acids, are considerably more toxic than those formed from the sugars alone (Adrian, 1973). Five such nitrogen-containing reductones formed from hexoses and secondary amines were studies by Ambrose *et al.* (1961) and found to be toxic to mice. The approximate LD_{50} values of dimethylaminohexosereductone (**III**) and morpholinohexosereductone (**IV**) administered intraperitoneally were 300 and 850 mg/kg body weight, respectively. The oral LD_{50} values of anhydrodimethylaminohexosereductone (**V**), dimethylaminohexosereductone, and anhydropiperidinohexosereductone (**VI**) were 300, 400, and 900 mg/kg, respectively, while in the case of morpholinohexosereductone and piperidinohexosereductone (**VII**) they were greater than 1200 mg/kg. Single intraperitoneal injections and single oral doses of dimethylaminohexosereductone in both mice and rats, as well as long-term feeding of dimethylaminohexosereductone and piperidinohexosereductone in rats, produced typical symptoms of hyperexcitability, elevation and nodding of the head, and whirling. The effect of

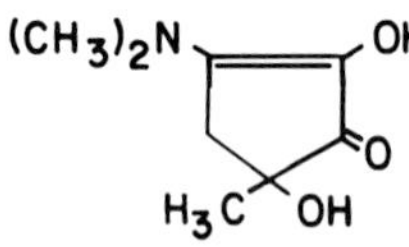

Dimethylaminohexose-reductone (III)

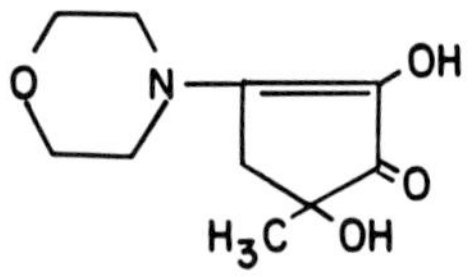

Morpholinohexose-furfural (IV)

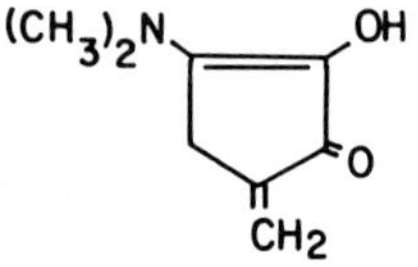

Anhydrodimethylamino-hexosereductone (V)

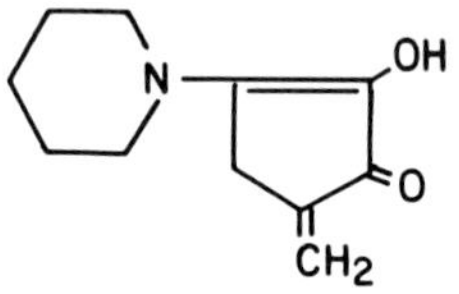

Anhydropiperidino-hexosereductone (VI)

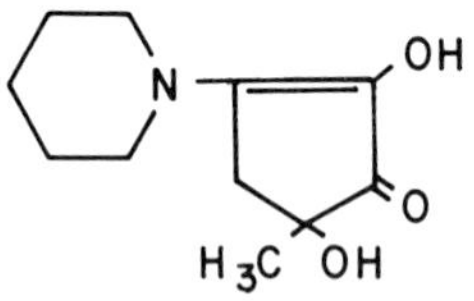

Piperidinohexosereductone (VII)

one of these compounds, dimethylaminohexosereductone, on mice was reported one year earlier by Cutting *et al.* (1960). The symptoms characterizing the circling syndrome induced by a single subcutaneous or intraperitoneal injection of the latter compound in distilled water or saline, or by administration of the reductone into the drinking water, were permanent. They included frequent head tossing and running in circles and a decreased ability to resist stress. A single dose of 250 mg/kg produced the maximum effect.

3. *Other Premelanoidins*

Other products result from the nonenzymatic browning reaction between amino acids and glyoxal (OHC—CHO). This aldehyde has been identified in a variety of foods [e.g., tomatoes, liquors, roasted barley, Kori tofu made from soybeans, and starch breakdown products (Davidek *et al.*, 1975)]. The reaction products formed by heating at various pH levels included amines and imines and a variety of nitrogen-containing heterocyclic compounds. The latter consisted, among other substances, of pyrazines and imidazoles, which were also identifed in ammoniated molasses and found to have toxic effects (Nishie *et al.*, 1969). The effect of ammoniation of food will be discussed in more detail in Section III. The presence of pyrazines in foods (e.g., potato chips) after being subjected to heat treatment was proved a few years earlier, and their formation has been attributed to fragmentation of sugars (Eskin *et al.*, 1971). Some of these compounds were identified by Dawes and Edwards (1966) in model systems consisting of aldoses and amino acids that had undergone heating, e.g., 2,5-dimethylpyrazine (**VIII**) and trimethylpyrazine (**IX**). The authors believe that

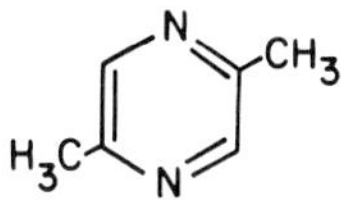

2,5- Dimethylpyrazine (VIII) 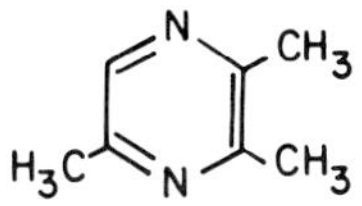 Trimethylpyrazine (IX)

these products are formed by the aldose–amino acid interaction, followed by the Strecker degradation. Koehler *et al.* (1969) support this view. They studied model systems containing amino acids and carbohydrates, all of which were low in water. Radioisotopic labeling indicated that sugars were the principal source of the carbon atoms, while the amino acids mostly furnished only nitrogen to the pyrazine molecule. Also, NH_4^+ ions served as a source of nitrogen. Fragmentation of hexoses into two- and three-carbon units provides intermediates which may then condense, thus forming pyrazines. The authors discuss possible pathways and suggest the manner in which different alkylated pyrazines develop.

In an early stage of the nonenzymatic browning reaction of the aminocarbonyl type involving amino acids and sugars, free radicals evolve. Namiki *et al.* (1973) heated a mixture of D-arabinose and β-alanine in distilled water and found that free radicals developed within a few minutes after the onset of heating; their quantity increased rapidly during the 90-min period reported in these authors' paper. The free radicals had appeared before considerable browning was observed. A similar result was obtained with α-alanine and D-arabinose and also with glucose, fructose, and other sugars, but no such radicals could be detected in a system of either sugar or amino acid alone. The free radicals are highly reactive and may interact with other components of the food, as, for example, through oxidation reactions, thus bringing about various changes in quality. Similar changes initiated by the free radicals may also take place in biological systems (Namiki *et al.*, 1975). The possible toxicological significance of free radicals formed in foods by processing will be discussed in Section IV.

B. Lipids

The most important chemical changes that may take place in fats and other lipids, through which potentially toxic substances may evolve, occur during a variety of industrial processing operations, such as refining, bleaching, deodorization, hydrogenation, frying, baking, and roasting. Some of these reactions also occur during prolonged storage under unfavorable conditions. When oil is heated in air, it is partially transformed into volatile chain-scission and oxidation products, nonvolatile oxidized derivatives, cyclic compounds, and dimers and higher polymers (Artman, 1969). Aizetmüller and Scharmann (1975) reported that 10–40% of the total amount of used frying fats consisted of oxidation products, glyceride hydrolysis products and polymers. Such reactions are undesirable, since some of the resulting substances are toxic (e.g., hydroperoxides, cyclic derivatives, and oxidized fatty acid dimers), and the oil also loses part of its nutritional value because it may contain substantial amounts of unabsorbable polymers. Fats altered in this way by processing operations, such as deep fat frying of foods, are organoleptically acceptable in the human diet, although they may have adverse effects (Perkins, 1960). Most of these compounds do not form adducts with urea. The question whether heated and oxidized fats may be toxic is much in dispute, and one can find many conflicting reports in the literature on this issue. In numerous papers the authors provide what appears to be ample evidence for deleterious effects of heated oils in experimental animals, whereas many other workers failed to reveal any toxicity in even severely heated or oxidized oils. Some of the investigators who could not detect evidence of toxicity attribute the adverse physiological effects reported by others to vitamin deficiencies brought about by oxidative reactions in the diets containing previously oxidized fats (Artman, 1969). The possible reasons for the conflicting data will be discussed in more detail later in this section.

It is obvious that the nature and extent of the changes that might take place upon heating and oxidation depend very much on the kind of fat used and the conditions under which it has been heated and stored. Regarding the nature of the fatty acids involved, the results of most studies demonstrated that in nearly all cases the unsaturated fatty acids were the most susceptible to changes induced by processing and storage, and the higher the degree of unsaturation the greater the susceptibility of the fatty acid in question to these effects. These findings seem to be logical, as—needless to say—unsaturated materials are more reactive and thus more likely to undergo oxidation and polymerization than saturated compounds. Yet some investigators reported contradictory findings, according to which the less saturated fats failed to show more pronounced changes induced by heat treatments than did the more saturated ones (Sahasrabudhe and Bhalerao, 1963; Simko *et al.*, 1964; Thompson *et al.*, 1967). It is also a well-recognized fact that the fats in stored foods of animal origin become rancid at a relatively rapid rate, in spite of the low concentration of unsaturated fatty acids in these fats. This phenomenon can be explained by the presence in animal foodstuffs of fairly large amounts of heme compounds (consisting mostly of myoglobin and hemoglobin), which act as powerful biological prooxidants, and may therefore enhance the fat deterioration process (Bishov *et al.*, 1960). Plant foodstuffs, on the other hand, often contain potent antioxidants, among which are tocopherols, phospholipids, ascorbic acid, and flavonoids, which may greatly retard the oxidative deterioration in plant oils, despite their high levels of unsaturated fatty acids (Berk, 1976; Bishov *et al.*, 1960).

Aaes-Jørgensen *et al.* (1956) found, as would be expected, that highly unsaturated linseed oil was very susceptible to heat treatment. When this oil was heated at 275°C for 12 hr in a CO_2 atmosphere and fed to rats, the animals grew poorly and all died within the first 18 weeks of the experiment. Yet Raju and Rajagopalan (1955) reported that not only groundnut and sesame but even coconut oils (the latter containing mostly saturated fatty acids), which had been heated in an open iron pan at 270°C for 8 hr, caused toxic symptoms in young rats when incorporated in their diets at a level of 15%. The animals exhibited growth depression and enlarged livers, which also contained an abnormally high fat level. When the level of the heated oils in the diet was 30%, all the rats died within 1 week. In another set of trials, when 15% of oil heated at 280°C was used in the diet, severe jaundice occurred and most of the rats treated died in the sixth week of the experiment.

Perkins (1960) defined three types of chemical changes that may be encountered in fats and oils because of storage and/or heat treatments: (a) autoxidation, (b) thermal oxidation, and (c) thermal polymerization. Autoxidation is defined as the oxidation of a fat or an oil at temperatures not exceeding 100°C. Thermal oxidation is defined as oxidation of fat at about 200°C in the presence of air or with aeration. Thermal polymerization is defined as oxidation of fat between 200° and 300°C, in the absence of oxygen. As shown by some of the earlier studies on this subject, fats that had undergone such changes were toxic to experimental animals (Kaunitz *et al.*, 1952; Crampton *et al.*, 1953; Raju and Rajagopalan, 1955; Aaes-Jørgensen *et al.*, 1956; Custot, 1959; Matsuo, 1962). The above three categories of fat degradative changes are also referred to in more recent papers dealing with this topic, and much more evidence on the formation and nature of physiologically harmful products belonging to all three of them has been published since then. Fortunately, the flavor of rancid oil that develops during autoxidation serves as a built-in warning factor, which prevents voluntary consumption of autoxidized fats. However, as mentioned above, such a warning factor does not exist in polymerized fats, since they do not have an objectionable flavor.

Endres *et al.* (1962) stressed that data from studies of a particular fat or oil cannot be extrapolated to predict the results of another fat because of the varied and complex nature of different fats. The position and type of fatty acid with respect to other fatty acids may have a marked influence on the course of degradative changes that may occur in it. Thus, the qualitative and quantitative results to be anticipated from thermal treatment of different fats are unpredictable.

It has also been shown that ill-controlled heating during domestic frying can cause much more extensive chemical changes in fat than those occurring during carefully controlled laboratory or commercial frying operations (Custot, 1959; Fleishman *et al.,* 1963).

More of the earlier studies are discussed in reviews by Bergström and Holman (1948), Holman (1954), Lea (1958), Custot (1959), Perkins (1960), Swern (1961), Uri (1961), Lundberg (1962a), Mitchell and Henick (1962), Aaes-Jørgensen (1962), Sims and Hoffman (1962), Kaunitz (1962), and Kummerow (1962). For more recent information, see the comprehensive articles by Lea (1965a), Kaunitz (1967), Perkins (1967), Anonymous (1968a), Artman (1969), Labuza (1971), Alim and Morton (1974), Lang (1973), Cooper (1975), Vergroesen and Gottenbos (1975), and Berk (1976).

1. Non-Urea Adduct-Forming Compounds

Non-urea adduct-forming (NAF) compounds probably contain most of the harmful materials formed during heat treatments and storage of fats and foods containing fats (Lea, 1962; Barrett and Henry, 1966). They consist of fatty acid derivatives that are incapable of forming a complex with urea, due to the conversion of their natural chain structure into cyclic fatty acid derivatives and cyclic or branched polymers or due to the formation of oxidation and scission products the chains of which contain less than seven carbon atoms (Johnson *et al.,* 1957; Mehta and Sharma, 1956, 1957; Crampton *et al.,* 1953; Common *et al.,* 1957; Friedman *et al.,* 1961; Firestone *et al.,* 1961a,b; Anonymous, 1962, 1968a; Kummerow, 1962; Sahasrabudhe and Bhalerao, 1963; Firestone, 1963; Michael *et al.,* 1966; Michael, 1966a,b; Kaunitz, 1967; Jacobson, 1967; Artman and Alexander, 1968).

Firestone *et al.* (1961a) heated cottonseed oil at approximately 205°C for 7–8 hr each day for 40 days. The oil was stirred for a few seconds evey 2 hr. Later the oil was fractionated, and each fraction was administered orally to weanling rats. The fraction containing non-urea-adducting monomers and dimers proved to be toxic, and the animals receiving them died within 3 days. The NAF fatty acids isolated from commercial oleic acid and other fatty acids were also found to be toxic to weanling rats, even after hydrogenation (Firestone *et al.,* 1961b). Similar findings were reported in various papers (Perkins, 1960; Nolen *et al.,* 1967; Kaunitz, 1967). It has also been implied that the NAF from heated corn oil has a carcinogenic or cocarcinogenic effect. The incidence of tumors induced by painting 20-methylcholanthrene on rats' skin was substantially increased by feeding heated vegetable oils. The carcinogenicity of 2-acetylaminofluorene was also enhanced when fed to rats together with heated oil, and particularly with the NAF isolated from heated oil (Anonymous, 1962; Sugai *et al.*, 1962).

The amount of NAF present in heated corn oil and hydrogenated vegetable shortening was found to be in direct relation to the duration of heating. However, the molecular weight of the NAF—indicative of polymerization—did not change considerably, despite the high temperature (200°) and fairly long duration (24 hr) of heating (Sahasrabudhe and Bhalerao, 1963). Results obtained in our laboratory are in agreement with the above data. Kantorowitz and Yannai (1974) heated samples of soybean oil and hydrogenated soybean oil of the same origin at 200°C for 10, 19, 30, and 43 hr in air with continuous stirring at a fixed speed. The percentage of NAF in the liquid oil increased gradually from 0.56 (in the unheated oil) to 23.2 after 43 hr of heating, whereas in the case of the "hardened" oil the corresponding values were 2.9 and 28.2, respectively. Yet the molecular weight of the total NAF and of the three NAF subfractions (isolated by molecular distillation under high vacuum) failed to show appreciable changes throughout the long heating period. On the other hand, the average molecular weight of both oils did increase gradually during the heat treatment, from 780 to 1429 in the case of the liquid oil and to 1338 in the case of the hardened oil. The viscosity of the oils, which also reflects their molecular weight, showed even greater changes. A possible explanation for this discrepancy is that the increase in molecular weight of some of the NAF components (due to polymerization) was counterbalanced by the decrease in molecular weight (due to scission) of other NAF components that underwent degradation. As to the toxicity of the treated samples, only the liquid oil was found to have an adverse effect on the rats exposed to it (by daily doses, administered by a stomach tube). The adductable fraction from heated oil is innocuous to rats (Common *et al.*, 1957).

2. *Autoxidation: Primary Products and Secondary Degradation Products*

Autoxidation, which is defined as the oxidation of fats by oxygen at temperatures up to 100°C (Perkins, 1960), is of an autocatalytic nature. Many studies have shown that autoxidized fats may exhibit untoward physiological effects when administered to experimental animals. Numerous reports showed that the harmful influence was related to one or more nutritional deficiencies and that the autoxidation of food fats was accompanied by destruction of essential dietary factors, such as susceptible vitamins (A, E, B complex), essential fatty acids, and certain essential amino acids (Perkins, 1960; Andrews *et al.*, 1960; Artman, 1969). Other studies showed, however, that the poor performance of animals fed autoxidized oil could not be remedied by supplementing their diet with appropriate amounts of the damaged nutrients (see, for example, Kaunitz *et al.*, 1956a; Andrews *et al.*, 1960). It seems likely, therefore, that one of the effects of oxidized fats is an increased requirement for certain nutrients, which must be present in excessive quantities in order to meet the greater needs for these nutrients of the animals exposed to abused fats and perhaps also to counteract, at least in part, the adverse effects of the latter on the animals' metabolic functions. A protective effect of added vitamin E against the harmful influence of autoxidized oil was reported by Privett and Cortesi (1972). Still, as the findings of numerous studies demonstrate, an increased demand for nutrients is certainly not the only effect exerted by autoxidized fats.

Most of the discussion below will be devoted to the biological effects of oxidized fats, while the chemical changes occurring in them will be described briefly and the appropriate references given.

a. Primary Products. The initial stage in the autoxidation of unsaturated fatty acids and the lipids containing them involves a free-radical chain mechanism, with hydroperoxides as the major products (Labuza, 1971). Swern (1961) described the free-radical mechanism that facilitates the chain reaction, the structures of the first oxidation products, and the physical and chemical factors that may have an effect on the nature and rate of the reactions involved. Autoxidation is a phenomenon that concerns mainly unsaturated fatty acids. Yet Brodnitz and his co-workers (1968a,b) reported that a highly purified methyl palmitate, free of all detectable impurities, was oxidized by aeration at 150°C. The appearance of monohydroperoxide as the first autoxidation product was verified by thin-layer chromatography, spot and spray test, and polarography (Brodnitz *et al.*, 1968a). Oxidation may occur at any carbon atom along the alkyl chain, but the preferred site is usually near the middle of the chain (Brodnitz *et al.*, 1968b).

In the past it was generally believed that the primary autoxidation products were heterocyclic peroxide-containing compounds, which were supposed to evolve from the attack of double bonds by oxygen.

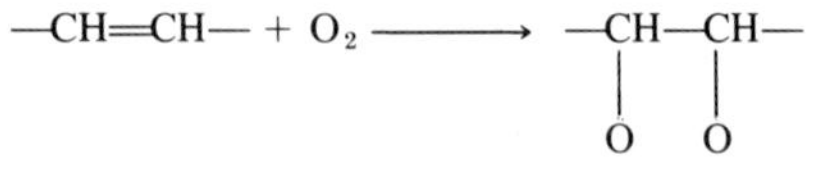

$$\text{—CH=CH—} + O_2 \longrightarrow \begin{array}{c}\text{—CH—CH—} \\ \;\;| \quad\;\; | \\ \;\;\text{O} \;\;\; \text{O}\end{array}$$

In fact, the existence of such cyclic peroxides has never been proved. A possible interpretation of this fact, which found wide acceptance, was that the cyclic peroxides are quite unstable and might therefore give rise to more stable epoxides, which possess the structure (Swern, 1961)

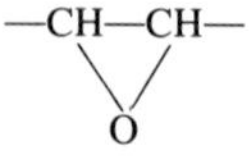

$$\begin{array}{c}\text{—CH—CH—} \\ \diagdown\;\diagup \\ \text{O}\end{array}$$

According to another theory epoxides are the first step in the autoxidation of ethylenic-group-containing materials and are formed by direct addition of oxygen to the double bonds (Fokin, 1909, cited by Swern, 1961).

$$\text{—CH=CH— } + \tfrac{1}{2}O_2 \longrightarrow \text{—CH—CH—} \text{ (bridged by O)}$$

Although epoxy compounds have been isolated from autoxidized materials, they do not seem to constitute major derivatives.

Farmer and his coinvestigators hypothesized that the primary products that appear during autoxidation of olefinic compounds are in fact hydroperoxides. The investigations of Farmer and his co-workers are probably the most significant ones in the development of modern concepts pertaining to lipid autoxidation. Much of the current knowledge of the reactions involved and the factors responsible for them stem from these authors' early studies conducted with olefinic substances. The initial free radicals formed in the autoxidation process, the appearance of peroxides, and the subsequent carbon-chain scission, which leads to the formation of carbonyl-group-containing derivatives, alcohols, and short-chain acids as secondary degradation products, were first described by them (Farmer *et al.,* 1942). One year later they developed a theory that during autoxidation there is a rearrangement of the double bonds, which results in the formation of new isomers (Farmer *et al.,* 1943).

Farmer's group was the first to suggest that the autoxidation process of all unconjugated olefinic-bond-containing compounds proceeds by a chain reaction in which molecular oxygen adds to a carbon atom adjacent to a double bond, namely, an α-methylene group, by a free-radical mechanism. The chain reaction is initiated by the removal of a hydrogen atom from an α-methylenic position, resulting in the formation of a free radical

$$\text{—CH}_2\text{—CH=CH—} \xrightarrow{-H} \text{—}\dot{\text{C}}\text{H—CH=CH—} \tag{1}$$

In the next step an oxygen molecule combines with the free radical, thereby creating a peroxide free radical [Eq. (2)]. The latter abstracts a hydrogen atom

$$\text{—}\dot{\text{C}}\text{H—CH=CH—} \xrightarrow{+O_2} \text{—CH(O—O}\cdot\text{)—CH=CH—} \tag{2}$$

from a methylene group adjacent to a double bond in another olefin molecule, thus forming a new free radical, available to react with oxygen, and the process becomes self-propagating [Eq. (3)]. As can be seen, the double bond remains

$$\text{—CH(O—O}\cdot\text{)—CH=CH— } + \text{ —CH}_2\text{—CH=CH—} \longrightarrow \text{—CH(O—OH)—CH=CH— } + \text{ —}\dot{\text{C}}\text{H—CH=CH—} \tag{3}$$

intact and is not saturated by oxygen, as was believed formerly. Formation of the above free radical involves a reaction that requires a very high activation energy, which is approximately 35–65 kcal/mole (Privett and Blank, 1962). Recent research, reviewed by Labuza (1971), suggests that singlet oxygen may be involved in the formation of free radicals. Singlet O_2 can be formed through photochemical reactions in the presence of a sensitizer such as metal, chlorophyll, etc. Further evidence of this effect was recently reported by Clements *et al.* (1973) and Carlsson *et al.* (1976). According to Heaton and Uri (1961, cited by Labuza, 1971) metals present in trace amounts are often responsible for the primary initiation, and metals that are oxidized by one-electron transfer are the most active. Accordingly, cobalt, copper, iron, nickel, manganese, and other such metals have been found to be potent lipid prooxidants (Ingold, 1962). These are transition metals, which possess two or more valency states, with a suitable oxidation–reduction potential between the states. They both shorten the induction period and increase the oxidation rate. Sometimes levels of as little as a few parts per billion of such metals in oil are sufficient to catalyze the autoxidation.

After initiation the chain cycle proceeds autocatalytically, but it can be greatly stimulated by illumination with ultraviolet light (Bergström and Holman, 1948; Swern, 1961) and to a lesser extent by visible and infrared radiation (Swern, 1961; Lundberg, 1962b; Artman, 1969; Carlsson *et al.*, 1976). Obviously, the amount of oxygen available plays a major role in the kinetics of this process (Mesrobian and Tobolsky, 1961; Lundberg, 1962b).

Another extraneous factor that influences the rate of autoxidation is, as with most other chemical reactions, the temperature. At ordinary temperatures, the effect of increasing temperature is greater than in most other chemical reactions, since not only the chain propagation, but also the peroxide decomposition process, is enhanced by it, and therefore a higher concentration of free radicals is available for the initiation and propagation of reaction chains (Lundberg, 1962b). Autoxidation takes place also during long cooling periods of fried fats (Artman, 1969).

Factors that retard lipid autoxidation are naturally occurring antioxidants (e.g., tocopherols and ascorbic acid) and added antioxidants of various kinds (the most widely used being phenolic compounds) and storage under an inert atmosphere or vacuum.

The rate of formation of hydroperoxides is not directly proportional to the degree of unsaturation. Rather, it increases drastically when the number of double bonds is greater. Thus, lineoleate oxidized ten times faster than oleate, and linolenate was 20 to 30 times faster. According to Labuza (1971) the reason for the increased rate of hydroperoxide formation with the more highly unsaturated fatty acids is that the methylene group adjacent to two double bonds is much more reactive than a group that is at an α position to one double bond. Labuza stresses, therefore, that the fat content of foods is of less importance in determining their susceptibility to lipid oxidation than are the amounts of unsaturated fatty acids present. Hence beef, which contains much more fat than potatoes, is less susceptible to oxidative rancidity when in the dehydrated state than are potatoes.

This fact can be attributed to the higher reactivity of potatoes, brought about by their relatively high level of unsaturated fatty acids in the lipid fraction. The relative tendency of lipids to become rancid (due to autoxidation) is tested by measuring the peroxide value (Kalbag *et al.*, 1955; Lea, 1962). A recent method for the assessment of oil rancidity based on the formation of hydroperoxides is that of Eskin (1976). Other workers, however, doubt the reliability of the peroxide value as a parameter for characterizing the degree of rancidity, for several reasons (Lundberg, 1962a; Akiya *et al.*, 1973).

The various hydroperoxides formed from the C_{18} polyunsaturated fatty acids, their concentration, the positions of the hydroperoxide groups, the different isomers formed in each case, and their characterization were described by Frankel (1962).

All of the physicochemical aspects of the autoxidation mechanism of formation, kinetics (initiation, propagation, and termination), thermodynamics, and activation energies of the various reactions that take place in the autoxidation process and the action of catalysts and inhibitors were described and discussed in the comprehensive article of Uri (1961). More recent studies and their practical implications were discussed by Labuza (1971).

It has been known for centuries that exposure of fats to air, especially while they are being heated, brings about rancidity. Such fats have been rejected because of their objectionable smell and taste. Until a few decades ago, however, little research had been conducted on the possible physiological effects of rancid fats. Early reports included observations of ophthalmia, digestive disturbances, reproductive failure, anemia, leukopenia, and dermatitis (Kaunitz, 1960). Rapid weight loss and a high death rate were often encountered (Kaunitz *et al.*, 1960). As mentioned above, most workers related these observations to the destruction of vitamins and other essential nutrients. Later, however, the results of many investigations indicated that rancid fats are truly toxic (Lundberg, 1962a) and that the toxicity correlated better with the peroxide concentration than with any other parameter of rancidity (Andrews *et al.*, 1960). When fats having a high peroxide value were fed to animals, toxicity symptoms often developed, and the severity of the changes observed depended on the concentration of peroxides occurring in it (Rao, 1960). This is also true for autoxidized fatty acids, the toxicity of which was found to be proportional to the amount of peroxide groups present (Matsuo, 1962). Greenberg and Frazer (1953) suggested that the adverse effects of rancid fats may be due to the following causes: (a) a condition of stress induced by the irritating action of the toxic factors, (b) interference with the normal nutritional properties of fat, (c) destruction of other nutrients, (d) interference with the function of intestinal flora, or (e) a lower food intake. The validity of the fourth possibility is supported by the observations of Roth and Halvorson (1952) that the germination of bacterial spores is inhibited by autoxidized linolenic acid, at a level of as low as 1 ppm. Frazer (1961) investigated the incidence of "tropical sprue" and suggested that it may well be associated with the consumption of excessive amounts of rancid fat, largely of plant origin. This chronic disease is characterized by diarrhea and ulceration of the mucous membrane of the digestive tract, which are symptoms that have been observed in animals fed autoxidized lipids, as will be described later. More recent publications show that autoxidized lipids interfere with many vital physiological functions and cause a variety of macroscopic and microscopic lesions in experimental animals exposed to them.

It should be realized that "the hydroperoxides themselves do not contribute directly, to any appreciable extent, to the undesirable flavors and odors of autoxidized food materials. Rather, the rancid flavors and odors are due to a host of secondary substances derived through various reactions and further oxidation of the peroxides and their degradation products" (Lundberg, 1962b). In other words, the rancid odors and flavors that develop in autoxidized fats and that are supposed to prevent their voluntary consumption are not the primary peroxide products themselves, but only their secondary degradation products. These will be discussed later in this section.

The results of older investigations, cited by Aaes-Jørgensen (1962), showed that rancid fats of both plant and animal origin caused weight loss, anemia, leukopenia, and eventually the death of rats fed diets containing these fats. Greenberg and Frazer (1953) added rancid soybean oil, with a peroxide value of 530–550 μmoles/gm, at a level of 10% in the diets of growing rats. They found that the rancid oil increased the protein requirement for optimal growth of the animals, but they were not certain whether the beneficial effect of the protein could be accounted for by a direct "neutralization" of a toxic factor present in the rancid oil or by a decreased biological availability of one or more of the essential amino acids in the protein of the diet containing rancid oil. The only organ weight seriously affected by the presence of the rancid fat was that of the intestinal tract, which increased, possibly because of an increased mucus secretion. Results obtained in the same laboratory showed that linoleate peroxide, reduced linoleate peroxide, and thermally decomposed linoleate peroxide were toxic to rats. The toxicity of ethyl oleate peroxide and ethyl linoleate peroxide, when administered to the rats intraperitoneally, was much higher than that of the peroxides of the same two fatty acid esters when administered orally, possibly due to their poor absorbability from the alimentary system (Aaes-Jørgensen, 1962). Similar findings were reported by other investigators as well.

Cortesi and Privett (1972) reported that methyl linoleate hydroperoxides administered to male rats by intubation caused enlargement of the intestines, intestinal irritation, loss of appetite, and diarrhea. However, there were no cases of mortality, as happened when the hydroperoxides were injected intravenously. It should be borne in mind, however, that intraperitoneal, or intravenous, injections can hardly be considered normal routes of intake for fat, or any food for that matter. Yet Matsuo (1962), who used ethyl esters of highly unsaturated fatty acids oxidized by air at room temperature, did observe marked toxicity when the esters were added to a rat's diet at a level of 5%. Untreated ethyl esters of the same fatty acids were given to another group of young rats, and ethyl oleate to a third group. The rats given the autoxidized ester lost weight rapidly and died within 8 days from the beginning of the experiment, while the performance of the other two groups appeared to be normal. When the experiment was repeated with mature rats, the facial fur of the animals that received the autoxidized esters became depilated, their mouth and hind legs were swollen, the color of their remaining fur changed to brown, and hemorrhages appeared in the skin. When the peroxides were removed from the autoxidized esters prior to feeding, the rats grew satisfactorily, as did those given the unoxidized esters. No toxicity appeared when the peroxide content was 100 mg/100 gm or less, as was also observed by Crampton *et al.* (1953). When Matsuo (1962) fed his autoxidized ester to rabbits the mucous membranes of their stomach and intestines deteriorated severely, and the intestines became less pliable and apt to break when stretched.

Unlike the findings of Greenberg and Frazer (1953), who failed to reveal changes in organ weights between the rats fed on diets containing treated oil and the controls except in the case of the intestine, Kaunitz *et al.* (1955) did find such differences in other organs as well. The latter workers used commercially available refined cottonseed oil, which they had aerated and heated at 90°–95°C for 50–300 hr. A dietary level of 15–20% caused diarrhea in rats; the animals lost weight rapidly and died within 3 weeks. The relative weight of the livers, kidneys, and adrenals were higher, and those of the spleens and thymuses smaller, than the relative weights of the respective organs in the control animals. The decrease in size of the thymus can be attributed to the nonspecific stressful stimulus brought about by the irritating effect of the autoxidized oil. Involution of the thymus and hypertrophy of the adrenals are two of the classical manifestations of the stereotyped response of the organism to a variety of widely different agents described by Selye (including infections, intoxication, trauma, nervous strain, heat, cold, muscular fatigue, X-irradiation, or exposure to allergens) and referred to as "stressors" (Selye, 1951). Greenberg and Frazer (1953) suggest that their above-mentioned failure to observe changes in the weight of the latter two organs indicates that the animals adapted themselves during the experimental period to such an extent that the early effects had been overcome. In the study of Kaunitz *et al.* (1955) no serious histopathological changes were observed, except occasional intestinal edema. The inclusion of fresh oil in the rats' diet provided protection against the harmful effect of the autoxidized oil, but no explanation for the latter observation was proposed by the authors. When the autoxidized oil constituted only 10% of the diet, few animals died, and the survivors were partly able to adjust to the treated oil in that they gradually lost their diarrhea and even gained weight slightly. The fact that the gain in weight was small even when the animals did not suffer from diarrhea suggests that the decrease in weight was not due primarily to diarrhea.

As mentioned earlier, most investigators agree that the higher the degree of unsaturation of the fat in question the greater its susceptibility to changes by oxidation and heating, yet some exceptions have been reported, and this is also true for autoxidized fats. Kaunitz and Johnson (1973) fed rats diets containing fats of different degrees of saturation which had been oxidized by aeration at 60°C for 40 hr. The toxicity of the oxidized fats, characterized by cardiac damage (mainly focal myocarditis and fibrosis), was most severe in animals fed corn oil, followed, in descending order of severity, by cottonseed, soybean, and olive oils. The conflicting results of different studies led Kaunitz (1967) to conclude his discussion on the safety of peroxidized lipids by suggesting that the toxicity of lipid peroxides may be related to their origin.

Although the toxicity of autoxidized fats has been extensively investigated, the exact mechanism of action has not yet been fully elucidated. There are reports implying that fat peroxides administered orally were unabsorbable. Glavind and Tryding (1960) gave male rats oral doses of peroxides and subsequently could not find any in the lymph of the animals. After testing *in vitro* systems in which pancreatic juice, bile, and lymph were shaken with the peroxides at suitable temperature and pH, the authors came to the conclusion "that the essential site of destruction of the lipoperoxides ingested with the food is the intestinal mucosa." Andrews *et al.* (1960) and Lea (1962) confirmed this observation and suggested that not only are the peroxides destroyed in the intestine, but they probably exert their action at that site also. Moreover, it has been suggested that their absorption may take place through damaged mucosa (O'Brien and Frazer, 1966).

Evidence based on several studies indicates that various fat autoxidation products elicit their effects by causing breakdown of the cellular membrane structure. Rats fed diets containing safflower oil or menhaden oil oxidized in air at room temperature exhibited retarded growth, increased relative weights of heart and lung, and decreased relative weights of the epididymal fat pads and vesicular glands (Privett and Cortesi, 1972). These macroscopic changes were accompanied by marked effects on the membrane properties of erythrocytes and liver mitochondria, both of which became prone to rupture. However, red cell hemolysis and liver mitochondrion fragility were not noted in animals the diets of which were supplemented with vitamin E or in animals raised on fat-free diets or on diets containing the unoxidized oils. The authors claim, therefore, that vitamin E has a protective effect against the adverse effects of autoxidized oils. According to Menzel (1967) cellular disruption results from the reaction between the free radicals arising from lipid oxidation and the membrane proteins. Macroscopic and microscopic observations were also reported by Nakamura *et al.* (1973), who used autoxidized safflower oil having a peroxide value of 465 mEq/kg oil. The oil was administered daily by intubation, in doses of 1 ml, to weanling male rats. The macroscopic effects reported included growth suppression and increased liver weight. Light microscopic examination revealed no significant changes in the liver, but electron microscopy showed marked changes in the endoplasmic reticulum and an increase in the number of microbodies in liver cells. The latter changes were manifested after 3 days, but not until 3 months after the beginning of the experiment was accumulation of lipofuscin-like substances in liver cells noted. After 3 months there was also a decrease in the animals' capacity to synthesize fatty acids in the liver under *in vitro* conditions, as evidenced by reduced rate of incorporation of [^{14}C]acetate by liver slices. Reduced capacity, and even complete inactivation, of enzymes brought about by the presence of lipid peroxides have been reported in several articles. There is ample evidence in the literature that autoxidizing fats may react with proteins and also with amino acids, and this fact may account for the above observations (Matsuo, 1962; Karel *et al.*, 1975). Matsuo (1962) found that when a dilute aqueous solution of ovalbumin was incubated with an ethyl ester of a highly unsaturated fatty acid, the solution showed little noticeable change. However, when an autoxidized ester was added to the ovalbumin solution, the mixture turned yellowish brown, a precipitate formed, and as time passed the color became darker. The author attributed this change to denaturation and precipitation of the protein by the autoxidized ester. The author's earlier work suggests that the greater part of ovalbumin precipitates by forming an adduct with the autoxidized ester (Matsuo, 1957a), possibly as a result of an attack on the disulfide bond of cystine to form cysteic acid (Matsuo, 1957b).

Karel *et al.* (1975) reported interactions of peroxidizing methyl linoleate with lysozyme and other proteins and amino acids. They found that histidine gave several reaction products, one of which was identified as histamine, and methionine was oxidized to methionine sulfone. The latter products may have harmful effects as is discussed in Section IV. It is not surpsising, therefore, that autoxidized lipids also react with enzymes. The inhibitory influence of oxidized fatty acids on certain oxidative enzymes was described by Bernheim *et al.* (1952) and Ottolenghi *et al.* (1955). Andrews *et al.* (1960) used soybean oil oxidized by air at 60°C for 1 week. The oil was incorporated into the diet of rats at a level of 15%. The authors found not only that the intestine is the primary site of autoxidized fatty acid toxicity, but that the intestinal xanthine oxidase is also inhibited by them. The authors believe that it is quite possible that other enzymes present in the intestine and other tissues are also inhibited by fatty acid peroxides. Matsuo (1957b) tested the influence of autoxidized fatty acid ethyl ester on the activity of rat muscle and liver succinodehydrogenase, on salivary amylase, and on sweet potato amylase. All three enzymes showed a decrease in activity as a result of interaction with the autoxidized esters.

On the other hand, the activity of the enzymes glutamic–pyruvic transaminase and glutamic–oxaloacetic transaminase in the serum was increased in animals receiving the autoxidized oil (Nakamura *et al.*, 1973). However, enhancement of the latter enzymes also results from exposure to other toxic agents, as is the case of browning reaction products (Lee *et al.*, 1975) mentioned in Section II,A.

The authors of the more recent studies discussed below conducted their experiments under *in vitro* conditions and investigated the direct interaction of lipid peroxides (and their secondary degradation products, which will be discussed later in this section) with enzymes and other proteins. Matsushita (1975) let linoleic acid and [1-^{14}C]linoleic acid autoxidize at 37°C. The hydroperoxides (referred to as LAHPO) formed were isolated from their secondary degradation products (designated SP) by column chromatography on silica gel and then incubated with RNase, pepsin, trypsin, and lipase. The author found that the LAHPO inhibited the RNase activity more than SP but had no inhibitory effect on the trypsin activity, whereas SP had a remarkable inhibitory action on the trypsin activity. In the case of pepsin, LAHPO caused a decrease in the enzyme activity, whereas SP enhanced it. The incorporation of LAHPO or SP into the proteins was determined by incubating the labeled products with the proteins. At certain intervals, the radioactivity incorporated into the enzyme proteins was determined. Both oxidized products were incorporated into RNase and pepsin, but only SP showed a considerable incorporation into trypsin. Certain amino acids in the enzymes were damaged by LAHPO and SP, especially cystine, methionine, lysine, histidine, and tyrosine in the case of SP. Only RNase also showed polymerization by the action of both autoxidation products. Triglyceride hydroperoxides (TGHPO) were also studied by Matsushita (1975) in terms of their effects on certain enzymes. The TGHPO were found by him to inhibit the activity of lipase, as was also the case with LAHPO (but not SP). They did not, however, show a significant effect on the activity of any of the other enzymes studied in this work, although their bacteriostatic action on the growth of *Escherichia coli* was as powerful as that of LAHPO. Kanner and Karel (1976) investigated the changes in lysozyme due to reaction with peroxidized methyl linoleate under *in vitro* conditions. The observed changes in solubility, fluorescence, enzyme activity, and electrophoretic pattern of the protein can be interpreted, in the authors' view, as being indicative of the appearance of lysozyme dimers and higher polymers. The polymerization was apparently due to formation of covalent bonds, which facilitated cross-linking of the protein molecules, thus bringing about aggregation. Disulfide bonds do not appear to be involved in these polymers, as evidenced by the absence of electron spin resonance (ESR) signals due to sulfur-centered radicals in the incubated lysosome/methyl linoleate system.

Sulfhydryl enzymes exhibited exceptional susceptibility to modification, through oxidation and addition reactions, by interactions with lipid peroxides (Gardner *et al.*, 1977). Cysteine reacts with linoleic acid to yield several products, some of which were later identified as fatty acid–cysteine adducts, and this addition was initiated by free-radical reactions. O'Brien and Frazer (1966) showed that linoleic acid hydroperoxide had a very marked effect on cytochrome *c* by causing damage to the hematin ring and also by polymerization of the protein, thereby preventing it from functioning efficiently in the mitochondrial electron transport system.

Finally, a few workers believe that, in addition to being toxic, hydroperoxides are also potentially carcinogenic. Cutler and Schneider (1973) tested the ability of oxidized linoleate in rats' diets, at levels of 10, 5, and 2.5%, to induce malignant tumors. These diets were given to the animals for three consecutive days per week, and the experimental duration was 56–75 weeks. All the doses induced one or more types of malignancy. There was an increase in the incidence of cervical carcinomas and mammary adenocarcinomas in the females and in the incidence of interstitial cell tumor of the testes in the males. Strangely enough, in the case of mammary tumor the lowest dose was the only one to induce it. In a more recent paper (Cutler and Schneider, 1974) the authors tested the carcinogenic potential of linoleate hydroperoxide injected subcutaneously to 6-week-old female rats. The treatment was repeated every 5 weeks. The incidence of malignant mammary tumors and of lymphosarcomas was increased among the rats that had received 200 mg of the fatty acid peroxide. In animals that had received 353 mg of the peroxide there was an increase in incidence of benign pituitary tumors and benign mammary tumors. The tumors appeared late in the life span of the rats. The incidence of total tumors of both organs increased as the dose of the linoleate peroxide was raised, but this increase was due to a greater number of benign rather than malignant tumors. No tumors developed at the site of injection, which proves that where neoplasms did appear they did not merely result from local irritation brought about by the treatment. Exposure to the linoleic acid hydroperoxide also caused pituitary gland hypertrophy and an increase in the proportion of eosinophilic cells in the pars anterior, in body weight and body length, as well as in the proportion of rats with follicular cystic ovaries. The authors suggested that the pituitary and mammary tumors arose as a consequence of an induced hypothalamic disturbance.

Fats can promote cancer in two ways: (1) by possessing a carcinogenic potential themselves or (2) by having the capacity to potentiate or promote other carcinogens. In the second case they act as cocarcinogens. The carcinogenic effect of fats may be due to the action of the fats per se or by their breakdown products or contaminants. The second possibility seems to be more likely (West and Redgrave, 1975). According to these authors, two of the most important fat products that have been blamed for causing malignancies are peroxides produced by autoxidation and polycyclic and aromatic hydrocarbons. The latter compounds and their significance will be discussed later in this section.

In 1945, Bolland and Koch found that as much as 70% of the primary oxidation products of linoleic acid contained conjugated double bonds. Later, the conjugated hydroperoxides were isolated by Cannon *et al.* (1952) and Privette *et al.* (1953a,b), who found nearly 90% conjugation in their peroxidized preparations. Infrared spectra showed that the autoxidation process induced the formation of cis, trans and trans, trans configurations in the dienoic acid preparation. Similar changes also occur in the more highly unsaturated fatty acids (Frankel, 1962), and this results from the conjugation of the double bonds in the polyunsaturated fatty acids (Alim and Morton, 1974). In most naturally occurring vegetable fats and oils only the unsaturated components contain isolated, that is to say nonconjugated, double bonds, in the cis configuration alone. However, a variety of processing operations that cause oxidation and hydrogenation promote isomerization from the naturally occurring cis to the trans isomers (American Oil Chemists' Society, 1975; Dutton, 1966; Alim and Morton, 1974). Autoxidation and also oxidation catalyzed by illumination (photooxidation) may yield trans derivatives from all-cis fatty acids (Clements *et al.*, 1973). Johnston *et al.* (1958a) found that while maternal tissues contain considerable amounts of trans fatty acids, these lipids did not occur to any measurable extent in placental, fetal, or baby fat. These results led them to the conclusion that trans fatty acids found in human tissue are not manufactured in the human body and must therefore originate solely from dietary fat. This conclusion was supported by the results of experiments with animals. When trans fatty acids, in the form of hydrogenated margarine stock, were fed to rats, they were later found to be deposited in the animals' tissues, but they were absent from the animals fed diets that did not contain trans fatty acids (Johnston *et al.*, 1958b). Most of the ingested trans fatty acids were metabolized, and little was excreted in the feces. When the trans fatty acids were removed from the diet, they gradually decreased in amount in the animals' tissues. The presence of these fatty acids in the diet did not retard the growth of the animals, and no untoward effects were reported. The results of a more recent study confirmed the above observation concerning the accumulation of dietary trans fatty acids in the tissues of the rat, but they also indicated that the consumption of these acids may not be safe (Sgoutas and Kummerow, 1970; Kummerow, 1974). The absorbed trans fatty acids were found to alter the permeability characteristics of cell membranes. The red blood cells of rats fed elaidic acid (the trans isomer of oleic acid) in olive oil became more susceptible to hemolysis, by a factor of 5, possibly due to increased osmotic fragility. It is likely that other biological functions of cell membranes are also impaired by such alteration of permeability.

b. Secondary Degradation Products. These are formed during autoxidation of unsaturated lipids, mostly by hydroperoxide decomposition. As the number of double bonds and their different positions along the fatty acid chains give way to a large number of possible hydroperoxides, and taking into account the fact that even a fatty acid possessing one double bond (e.g., oleic acid) may lead to the formation of four hydroperoxides (Badings, 1960), the number of different compounds that may evolve from them is very large (Keeney, 1962). Furthermore, the alcohols, hydroxy acids, aldehydes, epoxides, and unsaturated compounds formed by decay of the primary hydroperoxides are susceptible to further oxidative reactions, and thus a bewildering array of so-called secondary degradation products appears (Keeney, 1962). The compounds (and their mechanisms of formation from hydroperoxides) which are discussed by Keeney include alcohols, aldehydes, ketones, free short-chain fatty acids and esters, cyclic peroxides, and polymers. The latter products may be derived from interaction between peroxides and unsaturated species or by association of alkoxy and alkyl free radicals, with the resulting polymers containing carbon-oxygen-carbon linkages. Other types of polymers are formed by carbon-carbon bonds and probably originate from combination of alkyl free radicals. All these radicals may contain ring and branched-chain structures (Perkins, 1960).

As these reactions proceed and the secondary products accumulate, the level of the peroxides drops (Perkins, 1967; Karel *et al.*, 1975). Many of the secondary products are volatile, and they are believed to be primarily responsible for the characteristic flavor of rancid fats (Keeney, 1962; Kaunitz, 1967).

The results presented in various publications demonstrate that the most toxic compounds found in autoxidized lipids may be the secondary degradation products rather than the hydroperoxides from which they are derived (Kaunitz *et al.*, 1955, 1956a,b; Kaunitz, 1962; Yoshioka and Kaneda, 1975). This is why, as mentioned above, some investigators do not consider the peroxide value a satisfactory criterion for evaluating the anticipated toxicity of autoxidized fats. Akiya *et al.* (1973) proposed what appears to be a simple, rapid, and—as they claim—very sensitive method of estimating the degree of toxicity of deteriorated oils, which is not based on the occurrence of one component alone. This is an *in vivo* test, yet, it has obvious advantages over the long, tedious, and expensive conventional feeding experiments employing laboratory animals. In their test the authors injected oxidized soybean and rapeseed oils into the yolk sacs of chicken embryos. They found that the mortality rate of the embryos was a very sensitive parameter for the prediction of toxicity of the deteriorated oils.

Of the enormous number of lipid degradation products that have been isolated and characterized, only a few have been obtained in large quantities and subjected to toxicity tests. Epoxides may occur in damaged fats in amounts of several percent. Feeding of 5% methyl *cis*-epoxystearate to rats did not affect their growth or appearance (Kaunitz *et al.*, 1956a), but it did cause changes in mitochondrial oxidation. Feeding studies with rats have shown that 12,13-epoxyoleate is also toxic (Fioriti and Sims, 1970). Hydroxy acids have also been found to exert untoward physiological effects (Kaunitz, 1967). Yoshioka and Kaneda (1975) reported that the toxicity to mice, in terms of LD_{50}, of a mixture of hydroperoxyalkenals ranging in chain length from five to nine carbon atoms, and isolated from autoxidized methyl linoleate, was much greater (by a factor of 87) than the parent hydroperoxy compound. The aldehydes absorbed from the alimentary canal caused deterioration of tissue lipids. They also exhibited a marked hemolytic action and lipase, succinate dehydrogenase-, and thiokinase-inhibiting capability, which was much greater than that of methyl linoleate hydroperoxide. According to Menzel (1967), the aldehyde fraction derived from autoxidized lipids, which reacts with thiobarbituric acid (this fraction is mostly malonaldehyde), interacts with proteins and brings about their polymerization. The author used linolenic acid that was allowed to oxidize at 30°C and react with pancreatic ribonuclease at pH 7.6. The thiobarbituric acid-reactive product resulting from the linolenic acid autoxidation could be recovered from the RNase only by hydrolysis in hot 1 *N* HCl solution. The gel filtration pattern proved the formation of dimers and trimers of RNase. Kanner and Karel (1976) also pointed out that protein aggregation may take place by reaction with reactive compounds that evolve during degradation of peroxidized lipids.

Storage under ordinary conditions brings about three types of changes in fats and foods containing fats: (1) autoxidation at a relatively low rate, due to the mild ambient temperature; (2) hydrolysis resulting in an increase in the free fatty acid content of the lipid (Mitchell and Henick, 1962; Lea, 1965b); and (3) lipid oxidation reaction catalyzed by enzymes (Mitchell and Henick, 1962). The latter change will be discussed in Section II,B,3. In addition, storage (especially at elevated temperature and relative humidity) facilitates reactions between carbonyl-group-containing compounds, such as aldehydes and ketones formed during lipid degradation of hydroperoxide, and amino groups of proteins and free amino acids in the stored food. Thus, browning-type reactions may take place between oxidizing lipids and proteins or amino acids, resulting in loss of available lysine (Lea, 1958). Moreover, the carbonyl-containing secondary degradation products, obtained from oxidized lipids, react with lysine more readily, and under less rigorous time-temperature conditions, than do reducing sugars (Carpenter *et al.*, 1962). This observation probably results from the fact that the lipid secondary degradation products have much more reactive carbonyl groups than do sugars, since most of the latters' molecules have a ring structure (hemiacetal form), in which there is no active carbonyl present. The amino acids—free or as part of a protein molecule—that were found to react with the latter secondary degradation products by the Maillard reaction are lysine (Artman, 1969) and cysteine (Gardner *et al.*, 1977). When acetylcysteine was used instead of cysteine, the product mixture was not brown (Gardner *et al.*, 1977). Pokorny *et al.* (1975) reported that both primary and secondary amino groups of proteins react with the decomposition products formed by degradation of fatty acid peroxides. The protein structure apparently does not have a significant influence on the rate of the browning reaction (Pokorny *et al.*, 1975).

3. *Enzymatic Peroxidation*

Many plant foodstuffs become rancid during storage, even at a low temperature and in the dark, especially after mechanical disruption of the tissues, unless they are properly blanched. The main reason for this phenomenon is lipid oxidation catalyzed by enzymes called lipoxygenases (or lipoxidases), involving unsaturated fatty acids as well as atmospheric oxygen (Bergström and Holman, 1948). The primary oxidation products (hydroperoxides) and their secondary degradation derivatives are similar to those encountered during the lipid autoxidation process. Therefore, these reactions bring about formation of toxic compounds, as described above for autoxidized lipids. Evidence has been presented for the existence of several such lipoxygenase isoenzymes in a number of plants (Eskin and Henderson, 1975). The peculiar characteristic of these enzymes is that they catalyze only the oxidation of fatty acids containing the *cis,cis*-1,4-pentadiene system to form conjugated cis,trans-hydroperoxides (Lundberg, 1962a; Eskin and Henderson, 1975; Hadziyev and Haydar, 1975). According to Lea (1958) these enzymes may still be active at subzero temperatures and possibly also in dehydrated products. Gardner (1970) described the oxidation of linoleic acid by a sequential enzyme system isolated from extracts of defatted corn germ, which consisted of lipoxygenase and linoleate hydroperoxide isomerase, and characterized the products formed in the process.

4. *Thermal Oxidation and Polymerization*

These reactions occur in lipids during frying, baking, roasting, and similar processing operations, which are normally carried out at temperatures ranging from approximately 180° to 250°C. The higher part of the temperature range prevails more frequently during domestic cooking than in well-controlled industrial plants. With regards to the other processing conditions, in most cases they can be referred to as "thermal oxidation," since the oil is exposed to air, and frying causes continuous stirring, which greatly increases the surface area of the oil, even in deep fat frying operations. However, in certain cases the prevailing temperature is well above 200°C, and the foods being fried continuously supply a "steam blanket," which in turn shields the oil from the surrounding air, thus providing an anaerobic atmosphere in contact with the heated oil. The latter situation favors the so-called thermal polymerization process.

While many reports have demonstrated that fats subjected to high temperatures in the laboratory were definitely toxic when given orally to experimental animals, there is ample evidence of the apparent harmlessness of fats used in actual frying in food factories and restaurants, even after prolonged use. It is probably true that the heat treatment employed in the laboratory in many studies, which is intended to simulate a commercial frying practice, may be quite exaggerated, resulting in the development of compounds that do not form in appreciable quantities under the less severe practical frying conditions. In addition, some of the lipid degradation products reported to be present in considerable amounts in oils heated under laboratory conditions (without water) are fairly volatile and are distilled off with the steam released from the foods being fried. This also accounts for the absence from fried products of many compounds that have an objectionable flavor (Lea, 1962) and for the above-mentioned fact that fried fats may be organoleptically acceptable whereas rancid fats are not. In some cases the adverse effects on animals fed thermally oxidized fats can be ascribed to the presence of hydroperoxides formed during long cooling periods of the fat or after the fat was incorporated into the animals' diets (Artman, 1969). The wide variation in the handling of fats, in the frying practice employed (e.g., pan frying versus deep fat frying, continuous versus intermittent frying operations, fast versus slow turnover of the oil, the temperature used, the nature of the food commodity being fried) may all have a profound influence on the extent of damage suffered by the frying medium and on its contents of potentially harmful compounds. Therefore, it is not surprising that completely different and sometimes conflicting findings have been published by different workers who tested the toxicity of heated fats or fats used as an actual frying medium.

There is a marked difference between oxidation of a fat at a high temperature and oxidation at a low temperature (autoxidation). Not only are the chemical reactions involved accelerated at high temperatures, but some reactions occurring at the higher temperatures cannot take place under milder temperatures. The reason for this is that the initial oxidation products that form during heat treatments, as well as during autoxidation (i.e., while the oil is subjected to mild heat or storage), are too unstable to remain intact at high temperatures. The thermal instability of peroxides has been proved by many investigators, and the changes brought about in fats during thermal oxidation are generally believed to result from the presence of materials formed by the rapid degradation of hydroperoxides (Artman, 1969). Yet most of the reactions that may occur during autoxidation probably also take place at higher temperatures, but in the latter case they occur at highly elevated rates (Perkins, 1967). The water present in the food commodity being fried also plays a significant role in the series of events that take place in the frying medium. In contrast to the "steam blanket" effect mentioned above, which is supposed to protect the heated oil by preventing (or minimizing) the contact with oxygen, there is evidence suggesting just the opposite; namely, the water present in the food undergoing frying increases the extent of oxidation of the oil. The changes enhanced by the presence of water are mostly of a hydrolytic rather than an oxidative nature. However, at a high temperature the oxidative changes are the predominant ones (Artman, 1969; Perkins, 1967).

Simko *et al.* (1964) found that fats heated under conditions prevailing in the kitchen caused fatty necrosis in guinea pigs fed on them for 10 weeks. Fatty or calciferous lesions in the myocardium and aorta were noted, and changes in the coronary arteries were also exhibited. The fats used in this study were sunflower seed oil, hydrogenated vegetable fat, lard, and butter, which had been heated at 170°C for only 2 hr, that is, under conditions that can be considered mild, even when compared to common commercial practice, except that in this study the fats were stirred and aerated continuously during heating, and as a result the peroxide value of the heated fats was relatively high. An unexpected finding was that the hydrogenated vegetable fat was not markedly more resistant to heating damage than the sunflower oil. In another experiment, rats fed diets containing oils that had been blown with air for several hours at 180°C showed disturbance in the liver function, increased response of the central nervous system to stimuli, increased spontaneous moving about of the animals, alteration in kidney function, increased capillary permeability, changes in motor and secretory functions of the intestinal tract, microscopic changes in the thyroid gland, and deposits in the kidneys (Artman, 1969).

As mentioned above, the non-urea-adductable fraction contains the toxic factors, and this is also true for fats subjected to heat treatments (Huang and Firestone, 1969). Perkins and Kummerow (1959a) oxidized corn oil at 200° ± 5°C by bubbling air at a rate of 200 ml/min per kilogram oil for 48 hr. They found that 36% of the total amount of the fatty acids isolated from the thermally oxidized oil failed to form adducts with urea. Various fractions of fatty acids isolated from the same oil by urea adduction and molecular distillation under 3 μm at 150°C were incorporated into diets at a level of 14% and fed to weanling rats for 21 days (Perkins and Kummerow, 1959b). The animals fed the diet containing the nondistillable fraction isolated from the NAF (consisting of fatty acid polymers) lost weight, and all died within 7 days. The rats fed the total NAF gained weight during the first 12 days, although less than the controls, and then started losing weight; nonetheless, they survived the entire 21-day experimental period. The authors concluded that only the polymeric fraction formed from unsaturated fatty acids was toxic. Some workers in this field agreed with this view (Anonymous 1968a), but many other investigators are of a different opinion, as is discussed below.

French fries prepared continually in the same batch of oil for several 8-hr days caused an increase in the rate of degradation of the oil and in the percentage of high molecular weight products formed (Perkins, 1967). This was true for all five kinds of oils used in this study, including both plant and animals fats of different degrees of unsaturation. Sahasrabudhe and Bhalerao (1963) found that the NAF in some samples of used frying oils and in fats extracted from fried foods reached levels that were as much as ten times higher than those present in fresh oils. Nolen *et al.* (1967) pointed out that distillable compounds present in the NAF isolated from fried fat also contained a factor(s) that showed toxicity to rats, but its level was quite low. An important observation reported by some investigators is that the severity of toxicity of heat-abused oil greatly depends on the adequacy of the diet given to the animals. The levels and quality of the protein, minerals, and vitamins may have a marked influence on the susceptibility of the organism to the potentially toxic fat (Anonymous, 1968a). This fact may have important implications in the feeding of undernourished populations. In a review, Chang

(1967) described the toxicity symptoms in animals exposed to heat-oxidized fats. These included irritation of the digestive tract, organ enlargement, growth depression, and death. Highly oxidized and heated fats may even have carcinogenic and cocarcinogenic properties (Lea, 1965a). A carcinogenic effect of thermally oxidized oil was also described by Wurziger and Ostertag (1960), and additional reports describing compounds that are possibly responsible for the tumorigenicity of these oils will be cited below. The most frequently encountered toxicity symptoms are nonspecific and include decreased appetite, diarrhea, and alopecia (Artman, 1969). Heated fats may also promote atherosclerosis. It was found that rabbits fed a diet containing 2% cholesterol and 5% oil produced more florid lesions if the oil had been heated to 200°C for as little as 15 min. This effect was more severe when the heated oil was corn oil than when it was the more saturated olive oil (Anonymous, 1968a).

It has been suggested in numerous publications that cyclic fatty acid monomers, found in the distillable fraction, are the most toxic compounds occurring in thermally oxidized (and also thermally polymerized) fats. The formation of monomeric cyclic fatty acids was reported as early as 1876 by Cloëz (cited by Coenen *et al.*, 1967), who discovered them in heat-treated fatty acids obtained from tung oil. Firestone *et al.* (1961a) heated cottonseed oil at 225°C in a large aluminum pot for 170–195 hr, with continuous stirring. They found that 9% of the fatty acids were converted to a cyclic monomeric form (which did not form an adduct with urea) and to fatty acid polymers. Both monomeric and dimeric fractions, which are toxic to rats, have been recovered from the heated oil, and their oral administration to rats at a level of 0.4 ml caused death within 13 days. The monomeric fractions proved to be more toxic than the dimeric material. There was also a gradual increase in the concentration of trans isomers and in the percentage of conjugated unsaturated fatty acids. The biological significance of the trans forms and conjugated structure was stressed above. Nonetheless, there was no loss in toxicity of the cyclic monomers and dimers by hydrogenation, which proves that cyclization and polymerization are the more important changes from a toxicological viewpoint than conjugation of the double bonds and formation of trans isomers. The hydrogenation process itself does not bring about formation of cyclic fatty acids when carried out under normal conditions, but under extreme conditions they do appear (Coenen *et al.*, 1967). Firestone *et al.* (1961b) also demonstrated that urea filtrate monomers obtained from fats were toxic and that their toxicity remained the same after hydrogenation. They also showed that low-pressure hydrogenation failed to saturate the toxic fraction. This is a noteworthy observation, since these fats apparently did not contain high levels of aromatic compounds, as proved by Michael (1966b), who reported that thermally oxidized linoleate was converted to cyclic isomers and contained only one double bond in the ring structure. Artman and Smith (1972) heated cottonseed oil at 182°C for six 8-hr days. Later they isolated and characterized many of the products formed. They concluded that the cyclic fatty acids were probably the most important toxic products formed in the fats heated in the presence of air. Not only are the free cyclic fatty acid monomers the chief toxic products, but also the glycerides containing them, as the latter are hydrolyzed in the intestine, and the liberated cyclic fatty acids are absorbed (Crampton *et al.*, 1953). The toxicity of cyclic monomers, which are relatively stable, appears to be exerted systemically, whereas that of peroxides is of a local nature; i.e., it must occur within the intestinal lumen or cells (Mead, 1962). Mead believes that the fatty acid dimers and higher polymers, formed as either intraacid or intraglyceride derivatives, are nontoxic, and at any rate cannot be readily digested or absorbed. This view is in accord with the more recent findings of Kajimoto and Mukai (1970), who showed that fatty acid dimers formed during thermal oxidation were poorly absorbed by rats. Kaunitz (1967) believes, however, that in addition to the cyclic monomers the dimeric fraction evolved during thermal oxidation is also toxic. Michael *et al.* (1966) studied the biological effects of methyl linoleate that had been heated at 200°C for 200 hr. They observed that both the cyclic monomers and polar dimers were toxic to young rats when administered by stomach tube at a level of 0.5 ml/day for 3 consecutive days, while the nonpolar dimeric and polymeric fractions were not. The parameters used for evaluating toxicity were the body and thymus weights and the mortality ratio. The heated linoleate before fractionation was not toxic, probably because the concentrations of the toxic components were too low to manifest toxicity symptoms.

At temperatures higher than those customarily used, more severe changes occur, and even nearly saturated fats may become toxic. Raju and Rajagopalan (1955) heated groundnut, sesame, and coconut oils in an iron pan at 270°C for 8 hr and then incorporated them into synthetic diets for rats at a level of 15%. In all three cases the heated oil adversely affected the gain in weight and the feed efficiency, as compared with the respective unheated oil. The liver of the rats fed the heated oils underwent hypertrophy, and its fat content was almost twice that of the control animals. When the heated oils were fed at a level of 30%, all the rats died within a week. In another set of trials conducted by the same investigators, oil heated at 280°C caused severe jaundice, and most of the rats died in the sixth week of the experiment. However, most investigators agree that at temperatures normally used in frying operations the lower the degree of unsaturation of the fat in question the less the likelihood of its deterioration during heat processing (Kawada *et al.*, 1967) and consequently the less the chances that toxic compounds will be produced (Johnson *et al.*, 1957; Reddy *et al.*, 1968). Kantorowitz and Yannai (1974) studied the heat resistance of liquid and hydrogenated soybean oils of the same source and their toxicity to rats after being thermally oxidized for different periods of time. They found a 20-fold increase in the percentage of conjugated dienoic fatty acids and a considerable increase in the content of the monomeric nonurea-adductable fatty acid fraction in the heat-treated oil. No conjugated fatty acids were detected in the heated hydrogenated oil, and there was a smaller increase in its content of monomeric non-urea-adductable fatty acids. Only the liquid oil showed toxicity to rats when heated for more than 10 hr. The only effect of the hydrogenated oil was a temporary diarrhea in two animals, and only after the oil had been heated for 43 hr. These results, however, do not agree with the findings of some workers, who had previously reported that the degree of deterioration in thermally oxidized oils was independent of their degree of unsaturation (Sahasrabudhe and Bhalerao, 1963; Simko *et al.*, 1964; Thompson *et al.*, 1967).

In some papers more specific effects of thermally oxidized fats are described. Andia and Street (1975) studied the functional changes associated with the hepatomegaly observed as a result of feeding thermally oxidized fats (produced by heating at 180°C for 24 hr with continuous stirring). They found that thermally oxidized oil caused proliferation of the smooth endoplasmic reticulum and induction of certain microsomal enzymes in the liver. Miller and Landes (1975) reported that rats fed heated soybean oil showed, in addition to hepatomegaly, a decrease in the number of red blood cells and in their hemoglobin content. The rats suffered from depressed growth and appetite, and their serum levels of iron and copper, and of the respective binding proteins transferrin and ceruloplasmin, were also reduced by the use of heated fat in the diet. The animals' liver lipids content, however, decreased, whereas many other workers reported an increase in hepatic lipids after exposure to heated fats. The authors interpreted the decrease in the rate of production of hemoglobin, transferrin, and ceruloplasmin as indicating that an abnormally large portion of the dietary protein was being retained in the liver in order to cope with the metabolic effects of the damaged fat.

When fats are heated at temperatures in excess of 200°C in the absence of air, thermal polymerization occurs, in which the polyunsaturated fatty acids undergo conjugation and then cyclization. Cyclization may occur within the same molecule or between two (or more) such fatty acid molecules, resulting in the appearance of cyclic monomers or polymers, respectively (Matsuo, 1962; Perkins, 1967; Potteau *et al.*, 1970a). Such reactions occur mostly through the Diels–Alder type of condensation (Firestone, 1963; Johnson and Kummerow, 1957), and the higher the degree of unsaturation the greater the tendency of the oil to undergo such changes (Matsuo, 1962; Barrett and Henry, 1966). As discussed above, such products occur in thermally oxidized fats also, but here higher levels of these materials have been detected, and no oxygen has been found to participate in the new linkages formed during the reactions involved. It has been suggested that polymers produced by thermal polymerization may not be as toxic as the so-called oxypolymers formed during thermal oxidation (Kaunitz, 1967; Ohfugi and Kaneda, 1973). Yet Billek and Rost (1974) claimed that the dimeric fraction derived from safflower oil, which was thermally polymerized at 295°C for 25 hr in a CO_2 atmosphere, was the most toxic fraction to rats. A small amount of this fraction was absorbed and metabolized to carbon dioxide, and part of it was also incorporated into the body fat of rats. It should be noted that, in cases where all traces of oxygen have not been eliminated, the process taking place in the treated fat may actually be predominantly thermal oxidation rather than thermal polymerization, since traces of oxygen are sufficient to bring about oxidative changes (Artman, 1969). Thermally polymerized oils proved to be unwholesome in numerous studies, the most frequently reported symptoms being reduced appetite and growth depression. It was also reported that rats fed a diet containing thermally polymerized oils at a level as low as 10% of the diet exhibited a poor reproduction and lactation performance (Farmer *et al.*, 1951). Carcinogenic effects of such oils have also been reported (Chalmers, 1952; Perkins, 1960). Crampton *et al.* (1953) found that linseed oil polymerized at 275°C for 12 hr in a CO_2 atmosphere contained cyclic fatty acid monomers and polymers. The fatty acids were converted to ethyl esters by ethanolysis and subjected to fractionation by urea. The NAF was further fractionated by distillation. Both the monomeric and dimeric fractions obtained from the NAF were toxic to rats when included in their diets at a level of 20%. The animals failed to grow, had oily and matted coats, excreted dark and sticky feces, and displayed heavy mortality. The dimeric fraction caused low digestibility, which in turn brought about diarrhea, but it was less toxic than the cyclic monomers, probably due to its low absorbability. In another paper from the same laboratory, Wells and Common (1953) described in detail the procedure for the isolation of the different fractions from heated linseed oil. MacDonald (1956) claimed that the NAF isolated from linseed oil that had been polymerized as described by Crampton *et al.* (1953) contained a cyclic monomer possessing an unsaturated six-membered ring. Rats fed linseed oil that had been polymerized in a similar manner grew poorly, and all died during the first 18 weeks of feeding (Aaes-Jørgensen *et al.*, 1956). Cyclization also occurred in safflower oil during exposure to a temperature 300°C for 3 hr in an atmosphere of CO_2 (Mehta and Sharma, 1957). The effects of heat-polymerized soybean oil on the growth rate, body lipid composition, and liver hypertrophy and the effects of the cyclic monomers and polymers present in the heated oil were described by Potteau *et al.* (1970b). More information is given in a review by Artman (1969). The NAF's isolated from different oils, which were thermally polymerized under the same conditions, differed in their degree of toxicity. The NAF isolated from polymerized linseed oil displayed the highest toxicity, the one obtained from polymerized soybean oil was somewhat less toxic, and the one isolated from heated sunflower seed oil was only slightly injurious to the rats (Crampton *et al.*, 1956). Crampton *et al.* (1953), McInnes *et al.* (1961), and Firestone (1963) characterized the cyclic monomeric fatty acids present in the NAF isolated from thermally polymerized oils and described their general structure; Firestone (1963) and

Eisenhauer and Bell (1968) proposed a method for their determination. Mounts *et al.* (1970) suggested a mechanism by which dimerization of oxidized fatty acids by thermal polymerization takes place and provided experimental evidence for their theory. The higher polymers formed in heated fats, being unabsorbable, seem to be innocuous, except for a local irritating effect in the intestines reported by some workers (Artman, 1969).

Fats heated under conditions defined as thermal oxidation as well as thermal polymerization may also give way to volatile decomposition products, many of which are also formed during lipid autoxidation. These include aldehydes, ketones, hydroxy acids, dicarboxylic acids, epoxides, and other compounds, the presence of which has been shown by many investigators (e.g., Kawada *et al.*, 1967; Franzke *et al.*, 1970; Whitlock and Nawar, 1976) to be toxic to animals. Many of these products are also responsible for the various rancid flavors (Hoffman, 1962). In addition, a variety of compounds other than the types mentioned above are formed at high temperatures. The latter include hydrocarbons, aromatics, lactones, dibasic acids (Chang, 1967), and many other compounds which were found in trace amounts and some of which may be potentially toxic. In certain studies the nonsaponifiable fraction of the oil was involved. In one of the early reports on this subject (Torres *et al.*, 1956) it was implied that sterols may be altered by deep fat frying to form aromatic polycyclic compounds, many of which are known to be potent carcinogens, e.g., cholanthrenes (Torres *et al.*, 1956). The presence of the carcinogen benzpyrene in heated fats has also been detected (West and Redgrave, 1975). It is known, in fact, that most organic materials may form aromatic polycyclic hydrocarbons while undergoing pyrolysis (Artman, 1969). Another product of fat pyrolysis is acrolein, which constitutes the acid and lachrymatory vapor produced when fats are overheated, especially during pan frying operations. This commences at about 300°C by hydrolysis of the glycerides and liberation of the fatty acids and glycerol, and the latter is degraded to acrolein ($CH_2{=}CH{-}CHO$) (Crossley *et al.*, 1962). Exposure to this irritant by inhalation or injection induced, among other general toxicological effects, a marked increase in the activity of the liver alkaline phosphatase and tyrosine-α-ketoglutarate transaminase. Like many other irritants it stimulates the pituitary-adrenal system, thus leading to hypersecretion of glucocorticoids (Murphy, 1965). However, as has already been mentioned, a large proportion of these compounds are eliminated at the high temperatures prevailing during thermal oxidation and thermal polymerization.

As pointed out earlier, the authors of numerous publications claim that fried fats seem to be safe, since they failed to observe any untoward effects from ingestion of large amounts of fried fats or fried foods by experimental animals. Deuel *et al.* (1951) reported that margarine fat that had been heated at 205°C for 8 hr and used for deep fat frying had no adverse effects on the growth performance of male or female rats during the 10-week experimental period. This is also true for potato chips fried in that fat and fed to young rats for a long period (Deuel, 1954). Keane *et al.* (1959) investigated the biological effects of hydrogenated cottonseed oil used in a commercial deep fat fryer under actual production conditions for as long as 24 days (a day meaning an 8-hr period). The oil was incorporated into a rat diet at a level of 20%. Melnick (1957) failed to detect thermal polymers in any of the oils tested, all of which had been used as frying media for potato chips. Also Poling *et al.* (1960), who studied fats after use in commercial deep fat frying operations, stressed that such fats are harmless, whereas fats heated in the laboratory may not be safe, since in the latter case the heat treatment was more severe. The same investigators reviewed the available reports on this subject (Rice *et al.,* 1960) and pointed out that, according to the tests employed by them, no harmful substances occurred in fried foods or in the fats used in preparing such foods. They argued that the normally accepted good practice in current food preparation ensures the wholesomeness of fried foods and that often laboratory treatments are grossly exaggerated. Results obtained at the same laboratory, published a few years later, showed that other industrial heat treatments (alkali refining, adsorptive decolorizing, or deodorization), and even deodorization at 238°C, had no adverse influence on the safety of the treated oils (Wilding *et al.*, 1963). Nolen *et al.* (1967) tested five fats, used for frying potatoes, onions, and fish, at 182°C. Frying was done intermittently, with long heating periods without frying, and the fats were cooled overnight and reheated every day for 2–5 days. The used fats were fed to weanling rats at a dietary level of 15%, for 2 years. No increase in mortality or tumor incidence or in any other abnormality was noted. The small amount of distillable NAF isolated from the fried fats was toxic to rats when administered at high levels, but the authors believe that at the concentration found in these fats the NAF is harmless. Similar results were reported by Nolen (1973), who tested the safety of partially hydrogenated soybean oil used for deep fat frying under commercial conditions and later incorporated into a diet at a level of 15%. The female dogs consuming this diet for 54 weeks did not differ significantly in growth or in any other parameter from the control female dogs. However, in the case of the males the controls grew better than the animals maintained on the fried fat. Yet the clinical and histopathological tests employed failed to reveal any abnormalities in any of the groups. The author believes that his observations indicate that the reduced growth rate of the males is attributable to the lower absorbability of the treated fat, but not to any toxic effect. These findings are in accord with the many reports showing formation of appreciable amounts of unabsorbable fatty acid polymers during heat treatments applied to fats, but they conflict with the report mentioned above by Melnick (1957), who could not find any polymers in fried oils. The dimeric fatty acids formed in heated fats do not accumulate in the adipose tissue of the rats fed these fats for 2 months (Strauss and Billek, 1974). According to Lang and Krozingen (1972), not only are fried fats nontoxic, but their content of aromatic polycyclic hydrocarbons is lower than that of untreated oil. Moreover, Lang (1973) claims that the reported induction of microsomal enzymes in experimental animals exposed to heated fats may have a beneficial effect, because it provides a certain amount of protection against certain early disorders in the animal.

C. Miscellaneous Food Components

As mentioned earlier, most organic materials may form polycyclic aromatic hydrocarbons while undergoing pyrolysis (Artman, 1969), and some of these compounds are known to be carcinogenic (Lintas and Bernardini, 1975). For this reason the latter compounds have been detected in a variety of foods exposed to extreme temperatures in roasting and toasting operations, as well as in foods preserved and flavored by smoking. Perhaps the most prevalent among these compounds is benzo[*a*]pyrene (**X**), which is known to be a potent carcinogen

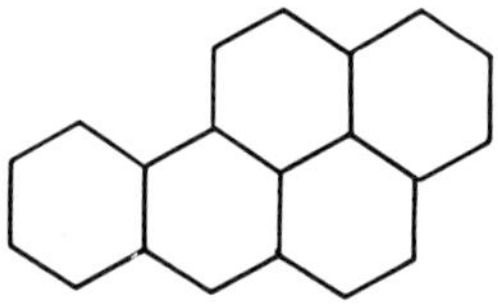

Benzo (*a*) pyrene (X)
(3,4-Benzopyrene)

(Wogan, 1966). Its concentration in smoked foods varies and may exceed 50 ppb (Tilgner and Daun, 1969). Some of the polycyclic aromatic hydrocarbons are also powerful inducers of the so-called drug-metabolizing enzymes, which may, in turn, create a major therapeutic problem (Pantuck *et al.*, 1976).

In one of the early studies on this subject the ether extracts of roasted foodstuffs were painted on the skin of mice. The treated animals showed a higher incidence of adenocarcinomas than the controls (Widmark, 1939). The most widely used foods subjected to processing at high temperatures, such as in roasting, are meat and fish (Joint FAO/WHO Expert Committee on Food Additives, 1975), but roasting and also smoking of plant foodstuffs is not uncommon (Sapeika, 1969). Polycyclic aromatic hydrocarbons are present in roasted coffee (Anonymous, 1964), bread and other baked and toasted cereal products (Lintas and Bernardini, 1975), vegetables, and oils (Grimmer, 1966; Fritz, 1975; Howard *et al.*, 1968; Siegfried, 1975). Information on the occurrence of these compounds in heated oils was given above. An example of a smoked plant foodstuff found to contain polycyclic aromatic hydrocarbons is torula yeast, in which benzo[*a*]pyrene levels ranging from less than 0.5 to 0.7 ppb have been detected (Malanoski and Greenfield, 1968). A review of the effects of smoking on the occurrence of polycyclic aromatic hydrocarbons in foods, and health hazards resulting from them, was written by Gilbert and Knowles (1975).

The soot and skin of coffee beans that had been roasted by direct contact with the combustion gases, were found to be very rich in benzo[*a*]pyrene. The values reported were 28 and 15 ppb, respectively (Anonymous, 1964). Fritz (1975) found that such roasting increased the benzo[*a*]pyrene content of coffee beans by a factor of 50. Also, Lintas and Bernardini (1975) observed that bread crust, which receives a more severe heating during baking than the crumb, contains considerably more benzo[*a*]pyrene than does the latter. The difference in the benzo[*a*]pyrene content between the crust and the crumb is much greater in the case of bread baked in a wood-fueled oven than in the case of bread baked in an electric oven (the levels in the crust being 7.67 and 2.27 ppb, respectively). This is probably due to direct contact of the loaf with the wood smoke. Masuda *et al.* (1966, cited by Haenni, 1968) reported the presence of polycyclic aromatic hydrocarbons in roasted barley, shoyu, and caramel. Even instant coffee was found to contain appreciable amounts of benzo[*a*]pyrene (Strobel, 1974). Drying processes involving exposure to open flame, which are often employed in certain countries in the manufacture of tea and copra and in the roasting of eggplants and other vegetables, may also bring about accumulation of polycyclic aromatic hydrocarbons (J. Ilany-Feigenbaum, personal communication, 1977).

It has been shown that several amino acids, especially tryptophan, glutamic acid, and lysine, may form potent mutagens when heated. Some of the tryptophan derivatives had been synthesized in sufficient quantities to permit animal testing and were later proved to be very active mutagens (Sugimura, cited by Fox, 1978).

III. ACID AND ALKALI TREATMENTS

A. Acid Treatment

Acids are used mostly for the control of food-spoiling microorganisms (Riemann, 1969) and of nonenzymatic browning (Eskin *et al.*, 1971). According to Riemann (1969) the effect of acids on microorganisms, which is to destroy or inhibit them, may result from the hydrogen ion concentration or from the toxicity of the undissociated acid or the anion. The author points out that no food-poisoning microorganisms, except toxin-producing fungi, can grow at pH levels below 4.5, and therefore most fruits and several fermented products are relatively safe and seldom cause food poisoning. Acidification of food for preservation is not a widely used practice in the industry, mainly because of the undesirable changes in organoleptic properties of the treated products. Yet in cases where such a treatment is applied, and especially when this is done while the food is hot, it may give rise to compounds that are definitely unwholesome. Despite the fact that the stomach itself contains a strong acid and the pH of the stomach juice is quite low, in some people the additional acid may be undesirable (Fricker, 1975). Oxidation and degradation of ascorbic acid may lead to the formation of furfural and other physiologically unfavorable compounds. Such reactions take place during heat treatment of ascorbic acid and dehydroascorbic acid, and they are greatly enhanced at low pH levels (Eskin *et al.*, 1971; Mooser and Neukom, 1975).

An acidic medium also enhances oxidation reactions of various kinds, which may give rise to toxic compounds. When methionine is oxidized by bleaching agents, its fate depends on the pH. When the medium is close to neutrality, or slightly acidic, it is only converted to methionine sulfoxide, which, as will be discussed later, is apparently innocuous. However, at a lower pH considerable amounts of the physiologically undesirable derivative methionine sulfone appear, and the lower the pH the higher the content of this compound (Njaa, 1962; Cuq *et al.*, 1977).

B. Alkali Treatment

Alkali treatment is performed on various foods in order to achieve different goals, the most important ones being (a) to dissolve proteins for the preparation of concentrates and isolates; (b) to produce proteins possessing special properties, such as foaming, emulsifying, stabilizing; (c) to destroy aflatoxins (Hollingsworth and Martin, 1972); (d) to prepare solutions suitable for spinning protein fibers, as is done in the production of "texturized vegetable protein" (Circle and Smith, 1972b; Cuq *et al.*, 1975); and (e) to chemically peel fruits (Circle and Smith, 1972a). Such treatments, however, may sometimes bring about the appearance of harmful compounds. A review dealing with many of the topics discussed in this and other sections of this chapter is that of Cheftel (1977).

De Groot and Slump (1969) observed that strong alkali treatment of several types of food proteins of both plant and animal origins caused the formation of an uncommon amino acid called lysinoalanine (LAL) (**XI**). The appearance of LAL

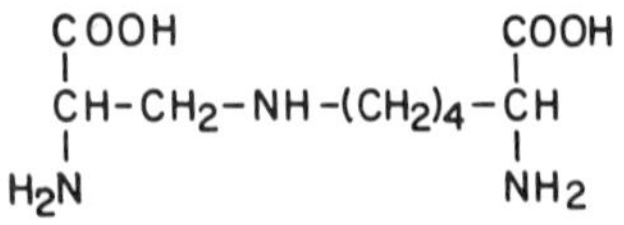

$$\begin{array}{ccc} COOH & & COOH \\ | & & | \\ CH-CH_2-NH-(CH_2)_4 & - & CH \\ | & & | \\ H_2N & & NH_2 \end{array}$$

Lysinoalanine (XI)

in alkali-treated soybean protein and its toxicity to rats were later demonstrated by Woodard and Short (1973). De Groot and Slump (1969) found that rats could tolerate a diet containing 20% of alkali-treated soybean protein for 90 days, and the only abnormal manifestation observed was an increased degree of nephrocalcinosis in females. However, Woodard and Short (1973) observed a nephrotoxic effect in rats fed semipurified diets containing alkali-treated soybean protein, which was designated renal cytomegalia. The causative agent of this lesion was suggested to be LAL. Van Beek *et al.* (1974) conducted an experiment which was designed in a manner similar to that of de Groot and Slump (1969) and also reported similar results. They failed to reveal renal cytomegalia or any pathological symptoms except for an increase in both relative kidney weights and degree of nephrocalcinosis in females. The two effects were dose-related. The latter symptom could be alleviated by adding calcium or magnesium to the diet. The authors believe that these changes in the kidneys resulted from the high level of available phosphorus in the diets containing the alkali-treated soybean protein. LAL is probably formed by the condensation—assumed to proceed via a nucleophilic attack by the ϵ-amino group of a lysine residue—of lysine with a dehydroalanyl residue (the latter evolving from cystine or glycosidically bound serine by a β-elimination reaction), and this in turn leads to cross-linking within or between polypeptide chains (Provansal *et al.,* 1975; Aymard *et al.,* 1978). More information on LAL formation was reviewed by Cheftel *et al.* (1976).

Since severe alkali treatments also include isomerization of amino acids (Provansal *et al.,* 1975) and since one stereoisomer of LAL may cause nephrocytomegalic changes while another does not (Tas and Kleipool, 1976), it is conceivable that alkali treatments differing in severity may or may not cause isomerization of the natural all-L form of the two amino acids involved, and thus in the configuration of the resulting LAL and in its toxicity. This may provide an explanation for the conflicting results of different investigators. Of much concern is a report by Sternberg *et al.* (1975), according to which LAL was identified even in food proteins that had not been subjected to alkali treatment. The authors claimed that proteins of various sources gave rise to LAL when heated in their laboratory under nonalkaline conditions. Yet only the rat showed nephrotoxic effects when exposed to as little as 100 ppm of LAL, whereas six other species fed much higher doses failed to exhibit nephrotoxicity (de Groot *et al.,* 1976; O'Donovan, 1976). Furthermore, even in the case of the rat 100 ppm of LAL proved to be nephrotoxic only when in the free state (as synthetic LAL or in hydrolyzed alkali-treated protein), while much higher levels (up to 6000 ppm) of LAL failed to show any toxicity when it was provided as the protein-bound compound in alkali-treated casein or soybean protein (de Groot *et al.,* 1976). Woodard and Short (1977) showed that synthetic LAL administered to young rats orally or intraperitoneally produced renal lesions after as little as 7 days of treatment. The tubular cells in the affected area had an increased nuclear size and DNA content, a condition that may be indicative of possible neoplastic transformation of cells (Woodard *et al.,* 1977). Comprehensive reviews on this subject have been published recently (Anonymous, 1976; de Groot *et al.,* 1977).

It is well known that combined or free arginine is degraded even by weak alkalis to form ornithine, which may in turn react with dehydroalanine (produced from cystine or serine, as explained above with regard to LAL). This condensation leads to the formation of ornithinoalanine (**XII**) (Ziegler *et al.,* 1967), the

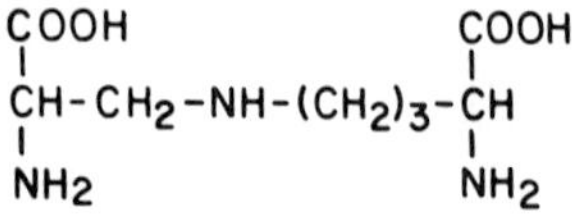

$$
\begin{array}{ll}
\text{COOH} & \text{COOH} \\
| & | \\
\text{CH-CH}_2\text{-NH-(CH}_2)_3\text{-} & \text{CH} \\
| & | \\
\text{NH}_2 & \text{NH}_2
\end{array}
$$

Ornithinoalanine (XII)

structure of which is similar to that of LAL. The toxicity of this compound, however, remains to be established.

Other processing operations in which an alkali is employed may also have physiologically adverse effects on animals or human beings who consume the products. Ammoniated molasses, used as feed supplements, were found to contain sugar–ammonia reaction compounds, which produce toxic effects in animals. The compounds identified included various imidazoles and pyrazine derivatives (Nishie *et al.,* 1969), some of which have been discussed in Section II,A,3. Farm animals that had consumed ammoniated molasses exhibited hysteria and convulsions. Beef cattle seem to be particularly susceptible to these effects (Anonymous, 1971).

A high pH value enhances nonenzymatic browning reactions involving sugars by catalyzing their conversion to the highly reactive enediol form (Eskin *et al.,* 1971). As discussed in Section II, many of the premelanoidins formed during this process are potentially harmful.

An alkaline medium also induces deesterification of pectin, resulting in the release of methanol, which may remain in the final products. This may be a real danger when pectin-rich plant materials are limed before fermentation, as is often done in the manufacture of citrus peel juice or apple cider (Berk, 1976).

IV. BLEACHING

This operation is carried out with various foods, but mainly with flours. The oxidizing agents used for this purpose are designated "aging agents" and include ammonium persulfate, potassium bromate, certain peroxides, chlorine dioxide, sulfur dioxide, and a variety of other compounds, some of which are also used for sterilization of foods. For many years the most important oxidizing chemical used was nitrogen trichloride (Agene). The first findings suggesting that the use of this chemical may not be safe began to appear quite early. However, they were attributed to viral or bacterial infections, nutritional deficiency, allergy, etc. (Campbell and Morrison, 1966). Later, Wagner and Elvehjem (1944, cited by

Campbell and Morrison, 1966) discovered that commercial wheat gluten rapidly produced in dogs a nervous disorder characterized by running fits. The investigators believed that the disorder resulted from a toxic factor present in the product. Later studies revealed that such fits occurred only in dogs fed flours that had been treated with Agene and that the affected dogs recovered when the Agenized flour was withdrawn from their diet and replaced by untreated flour of the same grist. Consequently, the use of nitrogen trichloride as a bleaching agent was abolished. Bentley *et al.* (1948) reported that proteins that readily become toxic when treated with Agene were relatively rich in methionine. Workers at the same laboratory (Bentley *et al.*, 1949, 1950) and also Misani and Reiner (1950) isolated the toxic compound and discovered its chemical structure. Misani and Reiner (1950) called it methionine sulfoximine; it is probably derived from methionine sulfoxide. This material is highly toxic to rabbits. It causes barking and running fits in dogs and later convulsive fits (Bentley *et al.*, 1949).

As a result of the unfortunate experience with Agene, other commonly used bleaching agents were also suspected of being unsafe and consequently subjected to careful testing. One of these, which has found many applications, is chlorine dioxide. Frazer *et al.* (1956a) found that wheat flour treated with chlorine dioxide at ten times normal levels of treatment does not have any adverse effects on rats consuming it. The only finding that may possibly give reason for concern is the fact that chlorine dioxide enhanced the oxidation process in the lipid fraction of wheat flour (Frazer *et al.*, 1956b). As discussed in the Section II,B, oxidized lipids may contain undesirable products. Daniels *et al.* (1963) suggested that the free chlorine occurring as a contaminant in chlorine dioxide may bring about chlorination of fatty acids present in the flour undergoing bleaching. They observed that the fatty acid fraction from treated flour contained a number of compounds, including dichlorostearic acid. A four-generation toxicity test with the lipid extracted from cake flour treated with chlorine showed that only the diet containing the highest concentration of the treated oil had an adverse effect on the animals. It should be remembered, however, that this level was much higher than that anticipated in human diets. The only effects encountered were impaired lactation in the females and a somewhat thin and rough fur in the males. The body fat of the latter animals contained 0.77% chlorine, compared with 0.075% in the case of animals maintained on a normal diet. Free chlorine is used as an oxidative improver in the production of certain speciality cake flour (Daniels, 1966). Yet the author believes that in view of the quantities of cake flour normally consumed, there is little or no toxicological hazard involved. However, according to Cunningham and Lawrence (1977) chlorinated cake flour, and lipids extracted from it, reduced growth rate and increased the relative liver size in growing rats. Moreover, even fat-free chlorinated wheat gluten induced similar effects.

Hydrogen peroxide is used in the food industry as a preservative, a bleaching agent, and an improver of the baking qualities of bread (Slump and Schreuder, 1973) and also for detoxifying the glucosinolates in rapeseed (Anderson *et al.*, 1975). It has been shown in several studies that this oxidizing agent may induce the formation of other methionine oxidative derivatives, methionine sulfoxide and methionine sulfone, which appear as a consequence of the reaction between methionine residues in food proteins and hydrogen peroxide. Tannenbaum *et al.* (1969) found that a similar reaction may also take place during storage at 37°C of protein and autoxidizing fatty acid. (Their system consisted of casein and autoxidizing methyl linoleate.) However, no such changes were observed in a system consisting of autoxidizing lipid and methionine, although the mixture was heated at 75° or 110°C, for 24 or 48 hr (Wedermeyer and Dollar, 1963). Methionine slfoxide was reported by Njaa (1962b), by other workers cited by him, and more recently by Slump and Schreuder (1973) to be utilized to the same extent as is methionine. Other workers, however, reported conflicting results (e.g., Ellinger and Palmer, 1969; Miller *et al.*, 1970). Yet even the latter two groups of investigators differ in their findings. Ellinger and Palmer (1969) reported that the protein-bound methionine sulfoxide had a lower biological availability than did the protein-bound methionine, whereas free methionine sulfoxide was nearly 100% available. However, Miller *et al.* (1970) found that even free methionine sulfoxide was less available than free reduced methionine and claimed that their results showed that the decreased availability of methionine sulfoxide was not due to reduced proteolysis of peptides containing this oxidized methionine, but rather to reduced utilization of the latter amino acid. Gjøen and Njaa (1977) attribute these conflicting results to the different contents of cystine in the test proteins in question and claim that methionine sulfoxide is poorly utilized only when cystine is not present in considerable amounts. Thus, when methionine sulfoxide is the sole source of sulfur-containing amino acids, synthesis of cystine is too slow to facilitate normal growth of the experimental animals. According to these authors when cystine is present in appropriate quantities, the conversion of methionine sulfoxide to methionine (by reduction) is given alone or with cystine. Yet methionine sulfoxide is apparently not toxic, whereas there is some evidence suggesting that methionine sulfone not only is unable to substitute for methionine but also is actually toxic (Njaa, 1962b; Slump and Schreuder, 1973). Methionine sulfone formed by hydrogen peroxide (used for detoxifying rapeseed) was found to depress the food intake, even when present in very small amounts (Anderson *et al.*, 1975). Supplementation of the diet with methionine failed to restore normal growth, possibly due to the decrease in food intake. The authors suggest that the decrease in protein nutritive value may have resulted from the appetite-depressing effect of the sulfone, but their results could also be indicative of impaired metabolism. This controversy could have been resolved had the investigators applied a paired-feeding test, which eliminates variability due to a difference in feed consumption. For a recent review of the subject see Cheftel (1977).

V. IRRADIATION

Food irradiation has been in practice for almost 30 years for various purposes. The main objectives are, in order of increasing dosage, sprout inhibition, insect disinfestation, disinfection, pasteurization, and sterilization (Welt, 1968). This method of food preservation makes it possible to store foodstuffs for long periods of time without refrigeration and is the only means of extending the shelf life of commodities that cannot be heat-treated or frozen. This technique is of particular importance for the military and for developing countries, where refrigeration is often impracticable. Therefore a large part of the research effort in this field is being conducted in developing countries (Diehl, 1977).

The health authorities in many countries consider food irradiation a food additive, since it may induce chemical changes in foods being treated. Since it was suspected that different products may evolve at different doses and with different food items, the regulations require that toxicity testing be carried out in each case. Diehl (1977) challenges this attitude, claiming that "enough animal feeding studies, enough teratogenicity, mutagenicity, cancerogenicity testing, enough microbiological studies, and enough chemical analyses have been carried out to classify all foods irradiated with a dose of 500 krad or less as generally recognized as safe (GRAS)." The Joint FAO/IAEA/WHO Expert Committee (1977) agreed more or less with this view and also accepted that data may be extrapolated from one food item to other items belonging to the same class of commodities, as long as the doses applied are below 1 Mrad. It also accepted the opinion of scientists who believe that when no toxic effects are obtained with a food treated with a high dose of radiation, there will be no effect when the same food is treated with a lower dose. This assumption is based on experimental evidence showing that the concentration of radiolytic products generally increases in proportion to radiation dosage, until it reaches a plateau with radiation doses of about 1 Mrad (Taub *et al.*, 1976). Numerous papers suggest that irradiated foods are apparently safe for experimental animals (e.g., Goldblith, 1970; Grünewald, 1973; Vidal, 1974). Furthermore, several studies show that they are also devoid of any mutagenic activity in mice (Renner, 1975; Levinsky and Wilson, 1975; Chauhan *et al.*, 1975) and rats (Hossain *et al.*, 1976).

There are, however, some indications in the literature that γ irradiation of foods may give rise to radiolytic products that may not be safe themselves, such as lipid peroxides, or that may participate in reactions leading to the formation of potentially harmful products, such as those from nonenzymatic browning. The undesirable compounds in both instances are discussed in the section II of this chapter.

Kawakishi *et al.* (1975) irradiated D-glucose and D-fructose solutions with cobalt γ rays. Some of the radiolysis products exhibited properties characteristic of reductones, which have been shown to have untoward effects (see above). Similar compounds are produced from starch breakdown derivatives, which may be formed, in turn, by γ irradiation (Davidek *et al.*, 1975). Much more information is available on the oxidative changes in lipids brought about by irradiation. The earlier papers were reviewed by Mead (1961), who also described briefly the main general chemical changes induced by high-energy radiation in various systems and dealt in more depth with the radiation-induced autoxidation of lipids. A more recent review on this subject is that of Nawar (1973), who described the peroxides formed via a free-radical mechanism and the great variety of secondary degradation products, including acrolein and crotonal, that are produced from irradiated fats. Irradiation-induced lipid–protein interaction was also suggested, and the possible unfavorable biological effects mentioned. The radiation-catalyzed formation of alkanes and alkenes from lipids was studied by Beke *et al.* (1975). The above-mentioned studies, in which irradiated food was found to be nonmutagenic, are also challenged by other findings that point to the contrary. Bhaskaram and Sadasivan (1975) fed children with irradiated wheat that had been stored for different time intervals. The children fed the wheat sample stored for 3 weeks or less after the irradiation had 3.8% abnormal lymphocytes, including 1.8% polyploid cells, whereas in the children fed wheat stored for 12 weeks before use the percentage of abnormal cells was lower, and no such cells were observed in the children fed nonirradiated wheat. The authors point to the fact that polyploidy may be associated with malignant transformation of the cells.

VI. SOLVENT EXTRACTION AND FUMIGATION

A. Solvent Extraction

Removal of oil, water, and various other components, including the elimination of toxic principles, is a widely used industrial operation (Frampton, 1960). This operation was considered to be without hazard to health as long as the toxic solvent residues were eliminated after the process. However, it became evident later that solvents are not completely inert and that under certain conditions they may interact with other components of the food being extracted to produce new compounds (Campbell and Morrison, 1966). Stockman (1916, cited by Campbell and Morrison, 1966) found that soybean meal produced by extraction with trichloroethylene was toxic to cattle and caused bleeding from body orifices, subcutaneous hemorrhages, and aplastic anemia. The toxic factor involved was not the solvent residues in the meal. Rather, it was associated with the protein component of the meal (McKinney *et al.*, 1957; Seto *et al.*, 1958), and later it was identified as a derivative of cystine residues in the polypeptide. McKinney *et al.* (1959) observed that calves fed synthetic *S*-(dichlorovinyl)-L-cysteine displayed the same symptoms. The same cysteine derivative may also be encountered in decaffeinized coffee produced by extraction of caffeine from the green coffee beans with trichloroethylene (Brandenberger and Bader, 1967; Sapeika, 1969). Extraction with dichloroethane caused interaction of the solvent with choline to form the toxic derivative chlorocholine (Campbell and Morrison, 1966). Meals produced by extraction of the oil with chloroform were also toxic to chicks (Carpenter *et al.*, 1963).

Another aspect of solvent extraction (and also of concentration by various methods), which is often overlooked, is the fact that when the water component of foods is extracted or evaporated the concentration of whatever toxic compounds they contain increases.

B. Fumigation

Fumigants are used for sterilization of foodstuffs that cannot be heat-sterilized (e.g., flours and many dehydrated products) and also to control pests that may infest foods during storage. Ethylene oxide belongs to the former class of fumigants. It is an effective disinfectant, but it has also been shown to react with several protein-bound amino acid residues, thereby rendering the protein toxic for experimental animals (Campbell and Morrison, 1966). Wesley *et al.* (1965) investigated the interactions between ethylene oxide and propylene oxide with water and with hydrochloric acid. They found that the reaction products with water were ethylene glycol and propylene glycol, respectively, while with hydrochloric acid the corresponding chlorohydrins were formed. The latter compounds are very toxic, and the authors believe that they also form in foods by reaction with chlorides.

Another fumigant, which has been used for many years for fumigating infested wheat and other cereals, is bromomethane (which is called in the literature, according to the old nomenclature, methyl bromide). Winteringham *et al.* (1955) studied whole-wheat flour that had been exposed to $^{14}CH_3Br$. The fat, starch, gluten, and water-soluble fractions were later separated, and their ^{14}C content was assayed. The protein (gluten) fraction was responsible for 80% of the decomposition of the absorbed fumigant. The ^{14}C was found to be bound mostly by methylation of N- and S-containing groups in the protein molecule. An evaluation of the possible toxicity for human beings from ingestion of wheat fumigated with bromomethane was also made in the same laboratory (Winteringham, 1955). On the basis of the nature of the products of the reaction between CH_3Br and the wheat ingredients and their metabolism, the author arrives at the conclusion that there is no evidence that the principal decomposition products of the fumigant are toxic or that their formation is associated with any significant reduction in essential food constituents. Information on other fumigants can be found in the recent review by Salunkhe and Wu (1977).

VII. CONTAMINANTS

Contamination of foodstuffs during processing and/or during storage may have various causes. The most frequently occurring contaminants belong to the following classes: microbial toxins, substances originating from food-packaging materials, and fermentation products. Detergents, lubricants, and many other contaminants may also be encountered (Oser, 1971). Although microbial toxins are the main cause of food poisoning, this subject is beyond the scope of the present discussion.

A. Packaging Materials

1. Metallic Cans

The metallic cans presently used for canning food are in most cases made of steel sheet plated with tin, and the alloy used for soldering the side seam of the can contains a large amount of lead. Since many canned products are corrosive media, especially such commodities as fruit juices (and even more so their concentrates) or tomato purée, or are rich in nitrate, oxygen, and sulfur-containing amino acids, tin ions start to appear in the product, and the higher the storage temperature the faster the rate of tin dissolution. Monier-Williams (1949) stressed that canned foods become contaminated with tin during storage even in laquered cans, probably because of the corrosion taking place in defects in the lacquer. Notwithstanding the fact that the toxicity of inorganic tin is quite low, there are numerous papers suggesting that this metal may not be entirely innocuous, even at levels commonly encountered in canned foods. Rats challenged with 5 ppm of tin, a level reported to be well within the range occurring in many canned foods, showed a shorter life span, as well as increased incidence of fatty degeneration of the liver, and an increased incidence of changes in the renal tubules (Schroeder *et al.*, 1968). The acute toxic effects observed in human beings and animals after consumption of canned foods may be ascribed to stomach irritation brought about by the dissolved tin (de Groot *et al.*, 1973). Using $^{113}SnCl_2$, Fritsch *et al.* (1977) found that very little of the amount of $SnCl_2$ ingested was absorbed from the alimentary canal and that the various food components had no influence on this result. These findings support the opinion that in all likelihood the toxic effects induced by tin in the body are essentially due to local irritation in the digestive system, which is manifested mostly in outbreaks of nausea, vomiting, and diarrhea. Many more papers reporting toxic effects caused by ingestion of canned foods contaminated with tin were cited by Knorr (1975). The author argues that the level of 250 ppm, which is in many countries the maximal permitted concentration for tin in canned foods, is not based on safety evaluation, but rather on the fact that rarely are higher values found in products prepared under "normal conditions of processing and storage."

As mentioned above, lead constitutes a major part (about 98%) of the solder used in the manufacture of cans. Whatever the type of can, the meniscus of the solder inevitably (albeit at a very small area) comes in contact with the stored products (Cheftel and Fourgoux, 1960) and may contaminate it. Lead accumulates in various tissues in the body, especially in the bones, until a toxic concentration may be attained. Children are considered a high-risk group in relation to lead exposure (Joint FAO/WHO Expert Committee on Food Additives, 1972).

2. *Plastic Containers*

Plastic packaging materials have been proved to be associated with health hazards that might be at least as severe as those caused by metallic cans, and sometimes they are appreciably more harmful. Some polymers used for making packaging materials exhibit toxicity to experimental animals, and some others are even carcinogenic. Toxicity may also occur from polymerization catalysts, fillers, stabilizers, curing agents, plasticizers, antioxidants, and also monomers, as a result of polymer disintegration in the digestive system, which involves processes of hydrolysis, oxidation/reduction, and degradation (Bischoff, 1972). Bischoff describes different types of polymers and their formation and discusses the carcinogenic mechanisms related to them. The migration of low molecular weight compounds from polymers to the food stored in them and the toxicological problems that may arise from exposure to them have been discussed by several investigators (e.g., Gilbert, 1975; Hausmann, 1974).

One of the serious problems related to this subject is the presence of vinyl chloride monomer in polyvinyl chloride (one of the most widely used polymers in the food industry) and its toxic and carcinogenic properties (Anonymous, 1975b; van Esch and van Logten, 1975). Many more papers on the toxicity and carcinogenicity of vinyl chloride, including clinical observations in workers exposed to the monomer during its industrial manufacture, studies with experimental animals, and epidemiological studies were presented in a conference devoted to this topic (Selikoff and Hammond, 1975).

Much information is available on health problems arising from exposure to a variety of plasticizers employed in the manufacture of polymers used as food-packaging materials. Polychlorinated biphenyls (PCB), widely used as plasticizers, caused in rats depressed body weight and an increase in liver weight and in the liver and plasma cholesterol contents (Kiriyama *et al.*, 1974). Prolonged exposure to a PCB called Aroclor 1242 induced lipid vacuolation and proliferation of smooth endoplasmic reticulum and an increase in liver weight and in the activity of hepatic microsomal hydroxylase, in urinary coproporphyrin, and in hepatic lipids (Bruckner *et al.*, 1974). The toxicity of plasticizers belonging to the group of phthalates has also been described in several papers (Nikonorow *et al.*, 1973; Gray *et al.*, 1977). Organic tin compounds are used as heat stabilizers and as catalysts in the manufacture of plastics, especially PVC. These tin derivatives are by far more toxic than the inorganic tin compounds discussed above. Rats fed di-*n*-octyltin dichloride suffered from atrophy of the thymus gland (Seinen and Willems, 1976).

There is not sufficient evidence to indicate whether the relatively high levels of the above compounds are also encountered in food. Thus, extensive surveys are needed to provide information on this subject.

B. Fermentation Products

During fermentation many metabolites are formed, although in most cases in minute amounts, that have been shown to be potentially toxic. They include alcohols, aldehydes, acids, and antibiotics. It is well known that when the brewing conditions are not well controlled certain alcohols, which are more toxic than ethanol (e.g., methanol), may be produced, and numerous cases of mortality have resulted from this fact. The aldehydes produced from partial oxidation of alcohols are even more toxic (Koerker *et al.*, 1976). Methanol can also be produced during fermentation of fruits containing large amounts of pectin, since many microorganisms participating in the process possess very potent pectin esterases, which release this toxic alcohol by hydrolysis of pectin (Berk, 1976). A small but significant amount (0.03–0.3 ppm) of phthalic acid esters are always present in raw materials for the production of vinegar. However, an increase in their contents in the vinegar during storage was observed due to accumulation by acetic acid bacteria and yeasts, which incorporate the phthalic acid esters, into their cells and possibly also synthesize them (Mori *et al.*, 1975). The latter authors found that old vinegars contained more phthalic acid esters than new ones of the same type.

References

Aaes-Jørgensen, E. (1962). *In* "Autoxidation and Antioxidants" (W. O. Lundberg, ed.), Vol. 2, pp. 1045–1094. Wiley, New York.

Aaes-Jørgensen, E., Funch, J. P., Engel, P. F., and Dam, H. (1956). *Br. J. Nutr.* **10,** 32–39.

Adrian, J. (1972). *Ind. Aliment. Agric.* **89,** 1281–1289.
Adrian, J. (1973). *Ind. Aliment. Agric.* **90,** 449–455.
Adrian, J., and Susbielle, H. (1975). *Ann. Nutr. Aliment.* **29,** 151–158.
Aizetmüller, K., and Scharmann, H. (1975). *Mitteilungsbl. GDCh-Fachgruppe Lebensmittelchem. Gerichtl. Chem.* **29,** 205–209.
Akiya, T., Araki, C., and Igarashi, K. (1973). *Lipids* **8,** 348–352.
Ambrose, A. M., Robbins, D. J., and DeEds, F. (1961). *Proc. Soc. Exp. Biol. Med.* **106,** 656–659.
Anderson, G. H., Li, G. S. K., Jones, J. D., and Bender, F. (1975). *J. Nutr.* **105,** 317–325.
Andia, A. M. G., and Street, J. C. (1975). *J. Agric. Food. Chem.* **23,** 173–177.
Andrews, J. S., Griffith, W. H., Mead, J. F., and Stein, R. A. (1960). *J. Nutr.* **70,** 199–210.
Anonymous (1962). *Nutr. Rev.* **20,** 346–348.
Anonymous (1964). *Food Cosmet. Toxicol.* **2,** 384.
Anonymous (1967). *Food Cosmet. Toxicol.* **5,** 719–721.
Anonymous (1968a). *Nutr. Rev.* **26,** 210–212.
Anonymous (1968b). *Nutr. Rev.* **26,** 244–245.
Anonymous (1971). "The Toxicology and Chemistry of Ammoniated Feed Supplements." Ad Hoc Tech. Caramel Comm. Ind. Manuf. Users.
Anonymous (1975a). *Nutr. Rev.* **33,** 123–126.
Anonymous (1975b). *Food Cosmet. Toxicol.* **13,** 275–276.
Anonymous (1976). *Nutr. Rev.* **34,** 120–122.
American Oil Chemists' Society (1975). *In* "Sampling and Analysis of Commercial Fats and Oils," 3rd ed. Tentative Method Cd. 14-61. A.O.C.S., Champaign, Illinois.
Aranjo Neto, J. S., Madi, K., and Panek, A. D. (1974). *J. Food Sci.* **39,** 643–644.
Artman, N. R. (1969). *Adv. Lipid Res.* **7,** 245–330.
Artman, N. R., and Alexander, J. C. (1968). *J. Am. Oil Chem. Soc.* **45,** 643–648.
Artman, N. R., and Smith, D. E. (1972). *J. Am. Oil Chem. Soc.* **49,** 318–326.
Aymard, C., Cuq, J. L., and Cheftel, J. C. (1979). *Food Chem.* (in press).
Badings, H. T. (1960). *Neth. Milk Dairy J.* **14,** 215–242.
Barrett, C. B., and Henry, C. M. (1966). *Proc. Nutr. Soc.* **25,** 4–9.
Beke, H., Tobback, P. P., and Maes, E. (1975). *Proc. Int. Congr. Food Sci. Technol., 4th, 1974* Vol. 1, pp. 194–198.
Bender, A. E. (1972). *PAG Bull.* **2,** 10–19.
Ben-Gera, I., and Zimmermann, G. (1964). *Nature (London)* **202,** 1007–1008.
Bentley, H. R., Booth, R. G., Greer, E. N., Heathcote, J. G., Hutchinson, J. B., and Moran, T. (1948). *Nature (London)* **161,** 126–127.
Bentley, H. R., Booth, R. G., Greer, E. N., Heathcote, J. G., Hutchinson, J. B., and Moran, T. (1949). *Nature (London)* **164,** 438–439.
Bentley, H. R., McDermott, E. E., Pace, J., Whitehead, J. K., and Moran, T. (1950). *Nature (London)* **165,** 150–151.
Bergström, S., and Holman, R. T. (1948). *Adv. Enzymol.* **8,** 425–457.
Berk, Z. (1976). "Braverman's Introduction to the Biochemistry of Foods." Elsevier, Amsterdam.
Bernheim, F., Wilbur, K. M., and Kenaston, C. B. (1952). *Arch. Biochem. Biophys.* **38,** 177–184.
Bhaskaram, C., and Sadasivan, G. (1975). *Am. J. Clin. Nutr.* **28,** 130–135.
Billek, G., and Rost, H. W. (1974). *Fette, Seifen, Anstrichm.* **76,** 436–439.
Bischoff, F. (1972). *Clin. Chem.* **18,** 869–894.
Bishov, S. J., Henick, A. S., and Koch, R. B. (1960). *Food Res.* **25,** 174–182.
Bolland, J. L., and Koch, H. P. (1945). *J. Chem. Soc.* pp. 445–447.
Brandenberger, H., and Bader, H. (1967). *Helv. Chim. Acta* **50,** 463–466.
Brodnitz, M. H., Nawar, W. W., and Fagerson, I. S. (1968a). *Lipids* **3,** 59–64.
Brodnitz, M. H., Nawar, W. W., and Fagerson, I. S. (1968b). *Lipids* **3,** 65–71.

Bruckner, J. V., Khanna, K. L., and Cornish, H. H. (1974). *Food Cosmet. Toxicol.* **12,** 323–330.
Burton, H. S., and McWeeny, D. J. (1964). *Chem. Ind. (London)* 462–463.
Calhoun, W. K., Hepburn, F. N., and Bradley, W. B. (1960). *J. Nutr.* **70,** 337–347.
Campbell, J. A., and Morrison, A. B. (1966). *Fed. Proc., Fed. Am. Soc. Exp. Biol.* **25,** 130–136.
Cannon, J. A., Zilch, K. T., Burcket, S. C., and Dutton, H. J. (1952). *J. Am. Oil Chem. Soc.* **29,** 447–452.
Carlsson, D. J., Suprunchuk, T., and Wiles, D. M. (1976). *J. Am. Oil Chem. Soc.* **53,** 656–660.
Carpenter, K. J., Morgan, C. B., Lea, C. H., and Parr, L. J. (1962). *Br. J. Nutr.* **16,** 451–465.
Carpenter, K. J., Lea, C. H., and Parr, L. J. (1963). *Br. J. Nutr.* **17,** 151–169.
Chalmers, J. G. (1952). *Biochem. J.* **52,** 31.
Chang, S. S. (1967). *Food Technol.* **21,** 33–34.
Chauhan, P. S., Aravindakshan, M., Aiyar, A. S., and Sundaram, K. (1975). *Food Cosmet. Toxicol.* **13,** 433–436.
Cheftel, C. (1977). *In* "Food Proteins" (J. R. Whitaker and S. R. Tannenbaum, eds.), pp. 401–455. Avi. Publ., Westport, Connecticut.
Cheftel, C., Cuq, J. L., Provansal, M., and Besançon, P. (1976). *Rev. Fr. Corps Gras* **23,** 7–13.
Cheftel, H., and Fourgoux, J. C. (1960). *Ann. Falsif. Expert. Chim.* **53,** 512–519.
Circle, S. J., and Smith, A. K. (1972a). *In* "Soybeans: Chemistry and Technology" (A. K. Smith and S. J. Circle, eds.), pp. 294–338. Avi Publ., Westport, Connecticut.
Circle, S. J., and Smith, A. K. (1972b). *In* "Soybeans: Chemistry and Technology" (A. K. Smith and S. J. Circle, eds.), pp. 339–388. Avi Publ., Westport, Connecticut.
Clements, A. H., Van Den Engh, R. H., Frost, D. J., and Hoogenhout, K. (1973). *J. Am. Oil Chem. Soc.* **50,** 325–330.
Coenen, J. W. E., Wieske, T., Gross, E. R., and Rinke, H. (1967). *J. Am. Oil Chem. Soc.* **44,** 344–349.
Common, R. H., Crampton, E. W., Farmer, F. A., and Defreitas, A. S. W. (1957). *J. Nutr.* **62,** 341–347.
Cooper, P. (1975). *Food Cosmet. Toxicol.* **13,** 571–579.
Cortesi, R., and Privett, O. S. (1972). *Lipids* **7,** 715–721.
Coulston, F. (1977). *Chem. & Eng. News* June 27.
Coulter, S. T., Jennes, R., and Geddes, W. F. (1951). *Adv. Food Res.* **3,** 45–118.
Crampton, E. W., Common, R. H., Farmer, F. A., Wells, A. F., and Crawford, D. (1953). *J. Nutr.* **49,** 333–346.
Crampton, E. W., Common, R. H., Pritchard, E. T., and Farmer, F. A. (1956). *J. Nutr.* **60,** 13–24.
Crossley, A., Heyes, T. D., and Hudson, B. J. F. (1962). *J. Am. Oil Chem. Soc.* **39,** 9–14.
Cunninghan, H. M., and Lawrence, G. A. (1977). *Food Cosmet. Toxicol.* **15,** 105–108.
Cuq, J. L., Provansal, M., Cheftel, C., and Besançon, P. (1975). *Proc. Int. Congr. Food Sci. Technol. 4th, 1974* Vol. I, pp. 578–586.
Cuq, J. L., Aymard, C., and Cheftel, C. (1977). *Food Chem.* **2,** 309–314.
Custot, F. (1959). *Ann. Nutr. Aliment.* **13,** 417–448.
Cutler, M. G., and Schneider, R. (1973). *Food Cosmet. Toxicol.* **11,** 443–457.
Cutler, M. G., and Schneider, R. (1974). *Food Cosmet. Toxicol.* **12,** 451–459.
Cutting, W., Furst, A., Read, D., Read, G., and Parkman, H. (1960). *Proc. Soc. Exp. Biol. Med.* **104,** 381–385.
Danehy, J. P., and Pigman, W. W. (1951). *Adv. Food Res.* **3,** 241–290.
Daniels, N. W. R. (1966). *Proc. Natr. Soc.* **25,** 51–60.
Daniels, N. W. R., Frape, D. L., Russell-Eggitt, P. W., and Coppock, J. B. M. (1963). *J. Sci. Food Agric.* **14,** 883–893.
Davidek, J., Velisek, J., and Janicek, G. (1975). *Proc. Int. Congr. Food Sci. Technol., 4th, 1974* Vol. I, pp. 298–305.
Dawes, I. W., and Edwards, R. A. (1966). *Chem. Ind. (London)* pp. 2203–2204.

de Groot, A. P., and Slump, P. (1969). *J. Nutr.* **98,** 45–56.

de Groot, A. P., Feron, V. J., and Til, H. P. (1973). *Food Cosmet. Toxicol.* **11,** 19–30.

de Groot, A. P., Slump, P., Feron, V. J., and van Beek, L. (1976). *J. Nutr.* **106,** 1527–1538.

de Groot, A. P., Slump, P., van Beek, L., and Feron, V. J. (1977). *In* "Evaluation of Proteins for Humans" (C. E. Bodwell, ed.), pp. 270–283. Avi Publ., Westport, Connecticut.

Deuel, H. J. (1954). *Prog. Chem. Fats Other Lipids* **2,** 99–192.

Deuel, H. J., Greenberg, S. M., Calbert, C. E., Baker, R., and Fisher, H. R. (1951). *Food Res.* **16,** 258–280.

Devik, O. G. (1967). *Acta Chem. Scand.* **21,** 2302–2303.

Diehl, J. F. (1977). *Radiat. Phys. Chem.* **9,** 193–206.

Dutton, H. J. (1966). *Prog. Chem. Fats Other Lipids* **9,** 351–375.

Eisenhauer, R. A., and Bell, R. E. (1968). *J. Am. Oil Chem. Soc.* **45,** 619–621.

Ellinger, G. M., and Palmer, R. (1969). *Proc. Nutr. Soc.* **28,** 42A.

Ellis, G. P. (1959). *Adv. Carbohyd. Chem.* **14,** 63–134.

Endres, J. G., Bhalerao, V. R., and Kummerow, F. A. (1962). *J. Am. Oil Chem. Soc.* **39,** 118–121.

Eskin, N. A. M. (1976). *J. Am. Oil Chem. Soc.* **53,** 746–747.

Eskin, N. A. M., and Henderson, H. M. (1975). *Proc. Int. Congr. Food Sci. Technol., 4th, 1974,* Vol. 1, pp. 263–269.

Eskin, N. A. M., Henderson, H. M., and Townsend, R. J. (1971). "Biochemistry of Foods," Chapter 3. Academic Press, New York.

FAO/IAEA/WHO Joint Expert Committee (1977). "Wholesomeness and Irradiated Foods." Food Agric. Org. U.N., Rome.

FAO/WHO Expert Group (1965). *FAO Nutr. Meet. Rep. Ser.* No. 37. Rome.

FAO/WHO Joint Expert Committee on Food Additives (1972). "Evaluation of Certain Food Additives and the Contaminants Mercury, Lead, and Cadmium," 16th Rep. orld Health Organ., Geneva.

FAO/WHO Joint Expert Committee on Food Additives (1975). "Evaluation of Certain Food Additives," 19th Rep. Food Agric. Organ. U.N., Rome.

Farmer, E. H., Bloomfield, G. F., Sundralingham, A., and Stutton, D. A. (1942). *Trans. Faraday Soc.* **38,** 348–361.

Farmer, E. H., Koch, H. P., and Stutton, D. A. (1943). *J. Chem. Soc.* pp. 541–547.

Farmer, F. A., Crampton, E. W., and Siddall, M. I. (1951). *Science* **113,** 408–410.

Feeney, R. E., Blankenhorn, G., and Dixon, H. B. F. (1975). *Adv. Protein Chem.* **29,** 136–203.

Feron, V. C. (1972). *Cancer Res.* **32,** 28–36.

Feuge, R. O. (1972). *In* "Nutritional Evaluation of Food Processing" (R. S. Harris and H. von Loesecke, eds.), pp. 244–260. Avi Publ., Westport, Connecticut.

Fioriti, J. A., and Sims, R. T. (1970). "Oxidation Products from Heated Vegetable Oils." Am. Oil Chem. Soc., Chicago, Illinois.

Firestone, D. (1963). *J. Am. Oil Chem. Soc.* **40,** 247–255.

Firestone, D., Horwitz, W., Friedman, L., and Shue, G. M. (1961a). *J. Am. Oil Chem. Soc.* **38,** 253–257.

Firestone, D., Horwitz, W., Friedman, L., and Shue, G. M. (1961b). *J. Am. Oil Chem. Soc.* **38,** 418–422.

Fleishman, A. I., Florin, A., Fitzgerald, I., Caldwel, A. B., and Eastwood, G. (1963). *J. Am. Diet. Assoc.* **42,** 394–398.

Ford, J. E., and Shorrok, C. (1971). *Br. J. Nutr.* **26,** 311–322.

Fox, J. L. (1978). *Chem. & Eng. News* January 2, p. 18.

Fragne, R., and Adrian, J. (1967). *Ann. Nutr. Aliment.* **21,** 129–147.

Frampton, V. L. (1960). *In* "Nutritional Evaluation of Food Processing" (R. S. Harris and H. von Loesecke, eds.), pp. 238–243. Avi Publ., Westport, Connecticut.

Frankel, E. N. (1962). *In* "Lipids and Their Oxidation" (H. W. Schultz, E. A. Day, and B. O. Sinnhuber, eds.), pp. 51–78. Avi Publ., Westport, Connecticut.
Franzke, C., Strobach, J., and Schilling, B. (1970). *Fette, Seifen, Anstrichm.* **72,** 629–635.
Frazer, A. C. (1961). *Chem. Ind. (London)* pp. 417–419.
Frazer, A. C., Hickman, J. R., Summons, H. G., and Sharratt, M. (1956a). *J. Sci. Food Agric.* **7,** 371–375.
Frazer, A. C., Hickman, J. R., Summons, H. G., and Sharratt, M. (1956b). *J. Sci. Food Agric.* **7,** 375–380.
Friedman, L., Horwitz, W., Shue, G. M., and Firestone, D. (1961). *J. Nutr.* **73,** 85–95.
Fritsch, P., de Saint Blanquat, G., and Derache, R. (1977). *Food Cosmet. Toxicol.* **15,** 147–149.
Fritz, W. (1975). *Ernaehrungsforschung* **20,** 49–51.
Gardner, H. W. (1970). *J. Lipid Res.* **11,** 311–321.
Gardner, H. W., Kleiman, R., Weisleder, D., and Inglett, G. E. (1977). *Lipids* **12,** 655–660.
Gilbert, S. G. (1975). *Environ. Health Perspect.* **11,** 47–52.
Gilbert, J., and Knowles, M. E. (1975). *J. Food Technol.* **10,** 245–261.
Gjøen, A. U., and Njaa, L. R. (1977). *Br. J. Nutr.* **37,** 93–105.
Glavind, J., and Tryding, N. (1960). *Acta Physiol. Scand.* **49,** 97–102.
Goldblith, S. A. (1970). *J. Food Technol.* **5,** 103–110.
Gonzalez del Cueto, A., Martinez, W. H., and Frampton, V. L. (1960). *Agric. Food Chem.* **8,** 331–332.
Gray, T. J. B., Butterworth, K. R., Gaunt, I. F., Grasso, P., and Gangolli, S. D. (1977). *Food Cosmet. Toxicol.* **15,** 389–399.
Greenberg, S. M., and Frazer, A. C. (1953). *J. Nutr.* **50,** 421–440.
Grimmer, G. (1966). *Erdoel Kohle, Erdgas, Petrochem.* **19,** 578–580.
Grünewald, T. (1973). *In* "Aspects of the Introduction of Food Irradiation in Developing Countries," pp. 7–11. IAEA, Vienna.
Hadziyev, D., and Haydar, M. (1975). *Proc. Int. Congr. Food Sci. Technol., 4th, 1974* Vol. I, pp. 199–207.
Haenni, E. O. (1968). *Residue Rev.* **24,** 41–78.
Hausmann, J. (1974). *Gordian* **74,** 390–393.
Henry, K. M., and Kon, S. K. (1958). *Proc. Nutr. Soc.* **17,** 78–85.
Hodge, J. E. (1953). *J. Agric. Food Chem.* **1,** 928–943.
Hoffman, G. (1962). *In* "Lipids and Their Oxidation" (H. W. Schultz, E. A. Day, and R. O. Sinnhuber, eds.), pp. 215–229. Avi Publ., Westport, Connecticut.
Hollingsworth, D. F., and Martin, P. E. (1972). *World Rev. Nutr. Diet.* **15,** 1–34.
Holman, R. T. (1954). *Prog. Chem. Fats Other Lipids* **2,** 51–98.
Hossain, M. M., Huismans, J. W., and Diehl, J. F. (1976). *Toxicology* **6,** 243–251.
Howard, J. W., Fazio, T., and White, R. H. (1968). *J. Assoc. Off. Anal. Chem.* **51,** 122–129.
Huang, A., and Firestone, D. (1969). *J. Assoc. Off. Anal. Chem.* **52,** 958–963.
Hurst, D. T. (1972). "Recent Developments in the Study of Non-enzymic Browning and Its Inhibition by SO_2," Sci. Tech. Surv., No. 75. Br. Food Manuf. Ind. Res. Assoc.
Ingold, K. U. (1962). *In* "Lipids and Their Oxidation" (H. W. Schultz, E. A. Day, and R. O. Sinnhuber, eds.), pp. 93–121. Avi Publ., Westport, Connecticut.
Jacobson, G. A. (1967). *Food Technol.* **21,** 147–152.
Johnson, O. C., and Kummerow, F. A. (1957). *J. Am. Oil Chem. Soc.* **34,** 407–409.
Johnson, O. C., Perkins, E. G., Sugai, M., and Kummerow, F. A. (1957). *J. Am. Oil Chem. Soc.* **34,** 594–597.
Johnston, P. V., Kummerow, F. A., and Walton, C. H. (1958a). *Proc. Soc. Exp. Biol. Med.* **99,** 735–736.
Johnston, P. V., Johnson, O. C., and Kummerow, F. A. (1958b). *J. Nutr.* **65,** 13–24.
Kajimoto, G., and Mukai, K. (1970). *J. Jpn. Oil Chem. Soc.* **19,** 66–70.

Kalbag, S., Narayan, K. A., Chang, S. S., and Kummerow, F. A. (1955). *J. Am. Oil Chem. Soc.* **30,** 271–274.

Kanner, J., and Karel, M. (1976). *J. Agric. Food Chem.* **24,** 468–472.

Kantorowitz, B., and Yannai, S. (1974). *Nutr. Rep. Int.* **9,** 331–341.

Karel, M., Schaich, K., and Roy, R. B. (1975). *J. Agric. Food Chem.* **23,** 159–163.

Kaunitz, H. (1960). *Exp. Med. Surg.* **18,** 59–69.

Kaunitz, H. (1962). *In* "Lipids and Their Oxidation" (H. W. Schultz, E. A. Day, and R. O. Sinnhuber, eds.), pp. 269–293. Avi Publ., Westport, Connecticut.

Kaunitz, H. (1967). *Food Technol.* **21,** 278–282.

Kaunitz, H., and Johnson, R. E. (1973). *Lipids* **8,** 329–336.

Kaunitz, H., Johnson, R. E., and Slanetz, C. A. (1952). *J. Nutr.* **46,** 151–159.

Kaunitz, H., Slanetz, C. A., and Johnson, R. E. (1955). *J. Nutr.* **55,** 577–587.

Kaunitz, H., Slanetz, C. A., Johnson, R. E., Knight, H. B., Saunders, D. H., and Swern, D. (1956a). *J. Am. Oil Chem. Soc.* **33,** 630–634.

Kaunitz, H., Slanetz, C. A., Johnson, R. E., and Guilmain, J. (1956b). *J. Nutr.* **60,** 237–244.

Kaunitz, H., Slanetz, C. A., Johnson, R. E., and Babayan, V. K. (1960). *J. Nutr.* **70,** 521–527.

Kawada, T., Krishnamurthy, R. G., Mookherjee, B. D., and Chang, S. S. (1967). *J. Am. Oil Chem. Soc.* **44,** 131–135.

Kawakishi, S., Kito, Y., and Namiki, M. (1975). *Proc. Int. Congr. Food Sci. Technol., 4th, 1974* Vol. I, pp. 306–314.

Keane, K. W., Jacobson, J. A., and Krieger, C. H. (1959). *J. Nutr.* **68,** 57–74.

Keeney, M. (1962). *In* "Lipids and Their Oxidation" (H. W. Schultz, E. A. Day, and R. O. Sinnhuber, eds.), pp. 79–89. Avi Publ., Westport, Connecticut.

Khan, N. A., Lundberg, W. O., and Holman, R. T. (1954). *J. Am. Chem. Soc.* **76,** 1779–1784.

Kiriyama, S., Banjo, M., and Matsushima, H. (1974). *Nutr. Rep. Int.* **10,** 79–88.

Knorr, D. (1975). *Lebensm.-Wiss.-Technol.* **8,** 51–56.

Koehler, P. E., Mason, M. E., and Newell, J. A. (1969). *J. Agric. Food Chem.* **17,** 393–396.

Koerker, R. L., Berlin, A. J., and Schneider, F. H. (1976). *Toxicol. Appl. Pharmacol.* **37,** 281–288.

Kritchevsky, D., Cottrell, M. C., Whitehouse, M. D., and Staple, E. (1961). *Fed. Proc., Fed. Am. Soc. Exp. Biol.* **20,** 283.

Kummerow, F. A. (1962). *In* "Lipids and Their Oxidation" (H. W. Schultz, E. A. Day, and R. O. Sinnhuber, eds.), pp.294–320. Avi Publ., Westport, Connecticut.

Kummerow, F. A. (1974). *J. Am. Oil Chem. Soc.* **51,** 255–259.

Labuza, T. R. (1971). *Crit. Rev. Food Technol.* **2,** 355–405.

Lang, K. (1973). *Fette, Seifen, Amstrichm.* **75,** 73–76.

Lang, K., and Krozingen, B. (1972). *Z. Ernaehrungswiss.* **11,** 177–199.

Lang, K., and Schaeffner, E. (1964). *Ernaehrungswiss.* **4,** 235–245.

Lea, C. H. (1958). "Fundamental Aspects of the Dehydration of Foodstuffs," pp. 178–196. Soc. Chem. Ind., London.

Lea, C. H. (1962). *In* "Lipids and Their Oxidation" (H. W. Schultz, E. A. Day, and R. O. Sinnhuber, eds.), pp. 3–28. Avi Publ., Westport, Connecticut.

Lea, C. H. (1965a). *Chem. Ind. (London)* pp. 244–248.

Lea, C. H. (1965b). *Nutr. Dieta* **1,** 88–105.

Lee, C. M., Lee, T.-C. and Chichester, C. O. (1975). *Proc. Int. Congr. Food Sci. Technol., 4th, 1974* Vol. 1, pp. 587–603.

Levinsky, H. V., and Wilson, M. A. (1975). *Food Cosmet. Toxicol.* **13,** 243–246.

Liener, I. E. (1958). *In* "Processed Plant Protein Foodstuffs" (A. M. Altschul, ed.), pp. 79–129. Academic Press, New York.

Lightbody, H. D., and Fevold, H. L. (1948). *Adv. Food Res.* **1,** 149–202.

Lijinsky, W. (1977). *Chem. & Eng. News* June 27, p. 25.

Lintas, C., and Bernardini, M. P. (1975). *Proc. Int. Congr. Food Sci. Technol., 4th, 1974* Vol. I, pp. 330–334.
Lundberg, W. O. (1962a). *In* "Autoxidation and Antioxidants" (W. O. Lundberg, ed.), Vol. 2, pp. 451–476. Wiley, New York.
Lundberg, W. O. (1962b). *In* "Lipids and Their Oxidation" (H. W. Schultz, E. A. Day, and R. O. Sinnhuber, eds.), pp. 31–50. Avi Publ., Westport, Connecticut.
MacDonald, J. A. (1956). *J. Am. Oil Chem. Soc.* **33,** 394–396.
McInnes, A. G., Cooper, F. P., and MacDonald, J. A. (1961). *Can. J. Chem.* **39,** 1906–1914.
McKinney, L. L., Weakley, F. B., Campbell, R. E., Eldridge, A. C., and Cowan, J. C. (1957). *J. Am. Oil Chem. Soc.* **34,** 461–466.
McKinney, L. L., Picken, J. C., Weakley, F. B., Eldridge, A. C., Campbell, R. E., Cowan, J. C., and Biester, H. E. (1959). *J. Am. Chem. Soc.* **81,** 909–915.
McWeeny, D. J., Biltcliffe, D. O., Powell, R. C. T., and Spark, A. A. (1969). *J. Food Sci.* **34,** 641–643.
Malanoski, A. J., and Greenfield, E. L. (1968). *J. Assoc. Off. Anal. Chem.* **51,** 114–121.
Matsushita, S. (1975). *J. Agric. Food Chem.* **23,** 150–154.
Matsuo, N. (1957a). *J. Jpn. Biochem. Soc.* **29,** 773–777.
Matsuo, N. (1957b). *Kagaku No Ryoiki* **12,** 64–73.
Matsuo, N. (1962). *In* "Lipids and Their Oxidation" (H. W. Schultz, E. A. Day, and R. O. Sinnhuber, eds.), pp. 321–359. Avi Publ., Westport, Connecticut.
Mauron, J. (1975). *Proc. Int. Congr. Food Sci. Technol., 4th, 1974* Vol. I, pp. 564–577.
Mead, J. F. (1961). *In* "Autoxidation and Antioxidants" (W. O. Lundberg, ed.), Vol. I, pp. 299–323. Wiley, New York.
Mead, J. F. (1962). *In* "Lipids and Their Oxidation" (H. W. Schultz, E. A. Day, and R. O. Sinnhuber, eds.), pp. 360–366. Avi Publ., Westport, Connecticut.
Mehta, T. M., and Sharma, S. A. (1956). *J. Am. Oil Chem. Soc.* **33,** 38–44.
Mehta, T. M., and Sharma, S. A. (1957). *J. Am. Oil Chem. Soc.* **34,** 448–450.
Melnick, D. (1957). *J. Am. Oil Chem. Soc.* **34,** 351–356.
Melnick, D., and Oser, B. L. (1949). *Food Technol.* **3,** 57–71.
Menzel, D. B. (1967). *Lipids* **2,** 83–84.
Mesrobian, R. B., and Tobolsky, A. V. (1961). *In* "Autoxidation and Antioxidants" (W. O. Lundberg, ed.), Vol. I, pp. 107–131. Wiley, New York.
Michael, W. R. (1966a). *Lipids* **1,** 359–364.
Michael, W. R. (1966b). *Lipids,* **1,** 365–368.
Michael, W. R., Alexander, J. C., and Artman, N. R. (1966). *Lipids,* **1,** 353–358.
Miller, J., and Landes, D. R. (1975). *J. Food Sci.* **40,** 545–548.
Miller, S. A., Tannenbaum, S. R., and Seitz, A. W. (1970). *J. Nutr.* **100,** 909–916.
Milner, M., and Thompson, J. B. (1954). *J. Agric. Food Chem.* **2,** 303–309.
Misani, F., and Reiner, L. (1950). *Arch. Biochem. Biophys.* **27,** 234–235.
Mitchell, J. H., and Henick, A. S. (1962). *In* "Autoxidation and Antioxidants" (W. O. Lundberg, ed.), Vol. 2, pp. 543–592. Wiley, New York.
Monier-Williams, G. W. (1949). "Trace Elements in Food." Wiley, New York.
Mooser, O., and Neukom, H. (1975). *Proc. Int. Congr. Food Sci. Technol., 4th, 1974* Vol. I, pp. 283–288.
Mori, A., Suneya, Y., Maruyama, Y., and Yasui, Y. (1975). *Proc. Int. Congr. Food Sci. Technol., 4th, 1974* Vol. 3, 298–304.
Mounts, T. L., McWeeny, D. J., Evans, C. D., and Dutton, H. J. (1970). *Chem. Phys. Lipids,* **4,** 197–202.
Murphy, S. D. (1965). *Toxicol. Appl. Pharmacol.* **7,** 833–843.
Nakamura, M., Tanaka, H., Hattori, Y., and Watanabe, M. (1973). *Lipids* **8,** 566–572.

Namiki, M., Hayashi, T., and Kawakishi, S. (1973). *Agric. Biol. Chem.* **37,** 2935–2936.

Namiki, M., Kawakishi, S., and Hayashi, T. (1975). *Proc. Int. Congr. Food Sci. Technol., 4th, 1974* Vol. I, pp. 208–216.

Nawar, W. W. (1973). *Prog. Chem. Fats Other Lipids* **13,** 91–118.

Nikonorow, M., Mazur, H., and Piekacz, H. (1973). *Toxicol. Appl. Pharmacol.* **26,** 253–259.

Nishie, K., Waiss, A. C., Jr., and Keyl, A. C. (1969). *Toxicol. Appl. Pharmacol.* **14,** 301–307.

Njaa, L. R. (1962a). *Acta Chem. Scand.* **16,** 1359–1362.

Njaa, L. R. (1962b). *Br. J. Nutr.* **16,** 571–577.

Nolen, G. A. (1973). *J. Nutr.* **103,** 1248–1255.

Nolen, G. A., Alexander, J. C., and Artman, N. R. (1967). *J. Nutr.* **93,** 337–348.

O'Brien, P. J., and Frazer, A. C. (1966). *Proc. Nutr. Soc.* **25,** 9–18.

O'Donovan, C. J. (1976). *Food Cosmet. Toxicol.* **14,** 483–489.

Ohfugi, T., and Kaneda, T. (1973). *Lipids* **8,** 353–359.

Oser, B. L. (1971). *Food Cosmet. Toxicol.* **9,** 245–250.

Ottolenghi, A., Bernheim, F., and Wilbur, K. M. (1955). *Arch. Biochem. Biophys.* **56,** 157–164.

Pantuck, E. J., Hsiao, K. C., Conney, A. H., Garland, W. A., Kappas, A., Anderson, K. E., and Alvares, A. P. (1976). *Science* **194,** 1055–1057.

Paulus, K. (1976). *Ernaehr-Umsch.* **23,** 116–123.

Perkins, E. G. (1960). *Food Technol.* **14,** 508–514.

Perkins, E. G. (1967). *Food Technol.* **21,** 611–616.

Perkins, E. G., and Kummerow, F. A. (1959a). *J. Am. Oil Chem. Soc.* **36,** 371–375.

Perkins, E. G., and Kummerow, F. A. (1959b). *J. Nutr.* **68,** 101–108.

Pokorny, J., El-Zeany, B. A., and Janicek, G. (1975). *Proc. Int. Congr. Food Sci. Technol., 4th, 1974* Vol. I, pp. 217–223.

Poling, C. P., Warner, W. D., Mone, P. E., and Rice, E. E. (1960). *J. Nutr.* **72,** 109–120.

Potteau, B., Lhuissier, M., Leclerc, J., Custot, F., Mezonnet, R., and Cluzan, R. (1970a). *Rev. Fr. Corps Gras* **17,** 143–153.

Potteau, B., Lhuissier, M., Leclerc, J., Custot, F., Mezonnet, R., and Cluzan, R. (1970b). *Rev. Fr. Corps Gras* **17,** 235–245.

Privett, O. S., and Blank, M. L. (1962). *J. Am. Oil Chem. Soc.* **39,** 465–469.

Privett, O. S., and Cortesi, R. (1972). *Lipids* **7,** 780–787.

Privett, O. S., Lundberg, W. O., and Nickel, C. (1953a). *J. Am. Oil Chem. Soc.* **30,** 17–21.

Privett, O. S., Lundberg, W. O., Khan, N. A., Tolberg, W. E., and Wheeler, D. H. (1953b). *J. Am. Oil Chem. Soc.* **30,** 61–66.

Provansal, M., Cuq, J. L., and Cheftel, C., (1975). *Agric. Food Chem.* **23,** 938–943.

Raju, N. V., and Rajagopalan, R. (1955). *Nature (London)* **176,** 513–514.

Rao, B. Y. (1960). *J. Sci. Ind. Res. Sect. A,* **9,** 430–437.

Rao, M. N., Srinivas, H., Swaminathan, S., Carpenter, K. J., and Morgan, C. B. (1963). *J. Sci. Food Agric.* **14,** 544–550.

Reddy, B. R., Yasuda, K., Krishnamurthy, R. G., and Chang, S. S. (1968). *J. Am. Oil Chem. Soc.* **45,** 629–631.

Renner, H. W. (1975). *Food Cosmet. Toxicol.* **13,** 427–431.

Reynolds, T. M. (1963). *Adv. Food Res.* **12,** 1–52.

Reynolds, T. M. (1965). *Adv. Food Res.* **14,** 167–263.

Rice, E. W. (1972). *Clin. Chem.* **18,** 1550–1551.

Rice, E. E., Poling, C. E., Mone, P. E., and Warner, W. D. (1960). *J. Am. Oil Chem. Soc.* **37,** 607–613.

Riemann, H. (1969). *In* "Food-Borne Infections and Intoxications" (H. Riemann, ed.), pp. 489–541. Academic Press, New York.

Roth, N. G., and Halvorson, H. O. (1952). *J. Bacteriol.* **63,** 429–435.

Sahasrabudhe, M. R., and Bhalerao, V. R. (1963). *J. Am. Oil Chem. Soc.* **40,** 711–712.
Salunkhe, D. K., and Wu, M. T. (1977). *Crit. Rev. Food Sci. Nutr.* **9,** 265–324.
Sapeika, N. (1969). "Food Pharmacology." Thomas, Springfield, Illinois.
Schroeder, H. A., Kanisawa, M., Frost, D. V., and Mitchener, M. (1968). *J. Nutr.* **96,** 37–45.
Seinen, W., and Willems, M. I. (1976). *Toxicol. Appl. Pharmacol.* **35,** 63–75.
Selikoff, I. J., and Hammond, E. C., eds. (1975). "Toxicity of Vinyl Chloride-Poly-vinyl Chloride," No. 246. N. Y. Acad. Sci., New York.
Selye, H. (1951). *Am. J. Med.* **10,** 549–555.
Seto, T. A., Schultze, M. O., Perman, V., Bates, F. W., and Sautter, J. H. (1958). *J. Agric. Food Chem.* **6,** 49–54.
Sgoutas, D., and Kummerow, F. A. (1970). *Am. J. Clin. Nutr.* **23,** 1111–1119.
Siegfried, R. (1975). *Naturwissenschaften* **62,** 576.
Simko, V., Bucko, A., Babaca, J., and Ondreicka, R. (1964). *Nutr. Dieta* **6,** 91–105.
Sims, R. P. A., and Hoffman, W. H. (1962). *In* "Autoxidation and Antioxidants" (W. O. Lundberg, ed.), Vol. I, pp. 629–682. Wiley, New York.
Slump, P., and Schreuder, H. A. W. (1973). *J. Sci. Food Agric.* **24,** 657–661.
Spark, A. A. (1969). *J. Sci. Food Agric.* **20,** 308–316.
Sternberg, M., Kim, C. Y., and Schwende, F. J. (1975). *Science* **190,** 992–994.
Strauss, H. J., and Billek, G. (1974). *Z. Ernaehrungswiss.* **13,** 81–88.
Strobel, R. G. K. (1974). *Int. Chim. Cafes Publ.* **6,** 128–134.
Sugai, A., Witting, L. A., Tsudiyama, H., and Kummerow, F. A. (1962). *Cancer Res.* **22,** 510–519.
Swern, D. (1961). *In* "Autoxidation and Antioxidants" (W. O. Lundberg, ed.), Vol. I, pp. 1–54. Wiley, New York.
Tannenbaum, S. R., Barth, H., and Le Roux, J. P. (1969). *J. Agric. Food Chem.* **17,** 1353–1354.
Tas, A. C., and Kleipool, R. J. C. (1976). *Lebensm.-Wiss.-Technol.* **9,** 360–362.
Taub, I., Angelini, P., and Merritt, C. (1976). *J. Food Sci.* **41,** 942–944.
Thompson, J. A., Paulose, M. M., Reddy, B. R., Krishnamurthy, R. G., and Chang, S. S. (1967). *Food Technol.* **21,** 405–407.
Tilgner, D. J., and Daun, H. (1969). *Residue Rev.* **27,** 19–41.
Torres, D., Trinchese, T., Foley, E., and Karabinos, J. V. (1956). *Trans. Ill. State Acad. Sci.* **49,** 205–206.
Uri, N. (1961). *In* "Autoxidation and Antioxidants" (W. O. Lundberg, ed.), Vol. I, pp. 55–106. Wiley, New York.
van Beek, L., Feron, V. J., and de Groot, A. P. (1974). *J. Nutr.* **104,** 1630–1636.
van Esch, G. J., and van Logten, M. (1975). *Food Cosmet. Toxicol.* **13,** 121–124.
Vergroesen, A. J., and Gottenbos, J. J. (1975). *In* "The Role of Fats in Human Nutrition" (A. J. Vergroesen, ed.), pp. 1–41. Academic Press, New York.
Vidal, P. (1974). *Ind. Aliment. Agric.* **91,** 11–25.
Wedermeyer, G. A., and Dollar, A. M. (1963). *J. Food Sci.* **28,** 537–540.
Wells, A. F., and Common, R. H. (1953). *J. Sci. Food Agric.* **4,** 233–237.
Welt, M. A. (1968). *Science* **160,** 483.
Wesley, F., Raurke, B., and Darbishire, O. (1965). *J. Food Sci.* **30,** 1037–1042.
West, C. E., and Redgrave, T. G. (1975). *Int. Lab.* March/April, pp. 45–50.
Whitlock, C. B., and Nawar, W. W. (1976). *J. Am. Oil Chem. Soc.* **53,** 586–591.
Widmark, E. M.P. (1939). *Nature (London)* **143,** 984.
Wilding, M. D., Rice, E. E., and Mattil, K. F. (1963). *J. Am. Oil Chem. Soc.* **40,** 55–57.
Winteringham, F. P. W. (1955). *J. Sci. Food Agric.* **6,** 269–274.
Winteringham, F. P. W., Harrison, A., Bridges, R. G., and Bridges, P. M. (1955). *J. Sci. Food Agric.* **6,** 251–261.

Wogan, G. N. (1966). *Fed. Proc., Fed. Am. Soc. Exp. Biol.* **25,** 124–129.

Woodard, J. C., and Short, D. D. (1973). *J. Nutr.* **103,** 569–574.

Woodard, J. C., and Short, D. D. (1977). *Food Cosmet. Toxicol.* **15,** 117–119.

Woodard, J. C., Short, D. D., Strattan, C. E., and Duncan, J. H. (1977). *Food Cosmet. Toxicol.* **15,** 109–115.

Wurziger, J., and Ostertag, H. (1960). *Fette, Seifen, Anstrichm.* **62,** 895–903.

Yoshioka, M., and Kaneda, T. (1975). *Proc. Int. Congr. Food Sci. Technol., 4th, 1974* Vol. I, pp. 276–282.

Ziegler, K., Melchert, I., and Lürken, C. (1967). *Nature (London)* **214,** 404–405.

CHAPTER 13

Miscellaneous Toxic Factors

Irvin E. Liener

TOXIC CONSTITUENTS OF PLANT FOODSTUFFS, SECOND EDITION

ISBN 0-12-449960-0

I. INTRODUCTION

The extremely varied nature of the chemical and physiological properties of naturally occurring toxic substances poses certain problems with respect to their organized presentation. This chapter represents an attempt to cover some of the toxic constituents of plants that do not legitimately fall within the framework of the individual chapters of this volume. No attempt is made, however, to provide a comprehensive coverage of the vast number of plants that are primarily of pharmacological interest or produce toxic effects if accidentally ingested by man or animals. Excellent treatises covering this aspect of the subject matter are available (Tanner and Tanner, 1953; Garner, 1961; Kingsbury, 1964; Radeleff, 1964) and should be consulted by the interested reader. The guiding principle here is to include only those plants that have present or potential value as a source of human food either by direct consumption or indirectly because they can serve as feed for animals whose products are intended for human consumption.

The manner in which the subject matter of this chapter has been organized is quite arbitrary and is based primarily on certain physiological effects or chemical properties that some plant materials have in common. In those few cases in which such common properties do not appear to exist, these plants are considered on an individual basis.

II. ESTROGENIC FACTORS

By definition, an estrogen is a substance capable of stimulating the growth of the vagina, uterus, and mammary gland and is responsible for the development of female secondary characteristics. Substances exhibiting estrogenic activity are widely distributed in the plant kingdom (Bradbury and White, 1951; Stob, 1973; Labov, 1977). Among the plants commonly used for human food, the following have been reported to contain substances that elicit an estrogenic response in an appropriate experimental animal: carrots (Ferrando *et al.*, 1961), soybeans (Carter *et al.*, 1955), wheat, rice, oats, barley, potatoes, apples, cherries, plums (Bradbury and White, 1951), and wheat bran, wheat germ, rice bran, and rice polish (Booth *et al.*, 1960). Estrogenic activity has also been reported in such vegetable oils as cottonseed, safflower, wheat germ, corn, linseed, peanut, olive, soybean, coconut, and refined or crude rice bran oil (Booth *et al.*, 1960).

The estrogenic principles of plants have been, in most cases, chemically identified as isoflavones, which occur naturally in the form of glucosides. The sugar moiety is attached to one or more of the hydroxyl groups located at various positions of the isoflavone nucleus. The chemical structures of some of the common estrogenic isoflavones are shown in Fig. 1.

Isoflavones have been isolated from the soybean (Carter *et al.*, 1955; Nilson, 1962; Naim *et al.*, 1974), and the content of isoflavones in defatted soybean meal was found to be 0.25%, of which 99% were present as genistein (64%) and daidzin (23%). Genistein has also been isolated from commercial soybean meal that has presumably been heated (Magee, 1963) so that it may be assumed that this substance is stable to the heat treatment involved in the toasting process. In addition to their estrogenic activity, the soybean isoflavones inhibit lipoxygenase activity and exert and antihemolytic effect on erythrocytes subjected to peroxidation (Naim *et al.*, 1976), although the physiological significance of these phenomena in the diet is not clear.

Biochanin A, another isoflavone derivative known to be estrogenic, has been isolated from Bengal gram (*Cicer arietinum*) and red clover (*Trifolium pratense* L.). In contrast to genistein, however, biochanin A appears to be heat sensitive (Siddiqui, 1945), which probably accounts for the observation that the drying of clover leads to a loss in its estrogenic activity (Alexander and Watson, 1951). Prunetin is found in species of *Prunus*, such as plums and cherries.

Another factor to be considered regarding isoflavones in plant products is their possible modification during processing. Pieterse and Andrews (1956a,b) found that moldy corn and alfalfa silage had increased estrogenic activity as a result of

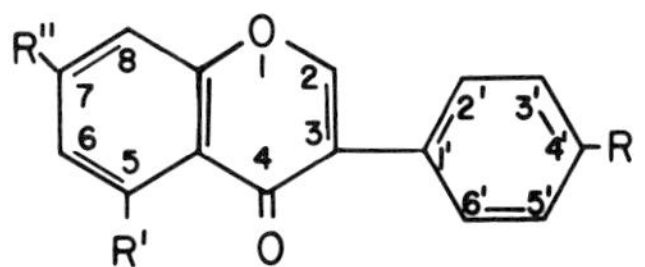

Estrogen	Substituents on isoflavone ring		
	R(4')	R'(5)	R''(7)
Genistein	OH	OH	OH
Biochanin A	OCH_3	OH	OH
Daidzein	OH	H	OH
Prunetin	OH	OH	OCH_3

FIG. 1. Structures of some common estrogenic isoflavones.

the fermentation process. The occurrence of yet another isoflavone, 6,7,4′-trihydroxyflavone, in the fermented soybean food, tempeh (György *et al.*, 1964), may be an example of isoflavone modification that may occur during the processing of soybeans.

It should be pointed out that not all of the estrogens found in plant materials are isoflavone derivatives. Estrone, the well-known steroidal sex hormone, was isolated from palm kernel (*Elaeis guincensis*) by Butenandt and Jacobi (1933). Anethole, which is chemically related to stilbene, was isolated from anise oil (Zondek and Bergmann, 1938), and coumestrol, a coumarin derivative, was isolated from alfalfa (Bickoff *et al.*, 1957; Knuckles *et al.*, 1977).

Soybean sprouts are also quite high in coumestrol, but the latter occurs in extremely low amounts in most other soybean products and other common vegetable food products (Knuckles *et al.*, 1977).

Since most of the naturally occurring estrogenic substances show only weak activity (Cheng *et al.*, 1955; Bickoff *et al.*, 1962), it is doubtful that the normal consumption of foods known to contain these substances would provide sufficient amounts of these substances to elicit a physiological response in human beings. Under unusual circumstances, there may be exceptions to this general conclusion. For example, Labov (1977) made reference to the fact that near the end of World War II, due to a severe shortage of food, the people in Holland consumed large quantities of tulip bulbs, which are high in estrogenic activities. Many women who ate them showed manifestations of estrogen imbalance including uterine bleeding and abnormalities of the menstrual cycle.

III. STIMULANTS AND DEPRESSANTS

A. *Datura*

Certain plants have been known from ancient times to contain substances that act as stimulants, depressants, or hallucinogenic agents. Most of these plants contain alkaloids that are useful in medicine and are not generally consumed as foods. It is nevertheless true that some of these plants have been the cause of numerous cases of poisoning in human beings and livestock and therefore merit some consideration here. Furthermore, because of the current interest in the so-called mind-expanding drugs, more attention should be given to the possible dangers of consuming such plants.

Seeds from the genus *Datura* (particularly *D. stramonium* and *D. sanguinea*) have been used among primitive societies for medicinal purposes as well as for their hallucinogenic properties. Such plants have been generally consumed in the form of a brew or tea made from the leaves and crushed seed of the plant (Blood and Rudolph, 1966). The accidental ingestion of the plant has led to numerous instances of poisoning, particularly among children (Raymond, 1944; Wright, 1944). Raymond (1944) described instances of *Datura* poisoning among Africans, which was traced to the consumption of bread made from grain that had become contaminated with *Datura* seed. Losses of all classes of livestock including horses, cattle, sheep, hogs, and chickens have been reported in the world literature, which has been reviewed by Kingsbury (1964). Most livestock find this plant distasteful and avoid it unless there is a shortage of more desirable forage.

The active principles of *Datura* are the alkaloids atropine, hyoscyamine, and hyoscine (scopolamine). For the reader interested in the chemistry of these alkaloids, the review by Fodor (1960) is recommended.

B. Nutmeg

The nutmeg (*Myristica fragrans*) has been used as a panacea for a wide variety of ailments, including toothache, dysentery, rheumatism, and halitosis, and for inducing abortions (Blood and Rudolph, 1966). In addition to its hallucinogenic effect (Fras and Friedman, 1969; Lovenberg, 1973), the ingestion of nutmeg is frequently accompanied by very serious side effects, such as nausea, constipation, tachycardia, and stupor (Weil, 1965; Painter *et al.*, 1971). The noxious ingredient of nutmeg is believed to be myristicin (Fig. 2), which makes up about 4% of the volatile oil (Hall, 1973). It is interesting that myristicin can also be found in the seed of fools parsley (*Aethusa cynapium* L.) (Small, 1949) and celery (*Apium graveolens* L.) (Karmazin, 1955), both of which have been known to kill livestock under certain conditions (Kingsbury, 1964).

C. Pressor Amines

A number of common foods contain appreciable levels of amines, such as serotonin, norephinephrin, tyramine, tryptamine, and dopamine (Fig. 3). These compounds are referred to collectively as pressor amines because they act as potent vasoconstrictors and thereby elevate the blood pressure. These pressor amines were found to be present at fairly high levels in such plants as pineapple, banana, plantain, and the avocado (Udenfriend *et al.*, 1959).

Although it is true that the direct introduction of these compounds into the bloodstream at the levels that would be obtained by eating normal quantities of these foods would no doubt produce serious effects, Strong (1966) is of the opinion that there is no real damage to be anticipated from the ingestion of such

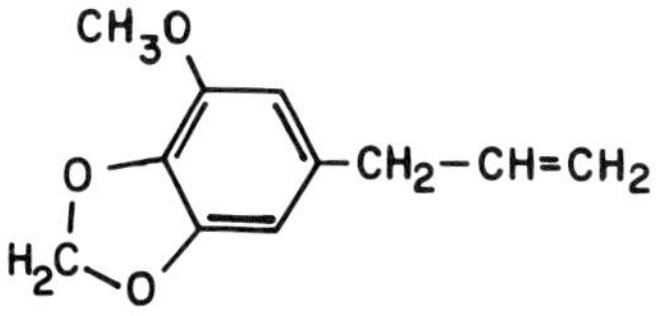

CH_3O

O — $CH_2-CH=CH_2$

H_2C — O

FIG. 2. Structure of myristicin, the active principle of nutmeg (*Myristica fragrans*).

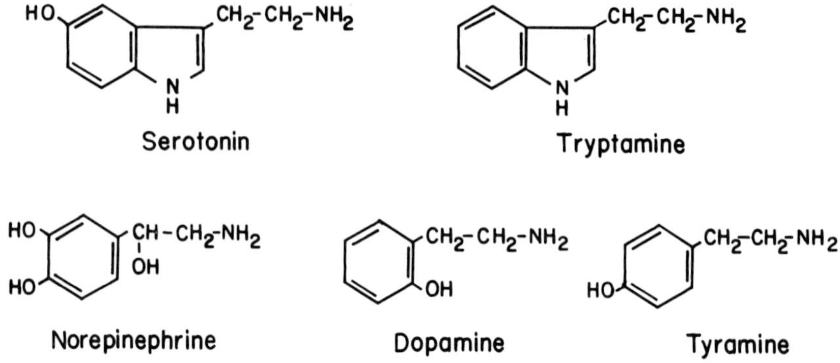

Serotonin Tryptamine

Norepinephrine Dopamine Tyramine

FIG. 3. Structures of some common pressor amines.

foods. In fact, the human body appears to be normally capable of coping with a constant influx of such compounds produced by the bacterial degradation of amino acids in the intestinal tract. A possible exception to this conclusion relates to the high incidence of carcinoid heart disease among Africans, a condition in which large amounts of 5-hydroxyindolylacetic acid, a measure of circulating serotonin, are produced endogenously (Crawford, 1962). It was suggested (Foy and Parratt, 1960, 1962) that one of the causative factors of this disease is the fact that the serotonin intake of many Africans may reach 200 mg/week due to the consumption of plantains or bananas as a major article of their diet.

Amines are normally rapidly deaminated after they enter the body through the action of the liver enzyme monoamine oxidase. Drugs that act as inhibitors of this enzyme are frequently used in the treatment of psychiatric depression. It has been observed that patients treated with such drugs are especially sensitive to certain types of cheeses that are particularly rich in tyramine. Hypertensive crisis, migraine headaches, and, in some instances, intercranial bleeding leading to death are some of the manifestations of this effect (Blackwell *et al.*, 1967). Presumably, the tyramine is not deaminated in the liver by the usual action of monoamine oxidase and subsequently displaces norepinephrin in nerve endings (Anthony and Lance, 1972). Although cheese has been the principal food implicated in this problem, Anet and Ingles (1976) made reference to the fact that certain foods of plant origin that are rich in amines—beans, wheat, nuts, and tomatoes—may also serve to trigger migraine in individuals using drugs that act as monoamine oxidase inhibitors.

IV. HYPOGLYCEMIC AGENTS

A. Hypoglycin

Consumption of the fruit of the plant *Blighia sapida* (known in Jamaica as *ackee* and in Nigeria as *isin*) has been linked to a disease of undernourished people, especially in Jamaica, known as "vomiting sickness." A detailed account of this disease and its cause can be found in the proceedings of a symposium (Kean, 1976) and a review article by Manchester (1974). The characteristic symptom of vomiting is followed by convulsions, coma, and even death in some instances (Tanaka *et al.*, 1976; Anonymous, 1976). Hypoglycemia is the principal feature of the blood picture, with sugar levels as low as 20 mg per 100 ml, as well as a gross depletion of liver glycogen (Jelliffe and Stuart, 1954). The causative principle has been identified as β-(methylenecyclopropyl)alanine (see Fig. 4) and is referred to as hypoglycin A. It is found in the seed both as the free amino acid and the γ-glutamyl dipeptide conjugate, hypoglycin B. Only hypoglycin A is found in the fruit (Hassal *et al.*, 1954). The unripe fruit has a higher content of hypoglycin A than the ripened fruit, which explains why most cases of poisoning have been associated with the consumption of the unripe fruit.

When injected into rats, hypoglycin follows the same pattern of degradation as branched-chain amino acids. It is first deaminated by transamination to yield β-(methylenecyclopropyl) pyruvate, which then undergoes oxidative decarboxylation to form β-(methylenecyclopropyl)acetyl-CoA (see Fig. 4). Formation of the latter is believed to somehow interfere with the transfer of long-chain fatty acyl residues from CoA to carnitine (Entman and Bressler, 1967). This interference with the synthesis of fatty acyl carnitine blocks β-oxidation. Since long-chain fatty acid oxidation is a requisite for gluconeogenesis, hypoglycemia results. Quite aside from this indirect hypoglycemic effect, hypoglycin, by virtue of its structural similarity to leucine, has been shown to inhibit the normal metabolism of this amino acid. The site of inhibition appears to be the enzyme isovaleryl-CoA dehydrogenase, thus leading to an accumulation of isovaleric acid and α-methylbutyric acid (Tanaka, 1972; Tanaka *et al.*, 1971,1972). Since short-chain branched fatty acids are known to act as depressants of the central nervous system (Borison *et al.*, 1975), the vomiting symptom would seem to be a direct result of the high levels of isovaleric acid and α-methylbutyric acid induced by

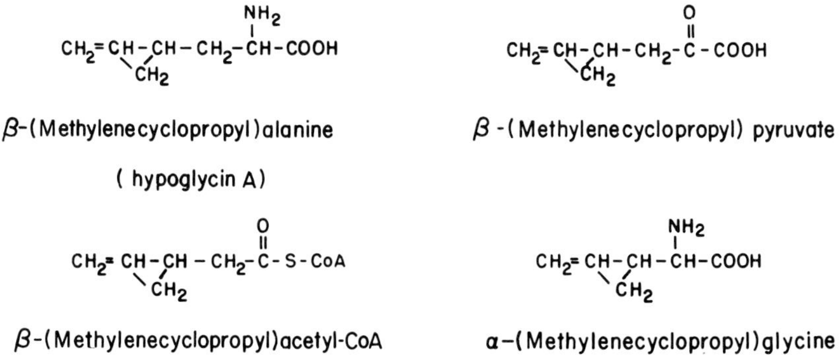

$CH_2{=}CH{-}CH(CH_2){-}CH_2{-}CH(NH_2){-}COOH$

β-(Methylenecyclopropyl)alanine

(hypoglycin A)

$CH_2{=}CH{-}CH(CH_2){-}CH_2{-}C({=}O){-}COOH$

β-(Methylenecyclopropyl) pyruvate

$CH_2{=}CH{-}CH(CH_2){-}CH_2{-}C({=}O){-}S{-}CoA$

β-(Methylenecyclopropyl)acetyl-CoA

$CH_2{=}CH{-}CH(CH_2){-}CH(NH_2){-}COOH$

α-(Methylenecyclopropyl)glycine

FIG. 4. Structure of hypoglycin A and some of its metabolic derivatives. Also shown is the structure of α-(methylenecyclopropyl)glycine, the lower homologue of hypoglycin A.

hypoglycin. The teratogenic effects of hypoglycin in rats and chick embryos (Persaud, 1968a,b; Persaud and Kaplan, 1970) may also be attributable to high concentrations of these abnormal fatty acids.

A lower homologue of hypoglycin A, α-(methylenecyclopropyl)glycine (Fig. 4), was isolated from litchi seeds (*Litchi chinensis*) and shown to produce hypoglycemia and depletion of liver glycogen reserves in mice (Gray and Fowden, 1962). It is doubtful, however, that this substance is of any significance as far as man is concerned, since there have been no reports of any toxicity associated with the consumption of litchi seeds. Since the inhibition of the growth of mung bean seeds by this compound could be reversed by leucine (Fowden, 1966), it is believed to function as a specific antagonist of this amino acid.

B. Atractyloside

The thistle, *Atractylis gummifera,* is well known for its toxicity in North Africa and the Mediterranean basin, and each year its accidental consumption by man and animals produces serious intoxications and sometimes death (Stanilas and Vignais, 1964). Animals that have been poisoned with this plant usually die in hypoglycemic convulsions. The toxic principle of this plant, potassium atractylate, was isolated over 100 years ago and has been shown to have the structure shown in Fig. 5. Potassium atractylate produces an *in vivo* depletion of glycogen reserves through an inhibition of glycogen synthesis. In addition to a drastic lowering of blood sugar, there is an increase in the concentration of lactic acid in the blood. The underlying mechanism responsible for these observations can be explained by *in vitro* studies, which have shown that potassium atractylate inhibits oxidative phosphorylation (Vignais *et al.*, 1962; Allman *et al.*, 1967) and energy transfer reactions in liver mitochondria (Bruni *et al.*, 1962).

V. TOXIC AMINO ACIDS

The subject of naturally occurring "uncommon" amino acids, some of which are known to be toxic, has been comprehensively reviewed by Fowden; *et al.* (1967), Hylin (1969), and Bell (1972, 1976). One of these, hypoglycin, has

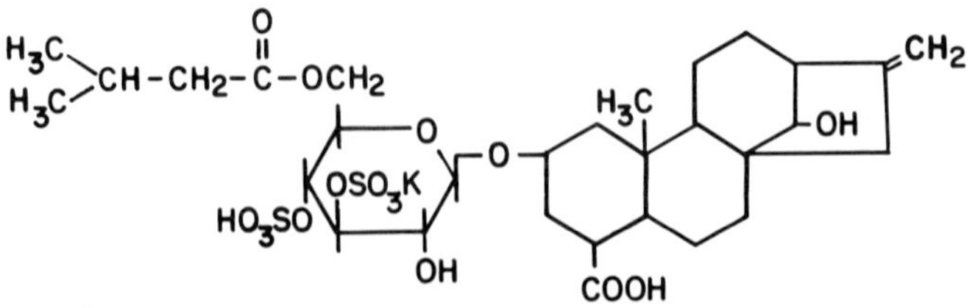

H_3C, H_3C >CH–CH$_2$–C(=O)–OCH$_2$; O ; O ; OSO_3K ; HO_3SO ; OH ; H_3C ; CH_2 ; OH ; COOH

FIG. 5. Structure of potassium atractylate, the toxic principle of *Atractylis gummifera*.

already been considered with respect to its action as a hypoglycemic agent (see this chapter, Section IV), and the neurotoxic amino acids that have been implicated in lathyrism have been discussed in Chapter 8. A brief survey will be made here of a number of other atypical amino acids found in plants that are of interest because of their toxic properties.

A. Mimosine

The National Academy of Sciences (1977) recently published a monograph that describes the high potential value of the legume *Leucaena leucocephala* (called *koa haole* in Hawaii) as a forage crop for livestock and human feeding. One of the principal factors limiting the use of this plant, however, is the fact that an unusual amino acid, mimosine (see Fig. 6), comprises 3–5% of the dry weight of the protein. This amino acid is responsible for the poor growth performance of cattle when leucaena makes up more than one-half of the diet. This adverse effect on growth has been traced to the underproduction of thyroxine presumably due to the fact that the rumen bacteria convert mimosine to 3,4-dihydroxypyridine, which acts as a goitrogenic agent (Hegarty *et al.*, 1976). In nonruminants, such as the horse, pig, and rabbit, the goitrogenic effect is not very marked, but the animals nevertheless do very poorly on diets containing leucaena, one of the characteristic symptoms being a loss of hair (Owen, 1958; Hegarty *et al.*, 1964). In fact, it has even been suggested that mimosine might be used as a defleecing agent in sheep (Reis *et al.*, 1975). Certain segments of the human population, particularly in Indonesia, are known to consume portions of the leucaena in their diet, and a loss of hair has been frequently observed among those individuals who have eaten the leaves, pods, and seeds in the form of a soup (Van Veen, 1973). As far as is known, mimosine has no effect on the meat or milk of ruminants that can be detrimental to human beings.

Although the goitrogenic effect of mimosine in ruminants seems to be well established, the precise mechanism of toxicity in other animals remains obscure. It can act as an inhibitor of pyridoxal-containing transaminases (Lin *et al.*, 1963), tyrosine decarboxylase (Crounse *et al.*, 1962), several metal-containing enzymes (Chang, 1960; Lin *et al.*, 1963; Tsai, 1961), and both cystathionine

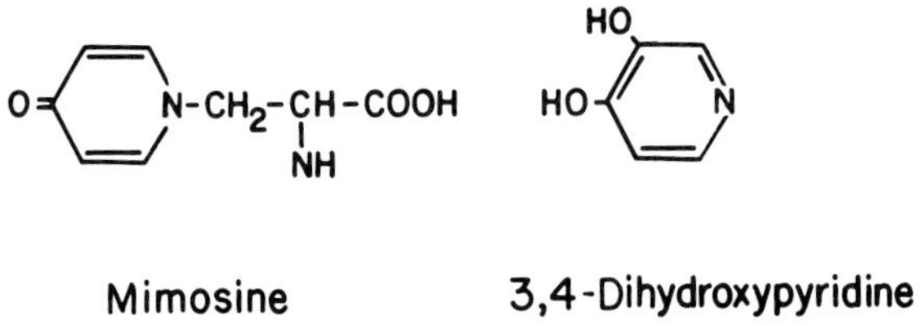

Mimosine 3,4-Dihydroxypyridine

FIG. 6. Structure of mimosine and its goitrogenic metabolite, 3,4-dihydroxypyridine.

synthetase and cystathionase (Hylin, 1969). The latter may be of particular significance, since an inhibition of the conversion of methionine to cysteine, a major component of hair protein, could account for the hair loss that is so characteristic of mimosine toxicity. Mimosine may also exert a more direct effect on hair growth since Montagna and Yun (1963) showed that leucaena extracts destroyed the matrix of the cells of the hair follicles of mice. This effect was reversible, however, since hair growth returned to normal once the animals were taken off the mimosine-containing diet.

The structural resemblance of mimosine to tyrosine would suggest that it might function *in vivo* as an antagonist of this amino acid. Lin *et al.* (1964) in fact reported that the growth inhibition produced by feeding 0.5% mimosine to rats could be partially reversed with phenylalanine and wholly reversed with tyrosine. Fowden (1966), however, was unable to reverse the mimosine-induced inhibition of the growth of mung bean seedlings with tyrosine, nor was he able to find any evidence for the incorporation of mimosine into the seedlings' protein.

Matsumoto *et al.* (1951) reported that the mimosine content of the seeds and leaves of leucaena could be decreased by storing the plant at temperatures in excess of 70°C in the presence of moisture. These authors also showed that ferrous sulfate added to the rations of rats containing unheated leucaena leaf meal was an effective means of reducing mimosine toxicity, due presumably to a decrease in the absorption of this amino acid from the gastrointestinal tract (Yoshida, 1944).

B. Djenkolic Acid

In certain parts of Sumatra, particularly in Java, the djenkol bean is a popular item of consumption (Van Veen, 1973). This bean is the seed of the leguminose tree, *Pithecolobium lobatum,* and resembles the horse chestnut in color and size. The bean is not particularly toxic except that it sometimes leads to kidney failure, which is accompanied by the appearance of blood and white needlelike clusters in the urine. The latter substance is a sulfur-containing amino acid known as djënkolic acid (Fig. 7), which is present in the free state in the bean to the extent of 1–4% (Van Veen and Hyman, 1933; Van Veen and Latuasan, 1949). In spite of its structural resemblance to cystine, it cannot replace cystine in the diet of

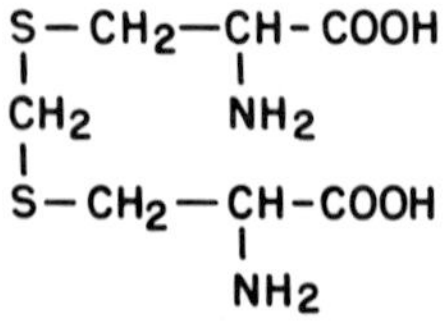

$$
\begin{array}{l}
\mathrm{S-CH_2-\underset{\displaystyle NH_2}{\underset{|}{CH}}-COOH} \\
| \\
\mathrm{CH_2} \\
| \\
\mathrm{S-CH_2-\underset{\displaystyle NH_2}{\underset{|}{CH}}-COOH}
\end{array}
$$

FIG. 7. Structure of djenkolic acid, an amino acid present in the djenkol bean, *Pithecolobium lobatum*.

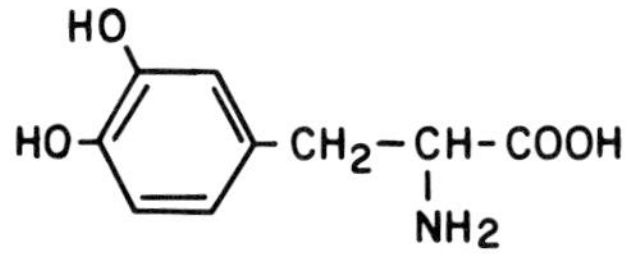

FIG. 8. Structure of 3,4-dihydroxyphenylalanine (dopa), an amino acid present in the faba bean (*Vicia faba*), the velvet bean (*Stizolobium derringianum*), and wheat and oats.

rats, although it can apparently be metabolized by the animal body (Van Veen and Hyman, 1933). Because of its relative insolubility, any djenkolic acid that escapes metabolic degradation tends to crystallize out in the kidney tubules and urine, hence the observations in human beings noted above.

C. Dihydroxyphenylalanine

The amino acid 3,4-dihydroxyphenylalanine (dopa) (Fig. 8) is present in fairly high concentrations in the fava bean, *Vicia faba* (Torquati, 1913; Guggenheim, 1913; Nagasawa *et al.*, 1961; Andrews and Pridham, 1965), the velvet bean, *Stizolobium deeringianum* (Miller, 1920), and wheat and oats (Hoeldtke *et al.*, 1972). Since the consumption of the fava bean is frequently associated with a disease in human beings known as favism (see Chapter 9), the question has been raised whether dopa might not play a causative role in this disease (Kosower and Kosower, 1967). Persons genetically deficient in the enzyme glucose-6-phosphate dehydrogenase appear to be particularly susceptible to favism, and one of the characteristic clinical features of this disease is a hemolytic type of anemia. The hemolysis of the red blood cells is believed to be due to a marked lowering of the glutathione content of the erythrocytes (Davies and Gower, 1964). In view of these facts, it may be pertinent to note that the *in vitro* addition of dopa to red blood cells from individuals deficient in glucose-6-phosphate dehydrogenase produced a significant lowering of the glutathione content of such cells (Kosower and Kosower, 1967). Because of its high content of dopa, it has been suggested that *Vicia faba* might be of therapeutic value in the treatment of Parkinson's disease (Natelson, 1969).

D. α-Amino-β-methylaminopropionic Acid

In addition to the carcinogenic and hepatoxic effects that accompany the ingestion of cycads (see Chapter 11), the consumption of this plant has also been reported to produce a neurological disorder in cattle involving the irreversible paralysis of the hindquarters (Mason and Whiting, 1966). Vega *et al.* (1968) were able to isolate from *Cycas circinalis* an amino acid, α-amino-β-methylaminopropionic acid (Fig. 9), which was markedly neurotoxic to chicks and rats (Nunn *et al.*, 1968). Judging from its structure and physiological

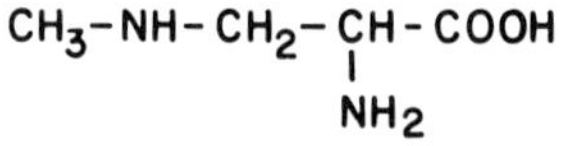

$$CH_3-NH-CH_2-\underset{\displaystyle NH_2}{\underset{|}{CH}}-COOH$$

FIG. 9. Structure of α-amino-β-methylaminopropionic acid, an amino acid with neurotoxic properties isolated from *Cycas circinalis*.

effects, it would appear that this amino acid is similar to those amino acids that have been implicated in lathyrism (see Chapter 8).

E. Selenoamino Acids

Selenium poisoning (''blind staggers'') in livestock is due to the consumption of selenium-accumulating plants of the genera *Astragalus* and *Machaeranthera* (Kingsbury, 1964; Hylin, 1969). The causative agents are believed to be the nonprotein amino acids methylselenocysteine and selenocystathionine (Fig. 10). Livestock intoxication can also occur from the consumption of plants that have grown in seleniferous soils. The selenium of such plants is present in protein in the form of selenomethionine and selenocystine residues (Fig. 10). Digestion of this protein in the digestive tract results in the liberation of these amino acids, which are absorbed and, because of their structural similarity to their sulfur analogues, compete for introduction into protein. Defective proteins thus formed could account for the loss of hair and sloughing of hoofs, which are characteristic features of selenium poisoning in livestock.

Since these selenoamino acids can be found in cereals grown in seleniferous soils, it is possible that human beings eating such cereals would also be affected. Chronic selenosis in human beings presumably caused by eating corn grown in

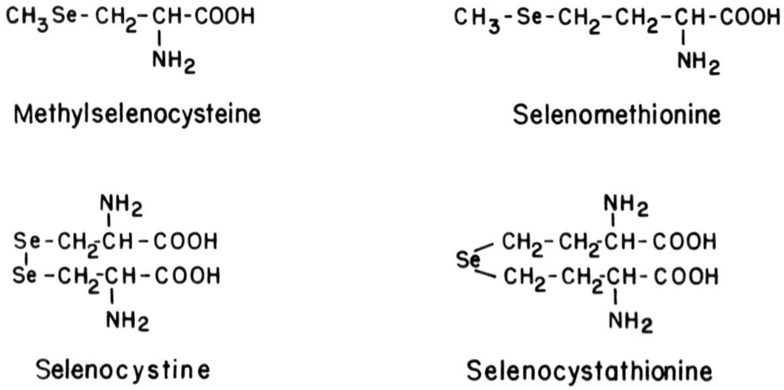

$$CH_3Se-CH_2-\underset{\displaystyle NH_2}{\underset{|}{CH}}-COOH$$

Methylselenocysteine

$$CH_3-Se-CH_2-CH_2-\underset{\displaystyle NH_2}{\underset{|}{CH}}-COOH$$

Selenomethionine

$$\begin{array}{l} NH_2 \\ \,| \\ Se-CH_2-CH-COOH \\ | \\ Se-CH_2-CH-COOH \\ \,| \\ NH_2 \end{array}$$

Selenocystine

$$Se\begin{array}{l} NH_2 \\ \,| \\ \diagup CH_2-CH_2-CH-COOH \\ \diagdown CH_2-CH_2-CH-COOH \\ \,| \\ NH_2 \end{array}$$

Selenocystathionine

FIG. 10. Structure of selenoamino acids.

seleniferous soil was reported in Colombia, South America (Rosenfeld and Beath, 1964). In certain parts of Venezuela, the ingestion of nuts from a tree known as coco de mono (*Lecythis ollaria*) produces a toxic syndrome in man characterized by abdominal distress, nausea, vomiting, diarrhea, and loss of scalp and body hair. The causative agent in this case is believed to be selenocystathionine (Arnow and Kerdel-Vegas, 1965; Kerdel-Vegas *et al.*, 1965). Direct proof for selenium toxicity in human beings in North America is lacking, although there appears to be a correlation between the selenium content in the food of rural populations in South Dakota and Nebraska in areas where there are seleniferous soils with disease symptoms specific for selenium toxicity (Smith *et al.*, 1936; Smith and Westfall, 1937; Smith and Lillie, 1940). The latter include higher than normal incidences of yellowish skin, dermatitis, chronic arthritis, bad teeth, and diseased nails on fingers and toes.

F. Indospicine

At one time, the trailing or creeping indigo (*Indigofera endecaphylla,* also known as *I. spicata*) showed great promise as a tropical forage crop. Subsequent studies, however, revealed that this plant was toxic to ruminants (Nordfeldt and Younge, 1949) as well as rabbits (Emmel and Ritchey, 1941), guinea pigs (Fuyre and Warmke, 1952), chicks (Hutton *et al.*, 1958a), mice (Hutton *et al.*, 1958b), and rats (Christie *et al.*, 1969,1975). The principal pathological feature in all cases is excessive damage to the liver, and, in some animals, a teratogenic effect has been demonstrated. Although β-nitropropionic acid was originally thought to be the hepatoxic agent (Cooke, 1955; Morris *et al.*, 1954), (see also Section XI,A), it now appears that the hepatoxic effect of this plant is largely due to an amino acid, indospicine, which, as shown in Fig. 11, bears a marked structure resemblance to arginine (Hegarty and Pound, 1968, 1970). This amino acid produces liver damage in sheep (Pearn and Hegarty, 1970) and in mice and rats (Christie *et al.*, 1969,1975) presumably by inhibiting the incorporation of arginine into protein (Madsen *et al.*, 1970). It is also antagonistic to arginine in bacteria (Leisinger *et al.*, 1972). Rather curiously, of 17 different species of *Indigofera* that were examined, only *I. endecaphylla* contained a detectable level of indospicine (Miller and Smith, 1973).

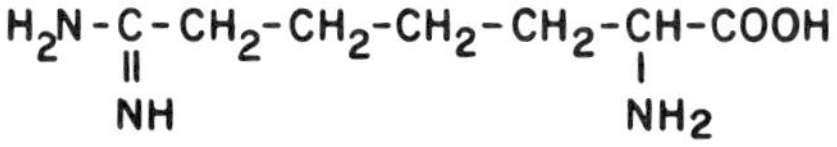

$$H_2N-\underset{\underset{NH}{\|}}{C}-CH_2-CH_2-CH_2-CH_2-\underset{\underset{NH_2}{|}}{CH}-COOH$$

FIG. 11. Structure of indospicine, an antimetabolite of arginine found in *Indigofera endecaphylla* (*spicata*).

VI. ANTIVITAMIN FACTORS

A. Antivitamin A

1. *Lipoxidase*

Raw soybeans are known to contain the enzyme lipoxidase, which oxidizes and destroys carotene (Sumner and Dounce, 1939). When 30% or more of ground soybeans were included in the diet of dairy calves, the levels of vitamin A and carotene in the blood plasma were markedly lowered (Shaw *et al.*, 1951). Although lipoxidase might logically be suspected as the causative agent, roasting soybeans at 100°C for 30 min did not prove beneficial in this respect (Ellmore and Shaw, 1954). Since lipoxidase measurements were not made in the latter study, it is not known whether this amount of heat treatment was in fact sufficient to inactivate the enzyme.

2. *Citral*

Citral (Fig. 12), a common constituent of orange oil, when administered subcutaneously or by mouth to rabbits and monkeys, caused damage to the vascular endothelia (Leach and Lloyd, 1956). Since this condition could be prevented by vitamin A, it was suggested by these authors that citral competes with retinene in the metabolism of the endothelial cells. As a constituent of orange oil, citral is present in such food products as marmalade, fruit juices flavored with orange oil, or orange drinks made by compression of the whole fruit. This raises the question whether the excessive consumption of such food products might not cause damage to blood vessels and hence be a contributory cause of cardiovascular disease.

B. Antivitamin D

Carlson *et al.* (1964a,b) reported that unheated soybean meal, or the protein isolated therefrom, caused rickets in turkey poults as judged by a decrease in bone ash. This rachitogenic effect was not observed when the soybean protein had been autoclaved and could be partially eliminated by supplementing the diet with an eight- to tenfold increase in vitamin D_3. Thompson *et al.* (1968) found

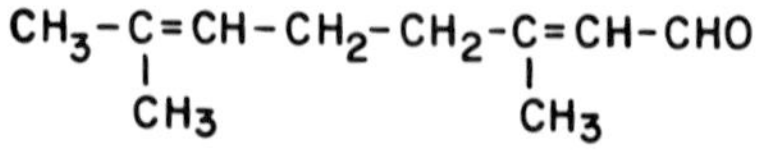

$$CH_3-\underset{\displaystyle CH_3}{\underset{|}{C}}=CH-CH_2-CH_2-\underset{\displaystyle CH_3}{\underset{|}{C}}=CH-CHO$$

FIG. 12. Structure of citral, a constituent of orange oil, believed to be an antagonist of vitamin A.

that autoclaving isolated soybean protein for 60 min caused the greatest destruction of the rachitogenic properties of the protein and was more effective in this respect than vitamin D_3. Calcium levels in the serum were increased by feeding the autoclaved protein, whereas increasing the vitamin D_3 levels did not affect serum calcium to any significant extent. For these reasons, these authors considered the possibility that the effect of autoclaving is mediated by a pathway different from that involved in the control of rickets by vitamin D_3 additions.

The rachitogenic activity of isolated soybean protein has also been reported for chicks (Jensen and Mraz, 1966a) and swine (Miller *et al.*, 1965). Jensen and Mraz (1966b) are of the opinion that isolated soybean protein simply interferes with the normal absorption of calcium or phosphorus or of both of these elements. These observations may perhaps be explained by the fact that isolated soybean protein is known to exhibit metal-binding properties (see Section VIII). If such is the case, however, it is difficult to understand why calcium and phosphorus supplementation does not overcome the rachitogenic activity of soy protein (Carlson *et al.*, 1964a,b).

C. Antivitamin E

1. Kidney Beans

Diets containing raw kidney beans (*Phaseolus vulgaris*) have been shown to increase the incidence of liver necrosis in weanling rats and muscular dystrophy in chicks and lambs (Hogue *et al.*, 1962; Hintz and Hogue, 1964, 1970). The muscular dystrophy in chicks was accompanied by a marked depression in plasma tocopherol levels (Desai, 1966). The incidence of muscular dystrophy in chicks could be largely eliminated by the administration of supplemental vitamin E or partially alleviated by autoclaving the bean for 60 min at 127°C. Supplementation of raw kidney bean diets with taurocholic acid also serves to alleviate the dystrophogenic effect in chicks (Hintz *et al.*, 1969; Banerjee and Hogue, 1975), possibly through an enhancement of the absorption of vitamin E. By means of alcohol extraction, two factors with antivitamin E activity could be demonstrated, one that was alcohol soluble and relatively heat stable and another that was heat labile and insoluble in alcohol. Hintz and Hogue (1964) believe that the alcohol-soluble factor is an unsaturated fat, but the chemical nature of the alcohol-insoluble antagonist is not known. Murillo and Gaunt (1975) recently purified an α-tocopherol oxidase from beans, alfalfa, and soybeans, which they feel might be responsible for an interference with the metabolism of this vitamin and hence could precipitate the symptoms characteristic of a deficiency of vitamin E. This enzyme may in fact be the heat-labile factor described by Hintz and Hogue (1964).

2. *Soybeans*

Fisher *et al.* (1969) reported that a soybean protein isolate increased the chick's requirement for α-tocopherol as measured by growth, mortality, exudative diathesis, and encephalomalacia. Whether this factor is the α-tocopherol oxidase described by Murillo and Gaunt (1975) remains to be proved.

3. *Alfalfa*

In studies on encephalomalacia in chicks, Singsen *et al.* (1955) noted an interference in the utilization of vitamin E from alfalfa (*Medicago sativa*). It was later shown by Pudelkiewicz and Matterson (1960) that alfalfa contained an ethanol-soluble fraction that interfered with the absorption of vitamin E. Whether this fraction is identical to the alcohol-soluble vitamin E antagonist present in kidney beans and soybeans remains to be demonstrated.

4. *Peas*

Sanyal (1953) reported that the pea, *Pisum sativum,* also contains a factor that interferes with the effectiveness of vitamin E for the prevention of fetal resorption in rats. The chemical nature of this factor does not appear to have been further investigated.

D. Antivitamin K

It is well known that farm animals, particularly cattle, that have eaten spoiled sweet clover (*Melilotus officinalis*) develop a fatal hemorrhagic condition known as sweet clover disease (Kingsbury, 1964). Link (1959) reviewed the classical studies which led to the identification and synthesis of the active principle, dicumerol, 3,3′-methylene-bis(4-hydroxycoumarin) (Fig. 13). This compound, when administered orally to man or animals, lowers the prothrombin level of the blood and thus leads to a breakdown in the chain of reactions involved in the clotting mechanism of the blood system. Dicumerol, in some still unknown way, interferes with the role that vitamin K plays in the production of thrombin by the

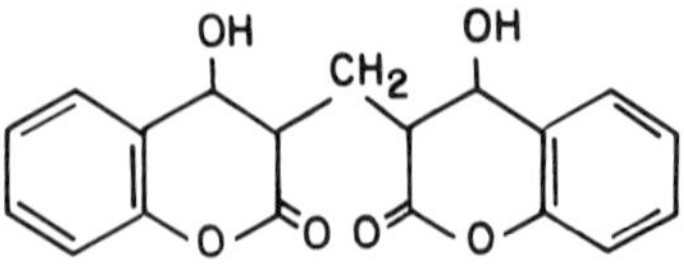

OH CH2 OH
O O O O

FIG. 13. Structure of dicumerol, the hemorrhagic principle responsible for the sweet clover disease in cattle.

liver (Collentine and Quick, 1951). Dicumerol has found important medical use as an anticoagulant and has served as a model for the synthesis of a related derivative known as "warfarin," which is widely used as a rat poison.

E. Antithiamine

The best known example of a naturally occurring factor in plants which is antagonistic toward thiamine is a thiaminase that occurs in bracken fern, *Pteridium aquilinum*. The carcinogenic properties of this plant have been considered in some detail in Chapter 11, Section V. The thiaminase present in the Australian nardoo fern (*Marsilea drummondii*) has been recently purified and partially characterized (McCleary and Chick, 1977). These plants have little value as forage, and losses in livestock, primarily horses and cattle, have been largely the result of animals eating this fern when other green forage is not available (Garner, 1961; Kingsbury, 1964).

Antithiamine activity has also been detected in many diverse plants, including ragi (*Eleusine coracana*), mung bean (*Phaseolus aureus*), rice bran, mustard seed, cottonseed, flax seed, blackberries, blueberries, black currants, beets, brussels sprouts, and red cabbage (Bhagvat and Devi, 1944a,b; Hilker, 1968). In only a few cases has the chemical nature of the antithiamine factor been identified. For example, Bhattacharya and Chaudhuri (1973) isolated an antithiamine factor from mustard seed, *Brassica juncea,* which they identified as the methyl ester of 3,5-dimethoxy-4-hydroxycinnamic acid (methyl sinapate). A related substance, 3,5-dimethoxysalicylic acid, which also had antithiamine properties, was also isolated from the cottonseed (*Bombex malabericum*) (Sarkar and Chaudhuri, 1976). (See Fig. 14 for structures.) The isolation of an antithiamine factor from rice bran which contains a benzene nucleus with *o*-dihydroxy groups was recently reported (De and Chaudhuri, 1976). This substance inhibited the utilization of ribose 5-phosphate by the thiamine pyrophsophate–transketolase system of human erythrocytes.

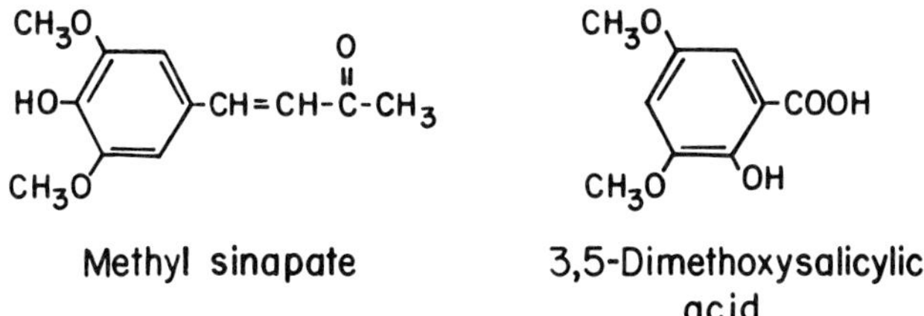

FIG. 14. Antithiamine factors, methyl sinapate and 3,5-dimethoxysalicylic acid, present in mustard seed and cottonseed, respectively.

F. Antiniacin

Belavady and Gopalan (1965) reported that "black tongue," a characteristic symptom of niacin deficiency in dogs, was observed when dogs were fed *Sorghum vulgare* (called jowar in India). This observation was quite unexpected since this plant is not particularly low in tryptophan and contains a fair quantity of nicotinic acid. Subsequent experiments (Belavady *et al.*, 1967) suggested that the high amount of leucine that is found in jowar may be responsible for the production of this niacin deficiency in dogs. These experiments may be of significance in human nutrition since pellagra is endemic in those parts of India where people consume large amounts of jowar.

G. Antipyridoxine

Flax or linseed (*Linum usitatissimum*) meal is considered a poor source of protein for chicks (Heuser *et al.*, 1946), but considerable improvement may be effected by extracting the meal with water and autoclaving for 30 min or by supplementation with pyridoxine (McGinnis and Polis, 1946; Kratzer, 1947; Kratzer *et al.*, 1954). Klosterman and co-workers (1967) reported the isolation of a pyridoxine antagonist from flaxseed, which was identified as a 1-amino-D-proline (Fig. 15) that occurs naturally in combination, via a peptide linkage, with glutamic acid. This peptide was given the name of linatine. 1-Amino-D-proline was actually four times as toxic as linatine when injected into chicks, and its toxicity could be effectively counteracted by simultaneous injections of pyridoxine. Sasaoka *et al.* (1976) carried out similar studies with rats and noted that 1-amino-D-proline causes marked changes in amino acid metabolism. These effects are no doubt related to its effect on enzyme systems that require pyridoxal phosphate as a cofactor. It has in fact been demonstrated that 1-amino-D-proline forms a hydrazone with pyridoxal phosphate (Klosterman *et al.*, 1967; Klosterman, 1974).

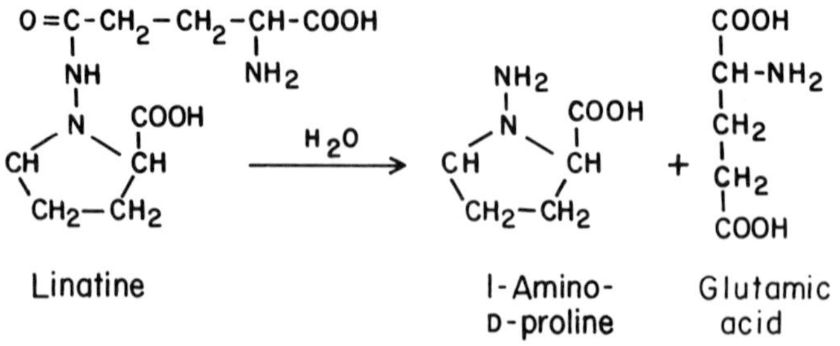

Linatine $\xrightarrow{H_2O}$ 1-Amino-D-proline + Glutamic acid

FIG. 15. Structure of 1-amino-D-proline, the antipyridoxine factor of linseed (*Linum usitatissimum*) and its natural precursor, linatine.

H. Antivitamin B_{12}

Since soybeans and other legumes are devoid of vitamin B_{12}, it is not surprising that diets containing soybean protein require supplementation with this vitamin in order to support normal growth of rats. Vitamin B_{12} supplementation, however, improves the growth of animals receiving raw soybeans to a greater extent than similar supplementation of diets containing heated soybeans (Baliga and Rajagopalan, 1954). If the excretion of methylmalonic acid, following a load of propionate, is used as an index of the vitamin B_{12} status of an experimental animal such as the rat, vitamin B_{12} is again much more effective in reducing the urinary excretion of methylmalonic acid on a raw soybean diet than on a diet containing heated soybeans (Edelstein and Guggenheim, 1969, 1970a,b; Williams and Spray, 1973). These results were interpreted to indicate that unheated soybeans contain a heat-labile substance that somehow accentuates the requirement for vitamin B_{12}. In addition, the vitamin B_{12} formed by the intestinal flora is, for some reason, less available for absorption when synthesized by rats subsisting on unheated soybeans (Edelstein and Guggenheim, 1970b).

VII. ANTIENZYMES

For general reviews on the subject of enzyme inhibitors in foods, the reader is referred to articles by Pressey (1972) and Whitaker and Feeney (1973).

A. Cholinesterase Inhibitors

Those common foods that are known to contain substances that inhibit the *in vitro* activity of cholinesterase are listed in Table I. It will be noted from this table that these inhibitors are found in various parts of the plant. The fact that different solvents are required for the extraction of the active substance from various plants would indicate that the active principles are not chemically identical. Little is actually known about the chemistry of these natural inhibitors of cholinesterase. In fact, the only cholinesterase inhibitor that has been identified with any degree of certainty is the glycoside, solanine (Fig. 16), which is present in highest concentration in the sprouts and skin (especially when green) of the potato. Although human fatalities due to the consumption of green potatoes have been reported from time to time (Hanson, 1925; Harris and Cockburn, 1918), proof that solanine was the causative agent is largely indirect. Values of 38–45 mg solanine per 100 gm have been obtained for potatoes implicated in fatal cases of human poisoning (Harris and Cockburn, 1918; Abbot *et al.*, 1960). These values may be compared to normal values of 3–6 mg per 100 gm, usually obtained with potatoes. The clinical symptoms described for isolated cases of potato poisoning in man and farm animals generally involve gastrointestinal

TABLE I

PLANTS DISPLAYING INHIBITION OF CHOLINESTERASE

Common name	Botanical name	Tissue tested	Reference
Broccoli	*Brassica oleracea*	Head[a]	Orgell *et al.* (1959)
Sugar beet	*Beta vulgaris*	Root[b]	Menn *et al.* (1964)
Asparagus	*Asparagus officinalis*	Stem[a]	Rohde (1960)
Eggplant	*Solanum melongena*	Fruit[a]	Orgell *et al.* (1959)
Potato	*Solanum tuberosum*	Tuber[a]	Orgell *et al.* (1959)
		Leaf[a]	Orgell (1963)
Turnip	*Brassica rapa*	Whole plant[c]	Crosby (1966)
Radish	*Raphanus sativus*	Whole plant[c]	Crosby (1966)
Celery	*Apium graveolens*	Whole plant[c]	Crosby (1966)
Carrot	*Daucus carota*	Whole plant[c]	Crosby (1966)
Horseradish	*Amoracia lapathifolia*	Leaf[a]	Orgell (1963)
Orange	*Citrus sinensis*	Fruit[a]	Menn *et al.* (1964)
Apple	*Malus sylvestris*	Fruit[a]	Menn *et al.* (1964)
Raspberry	*Rubus idaeus*	Leaf[a]	Beckett *et al.* (1954)

[a] Aqueous extract examined.
[b] Benzene extract examined.
[c] Ethyl acetate extract examined.

disturbances and certain neurological disorders (Willimot, 1933; Sapeika, 1969). It is significant that solanine is not destroyed by cooking (Baker *et al.*, 1955), and the poisoning of livestock has sometimes been observed even with cooked potatoes (Kingsbury, 1964).

An experimental variety of potatoes, Lenape, was found to have excellent properties for making potato chips (Akeley *et al.*, 1968) but was unacceptable because of an unusually high glycoalkaloid content (Zitnak and Johnson, 1970). Because of this varietal problem, all existing and newly developed varieties of potatoes are now monitored for alkaloid content. It is generally accepted that 20

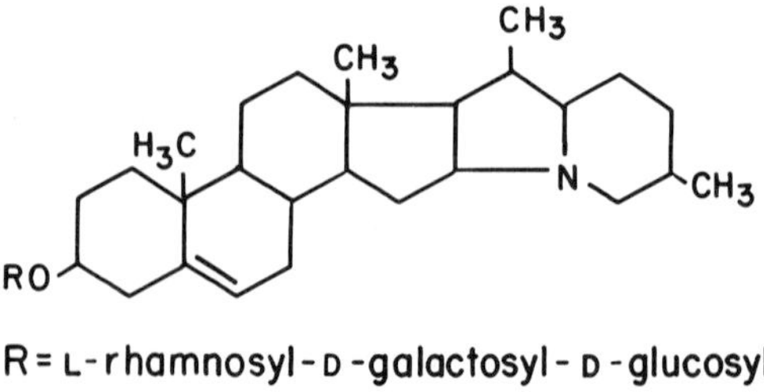

FIG. 16. Structure of solanine, an inhibitor of cholinesterase which is present in the potato (*Solanum tuberosum*).

mg of solanine per 100 gm fresh weight of potato tissue is the upper limit of safety (Bömer and Mattis, 1924; Oslage, 1956), or, to put it somewhat differently, an oral dose of 3 mg/kg body weight is generally considered a toxic level for man (Kuć, 1975).

For a more detailed treatment of the glycoalkaloids of potatoes, the reviews by Kuć (1975) and Jadhav and Salunkhe (1975) are recommended.

B. Amylase Inhibitors

1. Wheat

An inhibitor of α-amylase in aqueous extracts of wheat was first described by Kneen and Sandstedt (1943, 1946). This inhibitor was a nondialyzable, heat-labile protein that was active against salivary, pancreatic, and bacterial α-amylases (Militzer *et al.*, 1946), as well as α-amylases from various insects (Applebaum, 1964; Applebaum *et al.*, 1964; Applebaum and Konijn, 1965). A number of investigators have since demonstrated that over two-thirds of the albumin fractions of wheat is composed of multiple protein components capable of inhibiting α-amylases of diverse origin (Shainkin and Birk, 1970; Saunders and Lang, 1973; Silano *et al.*, 1973; Petrucci *et al.*, 1974,1976; Deponte *et al.*, 1976; Buonocore *et al.*, 1976; O'Donnell and McGeeney, 1976). Present evidence, recently reviewed by Buonocore *et al.* (1977), would indicate that these inhibitors can be classified into at least three major groups on the basis of molecular weight: 60,000, 24,000, and 12,500. All of these inhibitors are capable of inhibiting the α-amylase of the yellow mealworm, *Tenebrio molitor,* but only the one with a molecular weight of 24,000 can inhibit salivary and pancreatic amylases. All of these fractions, however, are still highly hetergeneous and can be resolved into multiple components. There is some indication that the various albumin fractions that exhibit α-amylase inhibition are actually composed of similar subunits having a molecular weight of about 12,500, which may be derived from a common ancestral gene (Deponte *et al.*, 1976).

The mechanism of interaction of the amylase inhibitors has received only limited attention to date. Buonocore *et al.* (1976) showed that the 24,000 molecular weight inhibitor inhibits the α-amylase of *T. molitor* in an uncompetitive fashion to form a 1:1 molar complex. The K_i was determined to be 3×10^{-7} *M*.

The fact that the wheat amylase inhibitors are so effective toward insect amylases would suggest that they are part of a defense mechanism of the seed against insect attack. Silano *et al.* (1975) have in fact shown that most of the insects that attack wheat grain and flour seem to have very high amylase activity which is inhibited by wheat albumins, whereas those insect species that do not normally feed on wheat have low amylase activity which is resistant to inhibition by wheat albumins. As far as the significance of the wheat amylase inhibitors in animal nutrition is concerned, it was shown with rats that these inhibitors can cause a marked decrease in the availability of starch (Lang *et al.*, 1974). It was also demonstrated in human subjects that the wheat amylase inhibitors can reduce hyperglycemia and hyperinsulinemia in diabetic patients (Puls and Keup, 1973). Macri *et al.* (1977), however, showed that the administration of amylase inhibitors, which had been enclosed in cellulose-coated microgranules in order to make them resistant to peptic digestion, to chickens depressed growth and caused pancreatic hypertrophy. This would cast doubt on the desirability of using oral administration of amylase inhibitors to control hyperglycemia in man. Strumeyer (1972) suggested that amylase inhibitors may be responsible for the sensitivity to wheat flour in coeliac disease by depressing the activity of an already deficient supply of pancreatic amylase, but this hypothesis was disputed by Auricchio *et al.* (1974).

Despite their susceptibility to heat inactivation, the wheat amylase inhibitors may persist through bread baking, being found in large amounts in the center of loaves (Bessho and Kurosawa, 1967) and in some wheat-based breakfast cereals (Marshall, 1975). In view of their possible action *in vivo,* the presence of amylase inhibitors in wheat-based food products should be considered undesirable, and their destruction during processing should be ensured. In particular, wheat amylase inhibitors should be excluded from the diet of infants and patients in which impaired proteolysis may not be adequate to destroy these inhibitors during digestion.

2. *Other Cereals*

α-Amylase inhibitors of a protein nature have also been reported to be present in oats (Elliot and Leopold, 1953) and rye (Kneen and Sandstedt, 1946; Strumeyer, 1972; Marshall, 1977). Marshall (1977) succeeded in isolating from extracts of rye an inhibitor which, in contrast to the wheat inhibitor that inhibits salivary α-amylase at least as strongly as the pancreatic enzyme (Strumeyer, 1972; O'Donnell and McGeeney, 1976), was at least 100 times more active toward salivary α-amylase than toward the pancreatic enzyme. This inhibitor, however, was not further characterized.

An inhibitor of the α-amylases of barley, wheat, and bacteria was also found in the bran and germ of certain varieties of sorghum (Miller and Kneen, 1947; Maxson *et al.*, 1973). In contrast to the amylase inhibitor of wheat and rye, the one from sorghum was dialyzable and heat stable. Studies by Sturmeyer and Malin (1969) on the chemical nature of the sorghum inhibitor suggest a carbohydrate-containing polyphenolic substance having a molecular weight of about 5000. The extent to which this substance is capable of interfering with the digestive processes of animals consuming sorghum remains to be assessed.

3. Legumes

Although the presence of an inhibitor of pancreatic amylase in navy beans (*Phaseolus vulgaris*) was first reported by Bowman (1945), only in recent years have attempts been made to purify this inhibitor (Hernandez and Jaffé, 1968; Marshall and Lauda, 1975; Powers and Whitaker, 1977). In a survey of 20 different legumes, 12 varieties of *Phaseolus* were found to have high levels of inhibitor activity toward porcine pancreatic amylase, whereas samples of lentils, chick-peas, and cowpeas showed only slight activity toward this enzyme (Jaffé *et al.*, 1973). No inhibitory activity was displayed toward plant or bacterial amylases by any of these legumes. The inhibitor purified from kidney beans by Marshall and Lauda (1975) proved to be a glycoprotein having a molecular weight of 45,000–50,000 and containing about 10% carbohydrate. It reacted noncompetitively with porcine pancreatic amylase to form a stoichimoetric 1:1 complex with an extremely small dissociation constant, $K_i = 3.5 \times 10^{-11}\ M$ (Powers and Whitaker, 1977). Rather curiously, oxidation of the carbohydrate moiety led to a complete loss of inhibitory activity.

The detection of undigested starch in rats consuming a diet containing raw beans with high amylase inhibitor activity led Jaffé and Vega (1968) to suggest that this factor may be active *in vivo*. If such an effect is indeed operative *in vivo*, it apparently does not affect growth, since Savaiano *et al.* (1977) did not observe any inhibition of the growth of rats fed a preparation of α-amylase inhibitor purified from the red kidney bean.

4. Other Plants

Amylase inhibitors have also been reported to be present in mango fruit, (Mattoo and Modi, 1970), taro root (Rao *et al.*, 1967,1970), and potatoes (Nilova and Guseva, 1975).

C. Invertase Inhibitor

Inhibitors of invertase have been found in a wide variety of plants, including the tuber of the potato (Schwimmer *et al.*, 1961; Pressey, 1966,1967; Pressey and Shaw, 1966), red beets and sugar beets (Pressey, 1968), sweet potato (Pressey, 1968; Winkenbach and Matile, 1970; Matsushita and Uritani, 1976), and corn (Jaynes and Nelson, 1971). The potato invertase inhibitor has been highly purified and characterized with respect to molecular weight (17,000), and amino acid composition, kinetic properties, and heat stability (Pressey, 1967). The invertase inhibitor isolated from red beet and sugar beets has properties similar to those of the potato inhibitor (Pressey, 1968), although the inhibitor from sweet potato appears to have a somewhat higher molecular weight (19,500)

(Matsushima and Uritani, 1976). Unlike many of the enzyme inhibitors present in plants, the invertase inhibitor is active against endogenous invertase and is believed to be responsible for the accumulation of sugar that occurs during the storage of potatoes (Pressey, 1972).

D. Other Enzyme Inhibitors

Brief mention should be made of other enzyme inhibitors that have been detected in plant tissue. Inhibitors of peroxidase and catalase have been partially purified from green mangoes and bananas (Mattoo and Modi, 1970). An inhibitor of microbial polygalacturonase was detected in extracts of onion, green bean, pepper, white cabbage, and cucumbers (Bock *et al.*, 1975), and an inhibitor of endogenous polygalacturonase was isolated from the green avocado (Reymond and Phaff, 1965). A derivative of chlorogenic acid in sunflower seeds was identified as an inhibitor of arginase (Muszynska and Reifer, 1970). An inhibitor of pancreatic lipase, fungal lipases, and castor bean lipase was isolated from soybeans (Mori *et al.*, 1973; Satouchi *et al.*, 1974; Satouchi and Matsushita, 1976), and an inhibitor of pancreatic and rice bran lipases was isolated from the green pepper (Kim *et al.*, 1977).

VIII. METAL-BINDING CONSTITUENTS

A. Phytate

Phytate, a cyclic compound (inositol) containing six phosphate groups (Fig. 17), is a common constituent of plant tissue. Because of its ability to chelate di- and trivalent metal ions such as calcium, magnesium, zinc, copper, and iron to form insoluble complexes that are not readily absorbed from the intestinal tract, it has been held responsible for the commonly observed interference that many plant sources of protein have on the availability of dietary minerals. The nutritional significance of phytate in the diet of man has assumed great importance in recent years because of its implication in the mineral deficiency so prevalent in those parts of the world, such as Iran, that are reliant on cereal proteins

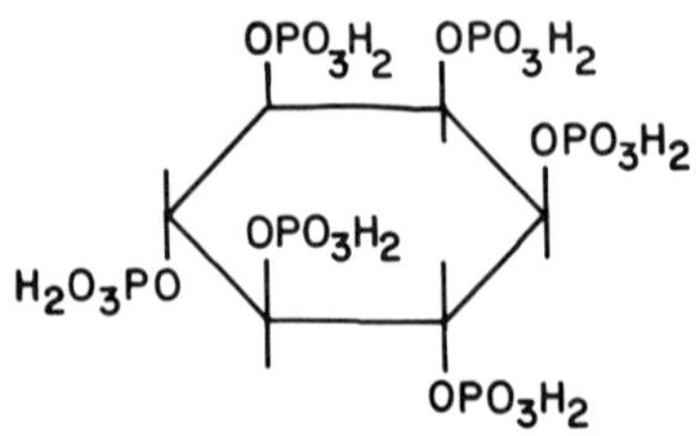

OPO_3H_2 OPO_3H_2 OPO_3H_2 OPO_3H_2 H_2O_3PO OPO_3H_2

FIG. 17. Structure of phytic acid (myoinositol 1,2,3,4,5,6-hexadihydrogen phosphate).

(Reinhold, 1971; Reinhold *et al.*, 1975). No attempt will be made here to review the vast amount of literature dealing with this subject. Several reviews on this subject (Liener, 1969,1977; Oberlas, 1973) are available to the interested reader.

B. Oxalate

Because certain plants, such as rhubarb (*Rheum rhaponticum*) and spinach (*Spinacea oleracea*), are known to contain rather high levels of oxalic acid, which has the ability to bind calcium, the general belief has arisen that such plants may be harmful by interfering with calcium metabolism. Oke (1969) and Fassett (1973) critically evaluated the literature pertaining to this problem and came to the conclusion that there is in fact very little danger associated with the ingestion of oxalate-containing plants. Since man shows a remarkable ability to adapt to very low levels of calcium, it would, according to Fassett, ''require a rather impossible combination of circumstances, a very high intake of oxalate-containing food plus a simultaneously low calcium and vitamin D intake over a prolonged period, for chronic effects to be noted.'' In those few instances in which ill effects and sometimes death have followed the ingestion of rhubarb, there has been little evidence to substantiate the claim that oxalate was the causative factor. Fassett is of the opinion that more attention should be given to the possible role of the toxic anthraquinone glycosides, which are known to be present in rhubarb and other related species of plants.

IX. TANNINS

As defined by Singleton and Kratzer (1973) in their review, a tannin is any polyphenolic substance that has a molecular weight greater than 500. The tannins may be classified as hydrolyzable, that is, degradable by enzymes to yield a sugar residue and a phenolcarboxylic acid, and condensed tannins, which are polymeric flavoids having the structure shown in Fig. 18. Among the cereals, certain selected strains of sorghum contain high levels of condensed tannins (about 5%), which make these strains of barley resistant to birds. These tannins are likewise believed to be responsible for the growth depression that has been frequently observed when sorghum is fed to experimental animals (Jambunathan and Mertz, 1973; Maxson *et al.*, 1973; Rostagno *et al.*, 1973a,b; Armstrong *et al.*, 1974,1975; Featherstone and Rogler, 1975; Nelson *et al.*, 1975; Featherstone, 1976; Martin-Tanguy *et al.*, 1976). Tannins have been implicated on the basis of three lines of evidence: (1) growth depression more or less parallels the tannin content of different strains of sorghum; (2) removal of the tannins by extraction serves to enhance the nutritive properties of sorghum protein; and (3) the addition of isolated tannins to animals diets devoid of tannins causes growth depression. There appears to be little doubt that the growth depression caused by

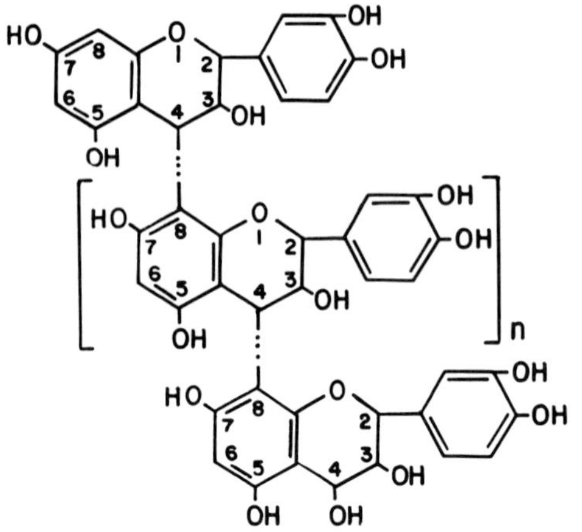

FIG. 18. Structure of condensed tannins.

sorghum tannins is due to an adverse effect on protein and dry matter digestibility (Glick and Joslyn, 1970; Harris *et al.*, 1970; Mitjavila, 1973; Schaffert *et al.*, 1974; Miller *et al.*, 1972; Rostagno *et al.*, 1973b). This effect may be related to the fact that tannins interfere with the digestive action of trypsin and α-amylase either by binding the enzymes themselves or by the binding of the dietary protein into an indigestible form (Feeney, 1969; Tamir and Alumot, 1969). Tannins can also complex with vitamin B_{12}, causing a decrease in the absorption of this vitamin in rats (Carrera *et al.*, 1974).

Condensed tannins have also been found in a number of other plants where they may exert an adverse biological response when included in the diet. The growth depression obtained in rats with carobs (*Ceratonia siliqua*) has been attributed to the inhibitory effect of tannins in protein digestion (Tamir and Alumot, 1969,1970). The growth of chicks may be adversely affected by the tannins present in rapeseed meal by causing a depression in metabolizable energy (Yapar and Clandinin, 1972). Over one-half of the heat-labile growth inhibitor responsible for the poor utilization of faba beans has been attributed to the presence of condensed tannins located primarily in the hull fraction of the bean (Marquardt *et al.*, 1977; Ward *et al.*, 1977). Tannins are also believed to be responsible for decreased laying rate, decreased efficiency of food utilization, and increased mortality of laying hens (Guillaume and Belec, 1977) and ducklings (Martin-Tanguy *et al.*, 1977) fed faba beans. Tannins also adversely affect the nutritive value of finger millet (*Eleusine coracana* Gaertn.) by reducing the digestibility of the glutelin fraction of the protein (Ramachandra *et al.*, 1977).

Chlorogenic acid (Fig. 19) can in a sense be considered related to tannins in that it is a polyphenolic compound. It is present in sunflower seed meal to the extent of 1.2% (Milić *et al.*, 1968) to 2.7% (Cater *et al.*, 1972). When fed to

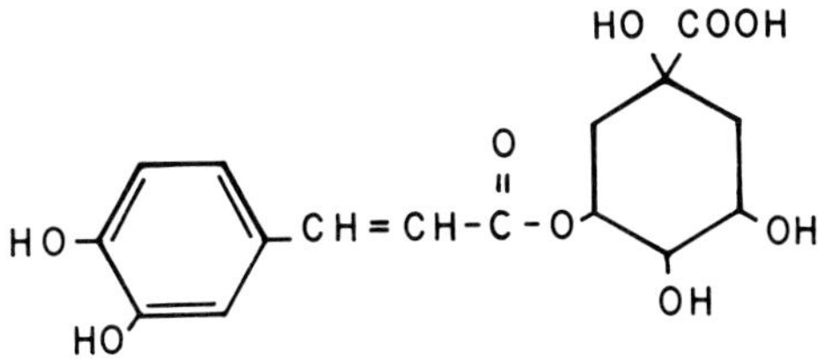

FIG. 19. Structure of chlorogenic acid.

mice at a level of 3% of the diet, growth and efficiency of feed utilization were depressed (Delić *et al.*, 1972). The exact mechanism whereby chlorogenic acid exerts its deleterious effect is not known, although this compound has been shown to inhibit the enzymatic activity of proteinases, amylases, and lipases (Delić *et al.*, 1972). The harmful effect of chlorogenic acid can also be counteracted by dietary supplementation with compounds having methyl donor groups, i.e., methionine and choline (Singleton and Kratzer, 1969; Delić *et al.*, 1975).

X. FLATUS-PRODUCING FACTORS

One of the major factors limiting the human consumption of legumes is their ability to produce gas in the gastrointestinal tract, also referred to as flatulence. This subject has been recently reviewed in several papers, where more detailed information can be found (Calloway *et al.*, 1971; Hellendorn, 1969,1973; Olson *et al.*, 1975; Rackis, 1975). It has been demonstrated by numerous investigators using rats and human subjects that the ingestion of various kinds of beans leads to the production of large quantities of intestinal gas composed largely of carbon dioxide, hydrogen, and, to a lesser extent, methane (Hedin and Adachi, 1962; Steggerda and Dimmick, 1966; Kurtzman and Halbrook, 1970; Rackis *et al.*, 1970a,b; Murphy *et al.*, 1972; Olson *et al.*, 1975; Wagner *et al.*, 1976). Such studies have clearly implicated the oligosaccharides raffinose and stachyose as the causative factors of flatus. As shown in Fig. 20, these oligosaccharides are related by having one or two α-D-galactopyranosyl groups attached to sucrose via α-1,6-galactosidic linkages. Owing to the absence of enzymes in the human intestinal mucosa that are capable of hydrolyzing this linkage (Gitzelmann and Auricchio, 1965), the intact oligosaccharides accumulate in the lower intestine, where they undergo fermentation by anaerobic bacteria (Rockland *et al.*, 1969; Rackis *et al.*, 1970a). The gases so produced are responsible for the characteristic features of flatulence, namely, nausea, cramps, diarrhea, abdominal rumbling, and the social discomfort associated with the ejection of rectal gas.

Such traditional soybean foods as tofu (soybean curd) and tempeh have little

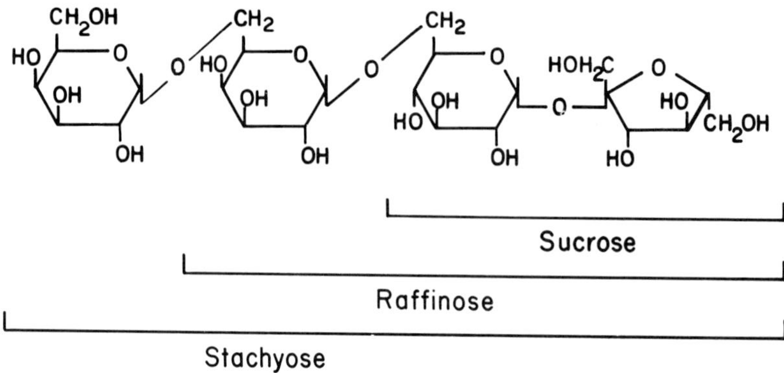

FIG. 20. Oligosaccharides believed to be responsible for flatulence-producing properties of legumes.

flatus activity (Calloway *et al.*, 1971). In the case of tofu, oligosaccharides are probably eliminated during the course of its preparation, and, in the case of tempeh, enzymes produced by the mold (*Rhizopus*) during fermentation probably hydrolyze the oligosaccharides. As might be expected because of their low carbohydrate content (<1%), soy protein isolates are devoid of flatus activity (Rackis, 1974). It follows that textured meat analogues made from soy protein isolates are most likely free of flatus activity.

Cooking has proved to be effective in causing some reduction in the gas-promoting activity of navy beans (Kakade and Borchers, 1967) and green gram, chick-pea, and cowpea (Venkataraman and Jaya, 1975). Toasted soybean meal still retains much of its flatus-producing activity (Rackis *et al.*, 1970a), although cooking soybeans in water causes a significant reduction in flatulence-related oligosaccharides (Ku *et al.*, 1976). Germination also has the effect of reducing the oligosaccharide content of soybeans (Kim *et al.*, 1973; Hun, 1974; Snauwert and Markakis, 1976) but has little or no effect on intestinal gas production by green gram, chick-pea, and cowpea (Shurpalekar *et al.*, 1973; Venkatamaran and Jaya, 1975). A number of studies have demonstrated that the treatment of legumes with enzymes of microbial origin (Sugimoto and van Buren, 1970; Sherba, 1970; Calloway *et al.*, 1971; Hansen, 1976), or by allowing the beans to undergo autolysis (Becker *et al.*, 1974), cause a significance reduction in flatulence activity. Cultivars of soybeans have been found which have a very low content of raffinose and stachyose (Hymowitz and Collins, 1974), suggesting that it may be possible to reduce the flatulence properties of legumes by selective breeding.

Legumes are not the only foods that contain flatus-producing factors. Intestinal gas production is also increased following the consumption of certain wheat cereals and the wheat milling fractions bran, red dog, and shorts (Hickey *et al.*,

1972a). Certain fruits and fruit juices, especially raisins, bananas, apple juice, and grape juice, have been shown to increase human intestinal gas production (Hickey *et al.*, 1972b). In the case of these nonleguminous foods, however, it remains to be established whether oligosaccharides are indeed the factors responsible for flatulence.

XI. OTHER PLANTS WITH TOXIC CONSTITUENTS

A. *Indigofera endecaphylla (spicata)* (Trailing or Creeping Indigo)

The hepatoxic principle of this plant, indospicine, has been discussed elsewhere in this chapter (see Section V,F). The identity of indospicine as the agent responsible for the hepatotoxic effects of *I. spicata* has been obscured by the presence of 3-nitropropionic acid (Fig. 21), which is also toxic to animals (Morris *et al.*, 1954; Cooke, 1955). Hutton *et al.* (1958a,b), however, reported that 3-nitropropionic acid could not be responsible for the liver damage produced by feeding this plant to animals since synthetic 3-nitropropionic acid had no hepatoxic effect on rabbits when force fed in amounts comparable to those found in the plant. The ingestion of 3-nitropropionic acid is believed to be toxic by virtue of its ability to inhibit cellular respiration (Lutton and Kopac, 1971). More recently, Alston *et al.* (1977) showed that the specific site of action is the enzyme succinate dehydrogenase, which 3-nitropropionate inhibits irreversibly by forming a covalent adduct with the active-site flavin moiety of this enzyme.

B. *Stizolobium deeringanium* (Velvet Bean)

The velvet bean (*Stizolobium deeringianum*) is grown as a soil-improving crop in the southern states and is often used as a livestock feed. Sure and Read (1921) observed that, although raw velvet beans were not particularly toxic to farm animals (Kingsbury, 1964), they were injurious to young rats when constituting 40% of the total ration. Autoclaving the seeds destroys most of the growth inhibitory activity of the beans (Borchers and Ackerson, 1950; Waterman and Jones, 1921; Read and Sure, 1923). Even the protein isolated from the velvet bean failed to support the growth of rats unless subjected to heat treatment (Finks and Johns, 1921). Experiments with chickens indicated the presence of some factor that inhibited growth and decreased the egg production of chickens (Harms *et al.*, 1961). The exact nature of the toxic constituent of the velvet bean remains

$$O_2N-CH_2-CH_2-COOH$$

FIG. 21. Structure of 3-nitropropionic acid, a toxic compound present in *Indigofera endecaphylla* (spicata).

unknown but could be possibly related to the known presence of dopa (see Section V,C).

Because of its prolific growth and resistance to disease the velvet bean has attracted experimental investigation to determine whether it would prove an acceptable food for man. Even boiled for an extended period of time, the beans were found to be unpalatable and produced symptoms of nausea and discomfort (Miller *et al.*, 1925). Despite these undesirable characteristics, the velvet bean has been suggested as a substitute for the broad bean, which is a stable food in some parts of the Mediterranean Basin (Kader, 1960).

C. Snake Root

A disease known as milk sickness has afflicted certain segments of the population in this country for over 150 years (see review by Hartman *et al.*, 1963; Kingsbury, 1964). This disease, which is frequently fatal (Lincoln's mother is said to have died of this disease), has been almost always traced to the consumption of milk from cows that have fed on the snake root (*Eupatorium rugosum*). The disease afflicting the cows themselves is known as trembles and generally occurs as a consequence of restricted pasturage when animals seek out this plant. The causative principle of milk sickness is attributed to an unsaturated alcohol, $C_{16}H_{22}O_3$, which has been given the name of tremetol (Couch, 1929).

D. Carrots

Crosby and Aharonson (1967) reported the isolation of a toxic substance from the ordinary carrot, *Daucus carota*. This toxic principle, called carotatoxin, was shown to be a polyacetylenic alcohol (Fig. 22). About 2 mg of carotatoxin could be isolated from 1 kg of carrot root, and, when injected into mice, it produced pronounced neurotoxic symptoms with an approximate LD_{50} value of 100 mg/kg. Based on these figures, it is extremely doubtful that the normal consumption of carrots would ever lead to toxic effects in human beings. It is interesting, however, that other highly toxic acetylene derivatives have been isolated from plants

$$(C_9H_{17})-C\equiv C-C\equiv C-CH_2-\underset{}{\overset{OH}{\overset{|}{C}}}H=CH_2$$

Carotatoxin

$$CH_3-CH_2-CH=CH-CH=CH-C\equiv C-C\equiv C-CH=CH-CH_3$$

Aethusin

FIG. 22. Structure of carotatoxin and aethusin, toxic agents present in carrots and fool's parsley, respectively.

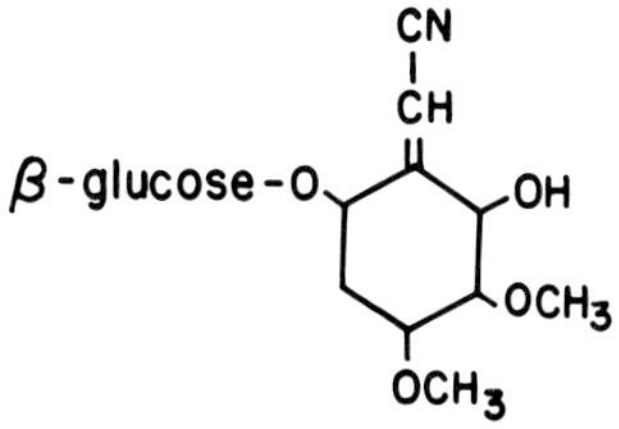

FIG. 23. Structure of toxic principle of the jojoba plant.

belonging to the same family, Umbelliferae, to which the carrot belongs (Clarke *et al.*, 1949; Anet *et al.*, 1953; Bohlmann *et al.*, 1960,1961). One of these, aethusin (Fig. 22), was isolated from fool's parsley (*Aethusa cynapium*), which was reported to have been responsible for many instances of human poisoning (Bohlmann *et al.*, 1960).

E. Jojoba Plant

The jojoba plant, *Simmondsia californica,* is an evergreen plant that grows wild in the arid regions of southwestern United States. Its economic importance lies in the fact that the female plants produce seeds that contain a liquid wax comparable in properties to sperm whale oil. The meal remaining after extraction of the oil contains about 27% protein but is very poorly utilized by animals. The toxic principle responsible for its poor growth-promoting property has been isolated and identified as a glucoside of a multiple-substituted cyclohexane system bearing a cyanomethylene substituent (Booth *et al.*, 1974). This compound has been given the name of simmondsin, and its structure is shown in Fig. 23. Its mechanism of action remains to be elucidated. Detoxification of the jojoba meal can be achieved by exposure of the meal to ammonia in a tightly sealed container (Elliger *et al.*, 1976).

F. Mint Plant

Perilla frutescens, known in the United States as purple mint plant, is widely distributed in this country and in Japan. In the latter country, intact leaves may be used as condiments of flavoring agents in a variety of human foods and are often included in preparations of tempura. The seeds are also used to flavor various foods, including pickled ginger. The volatile oil from the leaves is commonly used in Japan as a drug with reputed therapeutic properties. The seed cake which remains has been used as a high-protein animal feed. Wilson *et al.* (1974) reported the isolation of a ketone-substituted furan (see Fig. 24) from *P. frutescens,* which caused severe pulmonary lesions in mice, rats, and sheep. This compound is believed to be responsible for the acute pulmonary emphysema

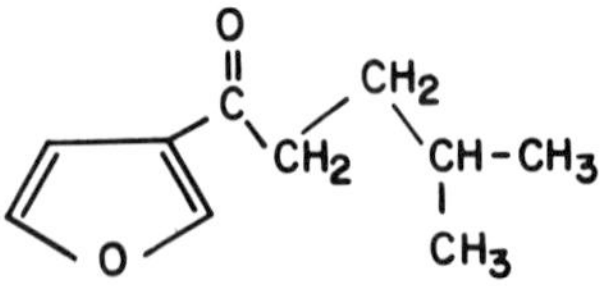

FIG. 24. Structure of 3-ketone-substituted furan, a potent lung toxin from the mint plant (*Perilla frutescens*).

observed in cattle ingesting this plant (Peterson, 1965; Selman *et al.*, 1976), but, more importantly, use of this plant in human foods in Oriental countries should be questioned because of the obvious potential hazard to health.

REFERENCES

Abbott, D. C., Field, K., and Johnson, E. I. (1960). *Analyst* **85,** 375–376.

Akeley, R. V., Mills, W. R., Cunningham, C. E., and Watts, J. (1968). *Am. Potato J.* **45,** 142–145.

Alexander, G., and Watson, R. H. (1951). *Austr. J. Agric. Res.* **2,** 480–493.

Allman, D. W., Harris, R. A., and Green, D. E. (1967). *Arch. Biochem. Biophys.* **122,** 766–782.

Alston, T. A., Mela, L., and Bright, H. G. (1977). *Proc. Natl. Acad. Sci. U.S.A.* **74,** 3767–3771.

Andrews, R. S., and Pridham, J. B. (1965). *Nature (London)* **205,** 1213–1214.

Anet, E. F. L. J., and Ingles, D. J. (1976). *CSIRO Food Res. Q.* **36,** 28–31.

Anet, E. J. L. J., Lythgoe, B., Silk, M. H., and Trippett, S. (1953). *J. Chem. Soc.* pp. 308–322.

Anonymous (1976). *Nutr. Rev.* **34,** 349–350.

Anthony, M., and Lance, J. W. (1972). *Drugs* **3,** 153–158.

Applebaum, S. W. (1964). *J. Insect Physiol.* **10,** 897–903.

Applebaum, S. W., and Konijn, A. M. (1965). *J. Nutr.* **85,** 275–281.

Applebaum, S. W., Harpaz, I., and Bondi, A. (1964). *Comp. Biochem. Physiol.* **13,** 107–111.

Armstrong, W. D., Rogler, J. C., and Featherstone, W. R. (1974). *Poult. Sci.* **53,** 714–720.

Armstrong, W. D., Featherstone, W. R., and Rogler, J. C. (1975). *Poult. Sci.* **53,** 2137–2142.

Aronow, L., and Kerdel-Vegas, F. (1965). *Nature (London)* **205,** 1185–1186.

Auricchio, S., de Viria, B., de Angelis, L. C., and Silano, V. (1974). *Lancet* , 98.

Baker, L. C., Lampitt, L. H., and Meredith, L. C. (1955). *J. Sci. Food Agric.* **6,** 197–202.

Baliga, B. R., and Rajagopalan, R. (1954). *Curr. Sci.* **23,** 51–52.

Banerjee, G. C., and Hogue, D. E. (1975). *Indian J. Anim. Sci.* **45,** 551–553.

Becker, R., Olson, A. C., Frederick, D. P., Kon, S., Gumbmann, M. R., and Wagner, J. R. (1974). *J. Food Sci.* **39,** 766–769.

Beckett, A., Belthie, F. W., Fell, K. R., and Lockett, M. F. (1954). *J. Pharm. Pharmacol.* **6,** 785–794.

Belavady, B., and Gopalan, C. (1965). *Lancet* **iii**, 1220–1221.

Belavady, B., Madhadwan, T. V., and Gopalan, C. (1967). *Gastroenterology* **53,** 749–753.

Bell, E. A. (1972). *In* "Phytochemical Ecology" (J. B. Harborne, ed.), pp. 163–167. Academic Press, New York.

Bell, E. A. (1976). *FEBS Lett.* **64,** 29–35.

Bessho, H., and Kurosawa, S. (1967). *Eiyo To Shokuryo* **20,** 317–319; *Chem. Abstr.* **68,** 113474e (1968).

Bhagat, K., and Devi, P. (1944a). *Indian J. Med. Res.* **32,** 123–125.

Bhagat, K., and Devi, P. (1944b). *Indian J. Med. Res.* **32,** 131–133.

Bhattacharya, J., and Chaudhuri, D. K. (1973). *Biochim. Biophys. Acta* **343,** 211–214.

Bickoff, E. M., Booth, A. N., Lyman, R. L., Livingston, A. L., Thompson, C. R., and DeEds, F. (1957). *Science* **126,** 969–970.

Bickoff, E. M., Livingston, A. L., Hendrickson, A. P., and Booth, A. N. (1962). *J. Agric. Food Chem.* **10,** 410–414.

Blackwell, B., Marley, E., Price, J., and Taylor, D. (1967). *Br. J. Psychiatry* **113,** 349—353.

Blood, F. R., and Rudolph, G. G. (1966). *N. A. S.-N. R. C. Publ.* **1354,** 62–71.

Bock, W., Dongowski, G., Goebel, H., and Krause, M. (1975). *Nahrung* **19,** 441–416.

Bohlmann, F., Arndt, C., Bornowski, H., and Herbst, P. (1960). *Chem. Ber.* **94,** 981–987.

Bohlmann, F., Arndt, C., Bornowski, H., and Kleine, K. M. (1961). *Chem. Ber.* **94,** 958–967.

Bömer, A., and Mattis, H. (1924). *Z. Unters. Nahr.-Genussm. Gebrauchsgegenstaende* **47,** 97–102.

Booth, A. N., Bickoff, E. M., and Kohler, G. O. (1960). *Science* **131,** 1807–1808.

Booth, A. N., Elliger, C. A., and Waiss, A. C., Jr. (1974). *Life Sci.* **15,** 1115–1120.

Borchers, R., and Ackerson, C. W. (1950). *J. Nutr.* **41,** 339–345.

Borison, H. L., Pendleton, J., Jr., and McCarthy, L. E. (1975). *J. Pharmacol. Exp. Ther.* **190,** 327–333.

Bowman, D. E. (1945). *Science* **98,** 358.

Bradbury, R. B., and White, D. E. (1951). *J. Chem. Soc.* pp. 3447–3449.

Bruni, A., Contessa, A. R., and Luciani, S. (1962). *Biochim. Biophys. Acta* **60,** 302–311.

Buonocore, V., Poerio, E., Pace, W., Petrucci, T., Silano, V., and Tomasi, M. (1976). *FEBS Lett.* **67,** 202–206.

Buonocore, V., Petrucci, T., and Silano, V. (1977). *Phytochemistry* **16,** 811–820.

Butenandt, A., and Jacobi, H. (1933). *Hoppe-Seyler's Z. Physiol. Chem.* **218,** 104–112.

Calloway, D. H., Hickey, C. A., and Murphy, E. L. (1971). *J. Food Sci.* **36,** 251–255.

Carlson, C. W., McGinnis, J., and Jensen, L. S. (1964a). *J. Nutr.* **82,** 366–370.

Carlson, C. W., Saxena, H. C., Jensen, L. S., and McGinnis, J. (1964b). *J. Nutr.* **82,** 507–511.

Carrera, G., Mitjavila, S., and Derache, R. (1974). *C. R. Hebd. Seances Acad. Sci., Ser. D* **278,** 153–156.

Carter, M. W., Matrone, G., and Smart, W. G., Jr. (1955). *J. Nutr.* **55,** 639–645.

Cater, C. M., Cheyasuddin, S., and Mattil, K. F. (1972). *Cereal Chem.* **49,** 5–12.

Chang, L. T. (1960). *J. Formosan Med. Assoc.* **59,** 882–891.

Cheng, E. W., Yoder, L., Story, C. D., and Burroughs, W. (1955). *Ann. N. Y. Acad. Sci.* **61,** 652–659.

Christie, G. S., Madsen, N. P., and Hegarty, M. P. (1969). *Biochem. Pharmacol.* **18,** 693–702.

Christie, G. S., Wilson, M., and Hegarty, M. P. (1975). *J. Pathol.* **117,** 195–205.

Clarke, E. G. C., Kidder, D. E., and Robertson, W. D. (1949). *J. Pharm. Pharmacol.* **1,** 377–381.

Collentine, G. C., and Quick, A. J. (1951). *Am. J. Med. Sci.* **222,** 7–12.

Cooke, A. R. (1955). *Arch. Biochem. Biophys.* **55,** 114–120.

Couch, J. F. (1929). *J. Am. Chem. Soc.* **51,** 3617–3619.

Crawford, M. A. (1962). *Lancet* , 352–353.

Crosby, D. G. (1966). *N. A. S.-N. R. C., Publ.* **1354,** 112–116.

Crosby, D. G., and Aharonson, N. (1967). *Tetrahedron* **23,** 465–472.

Crounse, R. G., Maxwell, J. D., and Blank, H. (1962). *Nature (London)* **194,** 694–695.

Davies, J. P. H., and Gower, D. B. (1964). *Nature (London)* **203,** 310–311.

De, B. K., and Chaudhuri, D. K. (1976). *Int. J. Vitam. Nutr. Res.* **46,** 154–159.

Delić, I., Vučurevic, N., and Stajanović, S. (1972). Cited by Delić *et al.* (1975).

Delić, I., Vučurevic, N., and Stajanović, S. (1975). *Acta Vet. (Belgrade)* **25,** 115–119.

Deponte, R., Parlamenti, R., Petrucci, T., Silano, V., and Tomassi, M. (1976). *Cereal Chem.* **53,** 805–820.

Desai, I. D. (1966). *Nature (London)* **209,** 810.

Edelstein, S., and Guggenheim, K. (1969). *Isr. J. Med. Sci.* **5,** 415–417.
Edelstein, S., and Guggenheim, K. (1970a). *Br. J. Nutr.* **24,** 735–740.
Edelstein, S., and Guggenheim, K. (1970b). *J. Nutr.* **100,** 1377–1382.
Elliger, C. A., Waiss, A. C., Jr., and Booth, A. N. (1974). U.S. Patent 3,919,432; *Chem. Abstr.* **84,** 42257d (1976).
Elliot, B. B., and Leopold, A. C. (1953). *Physiol. Plant.* **6,** 65–68.
Ellmore, M. F., and Shaw, J. C. (1954). *J. Dairy Sci.* **37,** 1269–1272.
Emmel, M. W., and Ritchey, G. E. (1941). *J. Am. Soc. Agron.* **33,** 675–678.
Entman, M., and Bressler, R. (1967). *Mol. Pharmacol.* **3,** 333–337.
Fassett, D. W. (1973). *In* "Toxicants Occurring Naturally in Foods," pp. 346–362. Natl. Acad. Sci., Natl. Res. Counc., Washington, D. C.
Featherstone, W. R. (1976). *Feed Manage.* **27,** 13–16.
Featherstone, W. R., and Rogler, J. C. (1975). *Nutr. Rep. Int.* **11,** 491–497.
Feeney, P. P. M. (1969). *Phytochemistry* **8,** 2119–2126.
Ferrando, R., Guilleux, M. M., and Guerilott-Vinet, A. (1961). *Nature (London)* **192,** 1205.
Finks, A. J., and Johns, C. O. (1921). *Am. J. Physiol.* **57,** 61–67.
Fisher, H., Griminger, P., and Budowski, P. (1969). *Z. Ennaehrungswiss.* **9,** 271–278.
Fodor, G. (1960). *In* "The Alkaloids" (R. H. F. Manske, ed.), Vol. 6, pp. 145–177. Academic Press, New York.
Fowden, L. (1966). *Proc. Int. Symp. Biochem. Physiol. Alkaloids, 3rd, 1963* pp. 35–42.
Fowden, L., Lewis, D., and Tristram, H. (1967). *Adv. Enzymol.* **29,** 89–163.
Foy, J. M., and Parratt, J. R. (1960). *J. Pharm. Pharmacol.* **12,** 360–364.
Foy, J. M., and Parratt, J. R. (1962). *Lancet* **ii,** 942–943.
Fras, I., and Friedman, J. J. (1969). *N. Y. State J. Med.* **69,** 463–465.
Fuyre, R. H., and Warmke, H. E. (1952). *P. R. Fed. Exp. Stn., Annu. Rep.*
Garner, R. J. (1961). "Veterinary Toxicology," 2nd ed. Williams & Wilkins, Baltimore, Maryland.
Gitzelmann, R., and Auricchio, S. (1965). *Pediatrics* **36,** 231–233.
Glick, Z., and Joslyn, M. A. (1970). *J. Nutr.* **100,** 516–522.
Gray, D. D., and Fowden, L. (1962). *Biochem. J.* **82,** 385–389.
Guggenheim, M. (1913). *Hoppe-Seyler's Z. Physiol. Chem.* **88,** 276–284.
Guillaume, J., and Belec, R. (1977). *Br. Poult. Sci.* **18,** 573–583.
György, P., Murata, K., and Ikehata, H. (1964). *Nature (London)* **203,** 870–872.
Hall, R. L. (1973). *In* "Toxicants Occurring Naturally in Foods," pp. 448–463. Natl. Acad. Sci.-Natl. Res. Counc., Washington, D. C.
Hanson, A. A. (1925). *Science* **61,** 340–341.
Hanson, O. K. (1975). German Patent 2,524,753; *Chem. Abstr.* **84,** 88299x (1976).
Harms, R. H., Simpson, C. F., and Waldroup, P. W. (1961). *J. Nutr.* **75,** 127–131.
Harris, F. W., and Cockburn, F. (1918). *Am. J. Pharm.* **90,** 722–726.
Harris, H. B., Cumins, D. G., and Burns, R. E. (1970). *Agron. J.* **62,** 633–635.
Hartmann, A. F., Sr., Hartmann, A. F., Jr., Purkerson, M. L., and Wesley, M. E. (1963). *J. Am. Med. Assoc.* **185,** 706–709.
Hassal, C. H., Reyle, K., and Feng, P. (1954). *Nature (London)* **173,** 356–357.
Hedin, P. A., and Adachi, R. A. (1962). *J. Nutr.* **77,** 229–236.
Hegarty, M. P., and Pound, A. W. (1968). *Nature (London)* **217,** 354–355.
Hegarty, M. P., and Pound, A. W. (1970). *Aust. J. Biol. Sci.* **23,** 831–834.
Hegarty, M. P., Schimkel, P. G., and Court, R. D. (1964). *Aust. J. Agric. Res.* **15,** 153–167.
Hegarty, M. P., Court, R. D., Christie, G. S., and Lee, C. P. (1976). *Austr. Vet. J.* **52,** 490–492.
Hellendorn, E. W. (1969). *Food Technol.* **23,** 87–92.
Hellendorn, E. W. (1973). *In* "Nutritional Aspects of Common Beans and Other Legume Seeds as

Animal and Human Foods" (W. G. Jaffé, ed.), pp. 261–272. Arch. Latinoam. Nutr., Caracas, Venezuela.

Hernandez, A., and Jaffé, W. G. (1968). *Acta Cient. Venez.* **19,** 183–185.

Heuser, G. J., Norris, L. C., and McGinnis, J. (1946). *Poult. Sci.* **25,** 130–136.

Hickey, C. A., Murphy, E. L., and Calloway, D. H. (1972a). *Cereal Chem.* **49,** 276–282.

Hickey, C. A., Murphy, E. L., and Calloway, D. H. (1972b). *Am. J. Dig. Dis.* **17,** 383–389.

Hilker, D. (1968). *Int. Z. Vitaminforsch.* **38,** 387–389.

Hintz, H. F., and Hogue, D. E. (1964). *J. Nutr.* **84,** 283–287.

Hintz, H. F., and Hogue, D. E. (1970). *Proc. Soc. Exp. Biol. Med.* **133,** 931–933.

Hintz, H. F., Hogue, D. E., and Walker, E. F., Jr. (1969). *Proc. Soc. Exp. Biol. Med.* **131,** 447–449.

Hoeldtke, R., Baliga, B. S., Issenberg, P., and Wurtman, R. J. (1972). *Science* **175,** 761–762.

Hogue, D. E., Proctor, J. F., Warner, R. G., and Loosli, J. K. (1962), *J. Anim. Sci.* **21,** 25–29.

Hun, K. E. (1974). *Kangwon Taehak Yongu Nonmunjip* **8,** 24–26; *Chem. Abstr.* **85,** 139816u (1976).

Hutton, E. M., Windrum, G. M., and Kratzing, C. C. (1958a). *J. Nutr.* **64,** 321–333.

Hutton, E. M., Windrum, G. M., and Kratzing, C. C. (1958b). *J. Nutr.* **65,** 429–440.

Hylin, J. W. (1969). *J. Agric. Food Chem.* **17,** 492–496.

Hymowitz, T., and Collins, F. I. (1974). *Agron. J.* **66,** 239–240.

Jadhav, S. J., and Salumkhe, D. K. (1975). *Adv. Food Res.* **21,** 308–354.

Jaffé, W. G., and Vega, C. L. (1968). *J. Nutr.* **94,** 203–210.

Jaffé, W. G., Moreno, R., and Wallis, V. (1973). *Nutr. Rep. Int.* **7,** 169–174.

Jambunathan, R., and Mertz, E. T. (1973). *J. Agric. Food Chem.* **21,** 692–696.

Jaynes, T. A., and Nelson, D. E. (1971). *Plant Physiol.* **47,** 629–633.

Jelliffe, D. B., and Stuart, K. L. (1954). *Br. Med. J.* **1,** 75–77.

Jensen, L. S., and Mraz, F. R. (1966a). *J. Nutr.* **88,** 249–253.

Jensen, L. S., and Mraz, F. R. (1966b). *J. Nutr.* **89,** 471–476.

Kader, M. M. A. (1960). *Alexandria Med. J.* **6,** 500; *Nutr. Abstr. Rev.* **31,** 435 (1961).

Kakade, M. L., and Borchers, R. (1967). *Proc. Soc. Exp. Biol. Med.* **124,** 1272–1275.

Karmazin, M. (1955). *Pharmazie* **10,** 57–59.

Kean, E. A., ed. (1976). "Hypoglycin," pp. 121–125. Academic Press, New York.

Kerdel-Vegas, F., Wagner, F., Russell, P. B., Grant, N. H., Alburn, H. E., Clark, D. E., and Miller, J. A. (1965). *Nature (London)* **205,** 1186–1187.

Kim, B., Shimoda, T., and Funatsu, M. (1977). *J. Fac. Agric., Kyushu Univ.* **21,** 1–8. *Chem. Abstr.* **86,** 117658n (1977).

Kim, W. J., Smit, C. J. B., and Nakayama, J. O. M. (1973). *Lebensm.-Wiss. t Technol.* **6,** 201–204.

Kingsbury, J. M. (1964). "Poisonous Plants of the United States and Canada." Prentice-Hall, Englewood Cliffs, New Jersey.

Klosterman, H. J. (1974). *J. Agric. Food Chem.* **22,** 13–19.

Klosterman, H. J., Lamoureux, G. L., and Parsons, J. L. (1967). *Biochemistry* **6,** 170–177.

Kneen, E., and Sandstedt, R. M. (1943). *J. Am. Chem. Soc.* **65,** 1247–1248.

Kneen, E., and Sandstedt, R. M.(1946). *Arch. Biochem. Biophys.* **9,** 235–249.

Knuckles, B. E., de Fremesy, D., and Kohler, G. O. (1977). *J. Agric. Food Chem.* **24,** 1177–1180.

Kosower, N. S., and Kosower, E. M. (1967). *Nature (London)* **215,** 285–286.

Kratzer, F. H. (1947). *Poult. Sci.* **26,** 90–91.

Kratzer, F. H., Williams, D. E., Marshall, B. J., and Davis, P. N. (1954). *J. Nutr.* **52,** 555–563.

Ku, S., Wei, L. S., Steinberg, M. P., Nelson, A. I., and Hymowitz, T. (1976). *J. Food Sci.* **41,** 361–364.

Kuć, J. (1975). *Recent Adv. Phytochem.* **9,** 139–150.

Kurtzman, R. H., and Halbrook, W. U. (1970). *Appl. Microbiol.* **20,** 715-719.
Labov, J. B. (1977). *Comp. Biochem. Physiol. A* **57,** 3-9.
Lang, J. A., Chang-Hum, L. E., Reyess, P. S., and Briggs, G. M. (1974). *Fed. Proc., Fed. Am. Soc. Exp. Biol.* **33,** 718.
Leach, E. H., and Lloyd, J. P. F. (1956). *Proc. Nutr. Soc.* **15,** xv-xvi.
Leisinger, T., Haas, D., and Hegarty, M. P. (1972). *Biochim. Biophys. Acta* **262,** 214-219.
Liener, I. E., ed. (1969). "Toxic Constituents of Plant Foodstuffs," 1st ed. Academic Press, New York.
Liener, I. E. (1977). *Adv. Chem. Ser.* **160,** 283-300.
Lin, D.-C., Lin, J.-H., and Tung, T.-C. (1964). *J. Formosan Med. Assoc.* **63,** 278-284.
Lin, K.-C., Kuang, T.-L., and Ling, K.-H. (1963). *J. Formosan Med. Assoc.* **62,** 587-593.
Link, K. P. (1959). *Circulation* **19,** 97-107.
Lovenberg, W. (1973). *In* "Toxicants Occurring Naturally in Foods," Natl. Acad. Sci.-Natl. Res. Counc., Washington, D. C.
Lutton, J. D., and Kopac, M. J. (1971). *Cancer Res.* **31,** 1564-1569.
McCleary, B. V., and Chick, B. F. (1977). *Phytochemistry* **16,** 207-211.
McGinnis, J. and Polis, H. L. (1946). *Poult. Sci.* **25,** 408-409.
Macri, A., Parlamenti, R., Silano, V., and Valfre, F. (1977). *Poult. Sci.* **56,** 434-441.
Madsen, N. P., Christie, G. S., and Hegarty, M. P. (1970). *Biochem. Pharmacol.* **19,** 853-860.
Magee, A. D. (1963). *J. Nutr.* **80,** 151-156.
Manchester, K. L. (1974). *FEBS Lett.* **40,** 5133-5139.
Marquardt, R. R., Ward, A. T., Campbell, L. D., and Cansfield, P. E. (1977). *J. Nutr.* **107,** 1313-1324.
Marshall, J. J. (1975). *Am. Chem. Soc. Symp. Ser.* **16,** 244.
Marshall, J. J. (1977). *Carbohyd. Res.* **57,** C27-C30.
Marshall, J. J., and Lauda, C. M. (1975). *J. Biol. Chem.* **250,** 8030-8037.
Martin-Tanguy, J., Vermorel, M., Lenoble, M., Martin, C., and Gallett, M. (1976). *Ann. Biol. Anim. Biochim., Biophys.* **16,** 879-890.
Martin-Tanguy, J., Guillaume, J., and Kossa, A. (1977). *J. Sci. Food Agric.* **28,** 757-765.
Mason, M. M., and Whiting, M. G. (1966). *Fed. Proc., Fed. Am. Soc. Exp. Biol.* **25,** 533.
Matsumoto, H., Smith, E. G., and Sherman, G. D. (1951). *Arch. Biochem. Biophys.* **33,** 201-211.
Matsushita, K., and Uritani, I. (1976). *J. Biochem. (Tokyo)* **79,** 633-639.
Mattoo, A. K., and Modi, V. V. (1970). *Enzymolgia* **39,** 237-247.
Maxson, E. D., Rooney, L. W., Lewis, R. W., Clark, L. E., and Johnson, J. W. (1973). *Nutr. Rep. Int.* **8,** 145-152.
Menn, J. M., McBain, J. B., and Dennis, M. J. (1964). *Nature (London)* **202,** 697-698.
Milić, B., Stojanovic, S., Vućurevic, N., and Turcić, M. (1968). *J. Sci. Food Agric.* **19,** 2-8.
Militzer, W., Ikeda, C., and Kneen, E. (1946). *Arch. Biochem. Biophys.* **9,** 309-320.
Miller, B. S., and Kneen, E. (1947). *Arch. Biochem. Biophys.* **15,** 251-264.
Miller, E. R. (1920). *J. Biol. Chem.* **44,** 481-486.
Miller, E. R., Massengale, O. N., and Barnes, M. A. (1925). *J. Am. Pharm. Assoc.* **14,** 1113-1114.
Miller, E. R., Ullrey, D. E., Zutant, C. L., Hoefer, S. H., and Luecke, R. L. (1965). *J. Nutr.* **85,** 347-354.
Miller, F. R., Lowrey, R. S., Monsen, W. G., and Burton, A. R. (1972). *Crop Sci.* **12,** 563-567.
Miller, R. W., and Smith, C. R., Jr. (1973). *J. Agric. Food Chem.* **21,** 909-912.
Mitjavila, S. (1973). *Bull. Laison, Groupe Polyphenols* **48,** 22-30; *Chem. Abstr.* **82,** 52414a (1975).
Montagna, W., and Yun, J. S. (1963). *J. Invest. Dermatol.* **40,** 325-327.
Mori, T., Satouchi, K., and Matsushita, S. (1973). *Agric. Biol. Chem.* **37,** 1225-1226.
Morris, M. P., Pagan, C., and Warmke, H. E. (1954). *Science* **119,** 322-323.

Murillo, E., and Gaunt, J. K. (1975). *1st Chem. Congr. North Am. Continent, 1975* Abstr. No. 155.
Murphy, E. L., Horsley, H., and Burr, H. K. (1972). *J. Agric. Food Chem.* **20,** 813-817.
Muszynaka, G., and Reifer, I. (1970). *Acta Biochim. Pol.* **17,** 247-252.
Nagasawa, T., Takagi, H., Kawakami, K., Suzuki, T., and Sahashi, Y. (1961). *Agric. Biol. Chem.* **25,** 441-447.
Naim, M., Gestetner, B., Zilkah, S., Birk, Y., and Bondi, A. (1974). *J. Agric. Food Chem.* **22,** 806-810.
Naim, M., Gestetner, B., Bondi, A., and Birk, Y. (1976). *J. Agric. Food Chem.* **24,** 1174-1180.
Natelson, B. H. (1969). *Lancet* **iii,** 640.
National Academy of Sciences (1977). "*Leucaena,* Promising Forage and Tree Crop for the Tropics." Natl. Acad. Sci., Washington, D. C.
Nelson, T. S., Stephenson, E. L., Burgos, A., Floyd, J., and York, J. O. (1975). *Poult. Sci.* **54,** 1619-1623.
Nilova, V. P., and Guseva, T. A. (1975). *Dokl. Vses. Akad. S'kh. Nauk* pp. 9-10; *Chem. Abstr.* **83,** 75563h (1975).
Nilson, R. (1962). *Acta Physiol. Scand.* **56,** 230-236.
Nordfeldt, S., and Younge, O. (1949). *Hawaii, Agric. Exp. Stn., Prog. Stn. Notes* **55,** 1-2.
Nunn, P. B., Vega, A., and Bell, E. A. (1968). *Biochem. J.* **106,** 15p.
Oberleas, D. (1973). *In* "Toxicants Occurring Naturally in Foods," Natl. Acad. Sci.—Natl. Res. Counc., Washington, D. C. pp. 363-371.
O'Donnell, M. D., and McGeeney, K. F. (1976). *Biochim. Biophys. Acta* **422,** 159-169.
Oke, O. L. (1969). *World Rev. Nutr. Diet.* **10,** 262-303.
Olson, A. C., Becker, R., Miers, J. C., Gumbmann, M. R., and Wagner, J. R. (1975). *In* "Protein Nutritional Quality of Foods and Feeds" (M. Friedman, ed.), Part 2, pp. 551-563. Dekker, New York.
Orgell, W. H. (1963). *Lloydia* **26,** 59-66.
Orgell, W. H., Vaidja, K. A., and Hamilton, E. W. (1959). *Proc. Iowa Acad. Sci.* **66,** 149-154.
Oslage, H. J. (1956). *Kartoffelbau* **7,** 204-211.
Owen, L. N. (1958). *Vet. Rec.* **70,** 454-456.
Painter, J. C., Shanor, S. P., and Winek, C. L. (1971). *Clin. Toxicol* **4,** 1-5.
Pearn, J. H., and Hegarty, P. M. (1970). *Br. J. Exp. Pathol.* **51,** 34-37.
Persaud, T. V. N. (1968a). *Nature (London)* **217,** 471.
Persaud, T. V. N. (1968b). *Naturwissenschaften* **55,** 39.
Persaud, T. V. N., and Kaplan, S. (1970). *Life Sci.* **9,** 1305-1308.
Peterson, D. T. (1965). *Proc Symp. Acute Bovine Pulmonary Emphysema, 1965* pp. R1-R13.
Petrucci, T., Tomasi, M., Cantagalli, P., and Silano, V. (1974). *Phytochemistry* **13,** 2487-2495.
Petrucci, T., Rab, A., Tomasi, M., and Silano, V. (1976). *Biochim. Biophys. Acta* **420,** 288-297.
Pieterse, P. J. S., and Andrews, F. N. (1956a). *J. Anim. Sci.* **15,** 25-36.
Pieterse, P. J. S., and Andrews, F. N. (1956b). *J. Dairy Sci.* **39,** 81-89.
Powers, J. R., and Whitaker, J. R. (1977). *J. Food Biochem.* **1,** 239-260.
Pressey, R. (1966). *Arch. Biochem. Biophys.* **113,** 667-674.
Pressey, R. (1967). *Plant Physiol.* **42,** 1780-1786.
Pressey, R. (1968). *Plant Physiol.* **43,** 1430-1436.
Pressey, R. (1972). *J. Food Sci.* **37,** 521-523.
Pressey, R., and Shaw, R. (1966). *Plant Physiol.* **41,** 1651-1661.
Pudelkiewicz, W. J., and Matterson, L. D. (1960). *J. Nutr.* **71,** 143-148.
Puls, W., and Keup, U. (1973). *Diabetologia* **9,** 97-101.
Rackis, J. J. (1975). *Am. Chem. Soc. Symp. Physiol. Effects Food Carbohyd., 1974* pp. 201-222.
Rackis, J. J., Sessa, D. J., Steggerda, F. R., Shimuzi, T., Anderson, J., and Pearl, S. L. (1970a). *J. Food Sci.* **35,** 634-639.

Rackis, J. J., Honig, D. H., Sessa, D. J., and Steggerda, F. R. (1970b). *J. Agric. Food Chem.* **18,** 977–982.

Radeleff, R. D. (1964). "Veterinary Toxicology." Lea & Febiger, Philadelphia, Pennsylvania.

Ramachandra, G., Virupaksha, T. K., and Shadaksharaswamy, M. (1977). *J. Agric. Food Chem.* **25,** 1101–1104.

Rao, M. N., Shurpalekar, K. S., and Sundaravalli, O. E. (1967). *Indian J. Biochem.* **4,** 185.

Rao, M. N., Shurpalekar, K. S., and Sundaravalli, O. E. (1970). *Indian J. Biochem.* **7,** 241–243.

Raymond, W. D. (1944). *East Afr. Med. J.* **21,** 362–364.

Read, J. W., and Sure, B. (1923). *J. Agric. Res.* **24,** 433–439.

Reinhold, J. G. (1971). *Am. J. Clin. Nutr.* **24,** 1204–1215.

Reinhold, J. G., Ismail-Beigi, F., and Faradji, B. (1975). *Nutr. Rep. Int.* **12,** 75–85.

Reis, P. J., Tunks, D. A., and Hegarty, M. P. (1975). *Aus. J. Biol. Sci.* **28,** 495–499.

Reymond, D., and Phaff, H. J. (1965). *J. Food Sci.* **30,** 226–230.

Rockland, L. B., Bardiner, B. L., and Pieczanka, D. (1969). *J. Food Sci.* **34,** 411–414.

Rohde, R. A. (1960). *Proc. Helminthol. Soc. Wash.* **27,** 121.

Rosenfeld, I., and Beath, O. (1964). "Selenium." Academic Press, New York.

Rostagno, H. S., Featherstone, W. R., and Rogler, J. C. (1973a). *Poult. Sci.* **52,** 765–772.

Rostagno, H. S., Rogler, J. C., and Featherstone, J. C. (1973b). *Poult. Sci.* **52,** 772–778.

Sanyal, S. N. (1953). *Calcutta Med. J.* **50,** 409–415.

Sapeika, N. (1969). *In* "Food Pharmacology," p. 67. Thomas, Springfield, Illinois.

Sarkar, L., and Chaudhuri, D. K. (1976). *Int. J. Vitam. Nutr. Res.* **46,** 417–421.

Sasaoka, K., Ogawa, T., Moritoki, K., and Komoto, M. (1976). *Biochim. Biophys. Acta* **428,** 396–402.

Satouchi, K., and Matsushita, S. (1976). *Agric. Biol. Chem.* **40,** 889–897.

Satouchi, K., Tomohiko, M., and Matsushita, S. (1974). *Agric. Biol. Chem.* **38,** 97–101.

Saunders, R. M., and Lang, J. A. (1973). *Phytochemistry* **12,** 1237–1241.

Savaiano, D. A., Powers, J. R., Costello, M. J., Whitaker, J. R., and Clifford, A. J. (1977). *Nutr. Rep. Int.* **15,** 443–449.

Schaffert, R. E., Lechtenberg, V. L., Oswalt, D. L., Axtell, J. D., Pickett, R. C., and Rhykerd, C. L. (1974). *Crop Sci.* **14,** 640–643.

Schwimmer, S., Makower, R. U., and Roum, E. S. (1961). *Nature (London)* **176,** 972.

Selman, I. E., Wiseman, A., Breeze, R. G., and Pirie, H. M. (1976). *Vet. Rec.* **99,** 181–183.

Shainkin, R., and Birk, Y. (1970). *Biochim. Biophys. Acta* **221,** 502–513.

Shaw, J. C., Moore, L. A., and Sykes, J. F. (1951). *J. Dairy Sci.* **34,** 176–180.

Sherba, S. E. (1970). South African Patent 6,902,053. *Chem. Abstr.* **74,** 2858n (1970).

Shurpalekar, K. S., Sunderavalli, O. E., and Desai, B. L. M. (1973). *In* "Nutritional Aspects of Common Beans and Other Legume Seeds as Animal and Human Foods" (W. G. Jaffé, ed.), pp. 133–138. Arch. Latinoam Nutr., Caracas, Venezuela.

Siddiqui, S. (1945). *J. Sci. Ind. Res.* **4,** 68–70.

Silano, V., Pocchiari, F., and Kasarda, D. D. (1973). *Biochim. Biophys. Acta* **317,** 139–148.

Silano, V., Furia, M., Gianfreda, L., Macri, A., Palescandolo, R., Rab, C., Scardi, V., Stella, E., and Valfré, F. (1975). *Biochim. Biophys. Acta* **391,** 170–175.

Singleton, V. L., and Kratzer, F. H. (1969). *J. Agric. Food Chem.* **17,** 497–503.

Singleton, V. L., and Kratzer, F. H. (1973). *In* "Toxicants Occurring Naturally in Foods," pp. 309–345. Natl. Acad. Sci.-Natl. Res. Counc., Washington, D. C.

Singsen, E. P., Potter, L. M., Bunnell, R. H., Matterson, L. D., Stinson, L., Amato, S. V., and Jungherr, E. L. (1955). *Poult. Sci.* **34,** 1234–1237.

Small, J. (1949). *Food* **18,** 268 (cited by Hall, 1973, p. 456).

Smith, M. I., and Lillie, R. D. (1940). *Natl. Inst. Health Bull.* **174,** 1–10.

Smith, M. I., and Westfall, B. B. (1937). *Public Health Rep.* **52,** 1375–1376.

Smith, M. I., Franke, K. W., and Westfall, B. B. (1936). *Public Health Rep.* **51,** 1496-1497.
Snauwert, F., and Markakis, P. (1976). *Lebensm.-Wiss.-Technol.* **9,** 93-95.
Stanilas, E., and Vignais, P. (1964). *C. R. Hebd. Seances Acad. Sci.* **259,** 4872-4875.
Steggerda, F. R., and Dimmick, J. F. (1966). *Am. J. Clin. Nutr.* **19,** 120-124.
Stob, M. (1973). *In* "Toxicants Occurring Naturally in Foods," pp. 550-557. Natl. Acad. Sci.—Natl. Res. Counc., Washington, D. C.
Strong, F. M. (1966). *Can. Med. Assoc. J.* **94,** 568-573.
Strumeyer, D. H. (1972). *Nutr. Rep. Int.* **5,** 45-52.
Strumeyer, D. H., and Malin, M. J. (1969). *Biochim. Biophys. Acta* **184,** 643-645.
Sugimoto, H., and van Buren, J. P. (1970). *J. Food Sci.* **35,** 655-660.
Sumner, J. B., and Dounce, A. L. (1939). *Enzymologia* **7,** 130-132.
Sure, B., and Read, J. W. (1921). *J. Agric. Res.* **22,** 5-15.
Tamir, M., and Alumot, E. (1969). *J. Sci. Food Agric.* **20,** 199-202.
Tamir, M., and Alumot, E. (1970). *J. Nutr.* **100,** 573-580.
Tanaka, K. (1972). *J. Biol. Chem.* **247,** 7465-7469.
Tanaka, K., Miller, E. M., and Isselbacher, K. J. (1971). *Proc. Natl. Acad. Sci. U.S.A.* **68,** 20-23.
Tanaka, K., Isselbacher, K. J., and Shih, V. (1972). *Science* **175,** 69-71.
Tanaka, K., Kean, E. A., and Johnson, B. (1976). *N. Engl. J. Med.* **295,** 461-463.
Tanner, F. W. and Tanner, L. P. (1953). "Food Borne Infections and Intoxications," 2nd ed. Garrard Press, Champaign, Illinois.
Thompson, O. J., Carlson, C. W., Palmer, I. S., and Olson, O. E. (1968). *J. Nutr.* **94,** 227-232.
Torquati, T. (1913). *Arch. Farmacol. Sper. Sci. Affini* **15,** 213-223.
Tsai, K. C. (1961). *J. Formosan Med. Assoc.* **60,** 58-61.
Udenfriend, S., Lovenberg, W., and Sjoerdsma, A. (1959). *Arch. Biochem. Biophys.* **85,** 487-492.
Van Veen, A. G. (1973). *In* "Toxicants Occurring Naturally in Foods," pp. 464-476. Natl. Acad. Sci.-Natl. Res. Counc., Washington, D. C.
Van Veen, A. G., and Hyman, A. J. (1933). *Geneeskd. Tijdschr. Ned.-Indie* **73,** 991-1001.
Van Veen, A. G., and Latuasan, H. E. (1949). *Chron. Nat.* **105,** 288 (cited by Van Veen, 1973, p. 466).
Vega, A., Bell, E. A., and Nunn, P. B. (1968). *Phytochemistry* **7,** 1885-1892.
Venkataraman, L. V., and Jaya, T. V. (1975). *Nutr. Rep. Int.* **12,** 387-398.
Vignais, P. V., Vignais, P. M., and Stanilas, E. (1962). *Biochim. Biophys. Acta* **60,** 284-300.
Wagner, J. R., Becker, R., Gumbmann, M. R., and Olson, A. C. (1976). *J. Nutr.* **106,** 466-470.
Ward, A. T., Marquardt, R. R., and Campbell, L. D. (1977). *J. Nutr.* **107,** 1325-1334.
Waterman, H. C., and Jones, D. B. (1921). *J. Biol. Chem.* **47,** 285-295.
Weil, T. (1965). *Econ. Bot.* **19,** 194-217.
Whitaker, J. R., and Feeney, R. E. (1973). *In* "Toxicants Occurring Naturally in Foods," pp. 276-298. Natl. Acad. Sci.-Natl. Res. Counc., Washington, D. C.
Williams, D. L., and Spray, G. H. (1973). *Br. J. Nutr.* **29,** 57-63.
Willimot, S. G. (1933). *Analyst* **58,** 431-435.
Wilson, B. J., Garst, J. E., Linnabary, R. D., and Channell, R. B. (1974). *Science* **197,** 573-574.
Winkenbach, F., and Matile, P. (1970). *Z. Pflanzenphysiol.* **63,** 292-295.
Wright, F. J. (1944). *East Afr. Med. J.* **21,** 365-367.
Yapar, Z., and Clandinin, D. R. (1972). *Poult. Sci.* **51,** 222-228.
Yoshida, R. K. (1944). Doctoral dissertation, University of Minnesota, St. Paul.
Zitnak, A., and Johnson, G. R. (1970). *Am. Potato J.* **47,** 256-260.
Zondek, B., and Bergmann, E. (1938). *Biochem. J.* **32,** 641-645.

Index

D

E

F

G

I

M

O

P

R

T

FOOD SCIENCE AND TECHNOLOGY

A SERIES OF MONOGRAPHS

Maynard A. Amerine, Rose Marie Pangborn, and Edward B. Roessler, PRINCIPLES OF SENSORY EVALUATION OF FOOD. 1965.

S. M. Herschdoerfer, QUALITY CONTROL IN THE FOOD INDUSTRY. Volume I – 1967. Volume II – 1968. Volume III – 1972.

Hans Riemann, FOOD-BORNE INFECTIONS AND INTOXICATIONS. 1969.

Irvin E. Liener, TOXIC CONSTITUENTS OF PLANT FOODSTUFFS. 1969.

Martin Glicksman, GUM TECHNOLOGY IN THE FOOD INDUSTRY. 1970.

L. A. Goldblatt, AFLATOXIN. 1970.

Maynard A. Joslyn, METHODS IN FOOD ANALYSIS, second edition. 1970.

A. C. Hulme (ed.), THE BIOCHEMISTRY OF FRUITS AND THEIR PRODUCTS. Volume 1 – 1970. Volume 2 – 1971.

G. Ohloff and A. F. Thomas, GUSTATION AND OLFACTION. 1971.

George F. Stewart and Maynard A. Amerine, INTRODUCTION TO FOOD SCIENCE AND TECHNOLOGY. 1973.

C. R. Stumbo, THERMOBACTERIOLOGY IN FOOD PROCESSING, second edition. 1973.

Irvin E. Liener (ed.), TOXIC CONSTITUENTS OF ANIMAL FOODSTUFFS, second edition. 1974.

Aaron M. Altschul (ed.), NEW PROTEIN FOODS: Volume 1, TECHNOLOGY, PART A – 1974. Volume 2, TECHNOLOGY, PART B — 1976. Volume 3, ANIMAL PROTEIN SUPPLIES, PART A — 1978.

S. A. Goldblith, L. Rey, and W. W. Rothmayr, FREEZE DRYING AND ADVANCED FOOD TECHNOLOGY. 1975.

R. B. Duckworth (ed.), WATER RELATIONS OF FOOD. 1975.

Gerald Reed (ed.), ENZYMES IN FOOD PROCESSING, second edition. 1975.

A. G. Ward and A. Courts (eds.), THE SCIENCE AND TECHNOLOGY OF GELATIN. 1976.

John A. Troller and J. H. B. Christian, WATER ACTIVITY AND FOOD. 1978.

A. E. Bender, FOOD PROCESSING AND NUTRITION. 1978.

D. R. Osborne and P. Voogt, THE ANALYSIS OF NUTRIENTS IN FOODS. 1978.

Marcel Loncin and R. L. Merson, FOOD ENGINEERING: PRINCIPLES AND SELECTED APPLICATIONS. 1979.

Hans Riemann and Frank L. Bryan (eds.), FOOD-BORNE INFECTIONS AND INTOXICATIONS, Second Edition. 1979.

N. A. Michael Eskin, PLANT PIGMENTS, FLAVORS AND TEXTURES: THE CHEMISTRY AND BIOCHEMISTRY OF SELECTED COMPOUNDS. 1979.

J. G. Vaughan (ed.), FOOD MICROSCOPY. 1979.

J. R. A. Pollock (ed.), BREWING SCIENCE. Volume 1. 1979.

Irvin E. Liener (ed.), TOXIC CONSTITUENTS OF PLANT FOODSTUFFS, Second Edition. 1980.

In preparation

INTERNATIONAL COMMISSION ON MICROBIOLOGICAL SPECIFICATIONS IN FOODS, MICROBIAL ECOLOGY OF FOODS, Volumes I and II